FOUNDATION MATHEMATICS FOR ENGINEERS

John Berry and Patrick Wainwright

MACMILLAN

First edition 1991

Published by
MACMILLAN EDUCATION LTD
Houndmills, Basingstoke, Hampshire RG21 2XS
and London
Companies and representatives
throughout the world

Printed in Hong Kong

ISBN 0–333–52717–8

A catalogue record for this book is
available from the British Library

CONTENTS

PREFACE

Mathematics is an essential tool for the engineer and applied scientist and mathematics is often up to one third of an engineering students' curriculum in the first year. To succeed in their studies in Higher Education, it is important for students to have a solid foundation of mathematics and science on which to build. In the past, the standard route into engineering was for students with GCE Advanced levels in mathematics and science or their BTEC equivalent. Now there is a chance for men and women, school leavers and older people without the necessary formal qualifications to move into engineering by joining one of the Foundation Engineering courses which are now part of many engineering degree courses. This book is the result of several years' experience teaching mathematics to engineering students at this Foundation level.

The mathematics syllabus for these Foundation courses is essentially a sound conceptual understanding of algebra, trigonometry, functions and calculus together with the confidence to apply the skills in engineering applications. With the arrival in the classroom of powerful computer algebra packages, such as DERIVE, the need for many hours' practice at standard exercise is now less important, and teachers of engineering mathematics can concentrate on a sound understanding of the concepts and applications of the theory. If the student has access to graphical packages and/or computer algebra packages then we would recommend that these packages are used in conjunction with the theory and problem solving activities in this book.

This book is intended to cover the mathematical content of Foundation Engineering courses (such as HITECC) in Higher Education. Each chapter contains applications of the mathematical theory so that students can see the relevance and need for the mathematics as well as have the opportunity to use their new mathematical skills. The applications are mainly drawn from science because at the Foundation level students will have little engineering background to use. In writing this book we have not made any distinction between the various areas of engineering since we consider that, no matter what discipline the student may go into, they all have the same common mathematical core syllabus.

Although primarily written for engineering students, the text should also be of value to science students whose mathematical requirements are similar to those of the engineer. For example, this book would be suitable for some of the new AS level courses which include calculus.

Each chapter identifies specific learning objectives for the student and these objectives are tested at the end of the chapter. The final chapter introduces the important topic of mathematical modelling. It is natural in a book containing applications to identify the problem solving skills needed for the prospective engineering student. Many courses now contain mathematical modelling as a part of the curriculum so that this chapter should provide a framework for these modules.

We would like to thank Sharon Ward for the expert typing of the manuscript and for her patience in the many corrections and amendments needed. We would also like to thank our colleagues within our Polytechnics for their advice during the preparation of this book and for the staff at Macmillans and their reviewers for their helpful suggestions, many of which have been incorporated into the text. We also thank the Controller of Her Majesty's Stationery Office for permission to use the 'shortest stopping distances' extract from the *Highway Code*, and Guinness Publishing for permission to reproduce one fact from the *The Guinness Book of Records 1984* (copyright © Guinness Publishing Ltd 1983).

Patrick Wainwright
John Berry

1 PROPORTIONALITY AND LINEAR LAWS

OBJECTIVES

When you have completed this chapter you should be able to

1. draw a straight-line graph to represent data
2. find the gradient and intercept of a straight-line graph and hence write down the line's equation
3. draw a straight-line graph from its equation
4. interpret the gradient as a rate of change
5. recognize whether variables are directly or inversely proportional and find the appropriate law between the variables
6. find and use a straight line of best fit as a model for making predictions

1.1 INTRODUCTION

There are many occasions in science and engineering where one quantity is to be related to another. Consider, as an example, an experiment in which different masses are suspended from the end of a spring and the extension of the spring measured (see Figure 1.1). Table 1.1 shows the results of such an experiment.

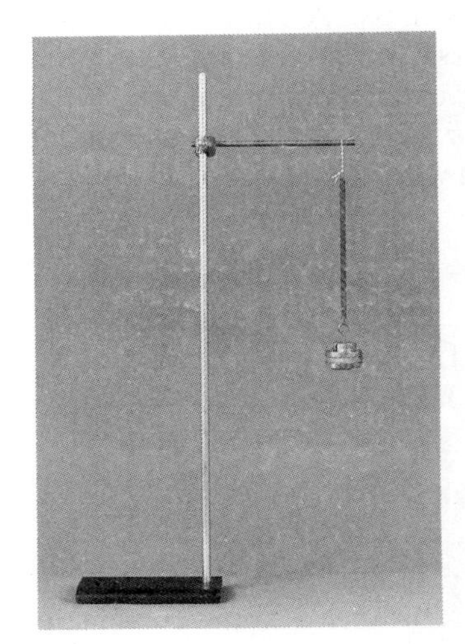

Figure 1.1

Table 1.1

Mass (g)	50	100	150	200	250
Extension (mm)	33	67	100	130	167

If we plot the measurements on a graph, we see that the points lie close to a straight line through the origin (Figure 1.2).

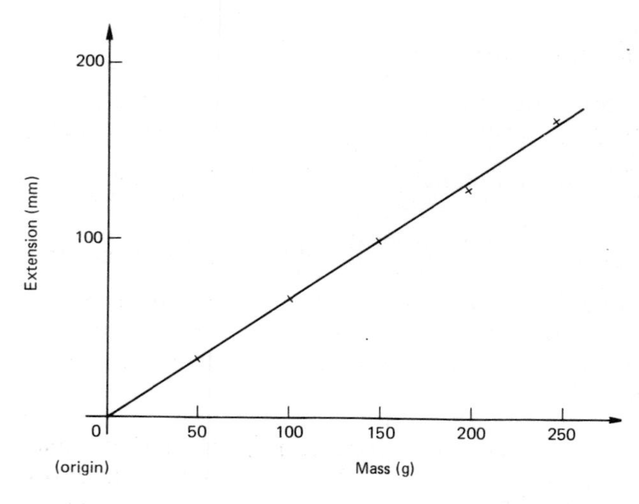

Figure 1.2

From this graph we can deduce that there must be a simple relationship between the extension of the spring and the mass on the end. The relationship is called a *linear relation* and in this case the extension is *directly proportional* to the mass. A feature of such a relationship is that if we double the mass, we double the extension and if we triple the mass we triple the extension, and so on.

As a second example, Table 1.2 shows the results of an experiment in physics in which the current through a wire is measured for wires of different resistance.

Table 1.2

Resistance (ohms)	0.9	1.2	1.5	1.8	2.1	2.4
Current (amperes)	5.56	4.17	3.33	2.78	2.38	2.08

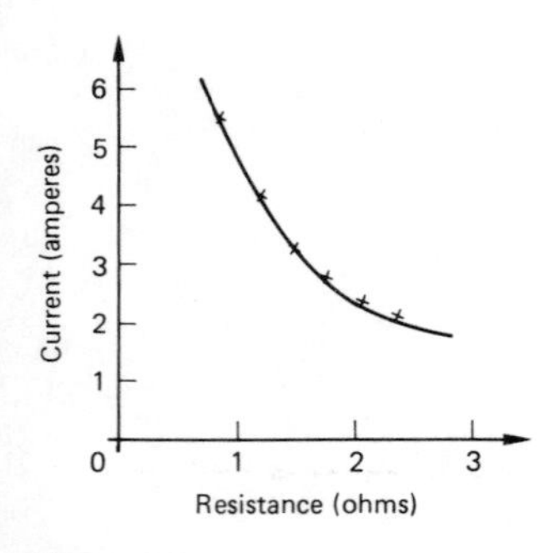

Figure 1.3

Drawing a graph of the measurements for this experiment suggests a more complicated relationship.

The fact that the points lie close to a smooth curve suggests that there is some relationship between the resistance and the current. If we look carefully at the graph, we can see that

(a) as the resistance increases the current decreases in a simple way; if the resistance is doubled the current is halved, if the resistance is tripled the current is reduced by a factor of 3, and
(b) the product of the resistance and the current for each pair of measurements is constant (approximately 5)

If these two properties hold true for two quantities we say that they are *inversely proportional* so that

$$\text{current is proportional to} \frac{1}{\text{resistance}}$$

This relationship becomes clearer when we draw a graph of current against 1/resistance. Such a graph is shown in Figure 1.4.

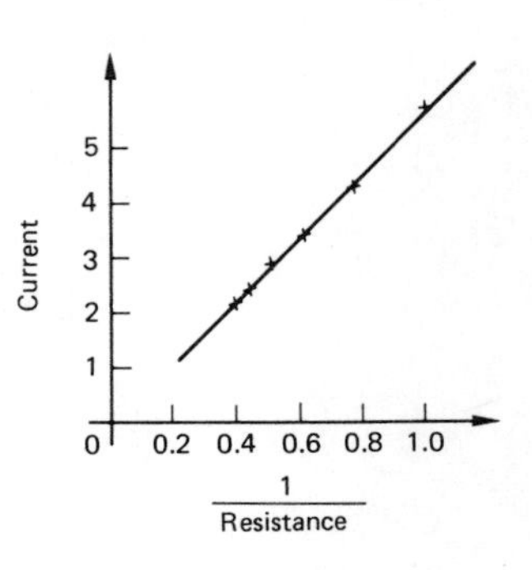

Figure 1.4

Now we have a straight line graph, which we shall see is easier to analyse than the curve in Figure 1.3.

In this chapter we look at proportionality in general and then concentrate on linear laws.

1.2 INTERVALS

In the first example (stretching a spring as in Table 1.1) the mass ranges in value between 50 grams and 250 grams and the extension ranges in value between 33 mm and 167 mm. These ranges of values are examples of *intervals* and it is important when finding a relationship between two quantities to specify the intervals over which the values of the quantities can be chosen.

More generally the set of real numbers represented geometrically by a straight line can be split into many sub-sets or intervals and we use the notation of inequalities to describe intervals algebraically.

For example, the set of numbers between -1 and 3 can be represented geometrically by a horizontal line as shown in Figure 1.5 and algebraically by $\{x: -1 < x < 3\}$ or simply $-1 < x < 3$.

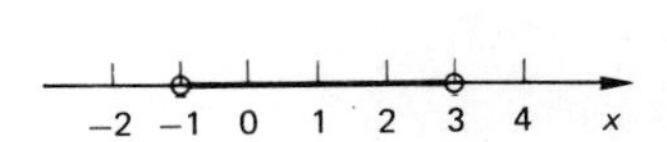

Figure 1.5

The 'curly brackets notation' is read as follows:

$\{x: -1 < x < 3\}$
'the set of x such that x lies between -1 and 3'.

In this example the end-points $x = -1$ and $x = 3$ are *not* included in the interval. This is denoted on the diagram by an open circle ○ and algebraically by $<$.

If the end-point $x = 3$ is included in the interval then we write

$x \leqslant 3$

which is read as 'x is less than or equal to 3' and on the diagram we use a solid circle ● (Figure 1.6).

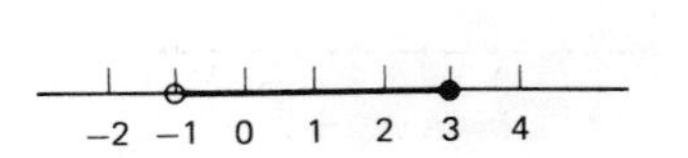

Figure 1.6

Exercise 1.2.1

1. Represent the following intervals on a diagram
   (a) $0 < x < 4$ (b) $0 \leqslant x \leqslant 4$
   (c) $-2 \leqslant x \leqslant 2$ (d) $-3 < x \leqslant 1$
   (e) $-4 \leqslant x < 3$ (f) $-2.5 \leqslant x \leqslant 1.3$

2. Consider the intervals shown in Figure 1.7. Write down the algebraic form for each interval.

(a)

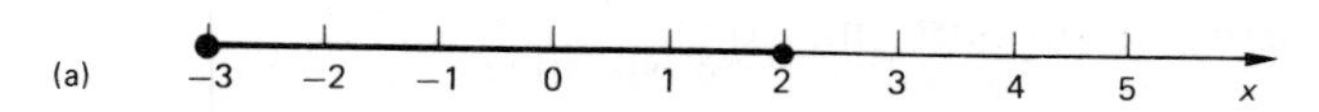

(b)

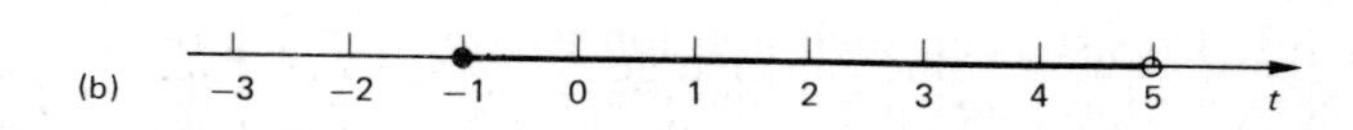

(c)

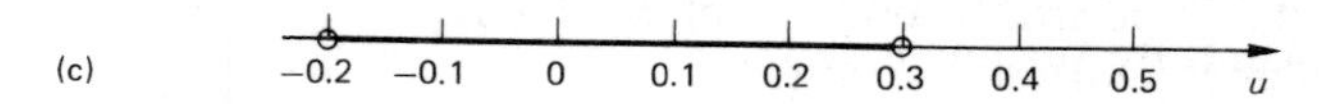

Figure 1.7

We can often use the diagrammatical representation of an interval to solve problems.

Example 1.2

Suppose that the quantity x satisfies the two inequalities $x \leqslant 1$ and $x + 2 \geqslant 0$. Show on a diagram the interval for x and write this interval algebraically.

Solution

First we rearrange the second inequality so that $x \geqslant -2$. Figure 1.8 shows the two inequalities.

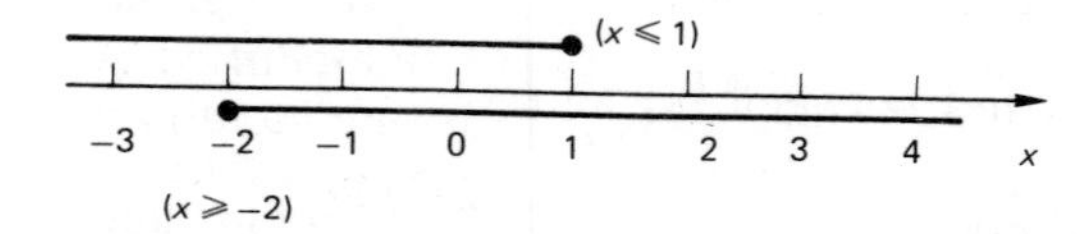

Figure 1.8

As x has to satisfy both of these inequalities, it must lie in the region where both of them are valid.

The interval of allowed values of x clearly lies between -2 and 1 so we can write

$$-2 \leqslant x \leqslant 1$$

Exercise 1.2.2

Find the range of values of x which satisfy both the inequalities in the following:

(a) $(x + 3) > 0$ $\quad (x - 1) < 0$
(b) $(x + 3) < 0$ $\quad (x + 6) > 0$
(c) $(x + 2) \geqslant 0$ $\quad (x + 3) \geqslant 0$

1.3 VARIABLES

A quantity whose value can change is called a *variable*. For example, in the experiment in Figure 1.1 in Section 1.1, the mass on the end of the spring and the extension of the spring are the variables. In the second experiment on the electric circuit, the variables are the resistance and the current.

In each of these experiments, we start by choosing one variable and then we find the other using an appropriate measuring device during the experiment. This is the case in science and engineering experiments. The variable whose values are preselected is called the *independent variable*. The other variable whose values are then found during the experiment is called the *dependent variable*, because its value depends on the value of the independent variable.

In the spring experiment the independent variable is mass and dependent variable is extension.

Exercise 1.3.1

Which is the dependent and independent variable for the electric circuit experiment?

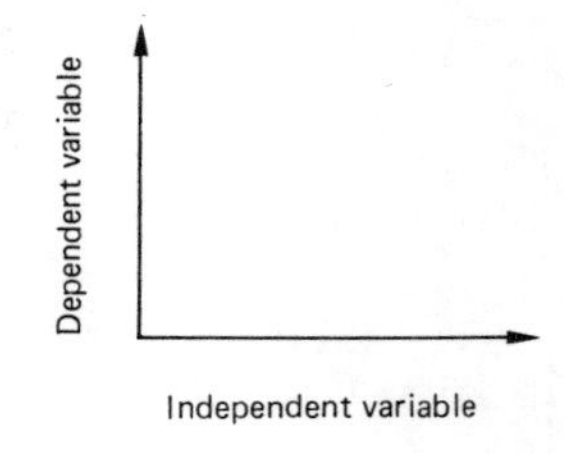

Figure 1.9

Having identified which is the dependent variable in an experiment or from a set of results from an experiment, we then draw a graph plotting the independent variable across the page (often identified as the x-axis) and the dependent variable up the page (often identified as the y-axis) (Figure 1.9). The shape of the resulting graph shows us how the variables are related.

Example 1.3.1

If a stone is thrown in the air it starts with a particular speed. Its speed then decreases to zero at the highest point before gaining speed as it falls back to the ground. Sketch a graph of speed against time showing how these quantities might be related.

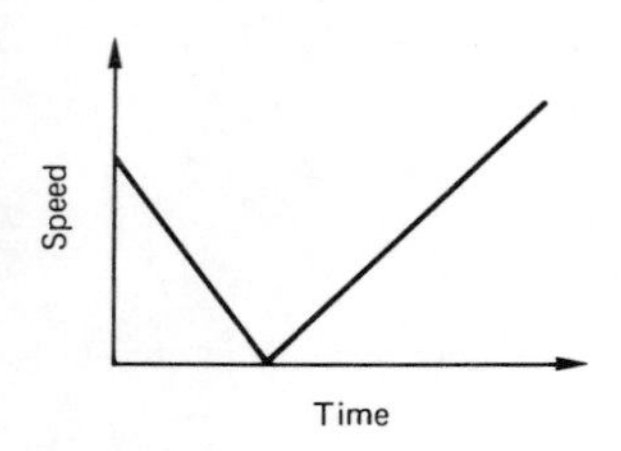

Figure 1.10

Solution

Figure 1.10 shows that the speed decreases to zero and then increases again. (As we have shown it here, an experiment would result in the actual graph consisting of two straight lines.)

Exercise 1.3.2

In the following situations choose appropriate dependent and independent variables and sketch graphs showing the sort of relationship that you would expect:

(a) the number of daylight hours in the day varies at different times of the year
(b) the amount of petrol that you use on a journey depends on the distance that you travel
(c) as you climb a mountain the atmospheric pressure goes down
(d) if you leave a cup of hot coffee its temperature gradually goes down to room temperature

---

In the last example and exercise we have suggested drawing sketches of graphs to illustrate the general form of the relationship between two variables. If we have a set of values from an experiment then we can draw (or plot) a more accurate graph. When graphing we locate points on a plane using *coordinates*.

To locate point (2, 5) on the plane, for example, we start at the origin and move 2 units to the right and then up 5 units. For $(-2, -5)$ we move 2 units to the left and then 5 units down. So + means right (if x) and up (if y) and – means left (if x) and down (if y).

Example 1.3.2

Using axes containing the intervals $-3 \leqslant x \leqslant 4$, $-2 \leqslant y \leqslant 5$, plot the following points: $(-3, 2)$, $(1, 5)$, $(-1, -2)$, $(4, -1)$.

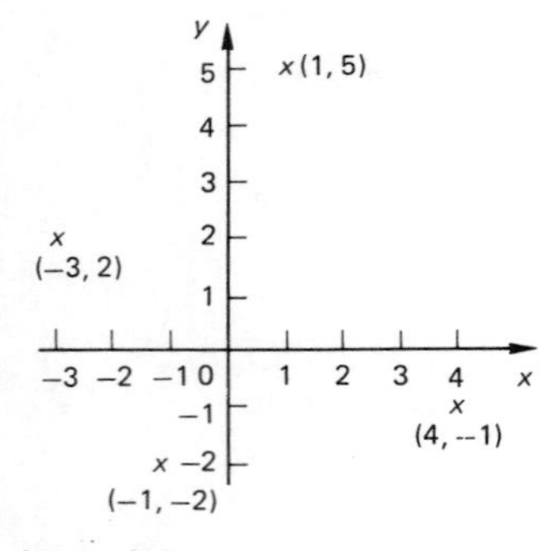

Figure 1.12

Solution

The four points are shown in Figure 1.12.

Exercise 1.3.3

Using axes containing the intervals $-4 \leqslant x \leqslant 4$, $-3 \leqslant y \leqslant 3$, plot the following points: $(-4, -3)$, $(0, 0)$, $(3, 3)$, $(-2, 2)$, $(2, -3)$, $(1.5, 2.5)$, $(-1.5, -0.5)$, $(2.5, 1.5)$.

Sometimes the x-axis is called the *abscissa* and the y-axis is called the *ordinate*. When choosing the scale for each axis it is important that the scale is easy to read and to use. Often graph paper is divided into large squares, each of which is in turn divided into 100 small squares (10 along and 10 up).

Figure 1.13 shows a sensible choice of scale, for the interval $40 \leqslant x \leqslant 100$. Each large square represents 10 units so each small square represents 1 unit.

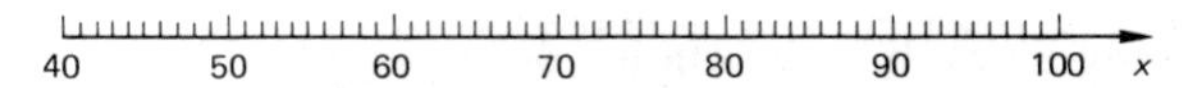

Figure 1.13

Figure 1.14 shows an unsatisfactory choice of scale, since it is not easy to read off values quickly.

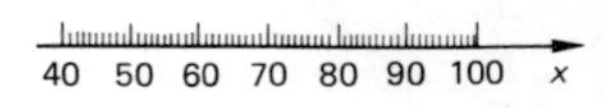

Figure 1.14

Another unsatisfactory choice of scales is shown in Figure 1.15. With such a choice it would be very difficult to plot points.

37 48 59 70 81 x

Figure 1.15

A simple general rule is to have each large square representing 10, 5, 2, 1 or 0.1 units so that each small square represents 1, 0.5, 0.2, 0.1 or 0.01 units respectively. Often having plotted the points on a piece of graph paper, we can identify a smooth curve that runs close to most of the points. This is the beginning of finding a relation between the variables. The following example and exercise deal with real data which do not lead to straight line graphs.

Example 1.3.3

The burning of fossil fuels adds carbon dioxide to the atmosphere around the Earth. This may be partly removed by biological reactions but the concentration of carbon dioxide is gradually increasing. This increase leads to a rise in the average temperature of the Earth. Table 1.3 shows this temperature rise over the last 100 years.

Table 1.3

Year	Temperature rise of the Earth over the 1860 figure (°C)
1880	0.01
1896	0.02
1900	0.03
1910	0.04
1920	0.06
1930	0.08
1940	0.1
1950	0.13
1960	0.18
1970	0.24
1980	0.32

If the average temperature of the Earth rises by about another 6 °C from the present value, this would have a dramatic effect on the polar ice caps, winter temperature, etc. If the ice caps melt there will be massive floods and a lot of the land mass would be submerged. Some countries might disappear except for the tops of the mountains!

Draw a graph representing this data.

Solution

The first step in drawing a graph is the choice of axes and scales. In this problem, the year is the independent variable and temperature rise is the dependent variable. Figure 1.16 shows the graph of temperature against year.

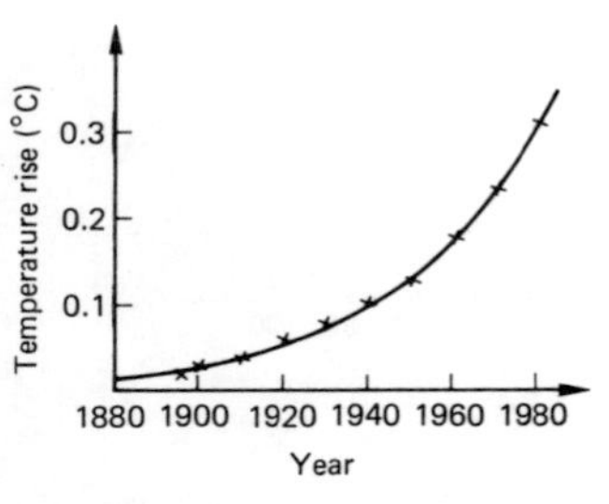

Figure 1.16

Exercise 1.3.4

The data in Table 1.4 shows the distance of each of the planets from the Sun (measured in millions of kilometres) and the time (measured in days) that it takes each planet to travel round the sun once, this time is called the *period*.

Table 1.4

Planet	Distance (millions of kilometres)	Period (days)
Mercury	57.9	88
Venus	108.2	225
Earth	149.6	365
Mars	227.9	687
Jupiter	778.3	4329
Saturn	1427	10753
Uranus	2870	30660
Neptune	4497	60150
Pluto	5907	90670

Draw a graph to show that there is a relationship between period and distance.

In the previous example and exercise, there is a smooth curve between the points suggesting a relationship between the variables. The equation relating the variables is not obvious. However, for straight line graphs finding an equation is much easier.

1.4 STRAIGHT LINE GRAPHS

The simplest, and perhaps one of the most important graphs, is the straight line graph. Consider the values shown in Table 1.5. A graph of these data is shown in Figure 1.17.

Table 1.5

x	0.1	0.3	0.5	0.7	0.9	1.1
y	1.2	1.7	2.2	2.7	3.2	3.7

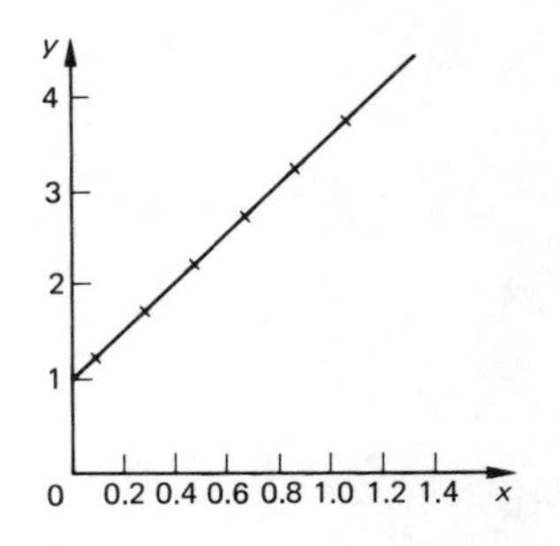

Figure 1.17

It is clear that in this figure the x-values increase in equal steps of 0.2 and the y-values increase in equal steps of 0.5. One of the identifying features of data that lie on a straight line is that given a step a say in x, there is a corresponding step b in y, wherever you start on the graph. In other words, there is a constant steepness or gradient.

Exercise 1.4.1

Given the following table of values plot a graph of the dependent variable (the second row) against the independent variable (the first row).

Table 1.6 (a)

x	0.4	0.7	1.0	1.3	1.6	1.9
y	2.1	2.7	3.3	3.9	4.5	5.1

(b)

Temperature, T(°C)	30	80	130	180	230	280
Volume, V (litres)	1.0	1.16	1.34	1.52	1.67	1.83

(c)

Mass (g)	50	100	150	200	250
Extension (mm)	33	67	100	130	167

(d)

Voltage (volts)	6.24	7.14	8.34	10.0	12.51
Current (amperes)	2.08	2.38	2.78	3.33	4.17

(e)

Time (s)	1	2	3	4	5	6
Speed (ms^{-1})	60.2	50.4	40.6	30.8	21.0	11.2

In this exercise you will have drawn straight line graphs to represent the data for each pair of variables. From each graph it would be possible to read off 'y-values' for given 'x-values'. Used in this way the graph is a *graphical model* describing the given situation. Often we are trying to find an algebraic model and drawing a graph is the first step in finding such a model.

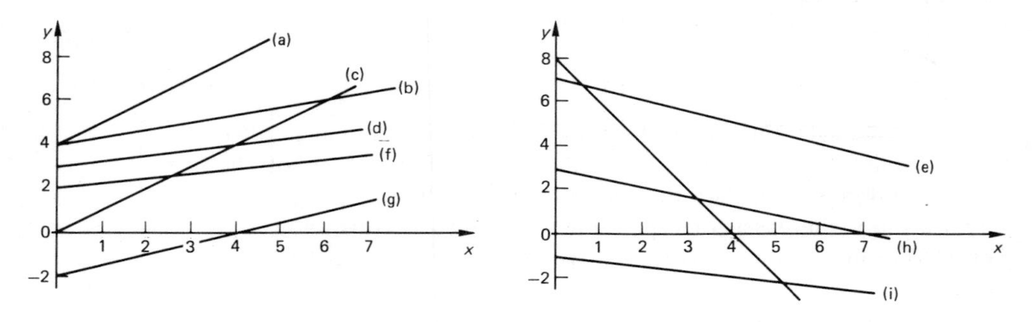

Figure 1.18

Consider the straight line graphs in Figure 1.18. Clearly they all have different features; some slope upwards (from left to right) and some slope downwards. Graph (a) has a steeper slope than graph (d). They cut the y-axis at different points. And so on.

We need a method of classifying straight-line graphs and of forming an equation for the relationship between the variables. There are two features that we use to classify the lines: the *gradient* or slope, and the *intercept* on the y axis.

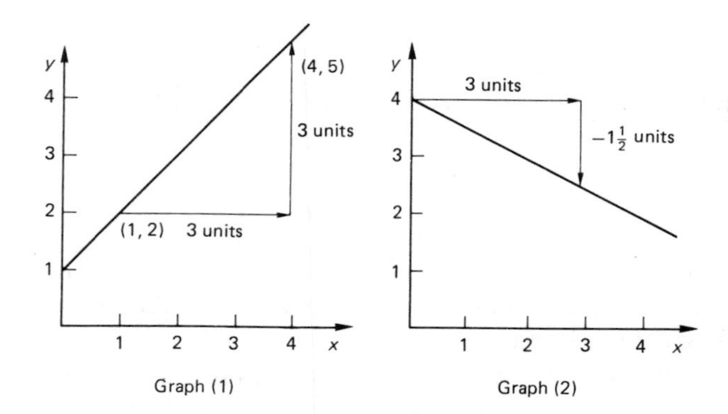

Figure 1.19

The gradient (*or slope*) is the steepness of the line. Consider the graphs in Figure 1.19. For graph (1), a step of 3 units in the x-direction leads to a step of 3 units in the y-direction. The gradient of graph (1) is defined as $3/3 = 1$. Now consider graph (2), a step of 3 units in the x-direction is followed by a step of $-1\frac{1}{2}$ units in the y-direction. The gradient of graph (b) is $-1\frac{1}{2}/3 = -\frac{1}{2}$.

Figure 1.20 shows the numerical definition of the gradient.

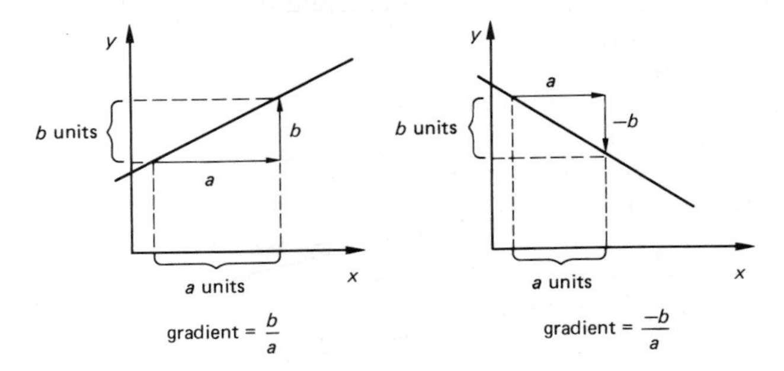

Figure 1.20

In calculating the gradient it is important *not* to count the squares on the graph paper but to use the scales on the axes.

Exercise 1.4.2

Find the gradient of all the straight-line graphs in Figure 1.18.

In the next example, we work the other way round. We are given the gradient and using it we have to draw the straight-line graph.

Example 1.4.1

Draw the straight lines of gradient 2 and -3 going through the point (2, 3).

Solution

For the straight line of gradient 2, we start at the point (2, 3) and move along 1 unit in the x-direction and then up 2 in the y-direction. Equally we could go along 2 and up 4; or along 3 and up 6; and so on.

For the straight line of gradient -3, we start at the point (2, 3) and move along 1 unit and then down 3 units (or along 2 units and down 6 units, etc).

Figure 1.21 shows these straight line graphs.

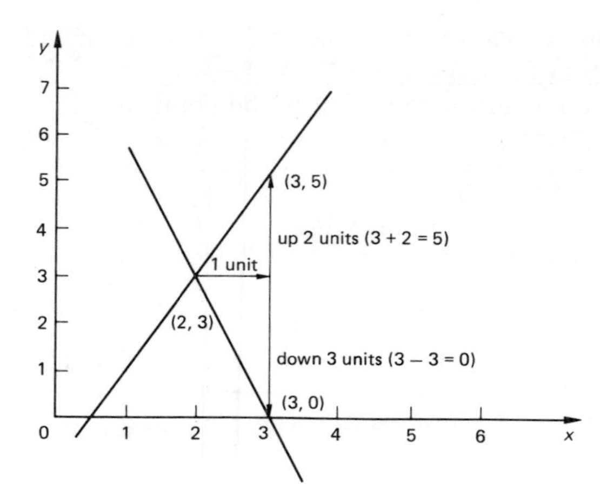

Figure 1.21

Exercise 1.4.3

On separate graphs draw the following:

(a) straight lines of gradient 1, 2, 3 starting at the point (0,1)

(b) straight lines of gradient $\frac{1}{2}$, $\frac{1}{3}$, $\frac{2}{3}$ starting at the point (0, 3)

(c) straight lines of gradient −1, −2, −3 starting at the point (0, 6)

(d) straight lines of gradient $-\frac{1}{3}$, $-\frac{1}{2}$, $-\frac{2}{3}$ starting at the point (0, 4)

The intercept on the y-axis is the point at which the graph crosses the y-axis. It is a measure of where the straight line lies in the plane. The larger the intercept the higher it cuts the y-axis.

For example, the intercepts of the graphs in Figure 1.18 are

(a) +4 (b) +4 (c) 0 (d) +3 (e) +7
(f) +2 (g) −2 (h) +3 (i) −1

Exercise 1.4.4

Write down the intercepts of all the straight line graphs in Exercise 1.4.3.

Example 1.4.2

Given the Table 1.7, plot a graph of the dependent variable (speed) against the independent variable (time).

Table 1.7

Time (s)	1	2	3	4	5
Speed (ms^{-1})	22.8	32.6	42.4	52.2	62.0

From the graph find the gradient and intercept.

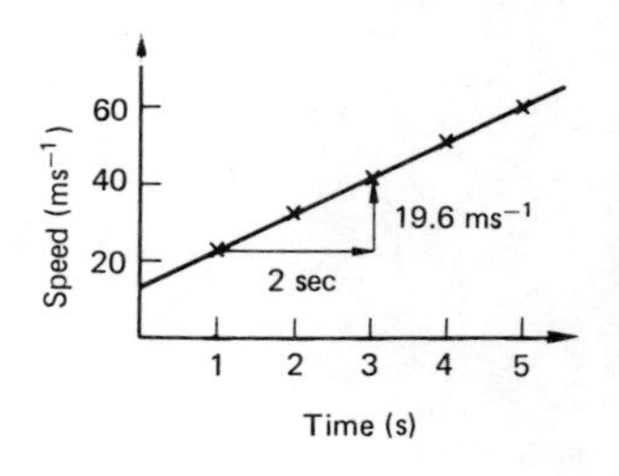

Figure 1.22

Solution

A graph of the data is shown in Figure 1.22. From the graph we can read off the intercept at 13 ms^{-1}.

To calculate the gradient, a step of 2 (seconds) across leads to a step of 19.6 (ms^{-1}) up. Hence

$$\text{Gradient} = \frac{19.6}{2} = 9.8$$

Exercise 1.4.5

Consider your graphs drawn in Exercise 1.4.1. Find the gradient and intercept of these graphs.

The gradient and intercept define a graph 'uniquely'. This means that if we know *both* the gradient and the intercept, we know *exactly* what the straight line looks like. We can now use these features to formulate the equation of a straight line. Consider the straight line graph in Figure 1.23.

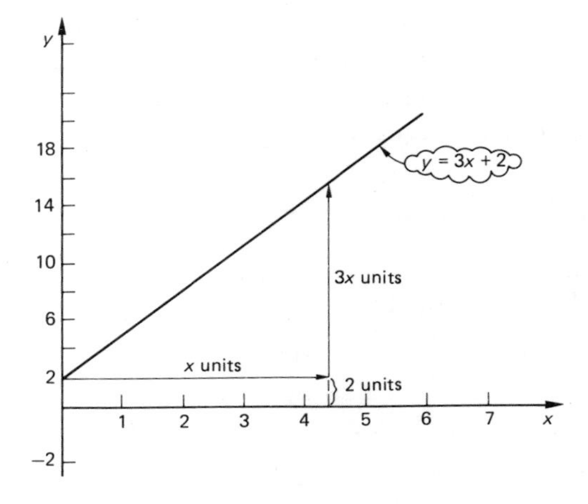

Figure 1.23

This line has a gradient of +3 and intercept 2. Suppose we start from the point (0, 2) and step x units in the x-direction, then the step in the y-direction is $3x$ units (because the gradient is 3). The y-value on the line is therefore given by $3x + 2$. Hence the equation of the line is $y = 3x + 2$. This leads us to the general result.

> The equation of a straight line law is
>
> $y = \text{gradient} \times x + \text{intercept}$
>
> and is often written as $y = mx + c$ where m is the *gradient* and c is the *intercept*.

An equation of this form describes a straight line and is called a *linear equation*.

Exercise 1.4.6

Write down the equations of the straight-line graphs in Figure 1.18 and those in Exercise 1.4.3.

Example 1.4.3

Draw the straight-line graph represented by the equation

$$2y - 4x = 3$$

Solution

The first step is to write the equation in the form $y =$ something, so that the coefficient of y is 1. In this example we need to rearrange the equation:

$$2y - 4x = 3$$

$$2y = 4x + 3 \quad \text{(add } 4x \text{ to each side)}$$

$$y = 2x + \frac{3}{2} \quad \text{(divide both sides by 2)}$$

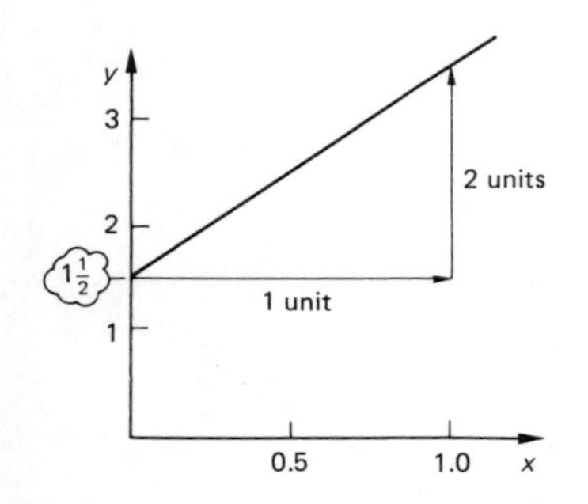

Figure 1.24

Now we can deduce that the straight line has gradient 2 and intercept 3/2. Figure 1.24 shows the graph of this straight line.

Exercise 1.4.7

1. Draw the straight lines represented by the following equations.
   (a) $y = 3x - 2$ (b) $2y + 1 = 3x$
   (c) $5x + 3y = 6$ (d) $x + 2y = 1$
   (e) $V = 20 - 9.8t$ (f) $T = 15.2e$

2. What is the gradient of the line connecting each of the following pairs of points?
   (a) (0, 0), (3, 5) (b) (0, 0), (2, 4)
   (c) (−2, 1), (1, 3) (d) (−1, −2), (2, 3)
   (e) (16, 7), (20, 19) (f) (1, 1), (4, 7)
3. Find the equation of each of the lines in problem 2 using variables x and y.

1.5 RATES

The gradient of a graph gives a measure of how a quantity is changing and can often be interpreted as a physical quantity. For example, the following graph shows the distance travelled by a car at various times.

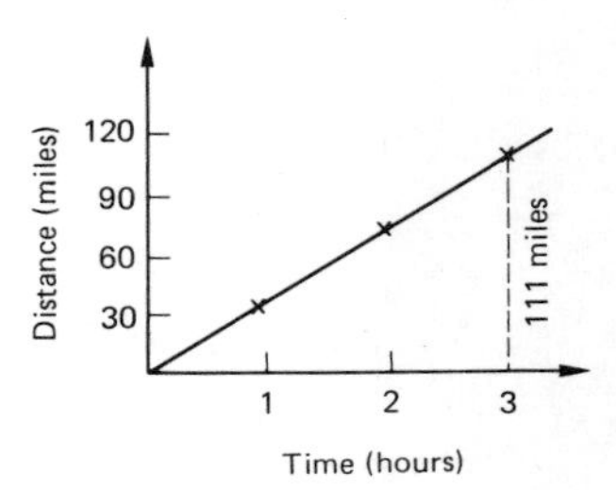

Figure 1.25

The gradient of the straight line graph is 37. In one hour the car travels 37 miles, in two hours 74 miles and so on. Clearly in this example the gradient of the graph represents the speed of the car (37 miles per hour). Note that the physical meaning of the gradient can be deduced using the units of the variables on the axes. We could have written

$$\text{Gradient} = \frac{111 \text{ miles}}{3 \text{ hours}} = 37 \text{ miles per hour}$$

Exercise 1.5.1

1. Water comes out of a bath tap at the rate of 15 litres per minute.
   (a) Copy and complete the following table showing the amount of water in the bath at different times.

Table 1.8

Time (min)	0	0.5	1.0	1.5	2.0	2.5	3.0
Amount (litres)	0	7.5					

   (b) Draw a graph showing this data.
   (c) Check that the gradient of the graph is the rate of flow of the water.
2. Figure 1.26 shows a graph representing a petrol tank being filled. Calculate the rate of filling of the tank.

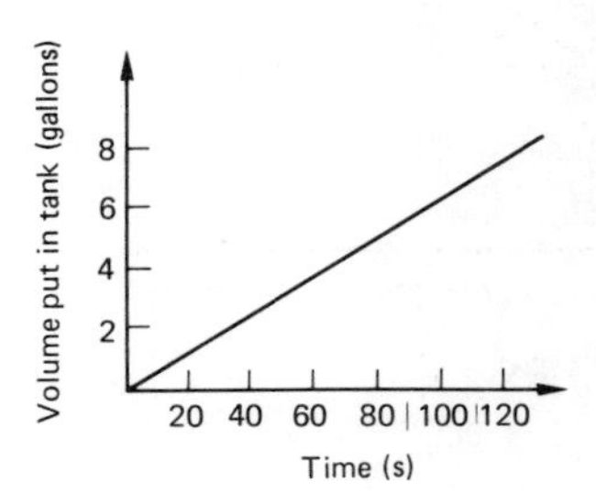

Figure 1.26

The gradient of a graph is often called *the rate of change* of the graph and describes the rate at which the y-variable changes relative to the x-variable. For example, for a graph of distance against time we have

$$\begin{aligned}\text{Gradient} &= \frac{\text{change in distance}}{\text{change in time}} \\ &= \text{rate of change of distance with respect to time} \\ &= \text{speed}\end{aligned}$$

Other examples with which you may be familiar are

gradient of speed/time graph = acceleration

gradient of voltage/current graph = resistance of wire

gradient of tension/extension graph = stiffness of spring (for an elastic spring)

gradient of force/acceleration graph = mass of object

There are of course many others!

At this stage the graphs that we have drawn are straight lines, i.e. lines with constant gradient. We are therefore dealing with constant rates of change. However in Chapter 7 we shall extend the ideas to more general situations.

Example 1.5

The graph below shows the position of a train travelling from Plymouth at various times throughout its journey.

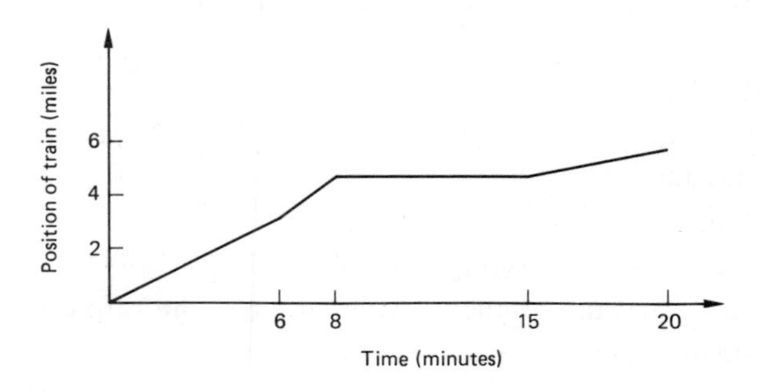

Figure 1.27

(a) What is the speed of the train during the following intervals of time: (0, 6), (6, 8), (8, 15), (15, 20)?
(b) Calculate the average speed for the journey.

Solution

(a) The following table shows the gradient during each time interval and hence the speed.

Table 1.9

Time interval	Gradient	Speed
(0, 6)	$\frac{3}{6} = 0.5$	0.5 miles per min = 30 mph
(6, 8)	$\frac{1\frac{1}{2}}{2} = 0.75$	0.75 miles per min = 45 mph
(8, 15)	$\frac{0}{7} = 0$	0 miles per min = 0 mph
(15, 20)	$\frac{1}{5} = 0.2$	0.2 miles per min = 12 mph

(b) The average speed for the journey is given by

$$\begin{aligned} \text{Average speed} &= \frac{\text{total distance travelled}}{\text{total time taken}} \\ &= \frac{5\frac{1}{2}\text{ miles}}{20\text{ minutes}} \\ &= 16.5\text{ miles per hour} \end{aligned}$$

Note that the average speed is not found by adding together the speeds for each part of the journey and dividing by 4. Be careful with such calculations.

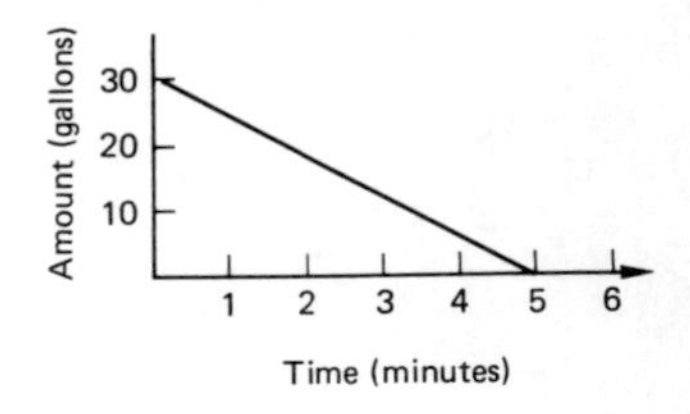

Figure 1.28

Exercise 1.5.2

1. The amount of water in a tank at various times t is shown in Figure 1.28.
   (a) Calculate the gradient of the graph.
   (b) Why does the graph have a negative gradient?

2. The graph in Figure 1.29 shows the amount of petrol in a car's tank during a journey.
   The journey consists of a mixture of town driving and motorways.
   (a) From the graph deduce the parts of the journey that are on motorways and the parts that are in towns. Explain how you decide.
   (b) What does the dotted line CD represent?

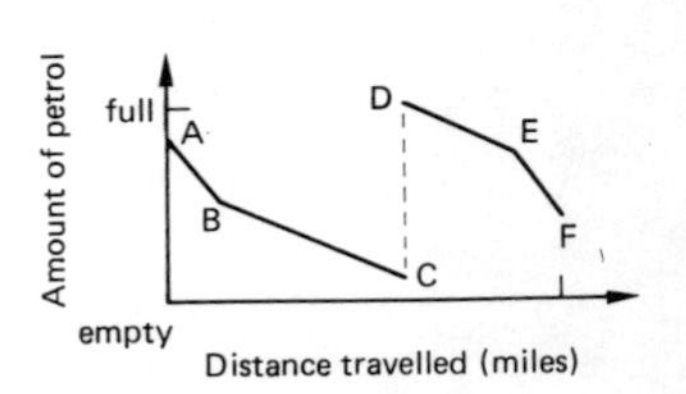

Figure 1.29

1.6 DIRECT PROPORTIONALITY

Consider the extension of a spring/mass system introduced at the beginning of this chapter (see Figure 1.1). Adding masses to the holder produces an extension in the spring. Results from such an experiment are shown in Table 1.10.

Table 1.10

Mass (g)	50	100	150	200	250
Extension (mm)	33	67	100	130	167

Clearly if there is no mass on the spring there is no extension so that a graph of the extension against mass is a straight line through the origin, as shown in Figure 1.30.

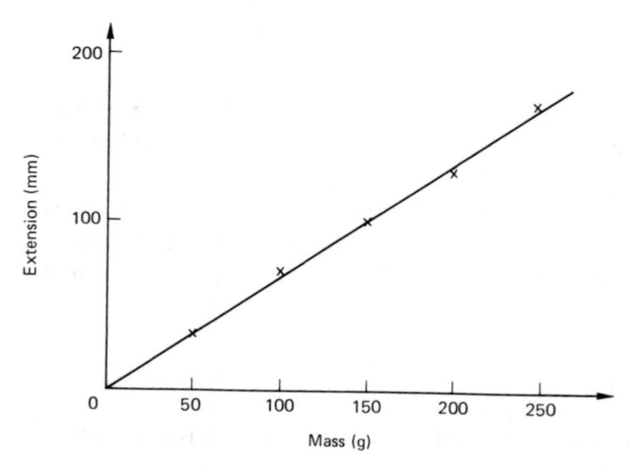

Figure 1.30

Exercise 1.6.1

For which of the following situations would you expect a straight line graph through the origin?

(a) The length l of the shadows of rugby posts of different heights h
(b) The area of a square of side length a
(c) The length l of a lighted candle against time t
(d) The distance d travelled against time for a car on a straight motorway at a constant speed of 50 km per hour if the car starts 30 km from home
(e) The number of seconds y in x minutes

Figure 1.31 shows another situation in which a straight line graph through the origin is obtained. The data shows the results of an experiment in which the current through a wire is measured for different applied voltages.

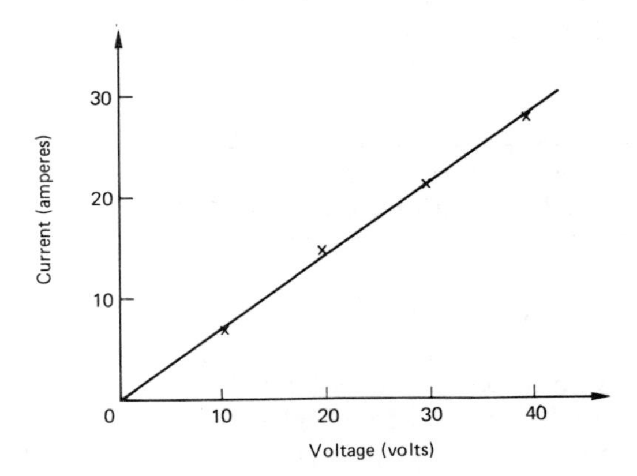

Figure 1.31

The results of this electric circuit experiment and the spring/mass system provide examples of a particular type of relationship between the variables.

If as in the case of the mass-spring and of the current/voltage experiment the graph of one variable, y say, against another, x say, is a straight line through the origin then the variables are said to be *directly proportional*.

Exercise 1.6.2

In which of the graphs shown in Figure 1.32 is one variable directly proportional to the other?

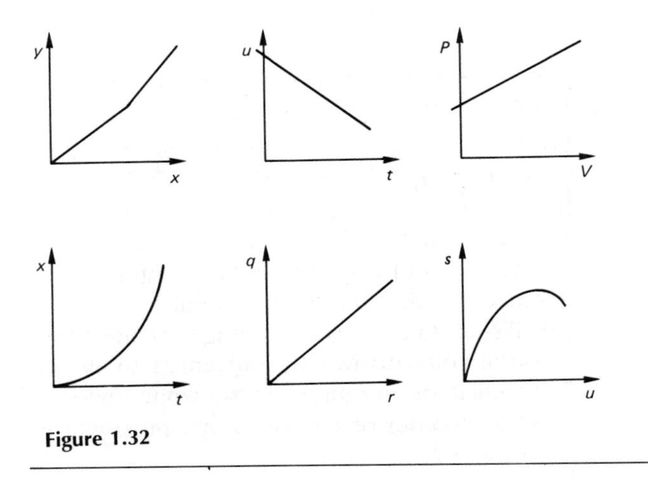

Figure 1.32

An important property of directly proportional variables is that if one variable is multiplied by a number, k say, then the other number is also multiplied by k. This provides a useful means of problem solving.

Example 1.6

The mass of a rope is directly proportional to its length. For a rope of length 100 cm its mass is 1 kg. What is the mass of a rope of length 135 cm? What is the length of a rope of mass 650 grams?

Solution

For a direct proportionality relationship, we look for the appropriate multiplier. For a rope of length 135 cm the multiplier from the given data is 1.35, i.e. we multiply 100 cm by 1.35 to give 135 cm. Now to find the mass of the rope, we multiply the mass 1000 grams (of the 100 cm rope) by 1.35 to give 1350 grams.

Similarly for the rope of mass 650 grams we need to multiply 1000 grams by 0.65, so to obtain the length we multiply 100 cm by 0.65 to give 65 cm.

Displaying the results, we have

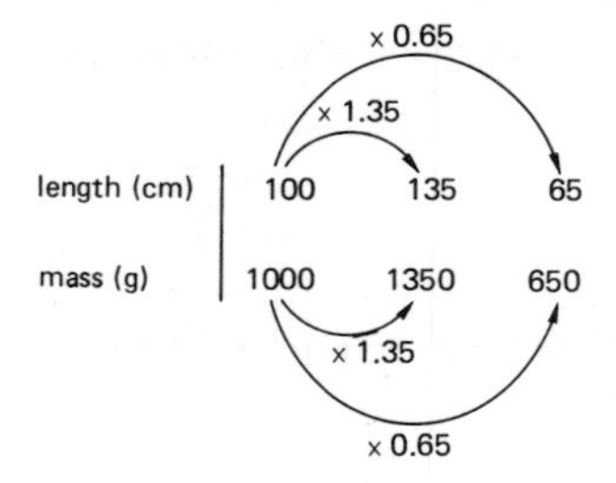

The mass of a rope of length 135 cm is 1350 g and the length of a rope of mass 650 g is 65 cm.

Exercise 1.6.3

1. The current and voltage across a piece of wire are directly proportional. For a voltage of 6 volts the current is 13 amperes.

   What is the current when the voltage is 8 volts? What is the voltage when the current is 16 amperes?
2. In which of the containers shown in Figure 1.33 will the volume of liquid be directly proportional to the depth of liquid?

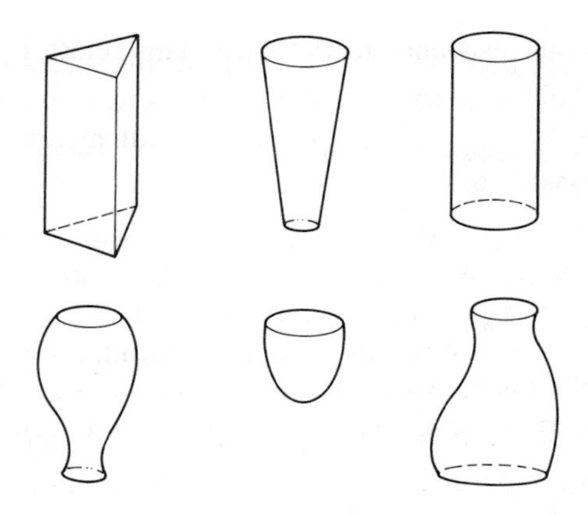

Figure 1.33

For those which the volume V is directly proportional to depth D, write down a formula relating V and D and identify the meaning of the constant.

Algebraically we write

$$y \propto x$$
↑
means proportional to

and since the graph of y against x is a straight line through (0, 0) we have $y = mx$, where m is called *the constant of proportionality*.

For the spring/mass system we have extension $\propto$ mass and for the electric circuit current $\propto$ voltage. The proportionality often leads to a physical law. For the spring/mass system the tension in the spring is given by mg where g is the acceleration due to gravity and instead of

$$m \propto \text{extension}$$

we write

$$\text{Tension} = K \times \text{extension}$$

where K is called the stiffness of the spring. The equation is called Hooke's Law.
For the electric circuit we have

$$\text{Current} \propto \text{voltage}$$

or more usually

$$\text{Voltage} = R \times \text{current} \qquad (\text{i.e. } V = RI)$$

and R is called the resistance of the wire. This relationship is called *Ohm's law*.

Exercise 1.6.4

1. The mass of a lump of wood is proportional to its volume. A wooden block of volume 400 cm^3 has a mass of 320 g. What is the mass of a block of volume 250 cm^3? Write down a law relating mass and volume. What is the value of the constant of proportionality and what name do we usually give to this constant?

2. An experiment to investigate the friction between two surfaces consisted of one block of wood sliding over another block. The frictional force and normal reaction were measured and the readings shown in Table 1.12 were noted.

Table 1.12

Normal reaction, N (in newtons)	24.6	27.1	43.1	52.9	62.7	71.3
Friction force, F (in newtons)	10.0	11.0	17.5	21.5	25.5	29.0

From a graph of this data, is it reasonable to assume that friction force is directly proportional to normal reaction? (Choose Normal Reaction as the independent variable). Find an equation which relates F to N and write down the value of the constant of proportionality (called the *coefficient of friction*).

In summary then if two variables are directly proportional

1. when one variable is multiplied by a number, the other is multiplied by the same number
2. the ratio of the two variables is constant
3. the graph of the variables is a straight line through the origin.

1.7 INVERSE PROPORTIONALITY

Consider the results of an experiment in which the volume and pressure of a quantity of gas are varied but the temperature remains constant (Table 1.13).

Table 1.13

Volume V (litres)	12	15	18	21	24
Pressure p (millibars)	1200	960	800	686	600

Clearly the variables are not directly proportional here; as volume increases the pressure decreases. There is a relationship though with a simple form. Notice that if V is multiplied by 2 then p is divided by 2.

Furthermore, the product of each pair of variables is the same value 14 400.

We say that *p is inversely proportional to V*. A law relating p and V is

$$pV = \text{constant} \quad \text{or} \quad p = \text{constant} \times \frac{1}{V}$$

Example 1.7

In music the frequency of a note is inversely proportional to the 'length' of the instrument. For 'middle C', for a particular string on a guitar, the frequency is 260 hertz and the length of the string is 60 cm.

What is the length of a guitar string to produce a note of frequency 110 hertz? What is the note produced by a guitar string of length 50 cm?

Solution

If the frequency f is inversely proportional to length L then we can write

$$f = \frac{k}{L}$$

for some constant k.

Knowing that $f = 260$ when $L = 60$ we calculate

$$k = 260 \times 60 = 15600$$

Hence

$$f = \frac{15600}{L}$$

When $f = 110$,

$$L = \frac{15600}{110} = 141.8 \text{ cm}$$

When $L = 50$,

$$f = \frac{15600}{50} = 312 \text{ hertz}$$

Exercise 1.7.1

1. A rod of circular cross section is supported at both ends. The mass W kg that the rod can support at its centre is inversely proportional to the length of the rod L cm.
   If $W = 1.5$ when $L = 30$, calculate
   (a) W when $L = 15$
   (b) W when $L = 40$
   (c) L when $W = 2.3$

2. The following table of results comes from an experiment in physics in which the current through a wire is measured for wires of different resistance.

Resistance (ohms)	0.9	1.2	1.5	1.8	2.1	2.4
Current (amperes)	5.56	4.17	3.33	2.78	2.38	2.08

   Show that it is reasonable to assume that current is inversely proportional to resistance. Write down the formula relating resistance and current.

3. The time for a journey is inversely proportional to the speed of the vehicle. Calculate the percentage increase or decrease in the journey time when the speed is
   (a) increased by 50%
   (b) decreased by 50%
   (c) increased from 50 mph to 60 mph
   (d) decreased from 50 mph to 40 mph

1.8 OTHER PROPORTIONALITY

There are many other laws in nature which lead to other types of proportionality. Here are just a few examples,

1. the volume of a sphere is proportional to the cube of the radius ($V = \frac{4}{3}\pi r^3$)
2. the wind resistance on a car is proportional to the square of the car's speed ($R = kv^2$)
3. The force of gravity is inversely proportional to the radius squared r^2, ($F_g = k/r^2$)
4. The distance travelled in time t by a ball dropped down a well is proportional to the square of the time, t^2 ($d = \frac{1}{2}gt^2$)
5. The power of an electric appliance is proportional to the square of the voltage applied to it ($P = E^2/R$)

The type of law given in example 3 is often called an *inverse square law*. It occurs often in science. Another example is the strength of light from a bulb falling on a flat surface which is inversely proportional to the square of the distance between the bulb and the surface.

Exercise 1.8.1

1. A ball is dropped from a window. The distance which it falls, d metres, in time, t seconds, is proportional to the square of the time.
   If $d = 44.1$ when $t = 3$, calculate
   (a) d when $t = 6$
   (b) d when $t = 1\frac{1}{2}$

   Hence deduce what happens to d when t is multiplied by a number k.

2. The power output of a windmill is proportional to the cube of the speed of the wind. Write this law as an equation relating power P and speed u.
   What happens to P if u is multiplied by 2? What happens to P if u is multiplied by any constant, say k? What percentage does P increase by if u increases from 30 knots to 40 knots?

Where there are more than two variables in a physical situation, we often need to combine various proportionality rules into one law, as shown in Example 1.8.

Example 1.8

According to Newton's law of universal gravitation, the force of gravity on an object of mass m due to an object of mass M is proportional to m, M and inversely proportional to the square of the distance between the centres of mass of the objects, r. Write down the law of Universal Gravitation.

Solution

We are told that

$$F \propto m \quad F \propto M \quad \text{and} \quad F \propto \frac{1}{r^2}$$

The first relation, $F \propto m$, means that if we do an experiment keeping M and r constant we find that $F = Cm$. The value of C will depend on the values of M and r chosen. Alternatively if we keep M and m constant then $F = k/r^2$ where k will depend on the values of m and M. Similarly if we keep m and r constant then $F = dM$.
In general if we write

$$F = \frac{GmM}{r^2}$$

where G is a constant, then this equation includes the three possibilities as Table 1.13 shows.

Table 1.13

Variable changing	Variables constant	Constant of proportionality
m	M, r	$c = GM/r^2$
r	m, M	$k = GMm$
M	m, r	$d = Gm/r^2$

Exercise 1.8.2

1. The pressure p of a quantity of gas is directly proportional to its temperature, T, and inversely proportional to its volume, V. Write down the equation describing the law between p, V and T.
2. An object of mass M is suspended from the ceiling by an elastic spring of stiffness k. When the object is pulled down the system vibrates with period P. If P is proportional to the square root of M and inversely proportional to the square root of k, write down the equation describing the law between P, k and M.

1.9 Modelling with Linear Laws

So far in this chapter the data from experiments have been carefully chosen so that the points are very close to a straight line. However in practice experimental data are not so convenient. In such cases we try to draw the 'line of best fit'. There are packages for producing a line of best fit and one is called the regression line. At this stage however we shall draw lines 'by eye'. A simple rule when trying to find a line of best fit in this way is to make sure that there are as many points on one side of the line as on the other. The following example illustrates the method.

Example 1.9.1

The shear strengths of electric welds of different thickness are given in Table 1.14.

Table 1.14

Thickness of weld (mm)	Shear strength (kg)
0.2	102
0.3	129
0.4	201
0.5	342
0.6	420
0.7	591
0.8	694
0.9	825
1.0	1014
1.1	1143
1.2	1219

Draw a graph to show that a linear law is reasonable for this data and from the graph find an equation relating thickness and shear strength.

Use the equation to find the value of the shear strength when (a) thickness = 0.75 mm, and (b) thickness = 1.4 mm.

Solution

A graph of the data is shown in Figure 1.34 and a straight line seems reasonable so we can assume a linear law.

From the graph we calculate the gradient as 1210 and the intercept as −240. The equation relating thickness, t, and shear strength, s, is thus

$$s = 1210t - 240$$

when $t = 0.75$, $s = 667.5$ and when $t = 1.4$, $s = 1454$.

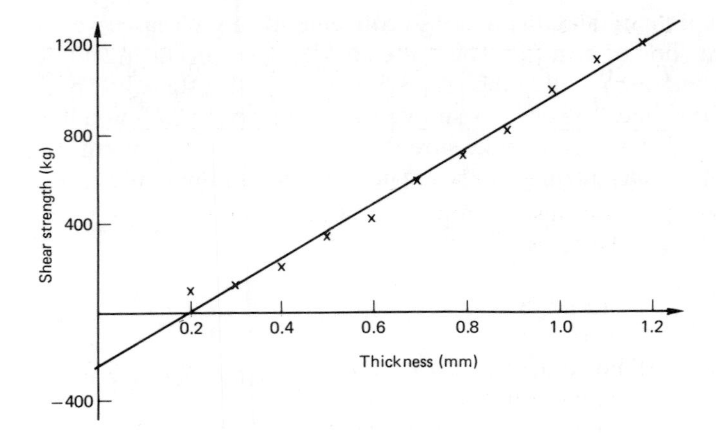

Figure 1.34

Exercise 1.9.1

1. The resistance of a component was measured at different temperatures and the readings shown in Table 1.15 were obtained.

Table 1.15

Temperature, T(°C)	100	120	130	150	170	180
Resistance, R (ohms)	61.8	64.8	66.0	70.0	73.4	75.2

(a) Draw a graph of the resistance (vertical axis) against temperature, draw the line of best fit, and hence find the equation which expresses R in terms of T.

(b) Using the equation determined above, find (on the assumption that this relationship holds true for all temperatures)
   (i) the resistance at a temperature of 20 °C,
   (ii) the temperature at which the component would have zero resistance, (i.e. the inferred zero-resistance temperature).

2. The data in Table 1.16 was gathered from an experiment in a laboratory where water was used to heat a copper bar and its length after heating was noted. The temperature T was measured in degrees centigrade and l the length of the bar in centimetres.

Table 1.16

T	20	30	40	50	60	70	80	90	100
l	7.4	7.5	7.6	7.7	7.8	7.9	8.0	8.1	8.2

Establish a linear equation relating l to T.

(a) Use this relation to find the length of the bar when the temperature is
  (i) 25 °C
  (ii) 55 °C
  (iii) 75 °C

(b) For obvious reasons it was not possible to heat the temperature of the bar to above 100 °C and room temperature would not allow its temperature to fall below 20 °C. Use the linear relation to estimate
  (i) the length of the bar at 0 °C
  (ii) the length of the bar at 200 °C
  (iii) the temperature at which the bar's length was double its value at 20 °C.

In Example 1.9.1 we used a linear equation to find the value of the shear strength for a thickness of 0.75 mm. This process of calculating values within the interval of the dependent variable is called *linear interpolation*. We can be fairly confident about values found from linear interpolation. We also found the value of the shear strength for a thickness of 1.4 mm. This point is outside the range of values in the experiment and the process is called *linear extrapolation*. It assumes that the linear model will be continued beyond the range of values used in the experiment. This method should be used with caution because there is no guarantee that the law will remain unchanged outside the range and it can lead to nonsense, as the following example illustrates.

Example 1.9.2

According to the Guinness Book of Records (Guinness Superlatives Ltd, 1984) "the tallest man of whom there is irrefutable evidence was ... Robert Pershing Wadlow, born ... on 22 February 1918 in Alton, Illinois, USA". The following figures for his height at various ages are given in Table 1.17. He died on 15 July 1940 at the age of 22.

Investigate whether Wadlow's growth could reasonably be described as linear during the period for which data are given. How tall might he have been at the age of 30 and of 40?

Table 1.17

Age (years)	Height (cm)
5	163
9	189
11	201
13	218
15	234
17	245
19	258
21	265

Solution

Figure 1.35 shows the graph of the data.

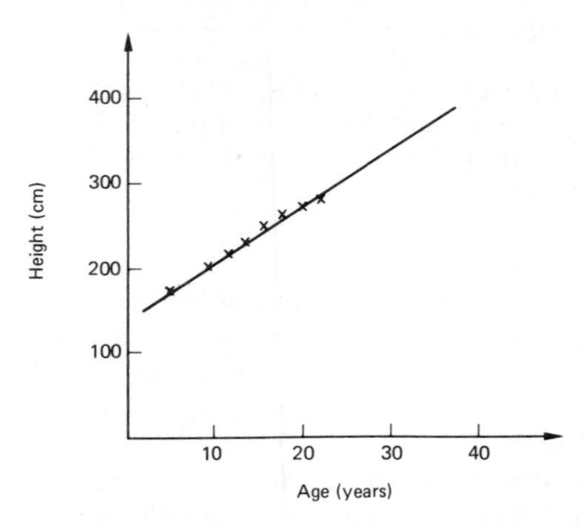

Figure 1.35

A straight line graph (i.e. a linear model) seems reasonable for the range of data given. Using the graph we could predict by extrapolation that at the age of 30 years Wadlow would be 337 cm tall and at 40 he would be 398 cm tall. Clearly this is highly unlikely and Wadlow's height would have been likely to level off to a constant value at some stage.

Exercise 1.9.2

In Exercise 1.9.1 identify whether the process of interpolation or extrapolation is being used.

OBJECTIVE TEST 1

1. Plot a graph of the data in Table 1.18. Find the gradient and intercept and hence write down the equation connecting the variables.

Table 1.18

t	0.1	0.5	1.0	1.4	1.7	2.1
x	−3.89	−2.65	−1.1	0.14	1.07	2.31

2. For each of the equations of the straight lines write down (a) the gradient, (b) the y-intercept and (c) the x-intercept (i.e. where it cuts the x-axis).

(i) $y = 2x + 5$ (ii) $y = 3x - 4$
(iii) $y = -4x + 8$ (iv) $x + 2y - 3 = 0$
Draw graphs of each of these equations.

3. Which of the following points lie on the line $y = -2x + 3$?
P(0, 1) Q(1, 1) R(−4, 11) S(3, 0)
A(1.5, 0) B(−1, 6) C(0.5, 4) D(2, −1)

4. For each of the following equations what quantities should be plotted to obtain straight line graphs?

(a) $T = 4x^2 + 5$ (b) $p = \frac{37}{V}$

(c) $W = \frac{2}{l^2} - 3$ (d) $T = 3\sqrt{L}$

5. In the isothermal compression (i.e. compression at constant temperature) of a gas, the pressure in the gas is directly proportional to the density. Complete Table 1.19 and find the equation relating pressure and density.

Table 1.19

Pressure, p (kNm^{-3})	2.4	3.1	
Density, ρ (kgm^{-3})	2.78×10^{-5}		5.01×10^{-5}

6. The conductance of a resistor is inversely proportional to its resistance. Complete Table 1.20 and find the equation relating conductance and resistance.

Table 1.20

Conductance, G	5		32
Resistance, R(ohms)	0.2	0.001	

7. The velocity of sound in a gas (V) is directly proportional to the square root of the pressure (P) and inversely proportional to the square root of the density (ρ). Suggest an equation between, V, P and ρ.

8. In an experiment the resistance of a wire is measured at different temperatures and Table 1.21 shows the results.

Table 1.21

Temperature, T(°C)	60	120	180	240	300
Resistance, R(ohms)	43.8	56.6	69.3	81.4	94.7

(a) Draw the line of best fit graph of resistance against temperature and hence find an equation relating R and T.
(b) What is the rate of change of resistance with temperature?
(c) Find the resistance at temperatures of (i) 100 °C, (ii) 400 °C.
(d) Find the temperature at which the resistance is zero.
(e) Which of your answers to (c) and (d) would you be most confident about?

2 QUADRATICS AND POLYNOMIALS

OBJECTIVES

When you have completed this chapter you should be able to

1. identify a polynomial and write down its degree
2. sketch the graph of a quadratic
3. factorize a quadratic
4. find the roots of a quadratic either from a sketch graph or by factorizing or by using 'the formula'
5. sketch the graph of a cubic polynomial
6. understand the difference between a local maximum, a local minimum and a point of inflexion
7. evaluate polynomials by nested multiplication
8. find the roots of a polynomial either from a sketch graph or by the interval bisection method

2.1 INTRODUCTION

In the previous chapter we discussed the properties and equation of a graph of a straight line and used the formula $y = mx + c$ as a linear law or rule between two variables occurring in various physical situations. Another law that we met was the inverse proportionality law $y = k/x$. Again we found several physical situations for which this law provides a good model. For this law, the graph is not a straight line (Figure 2.1).

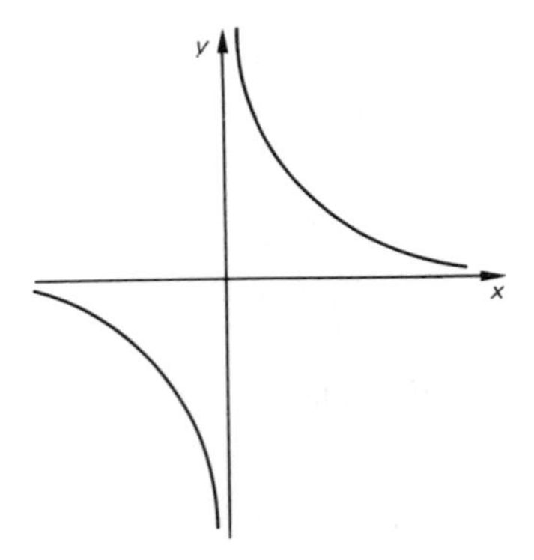

Figure 2.1

In this chapter we introduce another rule between two variables x and y, and investigate the graph and properties of this rule. The rule is called a *quadratic law* between x and y, and has the form $y = ax^2 + bx + c$.

As an example of a quadratic law consider the data in Table 2.1 on stopping distances taken from the Highway Code.

Table 2.1 Highway code stopping distances

Speed	Thinking distance		Braking distance		Overall stopping distance	
mph	m	ft	m	ft	m	ft
20	6	20	6	20	12	40
30	9	30	14	45	23	75
40	12	40	24	80	36	120
50	15	50	38	125	53	175
60	18	60	55	180	73	240
70	21	70	75	245	96	315

From Table 2.1 we can see that the overall stopping distance is the sum of the thinking distance and the braking distance. Each of these in fact depends on the speed. Let us take them in turn.

The thinking distance is the distance travelled while the driver reacts to any impending danger. From the data we can see that the thinking distance in feet is equal to the velocity in miles per hour, so we can write

$$D_T = V$$

where D_T is measured in feet and V is in mph.

For the braking distance the law is not quite so obvious. However it turns out to be a simple quadratic rule

$$D_B = \frac{1}{20} V^2$$

where D_B is in feet and V is in mph.

Hence the overall stopping distance obeys the rule

$$D = V + \frac{1}{20} V^2$$

The equation describing the path of a projectile is a quadratic (providing we neglect the effects of air resistance). For example, the quadratic rule

$$y = 2 + 0.5x - 0.049x^2$$

describes the path of a cricket ball thrown from a height of 2 metres with a speed of 20 ms^{-1} at an angle of 60 ° to the horizontal. Figure 2.2 shows a graph of this rule; the curve is called *a parabola*. A projectile will follow such a path, you can try it by throwing a stone or a ball across a field to check this.

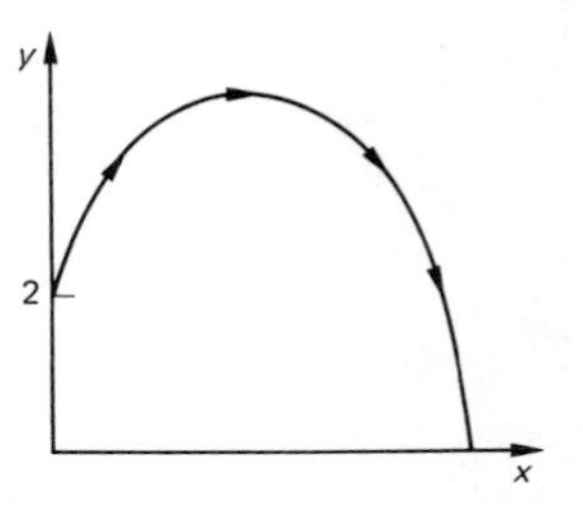

Figure 2.2

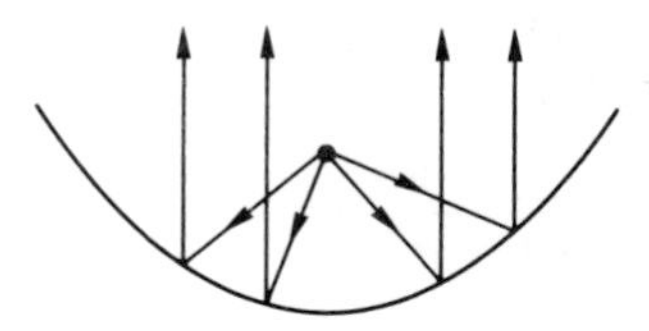

Figure 2.3

Reflectors of light in torches (Figure 2.3) are designed to be parabolic in shape. This is because the light rays from the bulb can be reflected off the parabolic surface (which is usually made of a reflective material) to form a parallel beam of light.

Quadratic relationships between two variables occur often in science and engineering and in this chapter we introduce some of their properties. Consider the quadratic

$$y = ax^2 + bx + c$$

a, b and c are constants called the coefficients of the quadratic. Their values do *not* depend on x or y. For example, consider the quadratic law

$$y = 3x^2 - 4x + 2$$

The coefficients are $a = 3$, $b = -4$ and $c = 2$. Sometimes the equation is written with the x^2 term as the last term instead of the first. For example, the path of a projectile can be written as

$$y = 2 + 0.5x - 0.049x^2$$

The coefficients are $a = -0.049$, $b = 0.5$ and $c = 2$. Although common variables are x and y, we often use other symbols whose meaning can be easily referred to the names of the variables. For example, the Highway Code data leads to the quadratic law

$$D = V + \frac{1}{20}V^2$$

and D and V are overall stopping distance (in feet) and speed (in mph). The coefficients are then $a = 1/20$, $b = 1$ and $c = 0$. As before, a is the coefficient of the 'variable squared' term etc.

Exercise 2.1.1

Complete Table 2.2 showing the coefficients of the quadratics.

Table 2.2

Quadratics	Coefficients		
	a	b	c
$2x^2 + 3x + 1$			
$3x^2 - 5x - 3$			
$7 + 4x + x^2$			
$-2x^2 + 5$			
$-9x^2 + x$			
$-x^2 - 3x + 4$			
$4 - 20t + 9.8t^2$			
$1 + v + 0.7v^2$			
$u - 3u^2$			

2.2 GRAPHS OF QUADRATICS

In this section we plot the graphs of various quadratics and discover the types of parabolas that can occur and how their positioning is governed by the coefficients that appear in the formula.

To plot the graph we draw up a table showing the value of each term in the quadratic and add them.

Example 2.2.1

Plot a graph of the equation $y = 2x^2 - 5x - 12$.

Solution

Table 2.3 shows how to lay out the x-values and corresponding y-values.

Table 2.3

x	-2	-1	0	1	2	3	4	5
$2x^2$	8	2	0	2	8	18	32	50
$-5x$	10	5	0	-5	-10	-15	-20	-25
-12	-12	-12	-12	-12	-12	-12	-12	-12
y	6	-5	-12	-15	-14	-9	0	13

Choosing intervals $-2 \leqslant x \leqslant 5$, $-16 \leqslant y \leqslant 16$ the graph is shown in Figure 2.4

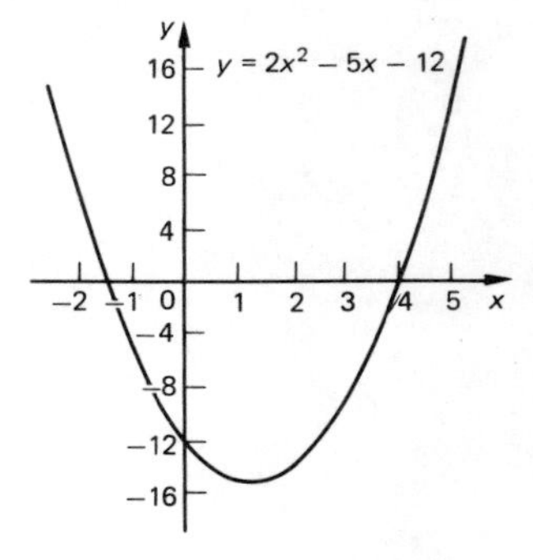

Figure 2.4 Graph of $y = 2x^2 - 5x - 12$

Example 2.2.2

Plot a graph of the equation $y = -2x^2 + 3x + 5$.

Solution

Table 2.4 gives values and the graph is shown in Figure 2.5.

Table 2.4

x	-1	$-\frac{1}{2}$	0	$\frac{1}{2}$	1	2	3
$-2x^2$	-2	$-\frac{1}{2}$	0	$-\frac{1}{2}$	-2	-8	-18
$3x$	-3	$-\frac{3}{2}$	0	$\frac{3}{2}$	3	6	9
5	5	5	5	5	5	5	5
y	0	3	5	6	6	3	-4

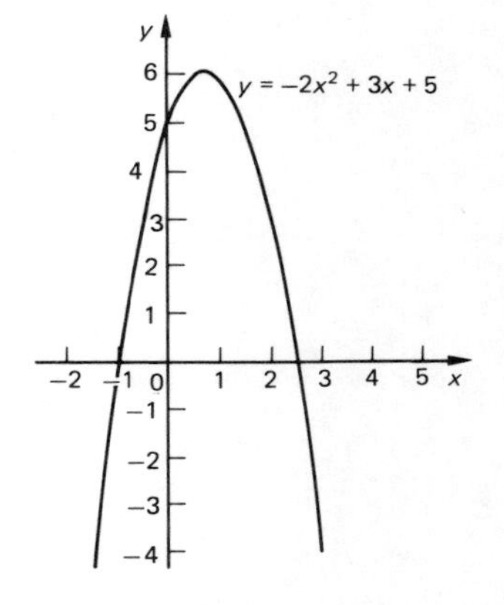

Figure 2.5 Graph of $y = -2x^2 + 3x + 5$

Exercise 2.2.1

Draw graphs for the following quadratics.

(a) $y = 2x^2 + 3x + 1$ (b) $y = -2x^2 + 5$
(c) $y = -x^2 - 3x + 4$ (d) $y = 3x^2 - 5x - 3$
(e) $y = 2x^2 - 4x + 2$ (f) $y = x^2 + 2x + 2$

The graph of $y = ax^2 + bx + c$ is called a parabola and any equation of this type has one of the two general shapes shown in Figure 2.6.

The orientation depends on the sign of the coefficient of x^2, i.e. the sign of a. If a is a positive number then the parabola has a *minimum value* (a trough) whereas if a is a negative number then the parabola has a *maximum value* (a peak). We will see in a later

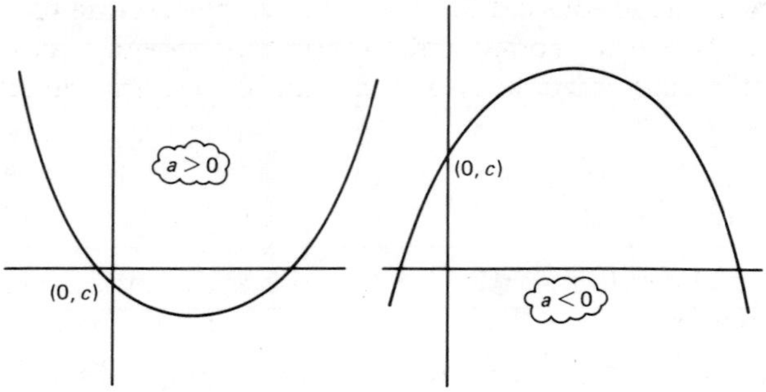

Figure 2.6

chapter that being able to identify these peaks and troughs is an important part of graph drawing. Where the graph cuts the y-axis is called *the intercept* and is given by the value of the coefficient c.

From the graphs of Examples 2.2.1, 2.2.2 and Exercise 2.2.1 you will see that the parabola may cut the x-axis in two places, may touch the axis or in some cases may miss the x-axis altogether.

Where the graph cuts the x-axis or touches the x-axis are special values of the quadratic equation

$$ax^2 + bx + c = 0$$

The x-values which are solutions of this equation are called the *roots* of the equation. There are three cases which may occur depending on the values of the coefficients a, b and c (Figure 2.7).

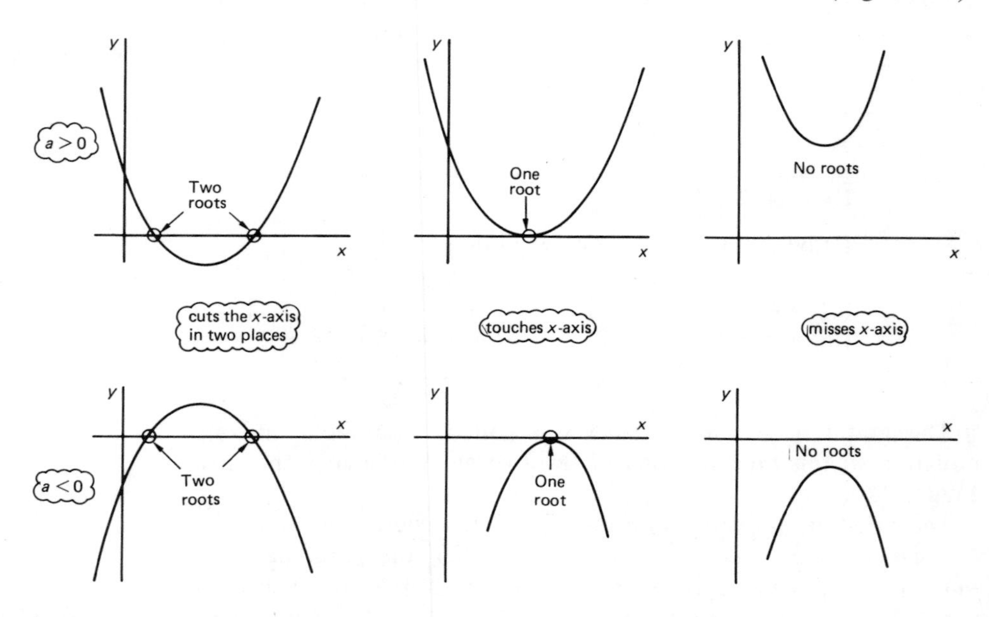

Figure 2.7

We have two roots if the graphs cut the x-axis in two places, one root (sometimes we say two equal roots) if the graph touches the x-axis. (Actually the latter case introduces a new number system called complex numbers.)

Example 2.2.3

Use the graphs in Figures 2.4 and 2.5 to find the roots of the quadratic equations $2x^2 - 5x - 12 = 0$ and $-2x^2 + 3x + 5 = 0$.

Use the graph to find the solution of the equation $2x^2 - 5x - 12 = 8$.

Solution

From Figure 2.4 we see that the graph cuts the x-axis at the points $x = -1.5$ and $x = 4$. Hence the roots are $x = -1.5$ and $x = 4$.

From Figure 2.5 we see that the graph cuts the x-axis at the points $x = -1$ and $x = 2.5$. Hence the roots are $x = -1$ and $x = 2.5$.

To solve the equation $2x^2 - 5x - 12 = 8$, we draw the line $y = 8$ (see Figure 2.8) and the solution of the equation is then the x-values of the points of intersection of the two graphs $y = 2x^2 - 5x - 12$ and $y = 8$. These x-values are $x = -2.2$ and $x = 4.6$.

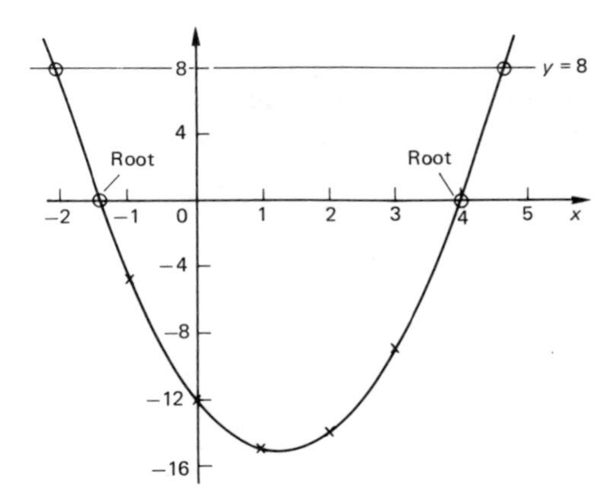

Figure 2.8

Exercise 2.2.2

1. Use your graphs in Exercise 2.2.1 to estimate the solutions of the quadratic equations
   (a) $2x^2 + 3x + 1 = 0$ (b) $-2x^2 + 5 = 0$
   (c) $-x^2 - 3x + 4 = 0$ (d) $3x^2 - 5x - 3 = 0$
   (e) $2x^2 - 4x. + 2 - 0$ (f) $x^2 + 2x + 12 = 0$

2. Use your graphs in Exercise 2.2.1 to estimate the solutions of the equations
(a) $2x^2 + 3x + 1 = 5$ (b) $-2x^2 + 5 = 1$
(c) $-x^2 - 3x + 4 = 0$ (d) $-2x^2 + 5 = 2x + 1$

3. The height s above the ground of a projectile at a time t seconds after release is given by $s = 2 + 20t - 4.9t^2$. Plot a graph of this quadratic and from your graph find (a) the height of release, (b) the time to reach the maximum height, and (c) the time for the projectile to reach the ground.

2.3 FORMING QUADRATICS FROM LINEAR FACTORS

Quadratics can be formed by multiplying two (linear) factors of the form $(x + \alpha)(x + \beta)$. For example,

$$\begin{aligned}(x + 2)(x - 1) &= x(x - 1) + 2(x - 1)\\ &= x^2 - x + 2x - 2\\ &= x^2 + x - 2\end{aligned}$$

The quadratic $x^2 + x - 2$ can be formed by multiplying out the brackets $(x + 2)$ and $(x - 1)$. These are called *linear factors* of $x^2 + x - 2$.

A useful way of displaying the multiplication is in the form of a table, as shown in Table 2.5.

Table 2.5

	x	-1
x	x^2	$-x$
2	$2x$	-2

The entries in Table 2.5 are formed by multiplying the headings of each row and column. The result of multiplying $(x + 2)$ by $(x - 1)$ is the sum of the entries in the table. So

$$(x + 2)(x - 1) = x^2 - x + 2x - 2 = x^2 + x - 2$$

Example 2.3

Expand the following as quadratics.

(a) $(x + 2)(x - 3)$ (b) $(x - 2)^2$ (c) $(x - 2)(x + 2)$

Solution

(a)

	x	-3
x	x^2	$-3x$
2	$2x$	-6

So $(x + 2)(x - 3) = x^2 - 3x + 2x - 6 = x^2 - x - 6$

(b)

	x	-2
x	x^2	$-2x$
-2	$-2x$	4

So $(x - 2)^2 = (x - 2)(x - 2) = x^2 - 4x + 4$

(c)

	x	2
x	x^2	$2x$
-2	$-2x$	-4

So $(x - 2)(x + 2) = x^2 - 4$

Exercise 2.3.1

Expand the following giving your answer as quadratics

(a) $(x + 1)(x + 3)$
(b) $(x - 1)(x - 2)$
(c) $(x + 1)^2$
(d) $(2x - 1)(3x + 2)$
(e) $(2x - 3)^2$
(f) $(x + 4)(2x - 1)$
(g) $(x + 5)(x - 7)$
(h) $(x - u)(x + v)$
(i) $(x - u)(x + u)$
(j) $x(x + 2)$
(k) $3x(2x - 4)$

The example and exercise have introduced two special cases which are worth noting and remembering.

Special case 1

$(rx + s)^2 \quad = \quad (rx + s)\,(rx + s) \quad = \quad r^2x^2 \quad + \quad 2rsx \quad + \quad s^2$

r^2x^2: (first term)^2

$2rsx$: (twice first term x second term)

s^2: (second term)^2

Special case 2

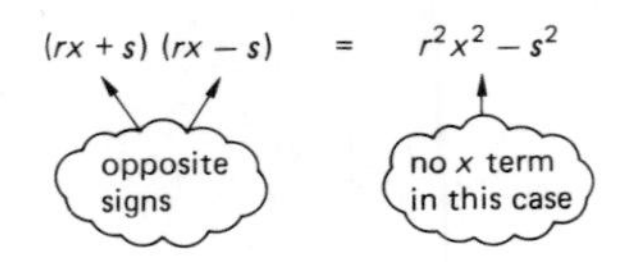

Exercise 2.3.2

Write down the quadratics in the following

(a) $(x - 1)^2$ (b) $(x + 4)^2$
(c) $(2x - 3)^2$ (d) $(x - 1)(x + 1)$
(e) $(x + 4)(x - 4)$ (f) $(2x - 3)(2x + 3)$

2.4 FACTORIZING QUADRATICS

An important skill in the algebra of quadratics is to be able to re-write quadratic rules in terms of linear factors. This is called *factorizing*.

For example, consider the quadratic $x^2 + 3x$. Finding the factors in this case is straightforward because we can spot that x is a factor of each term. We have

$$x^2 + 3x = x(x + 3)$$

It is not quite so obvious what the linear factors of $x^2 + 4x - 12$ are. Factorizing is the opposite of expanding brackets so we use the table to help us.

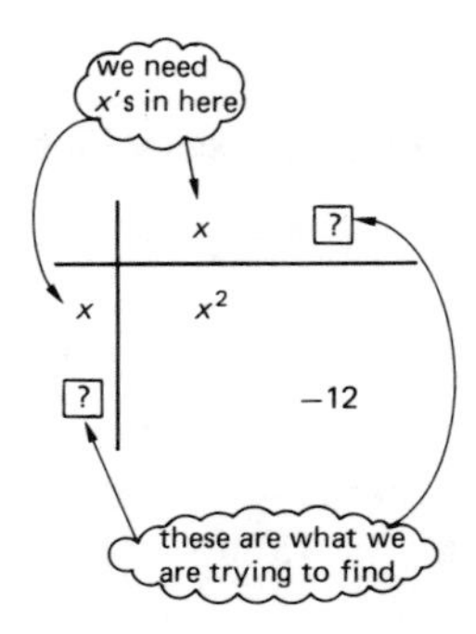

We can put the x^2 and -12 in the table. The two missing values must add to give $4x$.

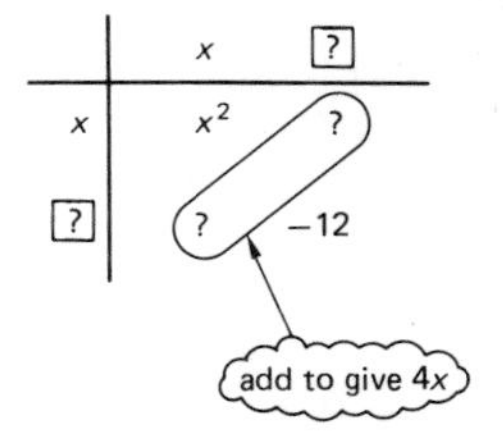

The product of the missing numbers in the boxes must be -12, so they are the factors of -12. The six pairs of factors of -12 are

$$(-4, 3) \quad (4, -3) \quad (-12, 1) \quad (12, 1) \quad (-6, 2) \quad (6, -2)$$

So we have six possible factors to try and only one pair will combine to give $4x$. The result of trying out each pair of factors is shown in Table 2.6.

Table 2.6

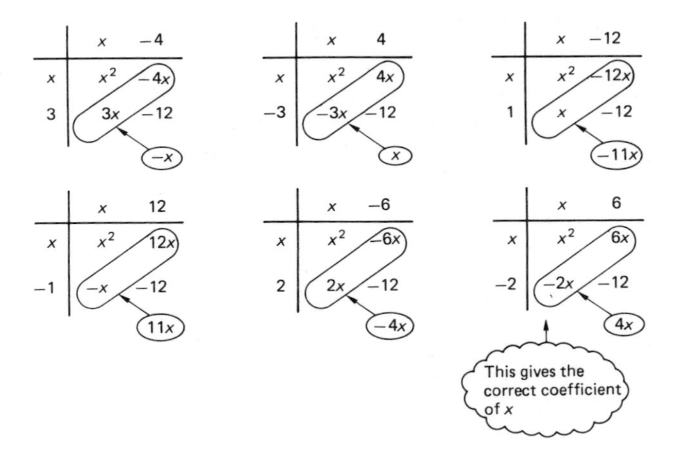

So the factors of $x^2 + 4x - 12$ are $(x - 2)$ and $(x + 6)$ and we can write

$$x^2 + 4x - 12 = (x - 2)(x + 6)$$

In practice we do not need to go through every case, it is usually fairly easy to reject some factors immediately. In the example above we could have rejected the factors involving 12 and 1, (and probably those involving 4 and 3) since we are looking for our factors to combine to give $4x$.

If the coefficient of x^2 is not equal to 1 the tabular method still works but we also have to consider the factors of the coefficient of x^2.

Example 2.4

Factorize $2x^2 + 5x - 3$.

Solution

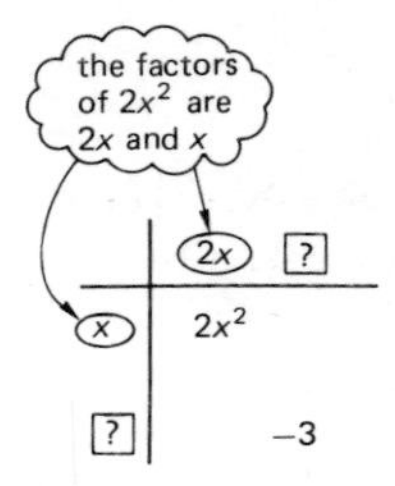

For the unknown boxes we need the factors of -3. These are $(-3, 1)$ or $(3, -1)$. This gives four possibilities (Table 2.7).

Table 2.7

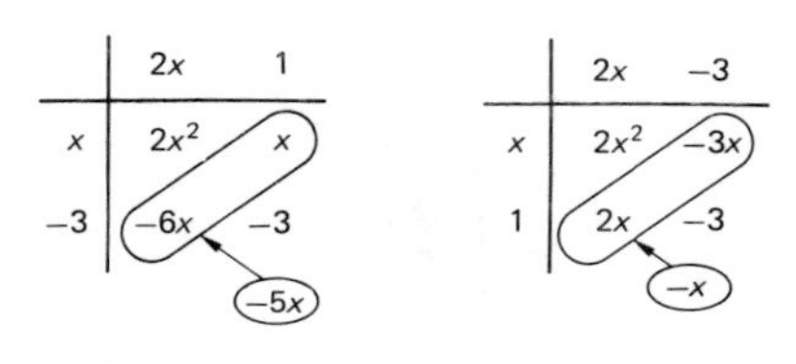

Clearly the last attempt gives $5x$ so we have

$$2x^2 + 5x - 3 = (x + 3)(2x - 1)$$

Exercise 2.4.1

Factorize the following quadratics

(a) $x^2 + 5x + 4$
(b) $x^2 - 3x - 4$
(c) $x^2 + x - 12$
(d) $x^2 + 7x + 10$
(e) $2x^2 + 7x + 5$
(f) $9x^2 - 1$
(g) $x^2 + 6x + 9$
(h) $x^2 - 8x + 16$
(i) $x^2 - u^2$
(j) $4x^2 + 4ux + u^2$
(k) $x^2 + 4x$
(l) $3x^2 - 10x$

2.5 FINDING ROOTS BY FACTORIZING

When a quadratic is written in terms of its linear factors then it is easy to find the roots of the quadratic equation. The method relies on the important rule:

> If $A \times B = 0$
> then either $A = 0$ or $B = 0$.

Consider the quadratic $2x^2 + 5x - 3$ and its linear factors $(x + 3)$ and $(2x - 1)$. The quadratic equation

$$2x^2 + 5x - 3 = 0$$

can be written as

$$(x + 3)(2x - 1) = 0$$

Hence using the rule

$$\text{either} \quad x + 3 = 0 \quad \text{or} \quad 2x - 1 = 0$$

$$\text{so} \quad x = -3 \quad \text{or} \quad x = \tfrac{1}{2}$$

The problem reduces to one of solving two linear equations, and is a more accurate and often easier method of finding the roots of a quadratic equation than drawing its graph.

Exercise 2.5.1

Use your factors from Exercise 2.4 to find the roots of the quadratic equations

(a) $x^2 + 5x + 4 = 0$ (b) $x^2 - 3x - 4 = 0$
(c) $x^2 + x - 12 = 0$ (d) $x^2 + 7x + 10 = 0$
(e) $2x^2 + 7x + 5 = 0$ (f) $9x^2 - 1 = 0$
(g) $x^2 + 6x + 9 = 0$ (h) $x^2 - 8x + 16 = 0$
(i) $x^2 - u^2 = 0$ (j) $4x^2 + 4u + u^2 = 0$
(k) $x^2 + 4x = 0$ (l) $3x^2 - 10x = 0$

2.6 FINDING ROOTS BY USING 'THE FORMULA'

Although factorizing provides an elegant method of solving quadratic equations, it is only easy to find the factors when they involve integers. An alternative method is to use a standard formula.

Before we develop a general formula, consider the quadratic equation

$$x^2 + 2x - 3 = 0$$

The two terms $x^2 + 2x$ can be combined as

$$x^2 + 2x = (x + 1)^2 - 1$$

This method is known as 'completing the square'. Hence

$$x^2 + 2x - 3 = [(x + 1)^2 - 1] - 3 = 0$$

so

$$(x + 1)^2 - 4 = 0$$
$$(x + 1)^2 = 4$$

Taking square roots of each side

$$x + 1 = \pm 2$$

Solving for x,

$$x = -1 + 2 \quad \text{or} \quad x = -1 - 2$$

i.e.

$$x = 1 \qquad \text{or} \quad x = -3$$

Example 2.6.1

For each of the following complete the square

(a) $x^2 + 16x$ (b) $x^2 + 16x + 14$

Solution

(a) We need to write $x^2 + 16x$ in the form $(x + s)^2$ for some numbers. Now we know that $(x + s)^2$ is $x^2 + 2s + s^2$.
Comparing this with $x^2 + 16x$ clearly we require $2s = 16$ so $s = 8$. Hence we write

$$x^2 + 16x = (x + 8)^2 - 64$$

(b) Similarly

$$x^2 + 16x + 14 = (x + 8)^2 - 50$$

The $(x + 8)^2$ gives the $x^2 + 16x$ and the -50 combines with the 8^2 to give 14.

Exercise 2.6.1

1. For each of the following 'complete the square'.
   (a) $x^2 + 6x$ (b) $x^2 + 10x$
   (c) $x^2 + 5x$ (d) $x^2 + 7x$

2. Hence solve the following quadratic equations.
   (a) $x^2 + 6x + 8 = 0$ (b) $x^2 + 10x + 9 = 0$
   (c) $x^2 + 5x + 1 = 0$ (d) $x^2 + 7x - 3 = 0$

More generally consider the quadratic

$$ax^2 + bx + c = 0$$

Our first step is to factorize by a, and rewrite the expression as

$$a\left(x^2 + \frac{b}{a}x + \frac{c}{a}\right)$$

Now to complete the square we attempt to write this as $a(x + s)^2 + r$ for some numbers s and r. Expanding we have

$$a(x^2 + 2xs + s^2) + r$$

so

$$2s = \frac{b}{a} \quad \text{and} \quad r = c - as^2$$

The important point here is that $s = b/2a$, so that the x terms have come together as a perfect square.

$$ax^2 + bx \rightarrow a\left(x + \frac{b}{2a}\right)^2$$

For example, consider the expression $3x^2 + 4x$. To complete the square we first write

$$3x^2 + 4x = 3\left(x^2 + \frac{4}{3}x\right)$$

Now 'completing the square' we have

$$3\left(x^2 + \frac{4}{3}x\right) = 3\left(x + \frac{2}{3}\right)^2 - \frac{4}{3}$$

Exercise 2.6.2

1. For each of the following 'complete the square'.
   (a) $2x^2 + 8x$ (b) $3x^2 + 7x$
   (c) $4x^2 + 12x - 3$ (d) $2x^2 - 5x + 1$

2. Now solve the following quadratic equations.
   (a) $4x^2 + 12x - 3 = 0$ (b) $2x^2 - 5x + 1$

A general formula exists to solve a quadratic equation, this can be derived using the completing square method but we have omitted the derivation. 'The Formula' is

The solution of the quadratic equation

$$ax^2 + bx + c = 0$$

is

$$x = \frac{-b \pm \sqrt{b^2 - 4ac}}{2a}$$

This is one of the most important and useful formulas in algebra and should be remembered.

Example 2.6.2

Use the formula to find the roots of the quadratic equation

$$x^2 - 2x - 8 = 0$$

Solution

Identifying the coefficients we have

$$a = 1 \quad b = -2 \quad c = -8$$

Substituting into the formula the roots are

$$x = \frac{-(-2) \pm \sqrt{(-2)^2 - 4(1)(-8)}}{2(1)}$$

$$= \frac{2 \pm \sqrt{4 + 32}}{2}$$

$$= \frac{2 \pm \sqrt{36}}{2}$$

$$= \frac{2 \pm 6}{2}$$

Hence

$$x = \frac{2 + 6}{2} = 4 \quad \text{or} \quad x = \frac{2 - 6}{2} = -2$$

Exercise 2.6.3

1. Use the formula to solve the quadratic equations in Exercise 2.10 and 2.11 and check your solutions to that exercise.

2. Use the formula to solve equations
   (a) $x^2 + x - 3 = 0$ (b) $2x^2 + x - 4 = 0$
   (c) $3x^2 - 4x + 1 = 0$ (d) $x^2 + 4x + 4 = 0$
   (e) $x^2 - 5 = 0$ (f) $x^2 + 5x - 3 = 0$

Clearly the value of $b^2 - 4ac$ is an important part of the formula. This quantity is often called the *discriminant* of the quadratic equation $ax^2 + bx + c = 0$. If $b^2 - 4ac$ is negative then we cannot take its square root.

From the formula it follows that

if $b^2 - 4ac > 0$, there are two distinct roots

if $b^2 - 4ac = 0$, there is one root (or two equal roots)

if $b^2 - 4ac < 0$, there are no (real) roots

When $b^2 - 4ac < 0$ we are trying to take the square roots of negative numbers which as yet we cannot do.

Exercise 2.6.4

1. Calculate the discriminant for each of the following quadratic equations. Hence state how many roots you would expect.
   (a) $3x^2 - 4x - 2 = 0$ (b) $3x^2 - 4x + 2 = 0$
   (c) $3x^2 - 6x + 1 = 0$ (d) $4x^2 + 1 = 0$
   (e) $x^2 + 6x + 5 = 0$ (f) $x^2 + 7x = 0$

2. Consider the quadratic equation

$$x^2 + Kx + 3 = 0$$

For what value of K will there be one root? Find the range of values for K so that the equation has two roots.

3. In mechanics it can be shown that the bending moment M at a distance s from one end of a uniform beam is given by

$$M = bs + as^2$$

Find the distance s from the end of the beam for the point at which the bending moment is 40 newton metre in each of the following cases
(a) $a = -3$, $b = 30$
(b) $a = -2.5$, $b = 25$
In each case find where the bending moment is zero.

4. The thermo-emf E (in microvolts) corresponding to the temperature T (in degrees centigrade) for a copper-iron thermocouple satisfies the quadratic law

$$E = -0.0192T^2 + 6.88T$$

The law is found to work well for temperatures between about 250–400 °C. This means that we can use the voltage reading across the thermocouple to calculate the temperature.
If the voltage is indicated as 300 μV find the temperature that is being measured.

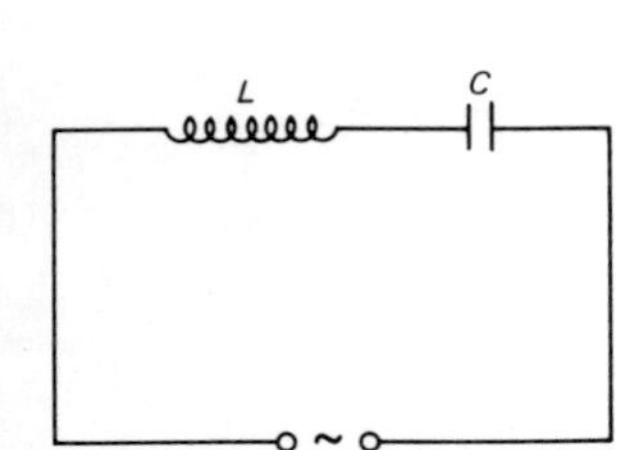

Figure 2.9

5. In the electric circuit shown in Figure 2.9 the reactance X is given by the formula

$$X = \omega L - \frac{1}{\omega C}$$

where $\omega = 2\pi f$ and f is the frequency of the a.c. supply.
In each of the following cases form and solve a quadratic equation for ω

(a) $L = 5$ H, $C = 0.05$ F, $X = 15$ ohms
(b) $L = 15$ μH, $C = 40$ μF, $X = -80$ ohms

2.7 POLYNOMIALS

The linear law $mx + c$ and quadratic law $ax^2 + bx + c$ are just two of a whole class of expressions called *polynomials*. A polynomial in x is an expression involving powers of x which are positive integers usually written in descending powers of x. For instance

$$p = 3x^4 + 7x^3 - 2x^2 + x - 11$$
$$q = 9.1x^5 - 3.7x^2 + 1.3x$$

are examples of polynomials. The numbers 3, 7, −2, 1 and −11 in p and 9.1, −3.7 and 1.3 in q are called *coefficients*. Note that although powers of x must be whole numbers the coefficients can be any number. Further it is not necessary to have all the powers of x. For example, q only contains x^5, x^2 and x.

The largest power occurring in the expression is called the *degree* of the polynomial. So that p has degree 4 and q has degree 5.

Exercise 2.7.1

Complete Table 2.8.

Table 2.8

Polynomial	Degree
$4x^5 + 3x^2 + 2x + 1$	
$9x^7 + 8x^6 - 4x^3 + 2x$	
$t^3 - 3t^2 + 4t + 1$	
$u^4 + 3u^3 - 2u^2 + u + 7$	
$0.1x^2 + 1.2x - 3.7$	
$3.1x^{11} + 4x^7$	

Polynomials of degree 0, 1 and 2 are already familiar and we have met their special names (Table 2.9). To these we add the polynomial of degree 3 called a *cubic* polynomial and the polynomial of degree 4 called a *quartic* polynomial.

Table 2.9

Degree	Name	Usual notation
0	constant polynomial	c
1	linear polynomial	$mx + c$
2	quadratic polynomial	$ax^2 + bx + c$
3	cubic polynomial	not a standard notation for these
4	quartic polynomial	

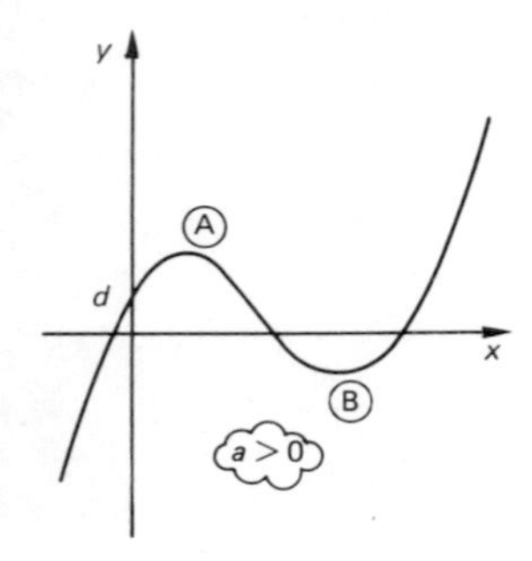

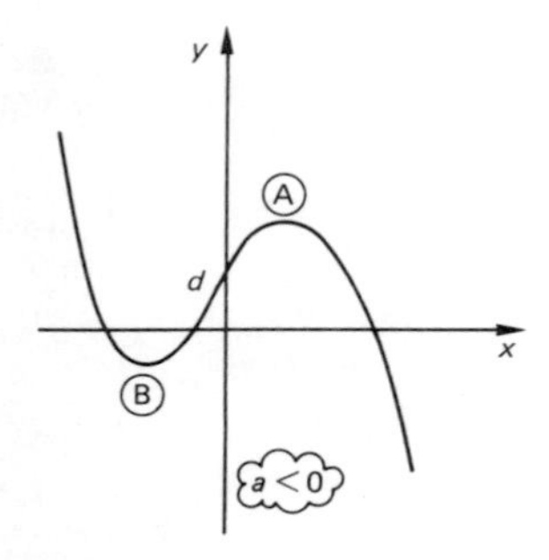

Figure 2.10

Exercise 2.7.2

On the same graph paper, plot the graphs of the polynomials

(a) $y = x^3$ (b) $y = x^4$ (c) $y = x^5$
(d) $y = x^3 + 2x^2 + 3x - 1$ (e) $y = 2 - x - 2x^2 + x^3$
(f) $y = 2 + 3x - 2x^3$

Of these graphs perhaps the most important one to recognize is the cubic polynomial. In general the graph of a cubic $y = ax^3 + bx^2 + cx + d$ may take two forms.

The sign of the leading coefficient, a (i.e. the coefficient of x^3), decides whether the graph comes up from $y = -\infty$ or down from $y = +\infty$.

The intercept on the y-axis is the coefficient d.

The graphs shown in Figure 2.10 illustrate two types of turning points that may occur. The feature (A) is called a *local maximum* because at the actual 'peak' the y-value is bigger than any other y-value close by. Feature (B) is called a *local minimum* because at the actual 'trough' the y-value is smaller than any y-value close by.

Another type of graph does occur as shown in Figure 2.11. This particular one is a graph of the cubic $y = x^3 - 3x^2 + 3x + 1$.

In case (a) the graph turns as if to become a local maximum but then turns back and continues on its upward path. At the point (C) a tangent to the curve is parallel to the x-axis.

In case (b) the tangent does not become horizontal.

In both cases the curve crosses its tangent at point (C). Such a point is called a *point of inflexion*.

It is knowing features such as (A), (B) and (C) that allow us to sketch graphs of functions quickly instead of plotting them accurately. (We will see how to do this in Chapter 8.)

The graph of a cubic polynomial cuts the x-axis at least once and at most three times. At such points the value of the cubic polynomial is zero and the corresponding values of x are called the *roots* of the cubic polynomial. For quadratics there is a formula for solving quadratic equations; for cubic and quartic polynomials there are equivalent formulas which are much more complicated. For polynomials of degree greater than four there is a theorem which says no such formula exists. (However such theorems and their proofs belong to the world of pure mathematics. We are interested in finding the roots of polynomials.)

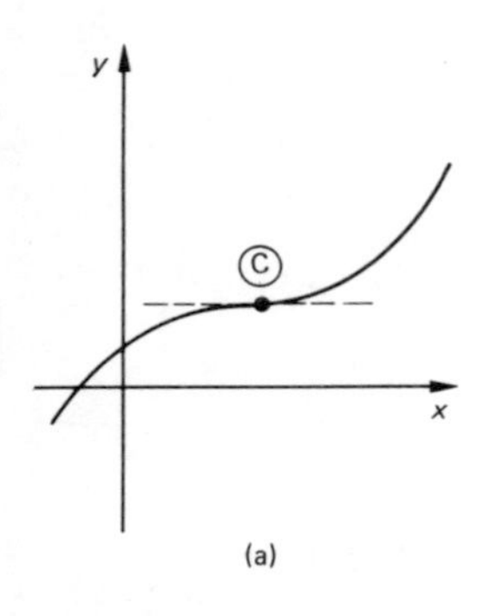

(a)

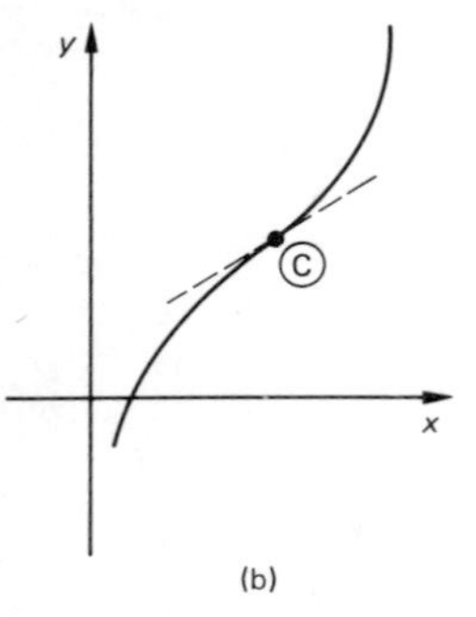

(b)

Figure 2.11

Clearly we could plot a graph of the polynomial or use a trial and error method to search for a value which gives $p = 0$. The next example illustrates this method for solving a cubic equation.

Example 2.7

Find the roots of the polynomial

$$2x^3 + x^2 - 6x + 2$$

(a) by plotting a graph, (b) by a decimal search method.

Solution

(a) Let the value of the polynomial be p. We can draw up a table (Table 2.9) to calculate p for different values of x.

Table 2.9

x	-3	-2	-1	0	1	2
$2x^3$	-54	-16	-2	0	2	16
x^2	9	4	1	0	1	4
$-6x$	18	12	6	0	-6	-12
2	2	2	2	2	2	2
p	-25	2	7	2	-1	10

The graph is shown drawn in Figure 2.12. The roots of the polynomial are values of x at which $p = 0$, i.e. where the graph crosses the x-axis.

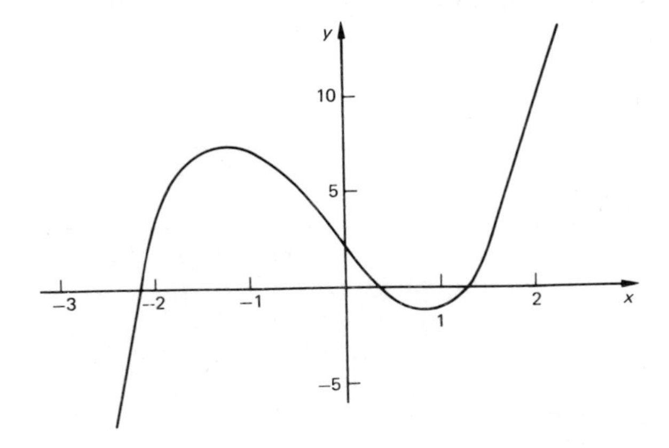

Figure 2.12

From the graph the roots are

$$x_1 = -2.1, \quad x_2 = 0.4 \quad \text{and} \quad x_3 = 1.3$$

(b) A decimal search method is one in which we guess where a root might be and then improve on the guess. Clearly a starting point is the table of values. Since the polynomial changes sign as we cross the x-axis we can see that the roots are between -3 and -2, between 0 and 1, and between 1 and 2.

Consider the root between -3 and -2. At $x = -3$, $p = -25$ and at $x = -2$, $p = 2$. So we might expect the root to be nearer $x = -2$ than $x = -3$. Table 2.10 shows the method. In each case we have used a calculator to obtain the value of p using the value of x.

Table 2.10

Trial value x	Value of p	Comment
-2.1	0.488	the root must be between -2.1 $(p > 0)$ and -3 $(p < 0)$, closer to -2.1
-2.15	-0.35425	between -2 $(p > 0)$ and -2.15 $(p < 0)$
-2.14	-0.181088	getting closer!
-2.13	-0.010294	
-2.12	0.158144	nearer -2.13

So, to two decimal places, one root is $x_1 = -2.13$.

For the second root which we have seen lies between 0 and 1 we start with $x = 0.5$ (Table 2.11)

Table 2.11

Trial value x	Value of p	Comment
0.5	-0.5	root between 0.5 $(p < 0)$ and 0 $(p > 0)$
0.4	-0.112	root between 0.4 and 0
0.3	0.344	nearer 0.4 $(p < 0)$ than 0.3 $(p > 0)$
0.36	0.062912	almost here!
0.37	0.018206	
0.38	-0.025856	

So $x_2 = 0.37$ (to two decimal places).

For the third root we start with $x = 1.1$ (Table 2.12).

Table 2.12

Trial value x	Value of p	Comment
1.1	−0.728	root between 1.1 and 2
1.2	−0.304	root between 1.2 and 2
1.3	0.284	root between 1.2 and 1.3
1.25	−0.03125	
1.26	0.028352	

To two decimal places, $x_3 = 1.26$.

Exercise 2.7.3

1. By plotting a graph find the roots of the polynomial

$$p = 6x^4 - 5x^3 - 14x^2 - x + 2$$

2. Use a decimal search method to find the roots of the cubic

$$p = 18x^3 - 27x^2 - 2x + 3$$

Neither of these methods is particularly elegant. The graphical method is unlikely to provide very accurate values of the roots and the decimal search method is somewhat tedious. A more systematic method which lends itself to an easy computation is introduced in the next section.

2.8 THE METHOD OF FORMULA ITERATION

In the previous section we used two inefficient methods to find the roots of a polynomial; the graphical solution provides a rough idea of the root and the decimal search method uses trial and error to gradually improve on an approximate value of the root. We now introduce a more systematic method of solution called *formula iteration*.

In general an iterative formula is a formula in the form

$$X_{\text{new}} = g(X_{\text{old}}) \tag{2.8.1}$$

where X_{new} is a label for a 'new' value of X which is calculated directly from an 'old' value of X labelled as X_{old} using function $g(X_{\text{old}})$. If formula (2.8.1) is to help solve $f(x) = 0$ then clearly the function $g(x)$ must be associated with $f(x)$. The aim of an iterative method is to find a function $g(x)$ so that by continually using the formula 2.8.1

$$X_{\text{new}} = X_{\text{old}} = r$$

where r is a solution of $f(x) = 0$.

To get an idea of how an iterative formula works, consider the iterative formula

$$X_{\text{new}} = \frac{1}{2}\left(X_{\text{old}} + \frac{2}{X_{\text{old}}}\right)$$

with the starting value $X_{\text{old}} = 1$. Then using $X_{\text{old}} = 1$

$$X_{\text{new}} = \frac{1}{2}\left(1 + \frac{2}{1}\right) = 1.5$$

This is said to be one iteration of the process. Now for a second iteration we chose $X_{\text{old}} = 1.5$. Substituting for this value into the formula gives

$$X_{\text{new}} = \frac{1}{2}\left(1.5 + \frac{2}{1.5}\right) = 1.4167 \text{ (to four decimal places)}$$

Now choosing $X_{\text{old}} = 1.4167$ the formula gives

$$X_{\text{new}} = \frac{1}{2}\left(1.4167 + \frac{2}{1.4167}\right) = 1.4142 \text{ (to four decimal places)}$$

Now choosing $X_{\text{old}} = 1.4142$ the formula gives

$$X_{\text{new}} = \frac{1}{2}\left(1.4142 + \frac{2}{1.4142}\right) = 1.4142$$

So we have reached the stage where the 'old' value of x and the 'new' value of x are equal to a chosen number of decimal places (in this case four decimal places). When this occurs we say that 'the iterative formula *converges* to the value 1.4142'.

Exercise 2.8.1

Solve the following iterative formulas stopping the calculation when the 'old' and 'new' values of x agree to three decimal places.

(a) $X_{\text{new}} = \frac{1}{2}\left(X_{\text{old}} + \frac{4}{X_{\text{old}}}\right)$ with starting value 1

(b) $X_{\text{new}} = 4 - \frac{1}{X_{\text{old}}}$ with starting value 2

A natural question to ask is 'what does the solution of an iterative formula have to do with solving $f(x) = 0$?'

Consider again the iterative formula

$$X_{\text{new}} = \frac{1}{2}\left(X_{\text{old}} + \frac{2}{X_{\text{old}}}\right)$$

We have found that after four iterations the new and old values of x are equal to four decimal places. Suppose when this occurs we let

$$X_{\text{new}} = X_{\text{old}} = x$$

then the iterative formula becomes an equation in x

$$x = \frac{1}{2}\left(x + \frac{2}{x}\right)$$

Rearranging this equation gives

$$x^2 = 2$$

so that the iterative formula has provided a method of solving the quadratic equation $x^2 - 2 = 0$. We can check that this is the case, since the iterative formula has solution $X = 1.4142$ (to four decimal places) and $(1.4142)^2$ is almost equal to 2. (The difference between $(1.4142)^2$ and 2 is due to the rounding in the solution of the iterative formula.)

Exercise 2.8.2

By rearranging the formulas in Exercise 2.8.1, show that the solutions to the iterative formulas are solutions of the equations

(a) $x^2 - 4 = 0$ (b) $x^2 - 4x + 1 = 0$

The problem that we have is one of solving $f(x) = 0$ by finding an appropriate iterative formula and starting value. There are often several rearrangements of $f(x) = 0$ which will lead to iterative formulas. For example, consider the problem of solving the equation

$$x^3 + 2x - 4 = 0 \tag{2.8.2}$$

The first step is to rewrite the formula in the form $x =$ something. There are two simple rearrangements of Equation (2.8.2)

$$x = \frac{4 - x^3}{2} \quad \text{and} \quad x = (4 - 2x)^{1/3}$$

which lead to two iterative formulas

$$X_{\text{new}} = \frac{4 - X_{\text{old}}^3}{2} \quad \text{and} \quad X_{\text{new}} = (4 - 2X_{\text{old}})^{1/3}$$

Now the original function $f = x^3 + 2x - 4$ is negative at $x = 1$ and positive at $x = 2$, so we can deduce that one solution of equation (2.8.2) lies between $x = 1$ and $x = 2$. Suppose we start off the iteration process with $X_0 = 1$. Table 2.13 shows the sequence of numbers formed by each iterative formula.

Table 2.13

X_{old}	$X_{\text{new}} = \dfrac{4 - X_{\text{old}}^3}{2}$	X_{old}	$X_{\text{new}} = (4 - 2X_{\text{old}})^{1/3}$
1	1.5	1	1.2599
1.15	0.3125	1.2599	1.1397
0.3125	1.9847	1.1397	1.1983
1.9847	−1.9089	1.1983	1.1704
−1.9089	5.4779	1.1704	1.1838
5.4779	−80.189	1.1838	1.1774

Clearly one iterative formula is a good one which is settling down to a definite value whereas the other one is jumping about wildly. We say that $X_{\text{new}} = (4 - 2X_{\text{old}})^{1/3}$ *converges* to the solution of $x^3 + 2x - 4 = 0$ and $X_{\text{new}} = \dfrac{4 - X_{\text{old}}^3}{2}$ *diverges*.

The convergence of numerical methods in problem solving is an important area in numerical analysis which we discuss in Chapter 10.

In summary then the formula iteration method to solve $f(x) = 0$ consists of

(a) a starting value, which is usually chosen as a value of x between two values where the function $f(x)$ changes sign
(b) a rule for generating new values, this is the iterative formula, found by rearranging the given equation $f(x) = 0$
(c) a condition for deciding when to stop the iterations, e.g. stop the iterations when X_{new} and X_{old} agree to a given number of decimal places

Example 2.8

Use the formula iteration method to find a solution correct to three decimal places of the equation

$$x^3 + 2x - 7 = 0$$

Solution

(a) To find a starting value we construct a table of values (Table 2.14).

Table 2.14

x	0	1	2
$x^3 + 2x - 7$	-7	-4	5

From this table we deduce that the solution lies between $x = 1$ and $x = 2$. Choose as a starting value $X_{old} = 1.5$.

(b) To find an iterative formula we arrange

$$x^3 + 2x - 7 = 0$$

Two rearrangements are fairly obvious

$$x = \frac{7 - x^3}{2} \quad \text{and} \quad x = (7 - 2x)^{1/3}$$

In Table 2.15 we try both of these.

Table 2.15

X_{old}	$X_{new} = \dfrac{7 - X_{old}^3}{2}$	X_{old}	$X_{new} = (7 - 2X_{old})^{1/3}$	
1.5	1.813	1.5	1.588	
1.813	0.523	1.588	1.564	
0.523	3.429	1.564	1.570	
3.429	-16.652	1.570	1.569	equal to
-16.652	2312.25	1.569	1.569	3 d.p.

Clearly the second rearrangement converges and a solution of $x^3 + 2x - 7 = 0$ correct to three decimal places is $x = 1.569$.

(Note that at this stage we need to try each possible iterative formula until we find one that converges. In Chapter 10 we use the method of calculus to decide on the appropriate formula.)

Exercise 2.8.3

1. The following iterative methods have been suggested in solving the equation $x^2 - 4x + 2 = 0$ near the solution $x = 3.0$.

(a) $X_{new} = \dfrac{X_{old}^2 + 2}{4}$ (b) $X_{new} = \sqrt{4X_{old} - 2}$

(c) $X_{new} = \dfrac{2}{4 - X_{old}}$ (d) $X_{new} = \dfrac{4X_{old} - 2}{X_{old}}$

Which method would you not be able to use to find the solution?

Which is the quickest method for each of the solutions?

2. By finding a rearrangement, choose an appropriate iterative method to solve each of the following equations.
(a) $x^2 - 3x + 1 = 0$
(b) $x^3 + x - 3 = 0$
(c) $x^3 - 2x^2 + 3x - 1 = 0$
In each case find the solution correct to two decimal places.

2.9 MODELLING WITH QUADRATICS AND POLYNOMIALS

The most common occurring models in science and engineering are linear models, applications of which were introduced in Chapter 1. However, polynomial laws do occur and in Table 2.16 we summarize some of them. These can all be verified by experiment.

Example 2.9

A ball is thrown from a height of 2 metres with a velocity of 20 ms^{-1} at an angle of 30 ° to the horizontal.

(a) Write down the equation of the path of the ball if it is modelled as a projectile. (Take $g \simeq 10$ ms^{-2}.)
(b) Find how far from the thrower the ball lands.

Solution

(a) The general equation of the path of a projectile launched from the origin is

$$y = \frac{v}{u}x - \frac{gx^2}{2u^2}$$

where $u = 20 \cos 30° = 10\sqrt{3}$; $v = 20 \sin 30° = 10$ and $g \simeq 10$ so that

$$y = \frac{1}{\sqrt{3}}x - \frac{1}{60}x^2 = 0.5774x - 0.0167x^2$$

Table 2.16

Physical situation	Variables	Model (and definition of constants)
Projectile motion	t = time y = height above origin	$y = vt - \frac{1}{2}gt^2$
Projectile path	x = horizontal distance from the origin y = height above origin	$y = \dfrac{v}{u}x - \dfrac{gx^2}{2u^2}$ u, v are components of initial velocity
Kinetic energy	v = speed E = kinetic energy	$E = \frac{1}{2}mv^2$ m = mass
Energy stored in an elastic spring	E = energy e = extension	$E = \frac{1}{2}ke^2$ k = stiffness
Energy stored in the magnetic field of the current in an inductor	E = energy i = current	$E = \frac{1}{2}Li^2$ L = inductance
Area of sphere	r = radius A = area	$A = 4\pi r^2$
Moment of inertia	I = moment of inertia k = radius of gyration	$I = mk^2$ m = mass
Thermo-couple	E = thermo-emf T = temperature	$E = aT^2 + bT$
Radiative heat loss (Stefan-Boltzmann law)	φ = net rate of heat loss by radiation T = temperature of body	$\varphi = \sigma A(T^4 - T_0^4)$ $\sigma = 5.67 \times 10^{-8}$ Wm^{-2}K^{-4} T_o = air temperature in K A = area
Bending of a beam supported at one end	y = deflection at distance x from support	$y = Ax^4$

(b) The ball hits the ground (i.e. it lands) when $y = -2$. Hence

$$-2 = 0.5774x - 0.0167x^2$$

$$0.0167x^2 - 0.5774x - 2 = 0$$

Using 'the formula' to solve for x we have

$$x = \frac{0.5774 \pm \sqrt{0.5774^2 - 4(-2)(0.0167)}}{2 \times 0.0167}$$

$x = 37.75$ or $x = -3.173$

The positive value gives the distance from the thrower of 37.75 metres. (The negative value gives a position behind the thrower.)

Exercise 2.9.1

1. Use the data given in Example 2.10 and the formula $y = vt - \frac{1}{2}gt^2$ to find the time for the ball to hit the ground.

2. An electric room heater consists of a square plate of material of area 1 square metres which radiates heat. It is supplied with 200 W in a room where the air temperature is 300 K. Find the temperature maintained by the material assuming that the heater is 100% efficient. (Hint: Use the formula for radiative heat loss in Table 2.16.)

3. The filament of a lamp maintains a temperature of 1500 K when it is supplied with 150 W. What temperature would it maintain if it were supplied with 200 W?

4. A scale pan for weighing objects consists of a spring of stiffness k and natural length l_0 as shown in Figure 2.13. The pan has mass m and a further object of mass M is placed on the pan and the system is released from rest with the spring unstretched.

   At any time in the subsequent motion

   $$\tfrac{1}{2}(m + M)v^2 + \tfrac{1}{2}k(x - l_0)^2 - (M + m)gx = \text{constant}$$

   where v is the speed of the pan and object and x is the length of the spring at time t.

   Find the maximum length of the spring in the ensuing motion. (Hint: At the maximum length of the spring $v = 0$.)

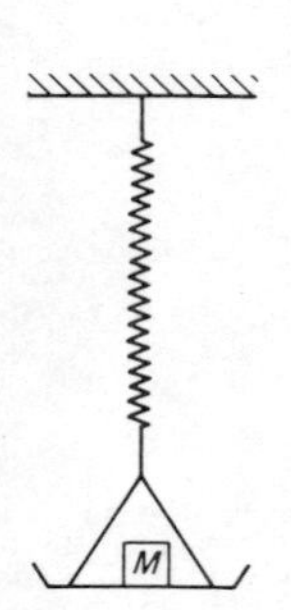

Figure 2.13

5. An electric circuit consists of a piece of iron wire and two pieces of copper wire, joined at junctions A and B. In an experiment, junction A was maintained at 0 °C while the temperature at junction B was raised to various temperatures T. A potentiometer was used to measure the potential difference V between the ends of

the copper wire. The data from the experiment leads to the relation

$$V = 1.2 \times 10^{-2}T - 2.1 \times 10^{-5}T^2$$

Find the values of T for V to be zero and the value of T for V to have its maximum value.

6. The electric circuit shown in Figure 2.14 consists of a battery of emf E volts and internal resistance r ohms and a load resistance of R ohms.

   The power P dissipated in the load resistance is given by

$$P = EI - rI^2$$

   where I is the current. Find the current in each of the following cases:

   (a) $P = 1.2 \quad E = 1.5 \quad r = 0.3$
   (b) $P = 5 \quad E = 11.8 \quad r = 5.5$

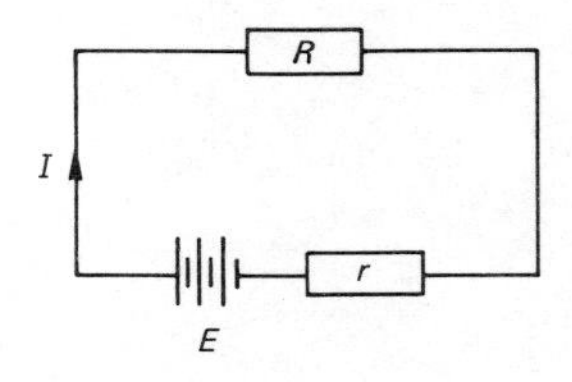

Figure 2.14

7. When any two resistors R and r are connected in series, the total resistance across them is $(R + r)$. When any two resistors are connected in parallel, the total resistance across them is $(Rr/(R + r))$.

   In an experiment using two resistors of unknown value, it is found that when combined in series their total resistance is 45 ohms and when combined in parallel their total resistance is 5 ohms.
   (a) Show that R satisfies the quadratic equation

$$R^2 - 45R + 225 = 0$$

   (b) Hence find the value of R and r.

8. A spherical buoy of diameter 1.2 metres and specific gravity 0.5 will sink to a depth x given by the cubic equation

$$x^3 - 1.8x^2 + 0.432 = 0$$

   (a) By drawing a graph estimate the solutions for x to one decimal place.
   (b) Use the method of interval bisection to find the values of x correct to three decimal places.

9. The deflection d of any point on a uniform beam supported at each end and uniformly loaded is given by

$$d = 0.01x^4 - 0.19x^3 + 1.08x^2 - 1.8x$$

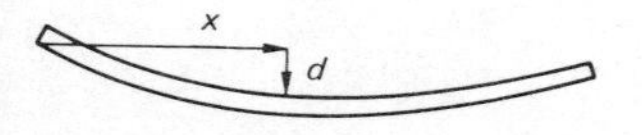

Figure 2.15

where x is the distance from one end of the beam (Figure 2.15).

(a) Sketch a graph of the polynomial d for x between 0 and 10.

(b) Find the values of x where the deflection is zero.

OBJECTIVE TEST 2

1. Write down the degree of the following polynomials.
   (a) $3x^2 - 4x + 1$ (b) $7x^8 + 6x^5 - 3x^2 + x$
   (c) $2x + 1$ (d) $2x^3 - x^2 + 3x + 5$

2. Label each of the graphs shown in Figure 2.16 with one of these equations.
   (a) $y = 4x^2$ (b) $y = -x^2 + 3x + 1$
   (c) $y = x - 3$ (d) $y = 2x^2 - 4x + 5$
   (e) $y = (x - 2)^2$ (f) $y = 2x^2 - 7x + 5$

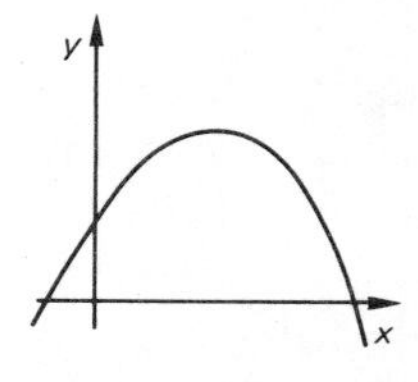

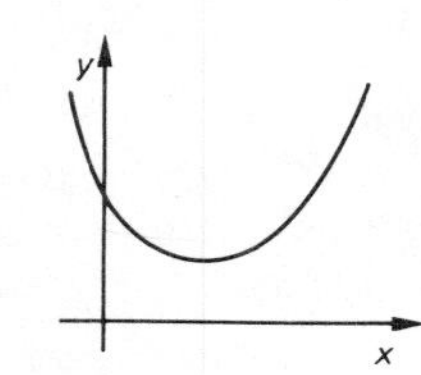

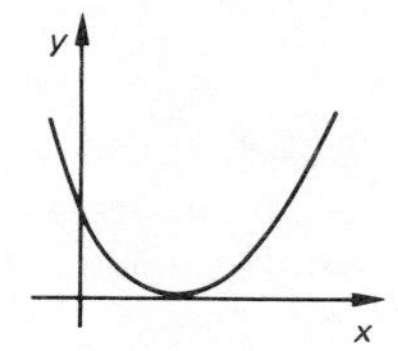

Figure 2.16

3. Factorize the following quadratics and hence find the roots.
   (a) $x^2 - 16$ (b) $x^2 - 3x + 2$ (c) $3x^2 + 5x - 2$

4. Sketch the graph of the quadratic $y = x^2 - 5x + 5$ and estimate the roots to one decimal place.

5. Solve the following quadratic equations.
   (a) $R^2 - 0.3R - 0.5 = 0$ (b) $2x^2 + 2x + 1 = 10$

6. Label each of the graphs shown in Figure 2.17 with one of these equations.
   (a) $y = x^3 - 4x$ (b) $y = -x^3 + 2x^2 + x - 2$
   (c) $y = x^3 - 3x^2 + 3x - 1$

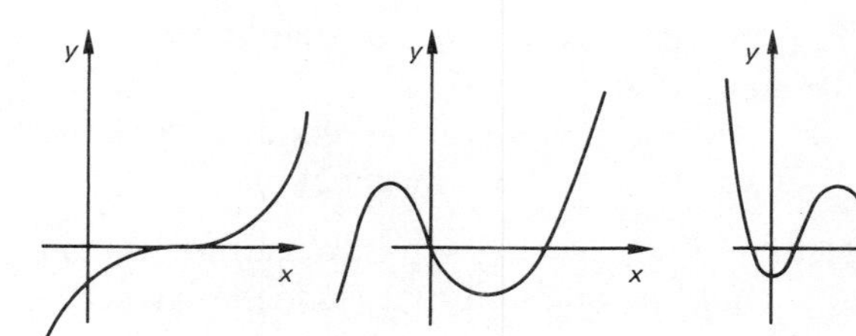

Figure 2.17

Identify the local maximum points, local minimum points and point of inflexion.

8. Sketch a graph of the polynomial

$$p = x^4 - 2x^3 + 2x^2 - 8$$

Estimate the roots correct to one decimal place.

Use the method of formula iteration to find the roots correct to three decimal places.

3 LOGARITHMS AND EXPONENTIAL FUNCTIONS

OBJECTIVES

When you have finished working through this chapter you should be able to

1. manipulate positive, negative and fractional indices in algebraic expressions
2. recognize and manipulate surds
3. understand the components of a function and functional notation
4. explain what is meant by an inverse function
5. define exponential functions to base 10 and base e and be able to sketch and recognize their graphs
6. understand the physical significance of the number e
7. define the logarithm functions to base 10 and base e and know their algebraic rules
8. find the half-life and decay constant of radioactive substances
9. find power laws and exponential laws between variables using appropriate logarithm graphs, log-log and log-linear graph paper

3.1 INTRODUCTION

In Chapter 3 we have seen how positive powers of an independent variable are often used to describe physical situations. For example, in Stefan's law the net rate of heat loss by radiation is proportional to T^4, i.e. the fourth power of temperature T. A natural question to ask now is can we have fractional (or decimal) powers or how about negative powers? In this chapter we see that such powers occur very often in the physical world.

As an example consider Table 3.1 which shows the period of vibration of a simple spring-mass oscillator (Figure 3.1), for different masses.

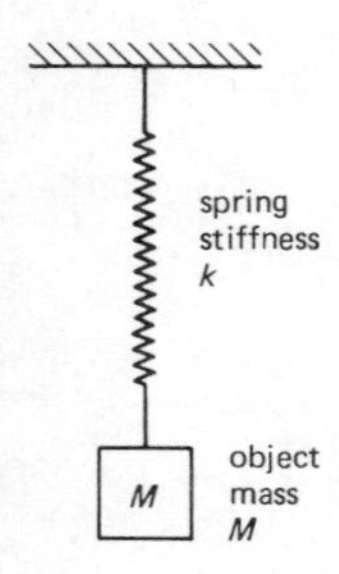

Figure 3.1

Table 3.1

Period, T (s)	0.36	0.51	0.61	0.71	0.80
Mass, M (kg)	0.1	0.2	0.3	0.4	0.5

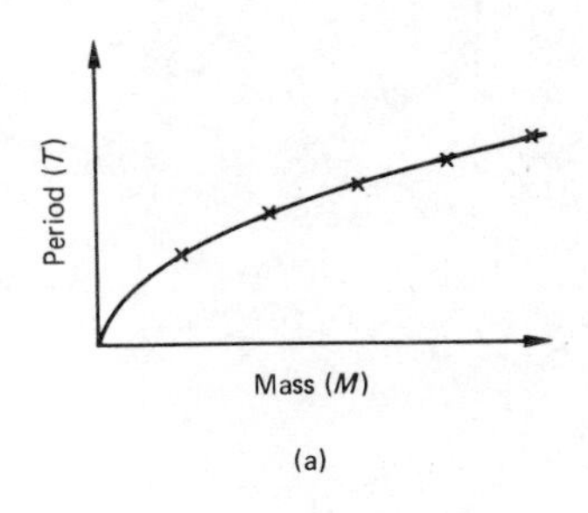

(a)

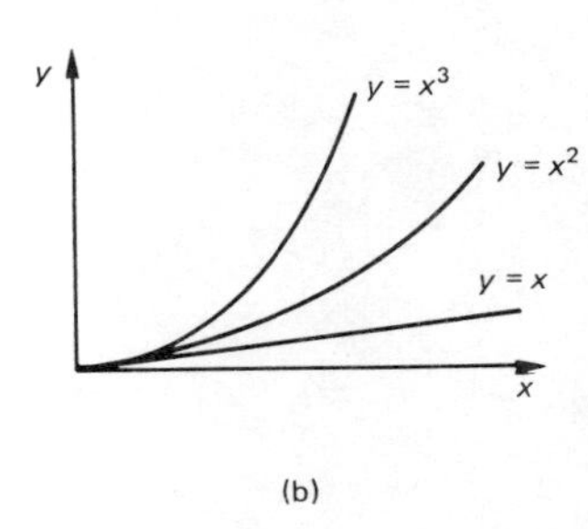

(b)

Figure 3.2

Figure 3.2(a) shows a graph of the period against the mass. Clearly the graph is not linear and further it does not have the general shape of a positive power of M, examples of which are shown in Figure 3.2(b).

There is in fact a simple law relating T and M, and it is

$$T^2 = 1.28M$$

Taking square roots of each side we have

$$T = 1.13\sqrt{M}$$

So, Figure 3.2(a) shows a new type of relationship involving a square root.

You are probably familiar with the square root sign. We can also express a square root as a fractional power

$$T = 1.13M^{1/2} \quad \text{or} \quad T = 1.13M^{0.5}$$

In this chapter we begin by investigating powers which are not integers and we will see that leads on to an important new function.

Exercise 3.1.1

The formula for the surface area of a sphere of radius r is $A = 4\pi r^2$. Rewrite the rule to give r in terms of a fractional power of A.

We can extend these ideas to other fractional powers. For example consider Stefan's law

$$E = K(T^4 - T_0^4)$$

where T and E are the temperature and the energy, and K and T_0 are constants. Rearranging for T we have

$$T^4 = \frac{E}{K} + T_0^4$$

Hence T is the fourth root of $E/K + T_0^4$. This is written in a power form using the fractional power $\frac{1}{4}$ or decimal power 0.25.

$$T = \left(\frac{E}{K} + T_0^4\right)^{1/4} \quad \text{or} \quad T = \left(\frac{E}{K} + T_0^4\right)^{0.25}$$

This process of 'undoing' a positive power is called the *inverse operation* and in this chapter we see how to evaluate expressions of the above form using a calculator. The laws associated with fractional and negative powers are often called *index laws*.

In general, if $a^n = b$ then $a = b^{1/n}$.

Exercise 3.1.2

The formula for the volume of a sphere in terms of its radius r is $V = \frac{4}{3}\pi r^3$. Rewrite the rule giving r in terms of a fractional power of V.

Negative powers can also be used. For example, the magnitude of the force on a satellite circulating the earth is given by

$$F = \frac{GMm}{r^2}$$

where G is the gravitational constant, M is the mass of the Earth, m is the mass of the satellite and r is the distance between the satellite and the centre of the earth. We can write this rule using a negative power

$$F = (GMm)r^{-2}$$

So in general if $a = 1/b^n$, then $a = b^{-n}$.

Exercise 3.1.3

The friction factor for the flow of water in a pipe is given by

$$f = \frac{0.316}{\sqrt[4]{R}}$$

where R is called Reynolds number. Rewrite this formula using a negative power of R.

Many quantities in science and engineering, when expressed in standard SI units, turn out to be very large or very small. For example, the mass of the earth is 5 980 000 000 000 000 000 000 000 kg and the mass of a proton is 0.000 000 000 000 000 000 000 000 001 67 kg. In order to write such quantities in a more compact form we adopt the scientific notation (or standard index form). This means writing the number as a multiple of a power of 10. For example,

$$\begin{aligned}\text{Mass of Earth} &= 5.98 \times 10^{24}\ \text{kg}\\ \text{Mass of proton} &= 1.67 \times 10^{-27}\ \text{kg}\end{aligned}$$

This is an example of using an integer as a 'base' (10 in this case) and another integer as a 'power' or 'index'. In the scientific notation a number is written as $a \times 10^n$ where a is between 1 and 10 and n is an integer.

Being able to represent quantities in index form and being able to manipulate powers is an important part of algebra.

3.2 INDEX LAWS

3.2.1 Integer Index Laws

When writing large numbers, such as the mass of the Earth., 5 980 000 000 000 000 000 000 000 kg, it is more convenient as we have seen to use the standard index form 5.98×10^{24} kg. We also write algebraic expressions using the power notation, for example, $a \times a \times a \times a \times a$ is written as a^5. This is called the *exponential form* of number; a is called the *base*, 5 is called the *exponent* or *index* or *power*.

Exercise 3.2.1

Complete Table 3.2

Table 3.2

Number	Base	Index
7^3	7	3
21^{13}		
a^4		
2^n		
a^m		
5^7		

When multiplying numbers written in standard index form, we have special rules for manipulating the powers of ten. For example, to multiply 10^3 by 10^4 we add the powers of 10 to get 10^7. This is clear if we write out the multiplication in full i.e.

$$10^3 \times 10^4 = (10 \times 10 \times 10) \times (10 \times 10 \times 10 \times 10) = 10^7$$

Exercise 3.2.2 Simplify

(a) $2^5 \times 2^4$ (b) $3^7 \div 3^2$ (c) $a^4 \times a^2$ (d) $x^6 \times x^3 \div x^4$

We now summarize the results for indices

Product rule $a^m \times a^n = a^{m+n}$

Quotient rule $a^m \div a^n = \dfrac{a^m}{a^n} = a^{m-n}$

Power rule $(a^m)^n = a^{mn}$

The quotient rule introduces a special case. Suppose that $m = n$ then we can write

$$a^m \div a^m = \frac{a^m}{a^m} = 1$$

By using the quotient rule $\dfrac{a^m}{a^m} = a^{m-m} = a^0$

Hence $a^0 = 1$

Exercise 3.2.3

1. Simplify, using the three rules for indices,
   (a) $x^5 \times x^2$ (b) $x^2 \times x^7 \times x^3$
   (c) $(3x)^2 \times (2x)^3$ (d) $a^5 \div a^3$
   (e) $6k^4 \div 2k^3$ (f) $x^{13} \div x^7$
   (g) $(4^3)^2$ (h) $(x^5)^3$
   (i) $(a^2b^3)^4$

2. Simplify, by removing the brackets,
   (a) $a^2(a^3 + a^5)$ (b) $2x^4(3x - 5x^7)$
   (c) $m^2(1 - m) - 2m(m + 2m^2)$ (d) $ab(a^2 + ab - b^2)$

3. Simplify, using the three rules for indices,
   (a) $\dfrac{a^4 \times a^3}{a^2}$ (b) $\dfrac{(3x)^3 \times (4x)^2}{2x^4}$
   (c) $\dfrac{e^n \times e^m}{e^p}$ (d) $\dfrac{6b(a^2b)^2}{t(2k)^2} \div \dfrac{3(ab)^2}{(2kt^2)^3}$

4. Evaluate,
   (a) $a^0 \times b^0$ (b) $x^0 + 2p^0 - a^0$
   (c) $\left(\dfrac{x^0}{y^0}\right)^{37}$ (d) $2a^0 \times 4b^0 \times 3c^0$

3.2.2 Negative Index Laws

So far we have manipulated indices which are positive integers but the rules also apply for negative indices. But what does a negative index, such as 2^{-3} mean?

Consider the quotient rule,

$$\frac{a^m}{a^n} = a^{m-n}$$

and suppose that $m = 0$. Then since $a^0 = 1$ we have

$$\frac{1}{a^n} = a^{-n}$$

So a negative index means 'one over' or the reciprocal. For example,

$$2^{-3} \text{ means } \frac{1}{2^3} = \frac{1}{8} \text{ and } 7^{-1} \text{ means } \frac{1}{7}$$

Example 3.2.1

Simplify the following:

(a) 3^{-2} (b) 5^{-1} (c) $x^{-2} \times x^{-3}$

(d) $a^3 \div a^4$ (e) $\dfrac{a^{-1} \times a^{-2}}{a^2 \times a^3}$ (f) $a^2b^{-3} \div ab^{-3}$

Solution

(a) $3^{-2} = \dfrac{1}{3^2} = \dfrac{1}{9}$

(b) $5^{-1} = \dfrac{1}{5}$

(c) $x^{-2} \times x^{-3} = x^{-5} = \dfrac{1}{x^5}$

Note that we add the indices as before and $(-2) + (-3) = -5$.

(d) $a^3 \div a^4 = a^{3-4} = a^{-1} = \dfrac{1}{a}$

(e) $\dfrac{a^{-1} \times a^{-2}}{a^2 \times a^3} = \dfrac{a^{-3}}{a^5} = a^{-8} = \dfrac{1}{a^8}$

(f) $a^2b^{-3} \div ab^{-3} = \dfrac{a^2b^{-3}}{ab^{-3}} = a^{2-1}b^{(-3)-(-3)} = a^1b^0 = a$

Exercise 3.2.4

1. Evaluate the following without using a calculator,
(a) 2^{-2} (b) 2^{-4} (c) 5^{-2} (d) 3^{-3} (e) 4^{-1}

2. Simplify the following,
(a) $a^{-2} \times a^{-3}$ (b) $x^{-5} \times x^2$ (c) $m^2 \div m^4$

(d) $p^{-1} \div p^{-3}$ (e) $\dfrac{4a^{-1} \times 3a^{-2}}{2a^{-3}}$ (f) $a^2b^{-3} \times a^{-1}b^2$

(g) $e^x(e^{2x} + e^{-2x})$(h) $3xy \times 3x^{-1}y^{-1}$ (i) $6x^{-2}y \div 2x^{-1}y^{-1}$

(j) $\left(\dfrac{x^2y}{a^3}\right)^{-1}$ (k) $\left(\dfrac{3a^{-1}b}{c}\right)^{-2}$ (l) $(m^px^{-n})^{-3}$

3.2.3 Fractional Indices

Again the same three index laws apply for fractional indices as for positive and negative indices. For example, $x^{1/2} \times x^{1/4} = x^{3/4}$, we add the indices so that $\frac{1}{2} + \frac{1}{4} = \frac{3}{4}$. But what does $x^{1/2}$ mean?

Consider $3^{1/2}$. Using the power rule for indices we can write

$$(3^{1/2})^2 = 3^1 = 3$$

since $\frac{1}{2} \times 2 = 1$. So taking square roots

$$3^{1/2} = \sqrt[2]{3}$$

Thus a fractional index is associated with the roots of a number: $x^{1/2}$ means square root of x, for example.

As a second example, consider $x^{1/3}$; we can write

$$x^{1/3} \times x^{1/3} \times x^{1/3} = x^{1/3+1/3+1/3} = x^{3(1/3)} = x^1$$

So the cube of $x^{1/3}$ is x, in other words $x^{1/3}$ means the cube root of x and so on.

Example 3.2.2

Evaluate $16^{1/2}$ and $27^{2/3}$.

Solution

Now $16^{1/2}$ means the square root of 16. Hence $16^{1/2} = 4$. Also $27^{2/3}$ can be written as $(27^{1/3})^2$ or $(27^2)^{1/3}$. Since the power $\frac{1}{3}$ means the cube root, we have $27^{1/3} = 3$ and $(27^{1/3})^2 = 3^2 = 9$. The alternative form leads to the cube root of 729 which is also 9. (Normally we take roots first and then calculate the powers.)

Exercise 3.2.5

1. Write each of the following in the form x^n.
   (a) $\sqrt{x}$ (b) $x\sqrt{x}$ (c) $x^3\sqrt{x}$ (d) $\sqrt{x^5}$
   (e) $\sqrt[3]{x}$ (f) $\dfrac{x^2\sqrt{x}}{x^3\sqrt{x}}$

2. Evaluate the following without using a calculator.
   (a) $4^{1/2}$ (b) $64^{1/2}$ (c) $27^{1/3}$ (d) $16^{1/4}$
   (e) $4^{5/2}$ (f) $64^{3/2}$ (g) $100^{5/2}$ (h) $(16x^6)^{1/2}$
   (i) $4^{-1/2}$ (j) $1^{-9/7}$ (k) $\dfrac{1}{8^{-1/3}}$ (l) $\dfrac{1}{1000^{-4/3}}$

You probably have a key on your calculator for working out powers of numbers. It usually looks like [x^y]. Try the following calculation using this key

press key [6]: press key [x^y] then press [2] [=]

Your calculator display should now show 36. You have evaluated 6^2. Now try evaluating 1.15^{-3}. The order in which you should have pressed the keys is:

[1] [.] [1] [5] [x^y] [−] [3] [=]

and you should have $1.15^{-3} = 0.658$ (to three significant figures). You can check this by evaluating $1 \div (1.15 \times 1.15 \times 1.15)$.

Exercise 3.2.6

Check your answers to Exercise 3.2.5 problem 2 using your calculator

---

3.2.4 Solving Exponential Equations

An equation that has an unknown as an index is called an *exponential equation*. An essential step in solving such equations is realizing that if $a^x = a^n$ then $x = n$.

Example 3.2.3

Solve for x,

(a) $2^x = 16$ (b) $16^{2-x} = 8$

Solution

(a) Writing 16 in terms of base 2, we have

$$2^x = 2^4$$

Hence $x = 4$.

(b) $16^{2-x} = 8$

This one is a bit more tricky; we have to find a base in terms of which both 16 and 8 can be written. In fact, 2 is such a base. Writing 16 and 8 in terms of base 2 we obtain

$$16 = 2^4 \quad \text{and} \quad 8 = 2^3$$

Substituting these in the above equations, we have:

$$(2^4)^{2-x} = 2^3$$

$$2^{8-4x} = 2^3 \quad \text{using the power rule}$$

Hence we can see that

$$8 - 4x = 3$$

Solving this for x,

$$x = \frac{5}{4}$$

Exercise 3.2.7

In each of the following cases find the unknown quantity,

(a) $5^k = 5^4$ (b) $3^k = 9$ (c) $2^{3r} = 32$
(d) $2^{a-1} = 2$ (e) $7^x = 1$ (f) $6^{3-x} = 1$

(g) $8^a = 4$ (h) $9^{-x} = \frac{1}{3}$ (i) $\left(\frac{1}{2}\right)^a = \frac{1}{16}$

(j) $\frac{1}{4^{a+1}} = 8$ (k) $3^x \div 3^{1-x} = 9$ (l) $\frac{1}{8^x} = 1$

3.3 FUNCTIONS

3.3.1 Basic Ideas

At this point we must step out of the mathematics of science and engineering and take a formal look at the question, what is a function?

The concept of variables being related to one another by rules is a familiar one. For example, the surface area of a sphere is related to the radius of the sphere by the rule $A = 4\pi r^2$. At the same time we must be careful with the values of the variables that we use. In this area example, the radius can clearly only take positive values, so r belongs to the set $0 \leqslant r < \infty$ and A will then belong to the set $0 \leqslant A < \infty$.

The sets are not always so straight forward. The rule for converting temperatures to degrees Fahrenheit from degrees Celsius is

$$F = \frac{9C}{5} + 32$$

If the substance under consideration is water then the set of values for T in °C is $0 \leqslant T \leqslant 100$ and the corresponding set of values in °F is $32 \leqslant T \leqslant 212$; (to find these extreme values we substitute for 0 and 100 in the rule). Essentially a function consists of three parts:

1. a *rule* relating a 'dependent' variable to an 'independent' variable
2. a set of values for the independent variable called its *domain*
3. a corresponding set of values for the dependent variable called its *codomain* or *range* or *image*

Table 3.3 shows the rule, domain and range for the two examples above.

Table 3.3

Variables	Rule	Domain	Range
Area, radius	$A = 4\pi r^2$	$0 \leqslant r \leqslant \infty$	$0 \leqslant A \leqslant \infty$
Temperature, C(°C) Temperature, F(°F)	$F = \frac{9}{5}C + 32$	$0 \leqslant C \leqslant 100$	$32 \leqslant F \leqslant 212$

Exercise 3.3.1

In the introduction we met several relations between variables for physical situations; complete Table 3.4.

Table 3.4

Physical situation	Variables	Rule	Domain	Range
Spring-mass oscillator				
Radiation heat loss rate				
Newton's law of gravitation				

When two variables, x and y say, are related by a rule so that for each value of the independent variable x there is only one value of the dependent variable y, we say that 'y is a function of x', and in general when no definite rule is specified we write

$$y = f(x)$$

The symbol $f(x)$ is read 'function of x' or 'f of x'.

One important requirement of a function is that the rule must be unambiguous, so that given a value of the independent variable the rule must specify just one corresponding value of the dependent variable. As an example of the care we need consider using the square root rule in a function. The possible values of $4^{1/2}$ are $+2$ and -2. Hence if we are to use the rule $y = x^{1/2}$ as a function then we must specify if we take the positive or negative values of y, i.e. $f(x) = +\sqrt{x}$ is a function and $f(x) = -\sqrt{x}$ is a function.

You should note that this idea of rules being unambiguous is a mathematical nicety and tends not to worry the engineer or applied scientist. The functions that describe physical situations usually contain allowable rules and their domains and ranges are easily established from an understanding of the problem.

As an example of the function notation consider the rule for converting degrees Celsius to degrees Fahrenheit. We could write the ingredients of the function as

1. the rule: $f(x) = \dfrac{9x}{5} + 32$
2. the domain: $\{x: 0 \leqslant x \leqslant 100\}$
3. the range: $\{y: 32 \leqslant y \leqslant 212\}$

When the x in brackets is replaced by a constant, such as $f(2.7)$, then the symbol $f(2.7)$ represents the value of the rule when x is replaced by 2.7. For example, for

$$f(x) = \frac{9x}{5} + 32$$

then

$$f(2.7) = \frac{9 \times 2.7}{5} + 32 = 36.86$$

Note that we are not trying to solve an equation here; $f(2.7)$ means the value of $f(x)$ when $x = 2.7$.

Exercise 3.3.2

(a) If $f(x) = x^4 - x^{1/2} + x^{-1}$ find $f(1)$, $f(3)$ and $f(0.61)$.
(b) If $f(T) = 0.7T^2 + 4.2T$ find $f(3.1)$ and $f(0)$.

It is not necessary in our function notation to stick rigidly to the symbol f. In Exercise 3.3.2(b) we could just have easily written $g(T) = 0.7T^2 + 4.2T$. In many problems in engineering we are likely to have several functions with different rules, so it is important to use different symbols. Often we use a symbol associated with the physical variables. For example, if we expect a temperature to be a function of time we might write

$$T = T(t) \quad \text{instead of} \quad T = f(t)$$

For the temperature conversion formula it is sensible to write

$$F = F(C) = \frac{9C}{5} + 32$$

3.3.2 Composite Functions

Many functions that we encounter in engineering and science are combinations of other functions. For example, in Einstein's theory of relativity the mass of a body varies with speed according to the rule

$$m = m_0\left(1 - \frac{v^2}{c^2}\right)^{-1/2}$$

where m_0 and c are constants. This rule for m can be seen as consisting of two rules

(a) a quadratic rule relating w and v, $w = 1 - v^2/c^2$. We could say $w = f(v)$
(b) a negative, fractional index rule relating m and w, $m = m_0 w^{1/2}$. We could say $m = g(w)$

We can see from this that m is a function of w, which is in turn a function of v. To calculate m for a given value of v we would first calculate w as a function of v and then m as a function of w.

Symbolically we can write the whole relationship in the form

$$m = g(f(v))$$

Such a combination is called *a composite function* or more often *a function of a function*. The latter reminds us of the process of following one function by another by successive applications of two rules.

Example 3.3.1

Given that $f(x) = x^2$ and $g(t) = t + 5$ find $g[f(x)]$, $f[g(t)]$, $f[f(x)]$ and $g[g(t)]$.

Solution

The notation $g[f(x)]$ means that 'f' is followed by 'g'

$$g[f(x)] = g(x^2) \quad \text{since} \quad f(x) = x^2$$

Replacing t by x^2 in the rule for g we have

$$g(x^2) = x^2 + 5$$

Hence

$$g[f(x)] = x^2 + 5$$

Similarly,

$$f[g(t)] = f(t + 5) = (t + 5)^2$$
$$f[f(x)] = f(x^2) = (x^2)^2 = x^4$$
$$g[g(t)] = g(t + 5) = (t + 5) + 5 = t + 10$$

Exercise 3.3.3

1. Given that $f(x) = 3x$ and $g(t) = t - 3$ find $g[f(x)]$, $f[g(t)]$, $f[f(x)]$ and $g[g(t)]$.

2. Given that $f(x) = x^3$, $g(x) = x - 2$ and $h(x) = \sqrt{x}$ find $f[g(x)]$, $g[h(x)]$, $f[g[h(x)]]$.
3. The function $f(x)$ is defined by $f(x) = \sqrt{x}$ for $x \geqslant 0$ and the function $g(x)$ is defined by $g(x) = x - 4$ for all x.
   (a) State the range of f and the range of g
   (b) Find the rule for $f[g(x)]$ and give its domain and range
   (c) Find the rule for $g[f(x)]$ and give its domain and range

---

3.3.3 Inverse of a Function

So far in our formal treatment of a function we have started with a value of the independent variable, x say, and used the rule $f(x)$ to find a corresponding value of the dependent variable, y say. Now suppose we start with y then we require a rule for finding x.

For example, the function relating temperatures in degrees Fahrenheit to degrees Celsius is

$$F = f(C) = \frac{9C}{5} + 32.$$

Given a value of C, 20 say, then

$$F = \frac{9 \times 20}{5} + 32 = 68$$

Suppose now we start with $F = 78$, what is the corresponding value of C? If we rearrange the rule $F = (9C/5) + 32$ then we have $C = 5(78 - 32)/9 = 25$.

The function from C to F is $f(C) = (9C/5) + 32$.

The function from F to C is $g(F) = 5(F - 32)/9$.

The function g is called the inverse of the function f (also f is the inverse of g). In a sense the inverse function 'undoes' the function: we use it to work backwards. This is clearer if we find the composite function $g[f(C)]$

$$g[f(C)] = g\left(\frac{9C}{5} + 32\right)$$

$$= 5\left[\left(\frac{9C}{5} + 32\right) - 32\right]/9$$

$$= C$$

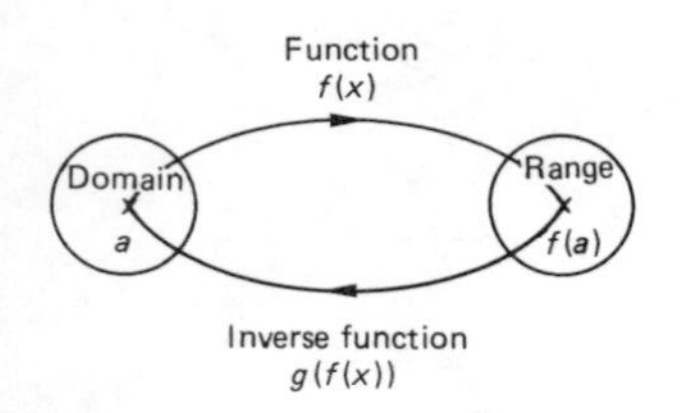

Figure 3.3

The composite function $g[f(C)] = c$ for all values of C. In general, for any function f, *the inverse function*, is the function g with the property

$$g[f(x)] = x \quad \text{for all } x \text{ in the domain of } f$$

Figure 3.3 illustrates this idea.

We can imagine the rule for the function processing the input and giving an output then the rule for the inverse function would process the output back into the input. This common idea is illustrated in the examples in Figure 3.4.

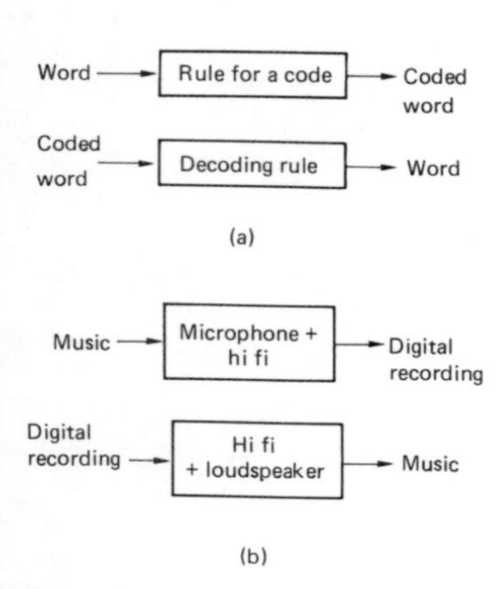

Figure 3.4

The special notation for the inverse function is $g = f^{-1}$ so that $f^{-1}[f(x)] = x$. Note that this does *not* mean one-over f, it is a special notation meaning the inverse function. You may have seen the notation before on your calculator and we meet it again in Chapter 4.

Suppose we know that $\sin(x) = 0.5$, then to find the unknown value of x we use the key labelled $\sin^{-1}$ on some calculators, or INVSIN, ARCSIN on others. (Which does your calculator use?) The answer is $x = \sin^{-1}(0.5) = 30°$ or $x =$ INVSIN$(0.5) = 30°$. This ties in with our discussion above; to find x we must 'undo' the sine function and the symbol on calculators uses one of our notations.

Exercise 3.3.4

1. Given that $f = 3x + 4$, find f^{-1}. Check that $f^{-1}[f(x)] = x$. Sketch the graph of $y = f(x)$ and $y = f^{-1}(x)$
2. Given that $f = (x + 1)^3 - 3$ find f^{-1}. Check that $f^{-1}[f(x)] = x$. Sketch the graphs of $y = f(x)$ and $y = f^{-1}(x)$.

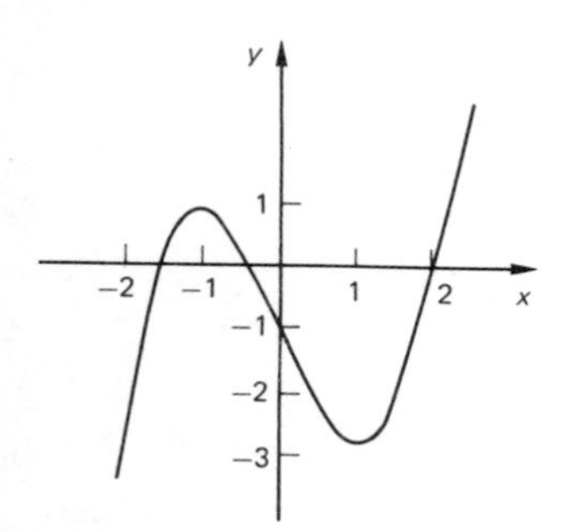

Figure 3.5

3.3.4 Asymptotes

A graph of a function provides a useful means of visualizing and recognizing the properties of the function. For example, if we draw a graph of the function $f(x) = x^3 - 3x - 1$ then the properties to describe might be the local maximum at $x = -1$ and the local minimum at $x = 1$ (Figure 3.5).

The graph drawn in Figure 3.5 consists of a smooth continuous curve. This is not always the case. Consider the graph of the function $f(x) = 1/x - 1$ shown in Figure 3.6.

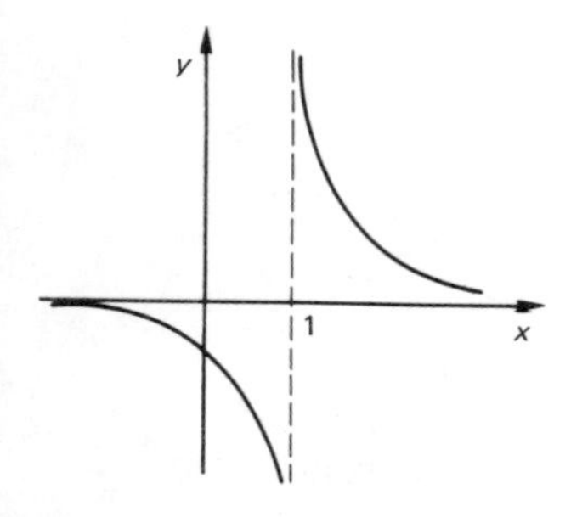

Figure 3.6

Clearly something strange happens when $x = 1$. For values of x just below 1 the function values are large and negative, for example, $f(0.9) = -10$, $f(0.99) = -100$, $f(0.999) = -1000$ and so on.

For values of x just above 1 the function values are large and positive, for example, $f(1.1) = 10$, $f(1.01) = 100$, $f(1.001) = 1000$ and so on.

When $x = 1$ we have $f(1) = 1/0$ which has no meaning. The graph of the function is not a continuous smooth curve. We say that the function $f(x) = 1/(x - 1)$ is *discontinuous* at $x = 1$ and on the graph the vertical line $x = 1$ is an example of a vertical *asymptote*. (The function $f(x) = x^3 - 3x - 1$ is said to be *continuous* for all values of x.) A vertical asymptote occurs when the function $g(x)/h(x)$ takes the form $k/0$ for some constant k. It is usually easy to find such asymptotes by solving the equation

$$\text{Denominator of } f(x) = h(x) = 0$$

For example, if

$$f(x) = \frac{(x + 3)}{(2x - 1)}$$

then a vertical asymptote occurs when $2x - 1 = 0$, i.e. when $x = \frac{1}{2}$ since $f(\frac{1}{2}) = 3.5/0$.

Exercise 3.3.5

Find the vertical asymptotes for the following functions:

(a) $f(x) = \dfrac{4}{2x - 5}$ (b) $f(x) = \dfrac{3x + 2}{7x + 15}$

(c) $f(x) = \dfrac{4}{x^2 - 3x + 4}$ (d) $g(t) = \dfrac{2t}{t^2 - 4}$

Asymptotes are not always vertical lines x = constant. Consider the function

$$f(x) = \frac{3(x + 2)}{(x - 1)}$$

Table 3.5 shows values of this function for large values of x.

Table 3.5

x	$f(x)$
10	4
100	3.0909
1000	3.009009
10000	3.00090009

If we continued this table for increasingly larger values of x, we would eventually find that $f(x)$ approaches the value 3. Figure 3.7 shows a graph of this function.

We say that $f(x)$ approaches the limit 3 as x approaches ∞. The idea of a limit and its importance in science and engineering is discussed more fully in Chapter 6.

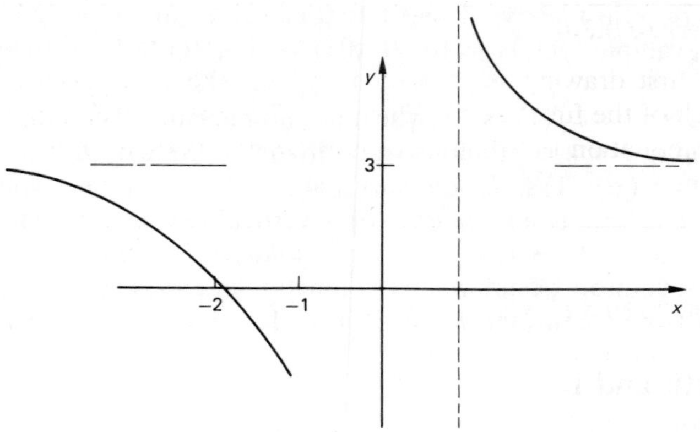

Figure 3.7

The horizontal line $y = 3$ is an example of an asymptote also. Notice that for this function there are two asymptotes $x = 1$ and $y = 3$.

In general an asymptote is defined in the following way:

> An *asymptote* of a function f is a line which is a tangent to the graph of f at an infinite distance from the origin.

Figure 3.8 illustrates this definition.

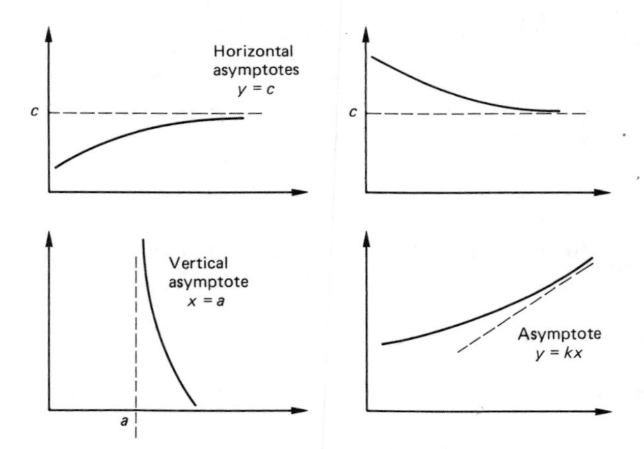

Figure 3.8

The most common asymptotes are the vertical lines x = constant and horizontal lines y = constant. Finding asymptotes is important because the function behaves awkwardly nearby. We need to investigate the physical system carefully if the mathematical model has an asymptote and hence a discontinuity.

Exercise 3.3.6

By first drawing up a table of values, sketch a graph of each of the functions in Exercise 3.3.5. In each case, give the equation of the horizontal asymptote for each function.

3.4 EXPONENTIAL FUNCTIONS

3.4.1 Growth and Decay Functions

Growth and decay processes such as the growth of bacteria, decay of radioactive material, cooling and uncontrolled population growth can be modelled by the function $f = a^t$ for some constant number a, where t is the time. Growth processes will have $a > 1$ and decay processes will have $0 < a < 1$. For $a = 1$ the function f takes constant values. (This is often called 'steady state'.) For example,

(a) if the mass of a bacterium doubles every minute then after t minutes the mass is proportional to 2^t
(b) if £1500 is invested at 8% per annum then after t years the investment grows to £1500 $(1.08)^t$

Of course, growth and decay processes may involve independent variables other than time t. For example, the pressure in the atmosphere at height h km above the sea level value is $p_0(0.86)^h$ where p_0 is the pressure at sea level.

The function defined by the index law $y = a^x$ is called an *exponential function* where $a > 0$, the domain of the function is $-\infty < x < \infty$ and the range is $0 < y < \infty$.

Example 3.4.1

Draw a graph of the exponential function $y = 2^t$.

Solution

Table 3.6 shows values of y corresponding to some integer values of t.

Table 3.6

t	-4	-3	-2	-1	0	1	2	3	4
$y = 2^t$	0.0625	0.125	0.25	0.5	1	2	4	8	16

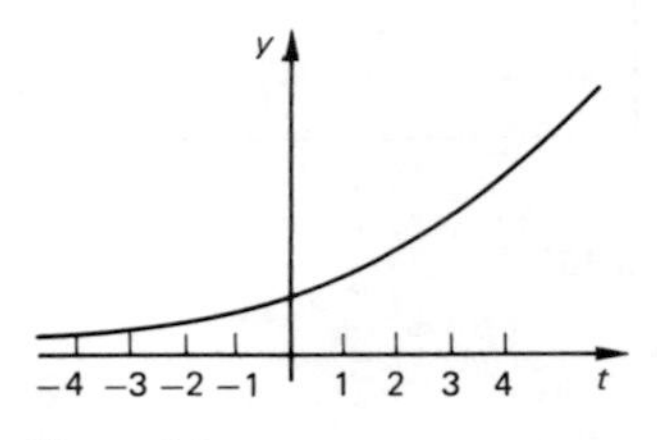

Figure 3.9

Plotting these values we obtain the graph shown in Figure 3.9.

This graph is typical of exponential growth functions, the shape of the graph suggests that the function increases at a faster rate (i.e. it

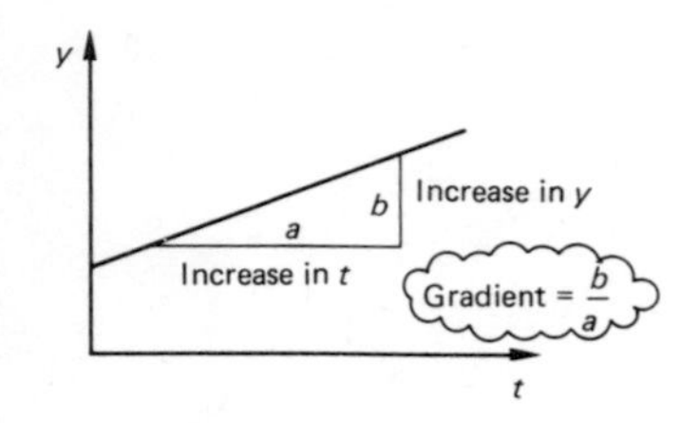

Figure 3.10

becomes more steep) as t increases. We can find the slope or gradient of a straight line graph easily by calculating the ratio of an increase in y to an increase in t (Figure 3.10).

For a curved graph it is not so straight forward. What do we mean by the gradient or rate of change at a particular point on the graph in Figure 3.9? To answer this question we draw a straight line at the point of interest that touches the graph once only. It is called the *tangent* to the graph. Then we define the *gradient* of the graph to be the slope of this tangent. There is only one tangent to the graph at each point.

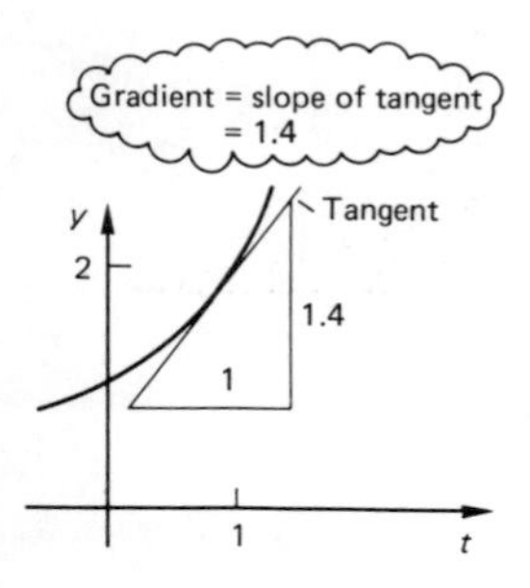

Figure 3.11

Consider the function $y = 2^t$ and the gradient at the point (1, 2). On the graph of $y = 2^t$ in Figure 3.11 is drawn a tangent at the point (1, 2). The slope of this tangent is 1.4 (to two significant figures) so that the gradient of the graph of $y = 2^t$ at the point (1, 2) is approximately 1.4. Clearly a graphical method is inaccurate and can only serve as a guide; an algebraic approach is introduced in Chapter 5. However, Table 3.7 shows the slope of the function $y = 2^t$ for various values of t.

Table 3.7

t	0	1	2	3
2^t	1	2	4	8
Rate of change of $y = 2^t$	0.7	1.4	2.8	5.6

What we can say is that the slopes of the tangents to the graph of $y = 2^t$ increase as t increases so that the rate of change of the function becomes greater.

Exercise 3.4.1

Draw graphs of the functions $y = 3^t$ and $y = 2.5^t$. Estimate the rate of change of these functions (i.e. the gradient of the graphs) when $t = 1$ and 2. (Give your answers to two significant figures).

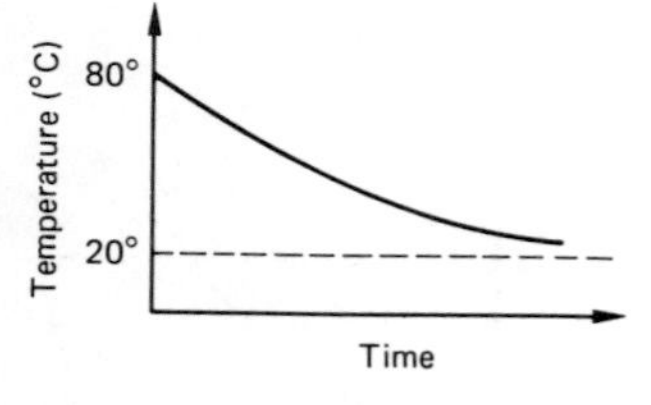

Figure 3.12

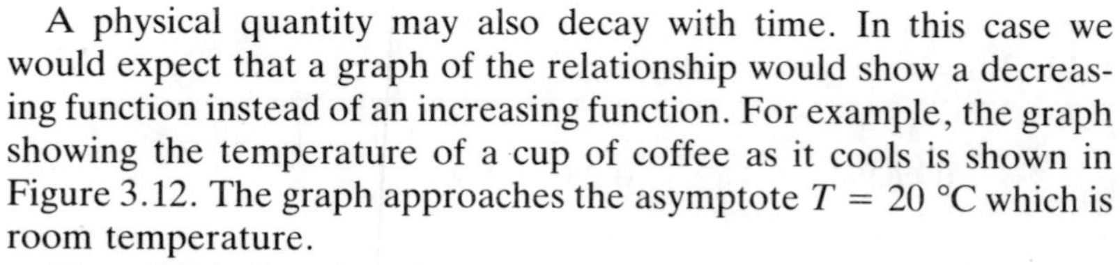
A physical quantity may also decay with time. In this case we would expect that a graph of the relationship would show a decreasing function instead of an increasing function. For example, the graph showing the temperature of a cup of coffee as it cools is shown in Figure 3.12. The graph approaches the asymptote $T = 20$ °C which is room temperature.

We said earlier that decay can be described by the function a^t where $a < 1$. In the following example and exercise we investigate two such values for a.

Example 3.4.2

Draw a graph of the function $y = 0.5^t$ and compare its general shape with Figure 3.12. Estimate the rate of change of the function when $x = 1$.

Solution

Calculating a table of values (Table 3.8) we have

Table 3.8

t	0	1	2	3	4
y	1	0.5	0.25	0.125	0.0625

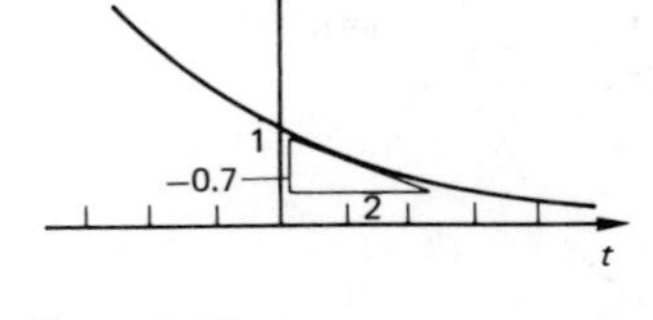

Figure 3.13

The graph of $y = 0.5^t$ is shown in Figure 3.13.

This graph has the same general shape as the cooling curve in Figure 3.12. As t increases the graph becomes flatter, its slope is decreasing. (However the rule 0.5^t does not in fact model the temperature/time relationship.)

To estimate the rate of change of the function where $t = 1$ we have drawn a tangent at the point (1, 0.5). The slope of this tangent is approximately −0.35, so that the rate of change of 0.5^t at $t = 1$ (or the gradient of the graph at (1, 0.5)) is roughly −0.35.

Exercise 3.4.2

Draw graphs of the function $y = 0.25^t$ and estimate the rate of change of the function at the points $t = 1$ and $t = 3$.

3.4.2 The Exponential Function e^t

In theory we could model any growth or decay process by a rule $y = a^t$ for different values of a, but this turns out to be very inconvenient. It is more effective to choose a particular base and try to describe all physical situations in terms of that base. There are two special bases that are chosen, base 10 and a base denoted by the letter e. We will consider base 10 in the next subsection, but first let us look at e.

The number e is an irrational number (like for example π) and to 12 decimal places is given by

$$e = 2.718281828459$$

All index laws can be expressed in terms of any base. So why choose e? There are essentially two reasons:

1. the rate of change of the function $y = e^t$ is equal to the same function e^t. So what this means is that at an instant $t = 2$ say the rate of change of y actually equals e^2. Compare this with the index laws in the previous subsection. For example the rate of change of $y = 2^t$ when $t = 2$ is 2.8 not $2^2 = 4$, similarly the rate of change of $y = 0.5^t$ when $t = 1$ is -0.35 not 0.5. Hence $y = e^t$ is a special function which obeys the rule

Rate of change of $y = y$

2. the second property of e is more mathematical. The number e is defined to be the limit of the expression $(1 + 1/n)^n$ as n grows large.

Example 3.4.3

Draw the graph of the function $y = e^t$ for $1 \leqslant t \leqslant 3$. Show that the rate of change of y when $t = 2$ is approximately e^2.

Solution

Scientific calculators contain the exponential function as a special key. It is usually labelled [e^x]. Calculating a table of values (Table 3.9) we have

Table 3.9

t	1	1.25	1.5	1.75	2	2.25	2.5	2.75	3
e^t	2.72	3.49	4.48	5.75	7.39	9.49	12.18	15.64	20.09

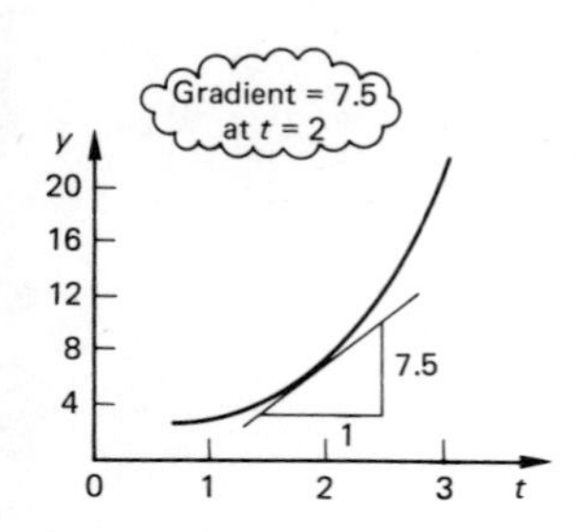

Figure 3.14

(Use your calculator to check these values, it will give you practice at using the [e^x] key.)

The graph of the function $y = e^t$ is shown in Figure 3.14.

Drawing a tangent at $t = 2$ and calculating its slope we have the rate of change of y equal to 7.5 which is approximately equal to e^2. (It is difficult to obtain the gradient of the graph exactly by this graphical means, but this example does illustrate Property 1 above.)

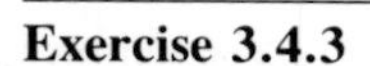

Exercise 3.4.3

Consider the expression $(1 + 1/n)^n$. Complete the following table (Table 3.10) thus showing Property 2 above. (Use the key [x^y] on your calculator.)

Table 3.10

n	$(1 + 1/n)^n$
1	2
1.5	2.1516574
2	
3	
4	
5	
10	
100	
1000	
10000	

In fact the two properties are not unrelated but we shall not prove the relationship here. For our purposes in science and engineering the use of e will provide a powerful function for modelling many real physical systems. In general we will show in Chapter 7 that

> In modelling a growth or decay process, if the rate of change of a quantity y is proportional to y then the rule relating the variables y and t is
>
> $$y = Ae^{kt}$$
>
> where A and k are constants.
> If $k > 0$ the process is *a growth process* and if $k < 0$ the process is *a decay process*.
> If $k = 0$ then y is constant and the process is often said to be in a *steady state*.

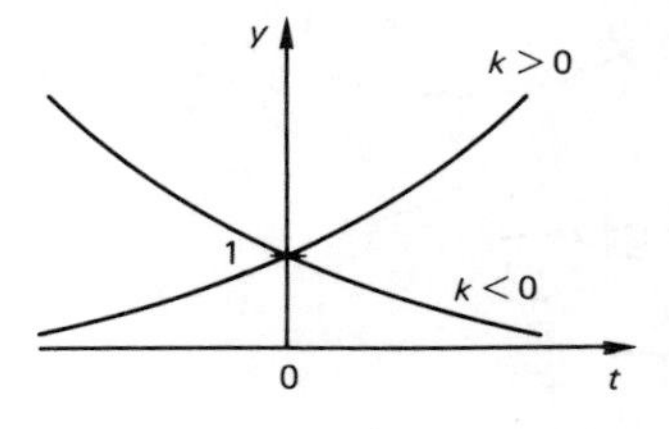

Figure 3.15

The general shape of the graph of the functions $y = Ae^{kt}$ for $k > 0$ and $k < 0$ are shown in Figure 3.15.

Exercise 3.4.4

1. Use your calculator to find the values of the following as decimals correct to four significant figures.
   (a) e^0 (b) e^{-2} (c) e^3 (d) $e^{1.7}$
   (e) $e^{0.31}$ (f) $e^{-0.42}$ (g) e^1 (h) $\sqrt{e}$
2. On the same graph paper draw graphs of the functions $y = e^x$ and $y = e^{-x}$ for the interval $-3 \leqslant x \leqslant 3$. In which line is the graph of e^x a reflection of the graph of e^{-x}?

3. On the same graph paper draw graphs of the following exponential functions

$$y = e^{2t}, \quad y = e^{3t}, \quad y = 5e^{-t}, \quad y = e^{-2t}$$

for the interval $-2 \leqslant t \leqslant 2$.

4. Use the formula iteration method and the suggested formula to solve the following equation to three decimal places.

$$e^{-x} - x = 0; \quad X_{\text{new}} = e^{-X_{\text{old}}} \text{ with starting value } X = 0.$$

3.5 LOGARITHMIC FUNCTIONS

3.5.1 Bases 10 and e

Consider the exponential function $y = 2^x$. If we are given a value of x, 1.3 say, then we can calculate $y = 2^{1.3} = 2.46$. However, suppose we are given a value of y, 3.7 say, then how do we find x? We need to solve the equation

$$3.7 = 2^x$$

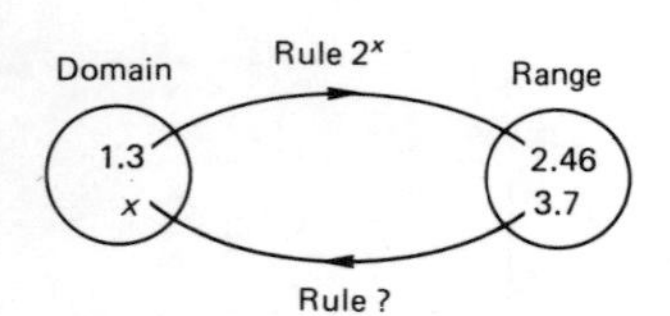

Figure 3.16

Figure 3.16 shows a diagram of the mathematical problem. The diagram suggests that we are looking for an inverse function for 2^x. The inverse functions for exponential functions are called *logarithmic functions* and defined in the following way:

$y = \log_a x \quad a > 0, \quad a \neq 1$ is the inverse function for the rule $x = a^y$

From this definition we see that logarithms are exponents in the sense that $x = a^{\log_a x}$. Since a^y is always positive then the domain of the function is $0 < x < \infty$ and the range is $-\infty < y < \infty$.

Example 3.5.1

Find y in each of the following equations.

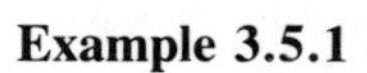

(a) $\log_4 16 = y$ (b) $\log_5 125 = y$ (c) $\log_2 2 = y$

Solution

In each case we convert the equation to its exponential form

(a) $\log_4 16 = y$ in exponential form is $16 = 4^y$. We recognize $16 = 4^2$ so that $y = 2$.
(b) $\log_5 125 = y$ in exponential form is $125 = 5^y$. Since $125 = 5^3$ we have $y = 3$.
(c) $\log_2 2 = y$ in exponential form is $2 = 2^y$. Clearly in this case $y = 1$.

Exercise 3.5.1

1. Find x in each of the following.
   (a) $\log_2 16 = x$ (b) $\log_3 (1/9) = x$
   (c) $\log_7 7 = x$ (d) $\log_{10} 1000 = x$
   (e) $\log_e e = x$ (f) $\log_{10} 0.01 = x$
   (g) $\log_6 36 = x$ (h) $\log_2 64 = x$

2. Express x in each of the following in logarithmic form.
   (a) $3^x = 7$ (b) $6^x = 4.1$ (c) $10^x = 12.2$
   (d) $e^x = 4.1$ (e) $2^x = 8.3$ (f) $5^x = 7.6$

Clearly we do not want to have to list the logarithms to all possible bases. There are two bases which are used most commonly. Logarithms to base 10 are called *common logarithms* and are written $y = \log(x)$. Logarithms to base e are called *natural logarithms* and are written $y = \ln(x)$. Each of these are available on your calculator.

Exercise 3.5.2

1. Use your calculator to find the values of the following logarithms.
   (a) $\log(25)$ (b) $\log(10)$ (c) $\log(7.3)$
   (d) $\log(1000)$ (e) $\ln(25)$ (f) $\ln(10)$
   (g) $\ln(e)$ (h) $\ln(7.3)$ (i) $\log(2.1)$
   (j) $\ln(1)$ (k) $\log(1)$ (l) $\ln(2)$

2. Draw a graph of the functions $y = \log(x)$ and $y = \ln(x)$ for the interval $0 < x < 10$.

3. Solve the following equations.
   (a) $2.7 = 10^x$ (b) $4.13 = e^x$
   (c) $7.1 = e^{3x}$ (d) $8.1 = e^{-2x}$
   (e) $11.2 = 10^x$ (f) $\log(x) = 3.2$
   (g) $\ln(x) = 1.5$ (h) $\ln(13x) = 6.2$
   (i) $\log(5x) = 1.5$

Logarithms to base 10 were introduced because our number system is based on 10. But they do have useful applications in themselves as relative measures of quantities.

For example, the magnitude of earthquakes is defined in terms of The Richter scale where the scale number R is related to the relative intensity of the shock, I, by the law

$$I = 10^R$$

The Richter scale number was 7 for the Italian earthquake in November 1980 and was 6 for the San Francisco earthquake of 1989. So the Italian earthquake was 10 times more intense ($I = 10^7$) than the San Francisco earthquake ($I = 10^6$).

Suppose we know the intensity I of an earthquake, then to find its Richter scale number we have $R = \log I$. Another use of common logarithms is as a measure of sound intensity. The *threshold of audibility* is the quietest sound that a 'good' human ear can hear and it is measured in watt/m^2. Its value is 10^{-12}. The sound level in decibels at a particular place is then defined by

$$S = 10 \log \left(\frac{P}{10^{-12}}\right)$$

where P is the power received per square metre.

Exercise 3.5.3

1. In normal conversation the power received per square metre is 10^{-6} watts. Find the sound level in dB.
2. At a pop concert the sound level is approximately 110 dB. Find the power received by a listener in watts per square metre.

Logarithms to base e are usually used in science and engineering applications and we concentrate on an important application in the next section.

We finish this subsection by using logarithms to base e to show that any index law can be expressed in terms of the exponential law e^x. Consider replacing the index law $y = Aa^x$ where $a > 0$ by $y = Ae^{bx}$ for some constant b. Then

$$y = A(e^b)^x = Aa^x$$

so that b is defined in such a way that $e^b = a$. Knowing the values of a we can use natural logarithms to find b, since

$$b = \ln(a)$$

For example, the atmospheric pressure at a height h km above sea level is $p_0(0.86)^h$ so that $a = 0.86$. If we let $b = \ln(0.86) = -0.15$ then we can write the rule for atmospheric pressure as $p_0 e^{-0.15h}$.

Hence we have proved a statement that we suggested earlier, that all index laws can be written in a standard way using the special exponential function e^x.

Exercise 3.5.4

1. The temperature of a large vat of hot liquid as a function of time is given by

$$T = 500(0.67)^t$$

Rewrite this law using the exponential function.

2. The current in a circuit is described by the law

$$i = i_0(1 - e^{-20t})$$

where i_0 is the equilibrium current and t is time in seconds. Find the value of t when $i = 2$ amperes if $i_0 = 2.5$ amperes.

3. The temperature (in °C) of a hot rod placed under running water is given by $T = 20 + 1180e^{-0.18t}$, where t is time in minutes. If you can hold the rod when its temperature is 40 °C how long will it take to cool to this temperature.

4. The work done in the isothermal expansion of a gas from pressure p_1 to p_2 is given by the formula

$$w = w_0 \ln(p_1/p_2)$$

If the initial pressure p_1 is 7000 pascals find the final pressure p_2 if $w = 3w_0$.

5. The atmospheric pressure at a height h (in km) above sea level obeys the rule

$$p = p_0 e^{-0.15h}$$

where p_0 the pressure at sea level = 100 000 pascals.

(a) Find the pressure at heights 1 km, 2 km, 5 km and 10 km.

(b) At what height is the pressure equal to half the value at sea level?

(c) At what height is the pressure equal to one-tenth of the value at sea level?

3.5.2 Rules for Logarithms

The following two properties of logarithms are worth remembering

$$\log_a 1 = 0$$
$$\log_a a = 1$$

in particular $\ln(e) = 1$ and $\log(10) = 1$. To prove these results we use the basic definition of logarithms from exponents.

Let x be a number such that $\log_a x = 0$ then $x = a^0 = 1$. Hence $\log_a 1 = 0$.

Let y be a number such that $\log_a y = 1$ then $y = a^1 = a$. Hence $\log_a a = 1$.

There are three important rules for logarithms which form the basis of algebraic manipulation with them. These rules are

$\log_a(xy) = \log_a(x) + \log_a(y)$	Product rule
$\log_a(x/y) = \log_a(x) - \log_a(y)$	Quotient rule
$\log_a(x^r) = r \log_a(x)$	Power rule

The product rule reduces manipulation to addition and the quotient rule reduces division to subtraction. Before the availability of cheap hand calculators some fifteen or so years ago, multiplication and division were often carried out using logarithms to base 10 and these two rules. To illustrate the process consider the problem of multiplying 3.042 by 16.96. Using the product rule and base 10 we have

$$\begin{aligned}\log(3.042 \times 16.96) &= \log(3.042) + \log(16.96)\\ &= 0.4831592 + 1.2294258\\ &= 1.712585\end{aligned}$$

To 'undo' this equation we use the exponential to base 10 so that

$$3.042 \times 16.96 = 10^{1.712585} = 51.59 \quad \text{(to four significant figures)}$$

This process was carried out using 'log tables', i.e. lists of logs of numbers given to four decimal places. Checking the product directly we have the same answer. Of course, these days we do not need to use logarithms to multiply numbers. Our electronic calculators are far more efficient. But at least this exercise demonstrates that the product rule gives the correct answer.

The proof of the three rules follows using the rules for exponents.

Rule 1: product rule

Let

$$\log_a(x) = p \quad \text{and} \quad \log_a(y) = q$$

then

$x = a^p$ and $y = a^q$ (using the definition of logs).

Multiplying x and y we have

$xy = a^p a^q = a^{p+q}$

then

$$\log_a(xy) = p + q$$
$$= \log_a(x) + \log_a(y)$$

Rule 2: the quotient rule
The proof is similar to that for the product rule.
Let

$\log_a(x) = p \qquad \log_a(y) = q$

then

$x = a^p, \qquad y = a^q$ and $x/y = a^{p-q}$

so

$\log_a(x/y) = p - q = \log_a(x) - \log_a(y).$

Rule 3: the power rule
Let

$\log_a(x) = p$ then $x = a^p$

Hence

$x^r = (a^p)^r = a^{pr}$ and then

$\log_a(x^r) = pr = r \log_a(x)$

We shall see in Sections 3.7 and 3.8 that Rules 1 and 3 are particularly important in finding exponential laws between two variables (where such a law exists).

Exercise 3.5.5

1. Expand the following expressions using the rules of logarithms.
   (a) $\log(p^2v^3)$ (b) $\ln(p/v)$
   (c) $\log_5(6p(a + b))$ (d) $\ln(0.21t/e)$
   (e) $\log\left(\dfrac{st}{\sqrt{3}}\right)$ (f) $\ln(0.1v^2)$
2. Write each of the following as a single logarithm, and simplify.
   (a) $\log(20) - \log(2)$ (b) $\ln(6e) - \ln(2e)$

(c) $\log(5w) + \log(v) - \log(u)$ (d) $3\ln(u) - 2\ln(v)$
(e) $\log(x) + 2\log(y) - 3\log(p)$ (f) $\frac{1}{2}\ln(e) - \ln(\sqrt{e})$

3. Use logarithms to base 10 to solve the following equations
(a) $3^{2x} = 5^{x+1}$ (b) $2.5^x = 0.62$
(c) $4^{3x-1} = 5^x$

3.6 THE HALF-LIFE

An important application of the exponential and natural logarithm functions is in radioactivity. Unstable atoms disintegrate to yield a new element and in doing so radiation is emitted. This phenomenon is called *radioactive decay*. Experiments show that the rate of decay of the number of atoms is directly proportional to this number of atoms. This is one of the key properties of the exponential function. So we can write

$$N = N_0 e^{-kt}$$

where N is the number of atoms present and N_0 is the original number of atoms present. k is a constant called the *decay constant*. The value of k varies from element to element.

The *half-life* of an element is the time taken for half the number of atoms to decay. For example, suppose that at some time T the number of atoms present is measured to be half the number N_0, i.e. when $t = T$, $N = N_0/2$. Substituting into the formula

$$\frac{N_0}{2} = N_0 e^{-kt}$$

Diving by N_0,

$$0.5 = e^{-kt}$$

Take logs to base e,

$$\ln(0.5) = \ln(e^{-kT}) = -kt$$

Since $\ln(0.5) = -0.6931$ we have the half-life T given by

$$T = \frac{0.6931}{k} \left(= \frac{\ln(2)}{k} \right)$$

An interesting consequence of this analysis is that the half-life of an element is a constant which depends only on k. Half-lives vary from millionths of a second to millions of years. For uranium it is 4.5×10^9 years and for xenon-133 it is 5 days.

Example 3.6

The decay constant for various elements is given below.

Uranium	$1.54 \times 10^{-10}\ \text{yr}^{-1}$
Carbon	$1.24 \times 10^{-4}\ \text{yr}^{-1}$
Strontium-90	$0.028\ \text{day}^{-1}$

(a) Find the half-life in each case.
(b) Find how long it takes for the amount of radiation to reduce to 1/4 and 3/4 of the original amount.

Solution

(a) The half-life of an element is given by

$$T = \frac{\ln(2)}{k}$$

where k is the decay constant. Substituting for each k we have

Uranium	4.5×10^9 years
Carbon	5590 years
Strontium-90	25 days

(b) For the radiation to reduce a fraction d of its original amount

$$d\,N_0 = N_0 e^{-kt}$$

so $$d = e^{-kt}$$

Take logs,

$$\ln(d) = -kt$$

$$t = \frac{\ln(d)}{-k}$$

Table 3.11 shows the results for the various elements.

Table 3.11

Element	Time for $\frac{1}{4}$	Time for $\frac{3}{4}$
Uranium	9×10^9 years	1.9×10^9 years
Carbon	11 180 years	2320 years
Strontium-90	49.5 days	10.3 days

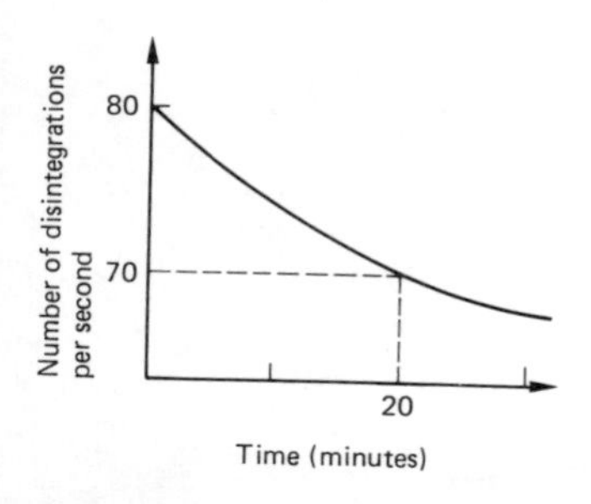

Figure 3.17

To find the half-life for an element we do not need to wait for the whole half-life period — fortunately! (Consider how long we would have to wait for uranium!) We carry out an experiment measuring the number of disintegrations per second (Becquerels) at different times, in such an experiment we would obtain a graph like Figure 3.17.

From the data and the exponential model $N = N_0 e^{-kt}$ we can find the value of k. When $t = 0$ and $N = 80$ so that $N_0 = 80$; when $t = 20$, $N = 70$ hence $70 = 80e^{-20k}$. Solving for k we have $k = -1/20 \ln(7/8) = 0.0067$.

Then the half-life $T = \ln(2)/0.0067 = 103$ minutes.

Exercise 3.6.1

1. The half-life of thoron is 52 seconds. Calculate its decay constant k in s^{-1}.

2. The unstable lead isotope ${}_{82}Pb^{212}$ has a decay constant of $0.0654\ h^{-1}$. Calculate its half-life.

3. A sample of thallium isotope has an initial activity of 703 kBq (kilobecquerels) and after nine minutes it has an activity of 159 kBq. Use the law $N = N_0 e^{-kt}$ to determine
   (a) the decay constant k
   (b) the half-life of thallium

4. Which of the curves in Figure 3.18 shows exponential changes?

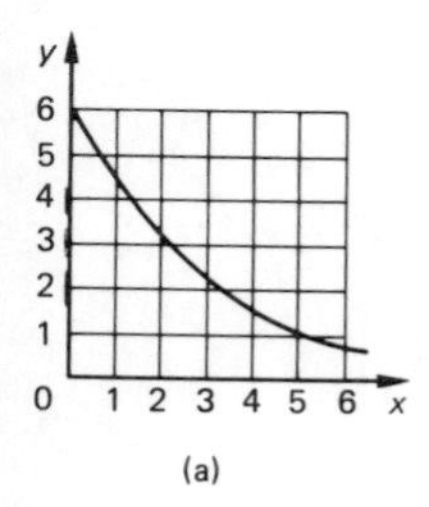

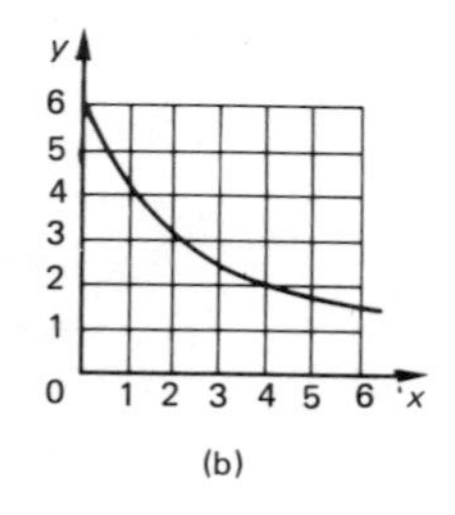

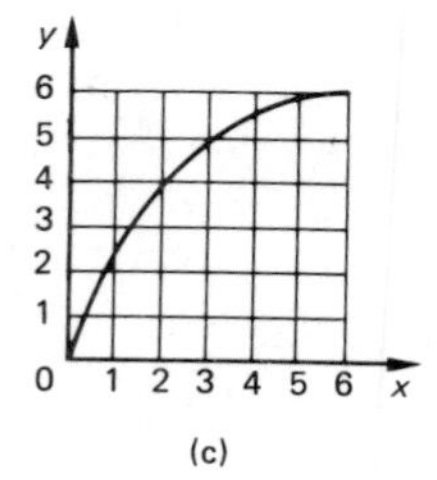

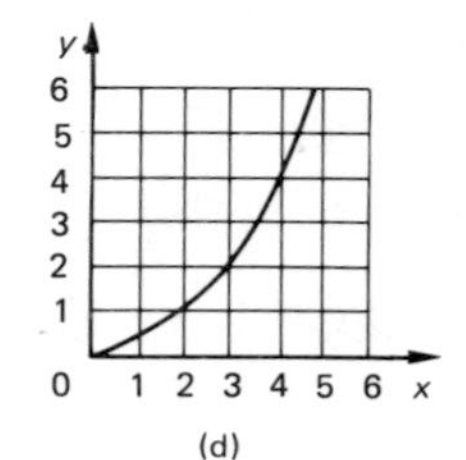

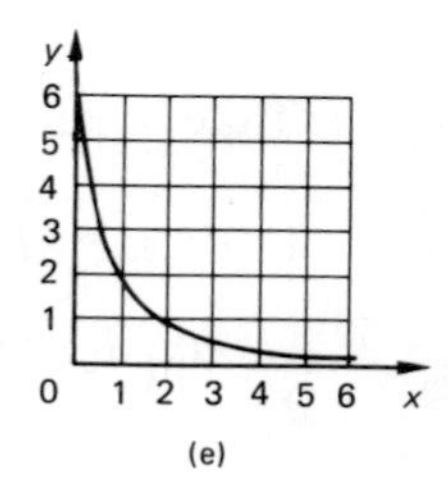

Figure 3.18

5. Barium-140 has a half-life of 13 days.
   (a) How many half-lives does a sample of barium-140 have in 13 weeks?
   (b) What fraction of the sample of barium-140 would remain after 13 weeks?
   (c) What fraction of the sample of barium-140 would have decayed after 26 weeks?

3.7 MODELLING WITH POWER LAWS

In Chapter 1 we saw that if given a set of data from an experiment that lies close to a straight line then we could find a linear law between the variables calculating the slope and intercept in the formula $y = mx + c$. But experimental data does not always give linear laws. Consider the data in Table 3.12 and associated graph (Figure 3.19) for the distance from the sun and period of the planets.

Table 3.12

Planet	Distance, R (millions of km)	Period, T (days)
Mercury	57.9	88
Venus	108.2	225
Earth	149.6	365
Mars	227.9	687
Jupiter	778.3	4 329
Saturn	1 427	10 753
Uranus	2 870	30 660
Neptune	4 497	60 150
Pluto	5 907	90 470

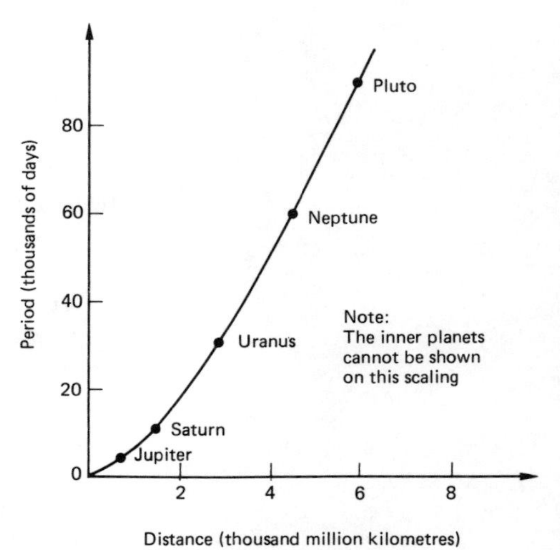

Figure 3.19

Clearly we would not be justified in proposing a linear model; however the points do appear to fit a smooth curve through the origin, so it is perhaps not unreasonable to suggest a relation $T = f(R)$. If the function f is a power law then logarithms can help us find it.

Suppose that the two variables are related by the rule

$$T = AR^m$$

where A and m are constants.

Taking logarithms (to base 10) yields

$$\log T = \log (AR^m) = \log A + m \log R$$

and $\log T$ is a linear function of $\log R$ since if we let $\log T = y$ and $\log R = x$ we have $y = mx + \log A$, i.e. a graph of y against x is a straight line of slope m and intercept $\log A$. Figure 3.20 shows a graph of $\log T$ against $\log R$.

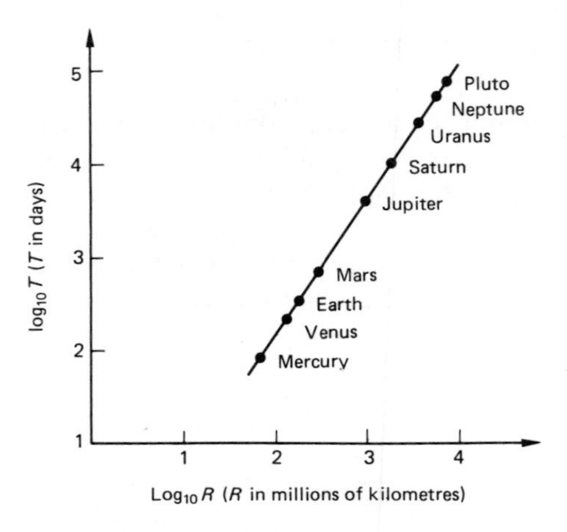

Figure 2.20

We see that the points do lie on a straight line of slope 1.5 ($=m$) and intercept -0.7 ($=\log A$); so $A = 10^{-0.7} = 0.2$. The relation between T and R is

$$T = 0.2R^{3/2}$$

This was first found by Kepler in the 17th century and is called Kepler's third law.

To summarize:

> If there is a power law between two variables x and y, i.e. if $y = Ax^n$ then a graph of $\log y$ against $\log x$ will be a straight line of slope n and intercept $\log A$

Exercise 3.7.1

1. Find the power law formula between the variables given in Table 3.13.

Table 3.13
(a)

x	1	2	3	4	5	6
y	3.4	22.1	66.0	143.6	262.2	429.0

(b)

u	0.2	1.4	2.6	3.8	5.0	6.2
v	13.78	1.10	0.49	0.30	0.21	0.16

2. Show that it is not reasonable to propose a power law for the data shown in Table 3.14 for the temperature of a hot rod against time.

Table 3.14

t(min)	2	4	6	8	10
T(°C)	816.4	564.8	390.7	270.3	187.0

3. The viscosity of a liquid is a measure of its resistance to flow. Its value is often found experimentally using a quantity of the liquid between two moving parallel plates and measuring the friction force on one plate. In such an experiment values of the coefficient of viscosity of lubricating oil against temperature were measured (Table 3.15). Show that a power law relation of the form $\mu = AT^m$ exists between the variables, and find values of the constants A and m.

Table 3.15

Temperature, T (°C)	20	40	60	80	100
Viscosity, μ ($\text{kgm}^{-1}\text{s}^{-1}$)	0.0986	0.0241	0.0110	0.0055	0.0036

4. Typical values for the number of counts per second, C, measured by a Geiger counter placed at various distances, x, from a radioactive source are given in Table 3.16. Show that a power law of the form $C = kx^n$ exists between C and x and find the values of the constants k and n.

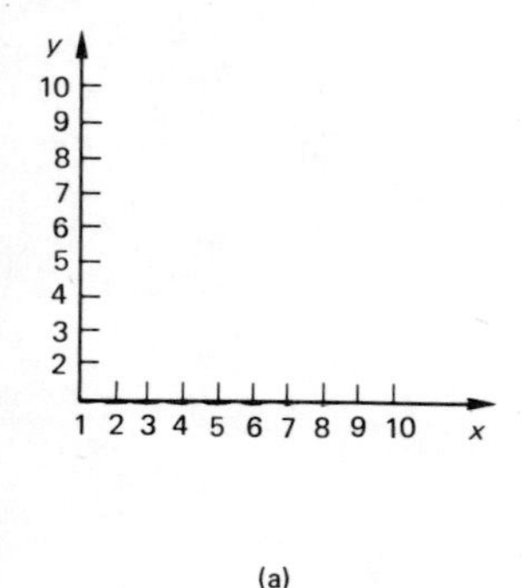

(a)

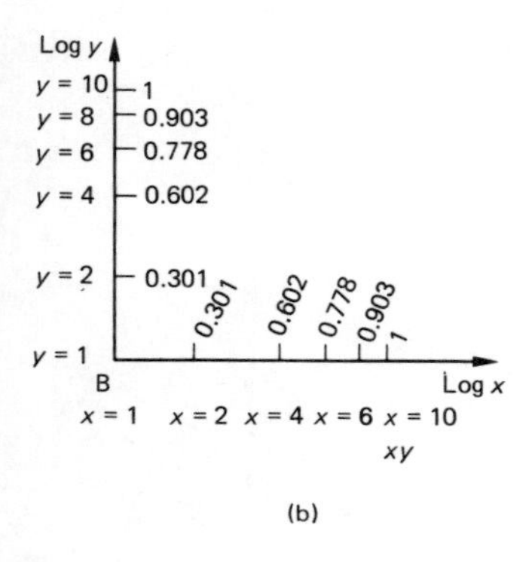

(b)

Figure 3.21

Table 3.16

x(m)	0.05	0.1	0.15	0.2	0.25	0.3
$C(s^{-1})$	48.0	12.0	5.3	3.0	1.9	1.3

It is often tedious to have to take logarithms of the pairs of data and plot a graph. Special graph paper exists which does the work for us. It is called *log-log graph paper* and looks rather strange compared with normal graph paper. To get an idea of how the scales are formed consider x and y values each between 1 and 10, and suppose that on normal paper we use 45 mm for the length of the scale between 1 and 10. Then the scales would be drawn equally spaced as shown in Figure 3.21(a).

Now if we use scales of $\log x$ against $\log y$ the spacing is non-uniform. Table 3.17 shows the distance from the point B of each number.

Table 3.17

Number	Logarithm	Distance from B (mm)
1	0	0
2	0.301	13.5
4	0.602	27.09
6	0.778	35.0
8	0.903	40.6
10	1.000	45.0

Figure 3.21(b) shows the shape of the scales on the resulting log-log graph paper.

Each axis on this graph represents *one cycle*. The numbers between 10 and 100 (whose logs are between 1 and 2) would form a second cycle, and so on. Figure 3.22 shows logarithmic graph paper with two

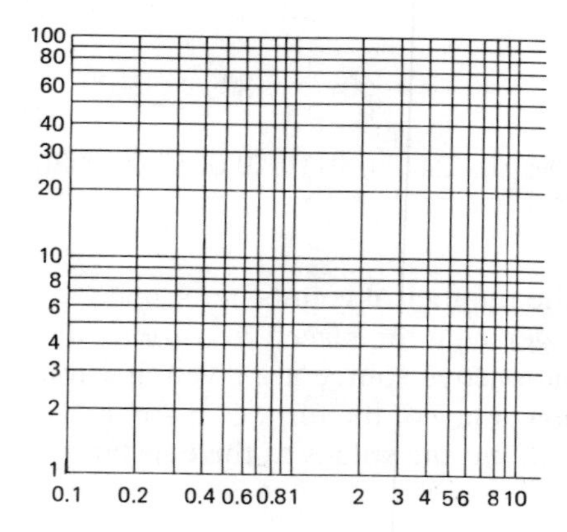

Figure 3.22

cycles along each axis. Along the x axis we have 0.1 to 10 for the x-values and along the y axis we have 1 to 100 for the y-values.

We use log-log paper if we expect a power law between two variables. To illustrate the use of log-log paper consider the data in Table 3.18.

Table 3.18

x	4	9	16	25	64
y	16	54	128	250	1024

Using log-log paper with scales 1–100 for x (two cycles) and 1–1000 for y (three cycle) as shown in Figure 3.23 we obtain a straight line graph.

We deduce from this graph that there is a power law between the variables x and y of the form $y = Ax^n$. It is important to note that the slope of the straight line graph in Figure 3.23 is *not* equal to n. We must calculate the value of n using two points on the graph.

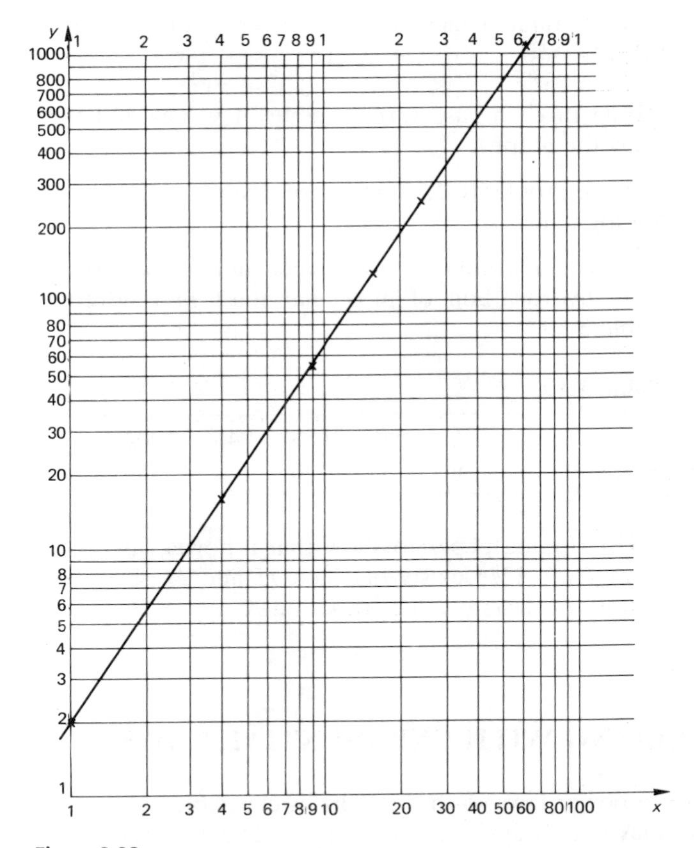

Figure 3.23

From the graph $y = 5.65$ when $x = 2$ and $y = 63$ when $x = 10$. Since $y = Ax^n$ we can write

$$\log y = n \log x + \log A$$

Substituting in values for x and y we have

$$\log 5.65 = n \log 2 + \log A \quad \text{or} \quad 0.75 = 0.3n + \log A$$

$$\log 63 = n \log 10 + \log A \quad \text{or} \quad 1.80 = n + \log A$$

Solving for n and A we have

$$n = 1.5 \quad \text{and} \quad A = 2.0$$

so the relationship between x and y is

$$y = 2x^{1.5}$$

Exercise 3.7.2

1. Draw a graph of each of the following on log-log paper.
   (a) $y = \sqrt{x}$ (b) $y = 0.1/x$
   (c) $y = 3x^{0.2}$ (d) $y = \pi x^2$
   (Hint: In each case form a table of values and then draw the graph).

2. Repeat Problems 2, 3 and 4 of Exercise 3.7.1 using log-log paper.

3. The heat of combustion H (joule mol^{-1}) for a petroleum hydrocarbon of molecular mass m is given in Table 3.19.

Table 3.19

H	213	373	530	688	845	1002	1159
M	16	30	44	58	72	86	100

Use log-log paper to show that a power law of the form $H = AM^k$ exists between H and M. Use your graph to estimate values of k and A.

3.8 MODELLING WITH EXPONENTIAL LAWS

Not all laws are power laws so that graphs of experimental data on log-log paper may not always give straight lines. The data in Problem

2 of Exercise 3.7.1 illustrates this fact. The law in this problem is actually exponential where the variable is the exponent.

Suppose that we have an exponential law $y = Ae^{ax}$. Taking logarithms to base e of each side yields

$$\ln(y) = \ln(Ae^{ax}) = \ln(A) + ax$$

So a graph of $\ln(y)$ against x will be a straight line with slope a and intercept $\ln(A)$. Consider the experimental data in Table 3.20.

Table 3.20 Cooling of a hot metal rod with time.

t(min)	2	4	6	8	10
T(°C)	816.4	564.8	390.7	270.3	187.0

Figure 3.24 shows a graph of $\ln(T)$ against t which is indeed a straight line. The slope of the straight line is -0.18 ($= a$) and the intercept is 7.05 ($= \ln(A)$ hence $A = e^{7.05} = 1150$). The relation between T and t is $T(t) = 1150e^{-0.18t}$.

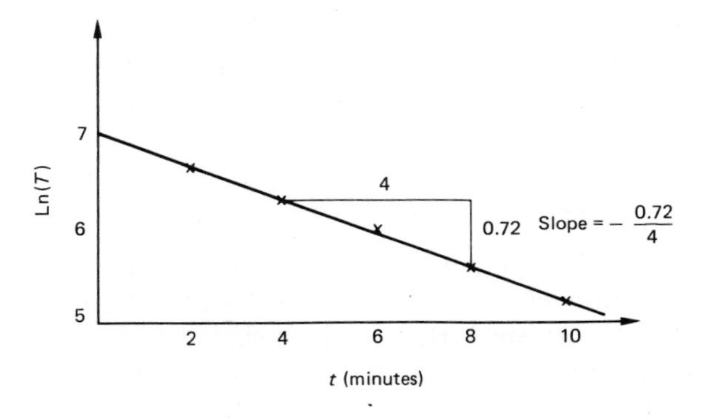

Figure 3.24

To summarize:

> If there is an exponential law between two variables x and y, i.e. if $y = Ae^{ax}$ then a graph of ln(y) against x will be a straight line of slope a and intercept $\ln(A)$

Exercise 3.8.1

1. A battery is discharging in a simple circuit and the voltage against time t is shown in Table 3.21. Find the exponential law between the variables.

Table 3.21

time t (s)	10	20	30	40	50
Voltage v (volts)	4.86	1.79	0.66	0.24	0.09

2. For most chemical reactions the velocity constant k is known to vary with temperature T according to the formula $k = Ae^{-\alpha/T}$ where A and α are constants. What graph could you plot to verify this fact given experimental data values of k and T?

   For the decomposition of nitrous oxide the data shown in Table 3.22 are available.

Table 3.22

Velocity constant, k (hr^{-1})	0.224	0.447	2.00	2.52	6.31
Temperature, T (°C)	985	1005	1058	1069	1105

From an appropriate graph estimate values of A and α for nitrous oxide and hence find the law relating k and T.

As before there is special graph paper available for graphing data for which an exponential model is expected. It is called *semilogarithmic* or *log-linear* graph paper; the vertical scale is similar to log-log paper being proportional to logarithms of numbers to base 10, while the horizontal axis is linear with the scale equally spaced. Figure 3.25 shows the data of Table 3.20 drawn on log-linear graph paper.

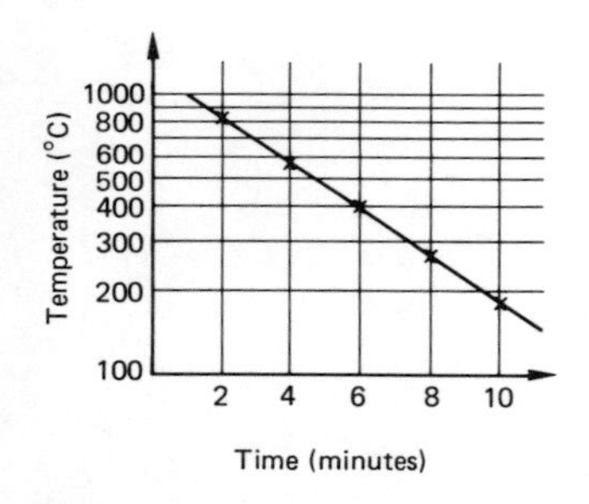

Figure 3.25

From the straight line graph we deduce a law of the form

$$T = Ae^{ax}$$

and taking logs (to base 10 because the graph paper uses base 10) we have

$$\log T = \log A + ax \log e.$$

Two points on the straight line graph are (2, 816.4) and (10, 187.0). Substituting these into the equation gives

$$\log 816.4 = \log A + a(2 \log e) \quad \text{or} \quad 2.91 = \log A + 0.87a$$

$$\log 187.0 = \log A + a(10 \log e) \quad \text{or} \quad 2.27 = \log A + 4.34a$$

Solving for a and A we have

$a = -0.18$ and $A = 1130$

The relation between t and T is $1130e^{-0.18t}$. (The numbers are slightly different from the previous formula for T because of small errors which occur in rounding the numbers).

Log-linear graph paper is useful as a guide to the existance of an exponential law between variables but we do not normally use it to find the constants and hence the form of the law.

Exercise 3.8.2

For the problems in Exercise 3.8.1 use log-linear graph paper to verify that appropriate exponential laws exist between the variables.

OBJECTIVE TEST 3

1. Simplify the following using positive indices.

   (a) $(x^2y^{-3})^{-1}$ (b) $\dfrac{ab^{-2}}{a^2b^2}$ (c) $(8x^3y^6)^{-1/3}$

2. The volume of a box, h metres high with a square base of side $(h - 3)$ metres is given by $V(h) = h(h - 3)^2$. State the domain and range of the function $V(h)$.

3. Consider the functions $f(x) = e^{3x}$ and $g(x) = 1 - 4x$ for values of x between 0 and 0.25. Find the rules for the inverse functions $f^{-1}(x)$ and $g^{-1}(x)$ and write down the domain and range of each inverse function. Write down the rule for the composite function $f[g(x)]$.

4. Find the unknown in these equations.

   (a) $e^{3x} = 4.12$ (b) $e^x + e^{-x} = 2$
   (c) $\ln(x) = 3$ (d) $3 \log(2x) - 2 \log(x) = 3.7$
   (e) $2^{x-2} = 47$ (f) $5^x = 30$

5. Radioactive carbon-14 decays at a rate of $k = 1.238 \times 10^{-4}\ \text{yr}^{-1}$. Calculate the half-life of carbon-14.

   The statue of Zeus at Olympia in Greece is made of gold and ivory. The ivory was found to have lost 35% of its carbon-14. Find the age of the statue.

6. In Table 3.23 showing three sets of experimental data, one pair of variables satisfies a power law, one pair satisfies an exponential law and the other pair does not satisfy either of these laws. Use appropriate graphs to find the relationships.

Table 3.23

x	0.1	0.4	0.7	1.0	1.3	1.6
y	2.010	0.606	0.182	0.055	0.017	0.005
u	0.1	0.3	0.6	0.9	1.2	1.5
v	1.00	0.96	0.83	0.62	0.36	0.07
t	0.1	0.5	0.9	1.3	1.7	2.1
s	0.084	1.293	3.511	6.561	10.35	14.83

4 TRIGONOMETRIC FUNCTIONS

OBJECTIVES

When you have completed this chapter you should be able to

1. state the convention for measuring angles
2. define sine and cosine as projections of a unit vector
3. sketch the graphs of sine, cosine and tangent
4. find the primary and secondary solutions to sine and cosine equations
5. state the period and the equations of the asymptotes of the tangent function
6. obtain the general solution to a trigonometric equation
7. obtain solutions within a specified range to a trigonometric equation
8. solve more complex trigonometric equations, including squares, multiple angles and phase angles
9. define sec, cosec and cot in terms of cos, sin and tan
10. sketch graphs of the form $a \sin(bx + c)$, $a \cos(bx + c)$ and $a \tan(bx + c)$ and determine their amplitude and period
11. convert from degrees to radians, and from radians to degrees
12. sketch graphs and solve equations using radians for angles
13. use trigonometric functions to model mechanical vibrations and alternating currents

4.1 INTRODUCTION

The diagram in Figure 4.1 shows a graph common in science. It occurs as the output of many vibrating systems. For example, it is

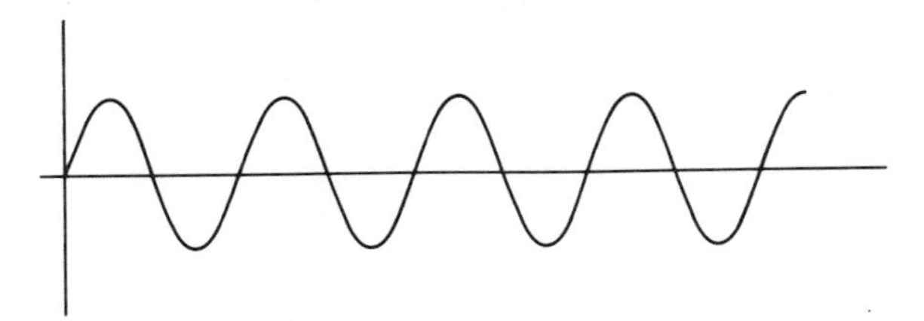

Figure 4.1

what we see on the screen of an oscilloscope when an alternating current is connected across the input terminals.

This is an example of the output of an electrical system but the graph also occurs in mechanical systems. Consider the oscillations of the simple mass/spring system shown in Figure 4.2.

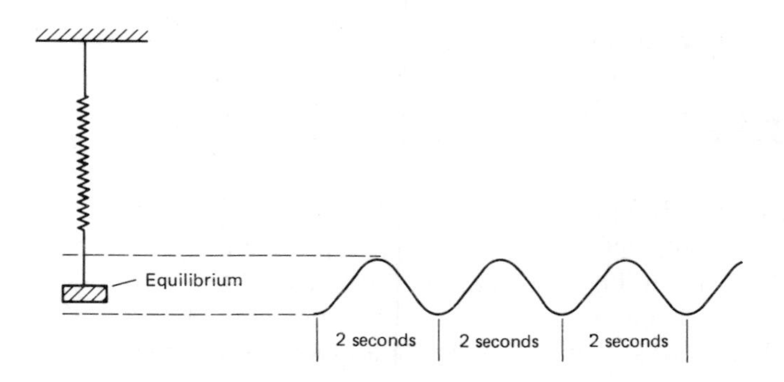

Figure 4.2

If we pull the mass down from its equilibrium position a distance of 50 mm and release it, then it will oscillate up and down in a regular pattern. The graph shows how the displacement of the mass varies with time. The graph repeats itself every two seconds, this is the time that it takes for the mass to travel from the lowest point to the highest point and back to the lowest point. This interval is called the *period* of the motion. The graph varies between -50 mm and $+50$ mm, which is the furthest that the mass travels from its equilibrium position. 50 mm is called the *amplitude* of the oscillation.

If we increase the mass suspended from the spring and release it from a position of 50 mm below the (new) equilibrium position, we find that the amplitude of the oscillations is still the same but the period is longer. So amplitude and period are independent quantities of the motion.

None of the functions that we have considered so far exhibit the periodic properties of the graphs in Figures 4.1 and 4.2. Polynomials and exponential functions are not periodic so it is clear that we need a new class of functions with periodic behaviour. The wave type behaviour is a very common feature of many systems in the physical world and the sine and cosine functions are extended from their trigonometric background to provide good models for oscillations.

Note: It is important to check that your calculator is in degree mode for the exercises in Sections 4.2 to 4.11.

4.2 ANGLES

You are probably familiar with angles in a triangle and angles drawn on a straight line. Most of the angles you have dealt with lie between

0° and 180°. Now we are going to introduce a convention for angles larger than 180° and negative angles.

For this we consider a line sweeping out an angle at a point so that 125° is the angle swept out by the line OP in an anti-clockwise rotation, shown in Figure 4.3. Figure 4.3 also shows the angles 232° and 340°.

The convention for measuring angles is to take anti-clockwise as positive and to measure from a horizontal datum. So it is possible to have negative angles (which represent clockwise rotations) and angles bigger than 360° (which represent more than one revolution) (see Figure 4.4).

Now, any angle is equivalent to an angle between 0° and 360°. For example, the angle 789° is made up of two complete rotations plus 69°. So the line OP in Figure 4.5 is in the same direction for angles of 69° and 789°.

We say that 69° and 789° are equivalent. In fact 69° is equivalent to many angles; 429° (360° + 69°), −291°(69° − 360°) are just two other examples.

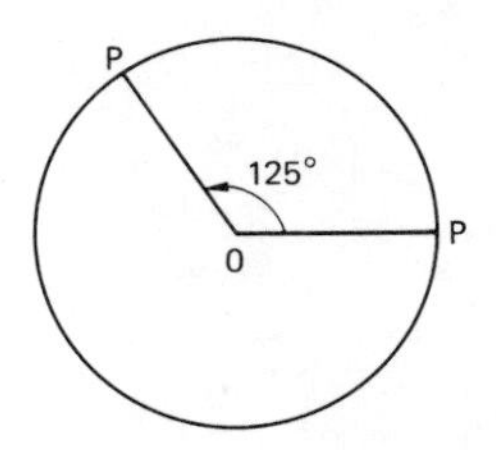

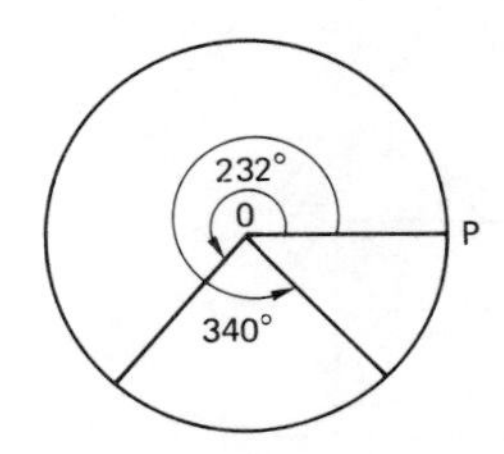

Figure 4.3

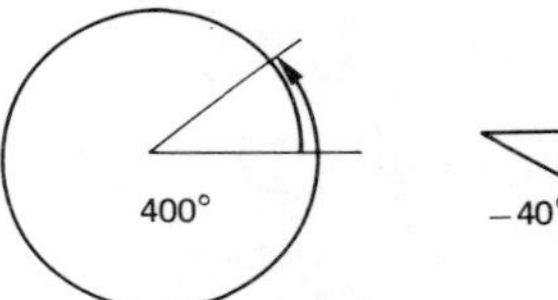

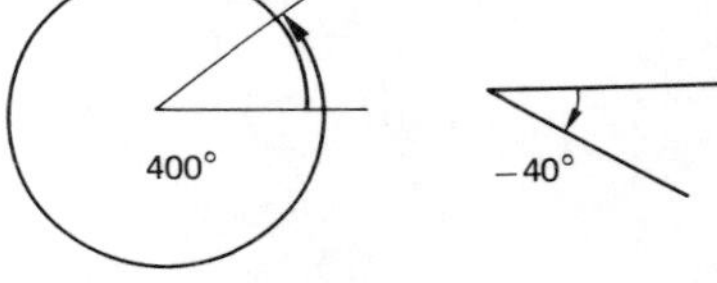

Figure 4.4

Exercise 4.2.1

1. Give the angle between 0° and 360° to which each of the following angles is equivalent.
   (a) 860° (b) −70° (c) −310° (d) −127° (e) 1020°
2. Give the angle between −180° and 180° to which each of the following angles is equivalent:
   (a) 200° (b) 559° (c) −382° (d) 315° (e) 1000°

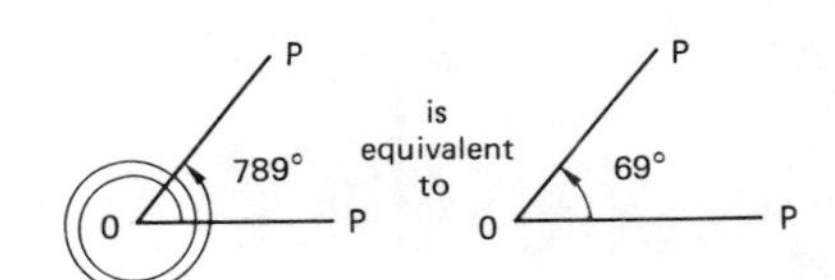

Figure 4.5

4.3 SINE AND COSINE AS PROJECTIONS

You will be familiar with the trigonometric rules sine, cosine and tangent for right-angled triangles.

For a right-angled triangle (Figure 4.6) we have

$$\sin A = \frac{\text{opposite}}{\text{hypotenuse}} \quad \cos A = \frac{\text{adjacent}}{\text{hypotenuse}}, \quad \tan A = \frac{\text{opposite}}{\text{adjacent}}$$

We will now introduce an alternative definition of these rules.

It is useful to think of sine and cosine as projections, especially when working with vectors in applied mathematics. The idea also leads to an easy definition of the sine and cosine of any angle.

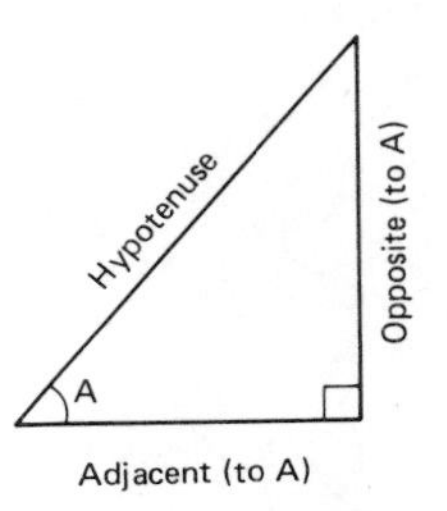

Figure 4.6

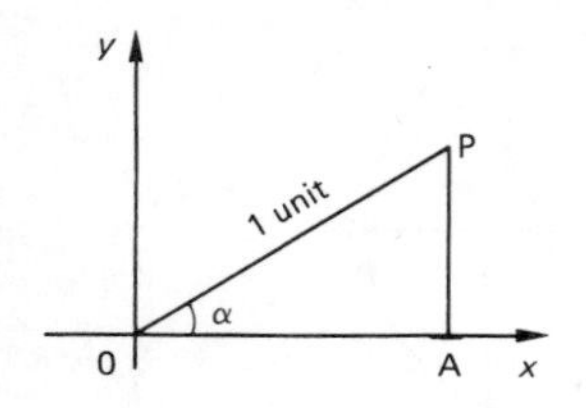

Figure 4.7

Take a right angled triangle OAP, in which the hypotenuse is one unit long (Figure 4.7).

Then

$$OA = OP \cos \alpha = 1 \cos \alpha = \cos \alpha$$

$$AP = OP \sin \alpha = 1 \sin \alpha = \sin \alpha$$

But OA is the projection of OP onto the x axis. This gives an important result:

The projection of OP onto the x axis = cos α.

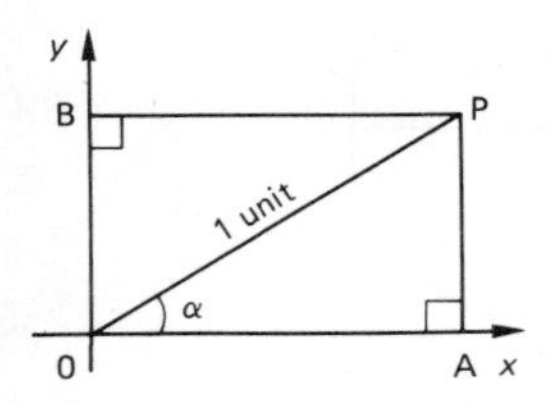

Figure 4.8

To project OP onto the y axis, draw PB perpendicular to the y axis (Figure 4.8). Then OB is the projection of OP onto the y axis.

From geometry OB = AP. But AP = $\sin \alpha$. This gives a second result

The projection of OP onto the y axis = sin α

Hence we may think of $\cos \alpha$ and $\sin \alpha$ as projections of a line 1 unit long onto the x and y axes.

However we can use these results to define cosine and sine for any angle:

> *Definition*: Let OP be 1 unit long and make an angle of α with the positive x axis. Then
>
> $\cos \alpha$ = projection of OP onto the x axis
> $\sin \alpha$ = projection of OP onto the y axis

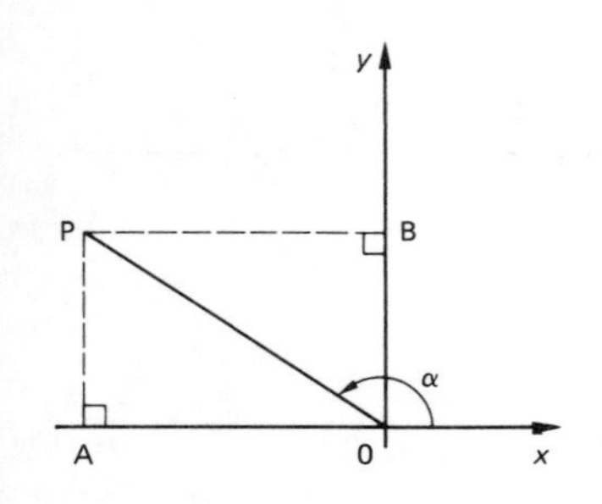

Figure 4.9

Example 4.3.1

Find the cosine and sine of an angle in the second quadrant (Figure 4.9).

Solution

$\sin \alpha$ = projection of OP onto the y axis = OB which is positive.
$\cos \alpha$ = projection of OP onto the x axis = OA which is negative.

Example 4.3.2

Which angle β in the first quadrant would have a sine and cosine with the same magnitude as $\sin \alpha$ and $\cos \alpha$?

Solution

Construct BR = PB.
Drop a perpendicular from R to meet the x axis at Q.

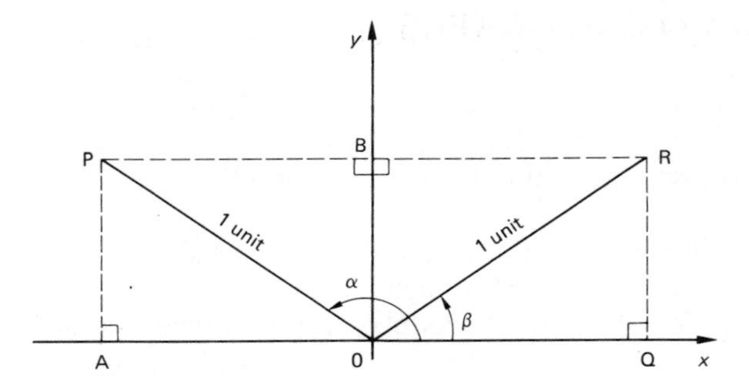

Figure 4.10

Then from the geometry of the diagram (Figure 4.10) we can see that β (= angle ROQ) will satisfy the criteria.

Note:

$$\cos\beta = -\cos\alpha \quad \text{and} \quad \sin\beta = \sin\alpha$$

Also:

$$\beta = 180° - \alpha$$

Example 4.3.3

Find the cosine and sine of an angle in the third quadrant (Figure 4.11)

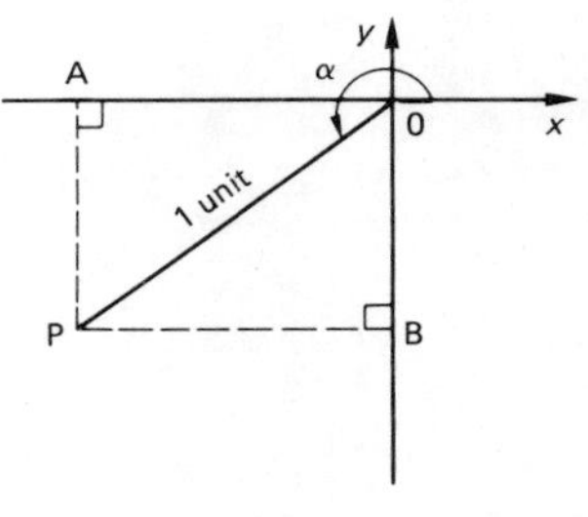

Figure 4.11

Solution

Sin α = projection of OP onto y axis = OB, which is negative.

Cos α = projection of OP onto x axis = OA, which is negative.

Exercise 4.3.1

1. Draw a diagram to show the values of the cosine and sine of an angle in the fourth quadrant.
2. Given that sin 30° = 0.5 and cos 30° = 0.8660, use the definitions of sine and cosine as projections to find the values of sin 150°, cos 150°, sin 210°, cos 210°, sin 330°, cos 330°, sin(−30°), cos(−30°).
3. Use the definitions of sine and cosine as projections to find the values of sin 180°, cos 180°, sin 270°, cos 270°, sin 360°, cos 360°.

4.4 SINE AND COSINE GRAPHS

First we draw the graph (Figure 4.12) of $y = \sin \theta$. Now

$\sin \theta = AP =$ vertical projection of the unit line OP.

As the line OP rotates through a circle the value of the height y ($= \sin \theta$) moves up and down and up again in a wave-like motion. Drawing a graph of height against θ produces a special type of graph called 'the sine graph'. The rotating line OP is often called a *phasor*.

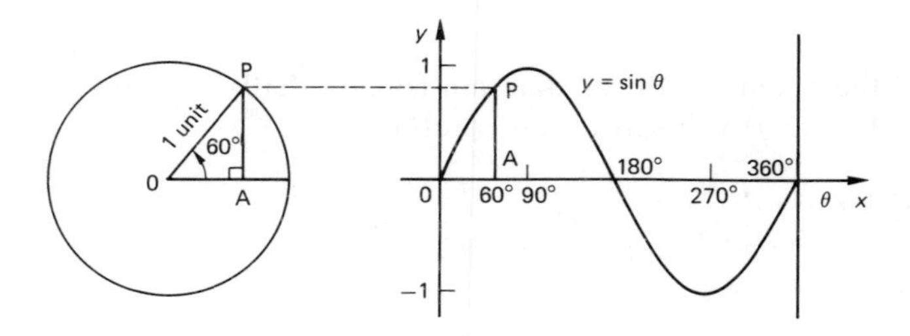

Figure 4.12

Next we consider the graph (Figure 4.13) of $y = \cos \theta$. Now

$\cos \theta = OA =$ horizontal projection of the unit line OP.

As the phasor OP rotates through a circle the value of the projection OA moves down and up in a wave-like motion, forming the cosine graph.

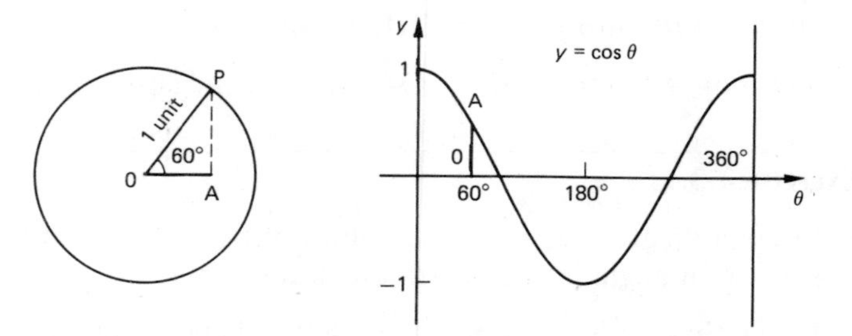

Figure 4.13

Exercise 4.4.1

1. On graph paper draw a circle of radius 4 cm. Let the radius be 1 unit.

(a) From your circle write down the sine of the following angles:
$0°, 30°, 60°, 90°, 120°, 150°, 180°, 210°, 240°, 270°, 300°, 330°, 360°$.
Check the values on your calculator.
(b) Repeat part (a) for the cosine of each angle.

2. Draw the sine and cosine curves for the range of values $-360°$ to $720°$. Note how the same basic shape is repeated. Explain this property.
Sketch the graph of the sine and cosine functions for the range $-720°$ to $1800°$. Check your sketch using the computer.

4.5 TANGENT OF ANY ANGLE

From elementary trigonometry we know that

$$\tan \theta = \frac{\sin \theta}{\cos \theta}$$

In the following exercise you can investigate some of the properties of the tangent function.

Exercise 4.5.1

(a) Use your calculator (or a computer) to evaluate $\tan \theta$ and draw a graph of $y = \tan \theta$ for θ between $0°$ and $360°$.
What happens for $\theta = 90°$ and $270°$?
(b) Although $\tan \theta$ is not defined for $\theta = 90°$ it is defined for values of θ close to $90°$. Use your calculator to find $\tan 89°$, $\tan 89.9°$, $\tan 89.99°$, $\tan 90.01°$, $\tan 90.1°$, $\tan 91°$.
(c) We say that at $\theta = 90°$, $\tan \theta$ has *an asymptote*. Where else does $\tan \theta$ have asymptotes?
(d) Draw the graph of $\tan \theta$ for the range of θ between $-720°$ and $720°$. Find all the asymptotes in this range.

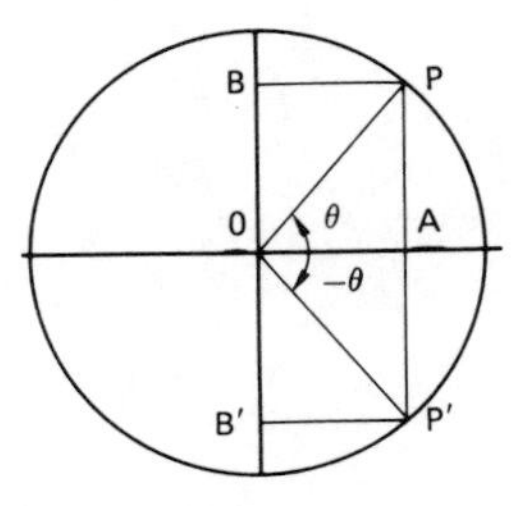

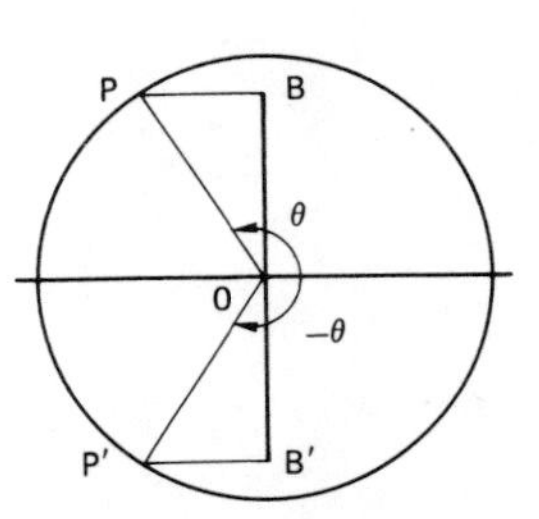

Figure 4.14

4.6 ODD AND EVEN FUNCTIONS

Since $\sin \theta$ can be considered as the vertical projection AP (= OB) of a unit line OP (Figure 4.14), we have

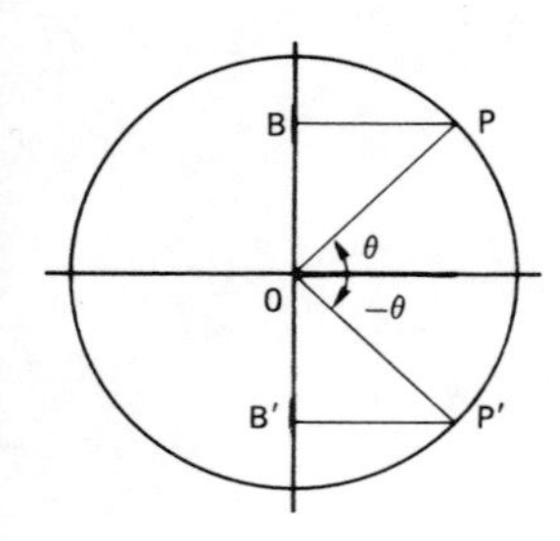

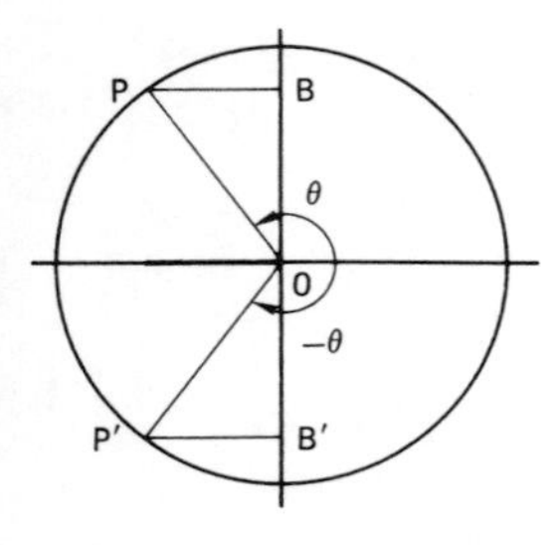

Figure 4.15

$$\text{OB} = \sin\theta \quad \text{and} \quad \text{OB}' = \sin(-\theta)$$

so that

$$\sin(-\theta) = -\sin\theta \quad \text{for any angle } \theta.$$

Since $\cos\theta$ can be considered as the horizontal projection of a unit line OP (Figure 4.15), we have

$$\text{BP} = \cos\theta \quad \text{and} \quad \text{BP} = \cos(-\theta)$$

so

$$\cos(-\theta) = \cos\theta \quad \text{for any angle } \theta.$$

Lastly since $\tan\theta = \sin\theta/\cos\theta$, we have

$$\tan(-\theta) = \frac{\sin(-\theta)}{\cos(-\theta)} = \frac{-\sin(\theta)}{\cos(\theta)} = -\tan\theta, \text{ for any angle } \theta.$$

Functions such as $\sin x$ and $\tan x$ with the property that

$$f(-x) = -f(x)$$

for all x are called *odd functions*. Their graphs have a half turn rotational symmetry about the origin.

Functions such as $\cos x$ for which

$$f(-x) = f(x)$$

for all x are called *even functions*. Their graphs have a reflectional symmetry in the y axis (vertical axis).

Exercise 4.6.1

1. Draw a graph of $y = \sin\theta$ for $-360° \leqslant \theta \leqslant 360°$. From the graph show that $\sin(-\theta) = -\sin\theta$ for all θ in this range. Confirm that the graph has rotational symmetry about the origin.

2. By sketching $\cos\theta$ for $-360° \leqslant \theta \leqslant 360°$, show that $\cos(-\theta) = \cos\theta$ for all θ in this range. Your graph should have reflective symmetry in the y axis since cosine is an even function.

3. By sketching $y = \tan\theta$ for $-180° \leqslant \theta \leqslant 180°$, show that $\tan(-\theta) = -\tan\theta$ for all θ in this range. Your graph (including asymptotes) should have rotational

symmetry about the origin as tangent is an odd function.

The following questions should be done without using a calculator.

4. If $\sin 20° = 0.342$, find $\sin(-20°)$.
5. If $\cos 61° = 0.485$, find $\cos(-61°)$.
6. If $\tan 50° = 1.192$, find $\tan(-50°)$.
7. If $\tan A = 0.7$, find $\tan(-A)$.
8. If $\cos B = -0.1$, find $\cos(-B)$.
9. If $\sin t = -0.39$, find $\sin(-t)$.
10. By means of a sketch, explain why the property $\cos(-\theta) = \cos\theta$ for all angles θ means that the cosine graph has reflective symmetry in the y axis.
11. By means of a sketch, explain why the property $\sin(-\theta) = -\sin\theta$ for all angles θ means that the sine graph has a half turn rotational symmetry about the origin.
12. Sketch the graphs of the following functions: $y = x$, $y = x^2$, $y = x^3$, $y = e^x$, $y = e^{-x}$, $y = 2$.

    Classify each function as (a) even, (b) odd, or (c) neither even nor odd.

4.7 PRIMARY AND SECONDARY SOLUTIONS OF SINE AND COSINE EQUATIONS

When solving an equation such as $\sin\theta = 0.4$, your calculator only gives one answer to the equation. Try it and you should obtain $\theta = 23.6°$. (To undo the sine you need to press the keys INV SIN.) This solution, $\theta = 23.6°$ is called the *primary solution* and labelled p in Figure 4.16.

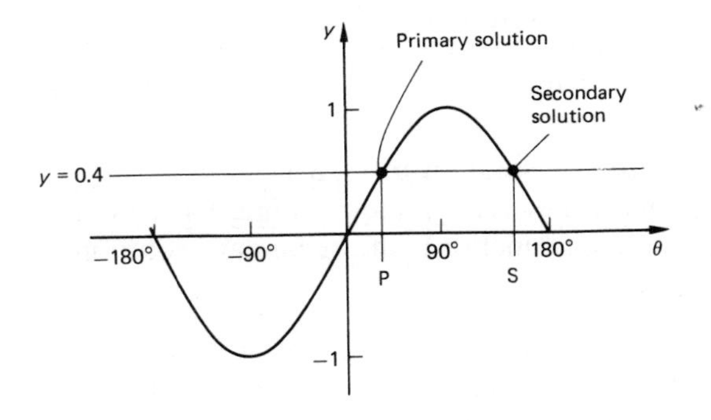

Figure 4.16

However the graph shows that there is another solution s, called the secondary solution. The sine curve has a symmetrical shape. So since p is 23.6° to the right of 0°, s is 23.6° to the left of 180°. Therefore $s = 180° - 23.6° = 156.4°$.

Example 4.7.1

Find the primary and secondary solutions to $\sin \theta = -0.4$.

Solution

The calculator gives the primary solution $p = -23.6°$.

From the sketch (Figure 4.17) and the symmetry of the sine curve, p is 23.6° to the left of 0°, so s is 23.6° to the right of $-180°$. Therefore $s = -180° + 23.6° = -156.4°$.

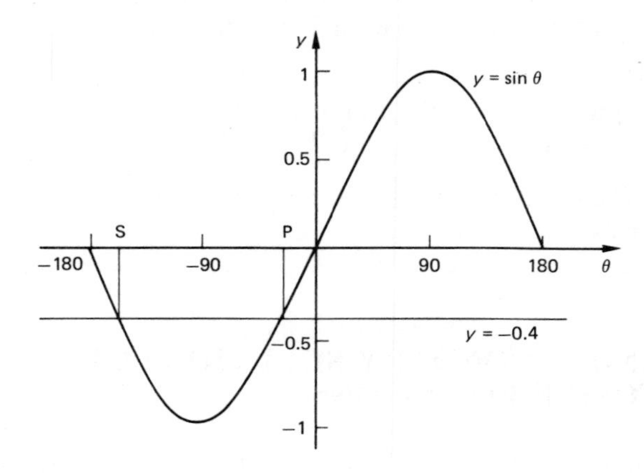

Figure 4.17

Example 4.7.2

Find the primary and secondary solutions to $\cos \theta = 0.3$.

Solution

The calculator gives the primary solution $p = 72.5°$.

Since the cosine curve has reflective symmetry (Figure 4.18) in the y axis (recall cosine is an even function), the secondary solution is $s = -p = -72.5°$.

Example 4.7.3

Find the primary and secondary solutions to $\cos \theta = -0.3$.

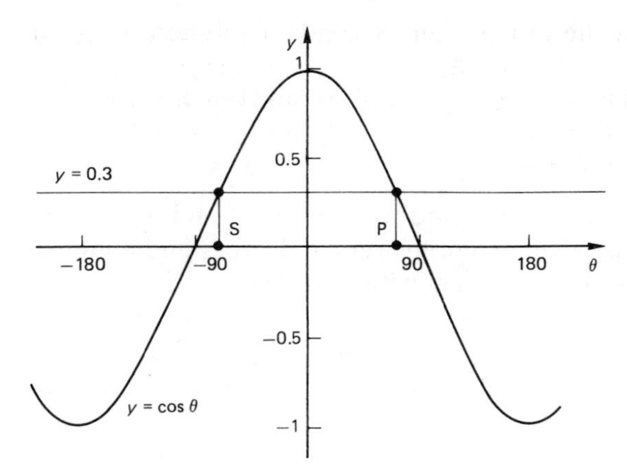

Figure 4.18

Solution

The calculator gives the primary solution $p = 107.5°$.

From the graph in Figure 4.19 we see that the secondary solution is $s = -p = -107.5°$.

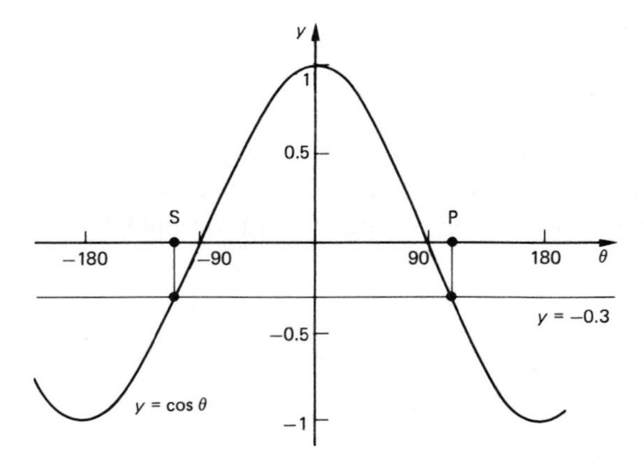

Figure 4.19

When solving equations of the type in these examples a quick sketch is often useful. Also note that for cosine equations $s = -p$. Therefore both solutions can be given in the compact for $\theta = \pm p$.

For tangent equations there is no secondary solution. These will be considered in more detail later.

Exercise 4.7.1

1. For each of the following problems sketch the sine or cosine graph for angles between $-180°$ and $+180°$.

Indicate the primary and secondary solutions on your sketch.

(a) $\cos \theta = -1/7$ (b) $\sin A = 1/8$
(c) $\cos B = 0.23$ (d) $\sin t = -0.09$
(e) $2 \sin \theta = -1$ (f) $\cos x = 0$

2. For the following find the primary and secondary solutions, giving your answers in degrees and minutes. Illustrate your answers with a sketch.

(a) $\sin t = 0.729$ (b) $\cos t = 0.395$
(c) $\sin \theta = -0.4$ (d) $\cos A + 3/8 = 0$

4.8 GENERAL SOLUTIONS OF TRIGONOMETRIC EQUATIONS

We have seen that the graph of sine repeats itself every 360°. So, in addition to the primary and secondary solutions, any sine equation will have an infinite number of solutions. These are related to the primary solution and secondary solution in a very simple way.

Consider the equation

$$\sin \theta = 0.6$$

Graphically this could be solved by drawing $y = \sin \theta$ and $y = 0.6$ and finding where they cut, as Figure 4.20 shows where P is the primary solution (36.9°) and s is the secondary solution (143.1°).

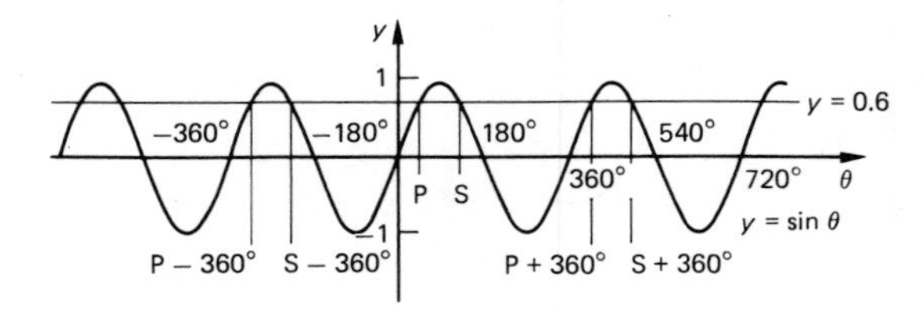

Figure 4.20

There are other solutions at $\theta = P + 360°$ (=396.6°) and $\theta = P - 360°$ (=−323.3°). Also there are solutions at $\theta = P + 2 \times 360°$ (=756.9°), $\theta = P + 3 \times 360°$ (=1116.9°), and so on. In fact from the primary solution we can generate an infinite number of other solutions by adding on multiples of 360°.

Further solutions can be generated from the secondary solution at $\theta = S + 360°$ (=503.1°) and so on.

The easiest way to give a formula for a multiple of 360° is to write $360°n$ where n is an integer.

From this we can see that the general solution is

$\theta = P + 360°n$, where n is an integer (positive or negative)

or

$\theta = S + 360°n$, where n is an integer (positive or negative)

so for $\sin\theta = 0.6$, $\theta = 36.9° + 360°n$ or $\theta = 143.1° + 360°n$.
A similar situation holds for cosine equations.

Example 4.8.1

Solve $\cos\theta = 0.6$

Solution

The cosine curve repeats itself every 360°.

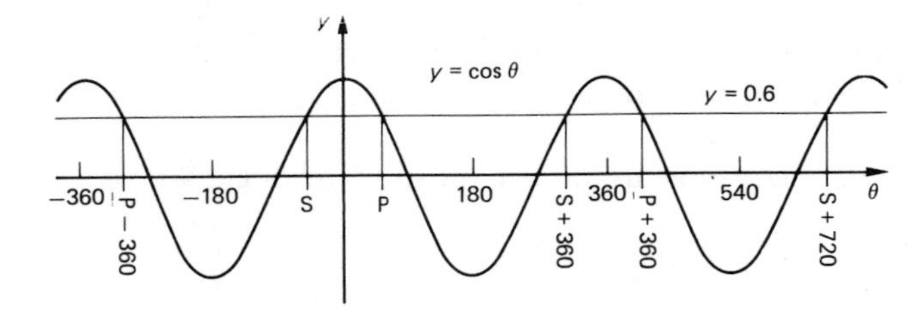

Figure 4.21

From a calculator we have P = primary solution (=53.1°) and then from the graph we deduce that S = secondary solution (= −53.1°). So the general solution generated from the primary solution is $\theta = 53.1° + 360°n$ and from the secondary solution $\theta = -53.1° + 360°n$.
In short we write the solution of $\cos\theta = 0.6$ as

$$\theta = \pm 53.1° + 360°n$$

The situation for tangent equations is simpler since we do not have secondary solutions. However one major difference is that the tangent graph repeats itself every 180°.

Example 4.8.2

Solve $\tan\theta = 0.6$.

Solution

From a calculator the primary solution $P = 31.0°$.

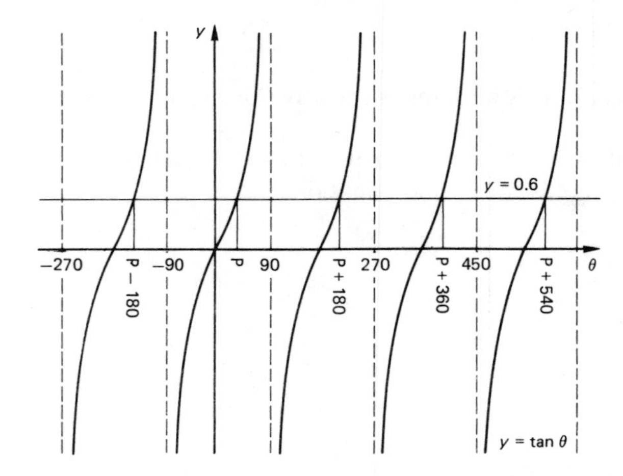

Figure 4.22

From the graph of $y = \tan\theta$ we see that the solution occurs at 180° intervals, so the general solution can be written as

$$\theta = P + 180°n, \text{ where } n \text{ is an integer (positive or negative)}$$

so for $\tan\theta = 0.6$, the general solution is $\theta = 31.0° + 180°n$.

Exercise 4.8.1

Give the general solutions in degrees to the following equations, correct to one decimal place.

(a) $\sin x = 1/4$ (b) $\tan y = 0.9$
(c) $3\cos R = -2$ (d) $3\cos t + 0.5 = 0$
(e) $(1/2)\sin B = -1/3$ (f) $3 - 2\tan A = 1$

4.9 PROPERTIES OF TRIGONOMETRIC FUNCTIONS

4.9 Period

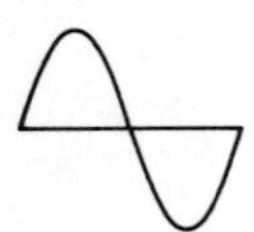

Figure 4.23

We have seen that the sine and cosine curves repeat the same basic shape every 360°. The basic shape is shown in Figure 4.23.

The graph of $y = \sin\theta$ goes through (0, 0) and the graph of $y = \cos\theta$ goes through (0, 1). However, in all other ways they are identical. So we only need to consider one function, sine say. Sine

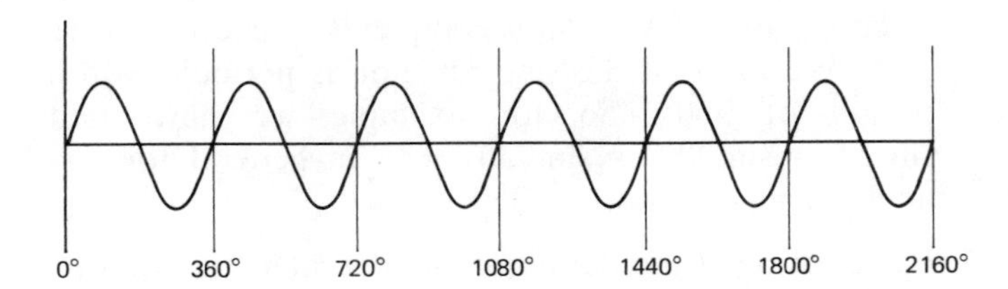

Figure 4.24

and cosine are often called *sinusoidal functions*. Figure 4.24 shows a graph of $y = \sin\theta$ between $\theta = 0°$ and $\theta = 2160°$.

Note the wave shape of the graph; this graph is called a *sine wave*. It repeats itself every 360°.

360° is called *the period* of the function. Any function with the property that a basic cycle repeats itself is called a *periodic function*. The tangent function is periodic, but its period is 180°.

Example 4.9.1

What is the period of the function shown in Figure 4.25?

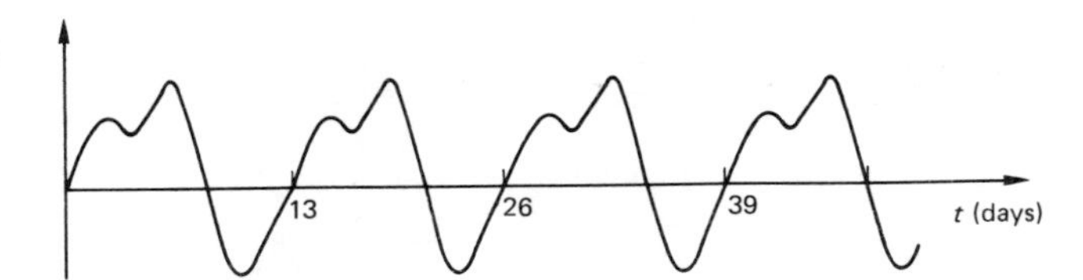

Figure 4.25

Solution

Clearly the basic cycle repeats itself every 13 days so the period is 13 days.

Exercise 4.9.1

1. Use a computer package to plot $y = \sin\theta$ for θ between $-360°$ and $360°$. Make sure you have chosen the 'degrees' option in the package. Note how the sine graph repeats itself. You should see two complete cycles over the chosen range.

   Now change the scale so that θ goes from $-720°$ to $720°$. How many complete cycles are there for this range? How many complete cycles would there be if the range for θ was $360°$ to $1080°$?

The graph of $y = \sin\theta$ completes one cycle every 360°. We say that the sine function is periodic with a period of 360°. So for example we have that $\sin 20° = \sin 380° = \sin 740° = \ldots$ In general for *any* angle θ, $\sin\theta = \sin(\theta + 360°)$.

2. By referring to the definition of general angles and the method of generating a sine graph by means of a rotating unit line, explain why $\sin\theta = \sin(\theta + 360°)$ for all angles θ.
3. Use a computer package to plot $y = \cos\theta$ for θ between $-360°$ and 360°. Note that the cosine function is periodic. What is its period?
4. Use a computer package to plot $y = \tan\theta$ for θ between $-360°$ and 360°. Note that the tangent function is periodic. What is its period?
5. What is the period of the periodic functions shown in Figure 4.26?

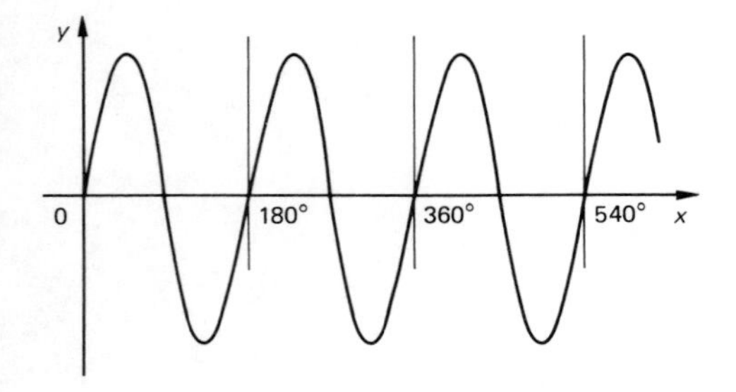

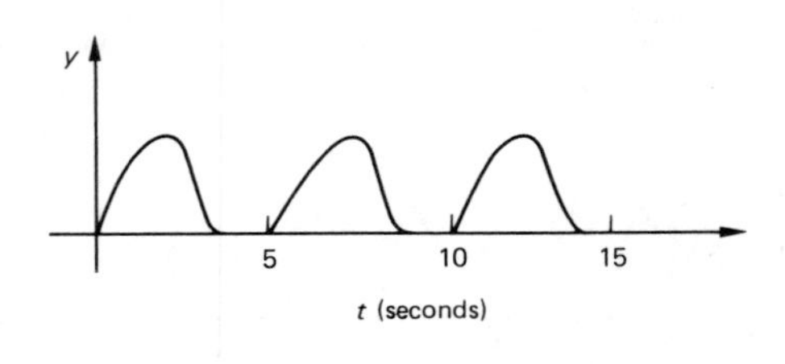

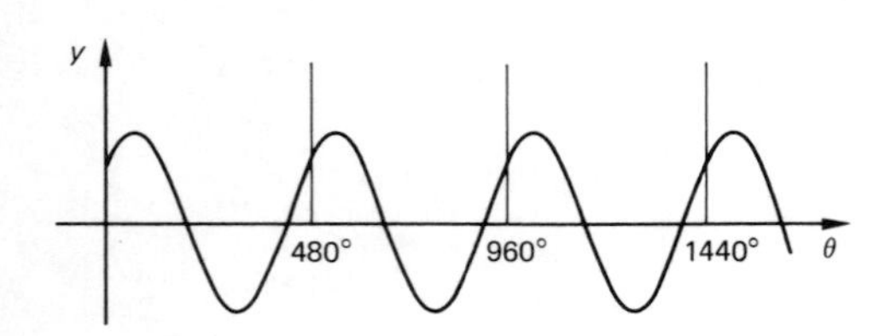

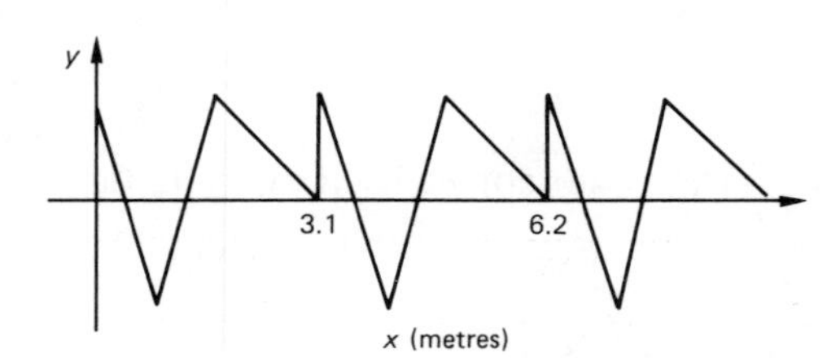

Figure 4.26

4.9.2 Angular Frequency

Consider the graph of $y = \sin 2\theta$ (Figure 4.27). (Note that $\sin 2\theta$ means double the angle θ and then take the sine of the result; e.g. if $\theta = 35°$ then $\sin 2\theta = \sin 70° = 0.94$.)

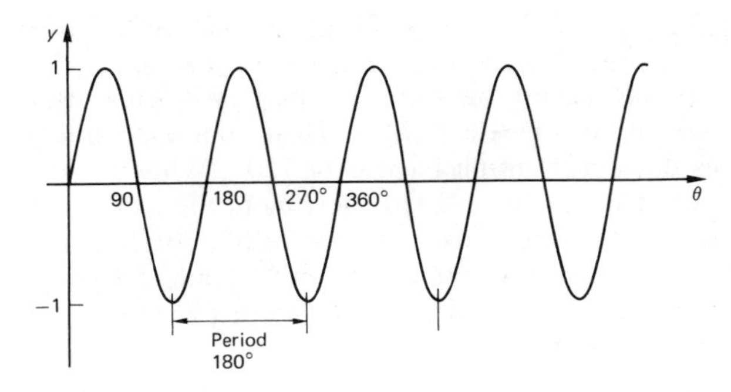

Figure 4.27

The period of this graph is 180° because it repeats itself every 180°. In other ways it has the same shape as the graph $y = \sin\theta$.

There are two cycles of $y = \sin 2\theta$ in every 360°. For $\sin 3\theta$ there are three cycles in every 360° and for $\sin(\omega\theta)$ there are ω cycles in every 360°. The number ω is called the *angular frequency* (or *circular frequency*).

Note that for a frequency of 2 the period is $360°/2 = 180°$. In general, for a frequency of ω the period is $360°/\omega$.

Example 4.9.2

For the sine wave shown in Figure 4.2.8, write down the period, calculate the angular frequency and hence write down a formula for the graph.

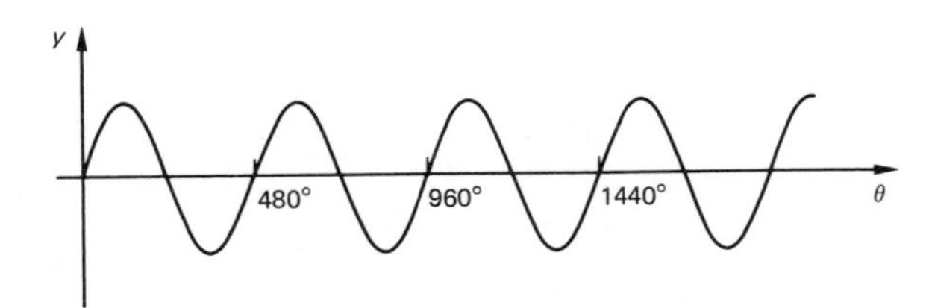

Figure 4.28

Solution

From the graph we see that the period is 480°. Hence the angular frequency is $\omega = 360°/480° = 0.75$. The graph has equation

$$y = \sin(0.75\theta)$$

Exercise 4.9.2

1. Use a computer package to plot $y = \sin \theta$ and $y = \sin 2\theta$ for $0° \leqslant \theta \leqslant 720°$. How many complete cycles does each graph have over 720°? What are the periods of $\sin \theta$ and $\sin 2\theta$? Referring to the method of generating the sine function by means of a rotating unit line, observe that the function $\sin 2\theta$ could be generated by a line which rotates twice as fast as the vector generating $\sin \theta$.

2. Use a computer package to plot $y = \cos \theta$ and $y = \cos(0.5\theta)$ for θ between $-360°$ and 360°. How many complete cycles does each graph complete over this 720° 'window'? What is the period and angular frequency of each function?

3. Use a computer package to obtain approximate solutions to the equation $\sin 2\theta = 0.75$ for angles between 0° and 360°. (Hint: plot $y = \sin 2\theta$ and $y = 0.75$.)

4. Use a computer package to plot $y = \cos \theta$ and $y = \cos(3\theta/2)$ for $0° \leqslant \theta \leqslant 360°$. Note that $\cos(3\theta/2)$ does not have a whole number of cycles over this 'window' of 360°. What is its period and angular frequency? Check that period = 360°/angular frequency.

5. Write down the period and angular frequency of each of the following functions.
   (a) $\sin(5x)$
   (b) $\cos(\theta/4)$
   (c) $\sin(3t/5)$
   (d) $\sin(k\theta)$ where k is an integer constant
   (e) $\cos(kx)$ where k is an integer constant.

4.9.3 Amplitude

The graph of $y = \sin \theta$ varies about a mean value of zero and has maximum values of $+1$ and minimum values of -1.

Consider the graph of $y = 4 \sin \theta$ (Figure 4.29) (Note $4 \sin\theta$ means 'find $\sin \theta$ and then multiply by 4', e.g. if $\theta = 55°$ then $4 \sin 55° = 4 \times 0.8192 = 3.277$. This is *not* the same as $\sin 4\theta$ which means sine of 4 times θ, e.g. if $\theta = 55°$, $\sin 4\theta = \sin 220° = -0.6428$.

The graph $y = 4 \sin \theta$ has the same shape as $y = \sin \theta$, a period of 360°, but varies between -4 and $+4$.

The 'size of the function' is called the *amplitude*. For the graph in Figure 4.24 the amplitude is 1 and for the graph in Figure 4.29 the amplitude is 4.

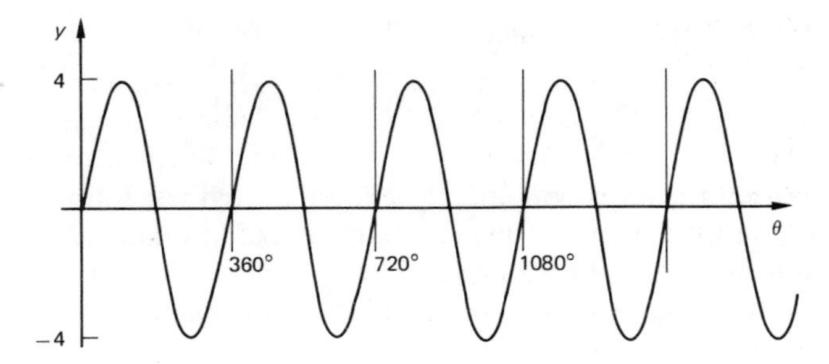

Figure 4.29

Example 4.9.3

Write down the amplitude, period and angular frequency of the sine wave shown in Figure 4.30. Write down the equation of the sine wave.

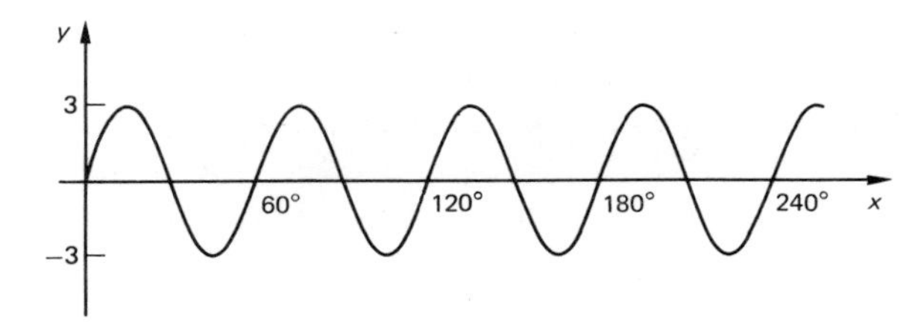

Figure 4.30

Solution

From the graph we have amplitude $= 3$, period $= 60°$ and hence the angular frequency $= 360°/60° = 6$.

The equation of the sine wave is

$$y = 3 \sin(6x)$$

Exercise 4.9.3

1. Use a computer package to plot the following functions. Make sure that you are in degrees mode and take the range for θ as $-360°$ to $360°$.

   In each case state the amplitude of the function. For sine and cosine graphs, note maximum and minimum values of the function, and the angles at which they occur. Also note where the graph cuts the θ axis.

   Obtain a print out of each graph (or make a sketch of it).

(a) $y = \sin\theta$ (b) $y = 2 \sin \theta$
(c) $y = 0.5 \sin \theta$ (d) $y = \cos \theta$
(e) $y = 3 \cos \theta$ (f) $y = (1/4) \cos \theta$
(g) $y = 2 \tan \theta$

2. Without using a computer package, sketch the following graphs, taking θ from $-360°$ to $360°$. In each case state their amplitude and period.
(a) $y = (1/3) \sin \theta$ (b) $y = 5 \cos \theta$
(c) $y = 0.5 \tan \theta$

3. Write down the amplitude and period of the sine waves shown in Figure 4.31. For each sine wave write down its equation.

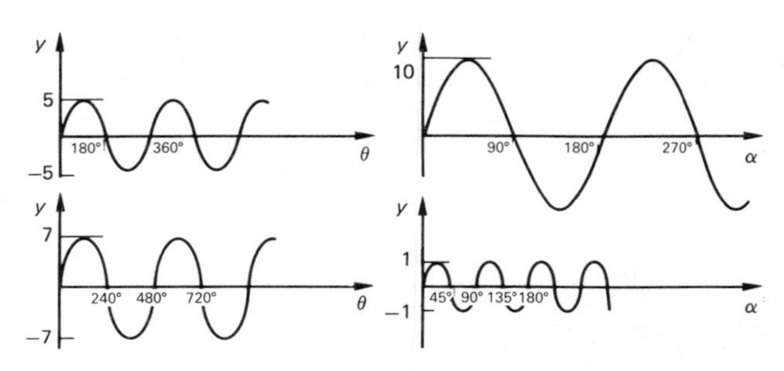

Figure 4.31

4. Write down the amplitude, angular frequency and period of the following 'sinusoidal functions'.
(a) $y = 2 \sin(3\theta)$ (b) $y = 4 \sin(2\theta)$
(c) $y = 4.3 \sin(3x/4)$ (d) $y = 0.73 \cos(0.21t)$

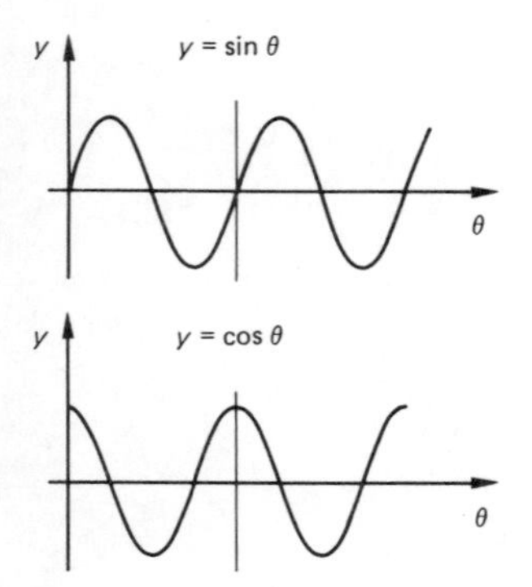

Figure 4.32

4.9.4 Phase Angles

Finally consider the graphs of $\sin \theta$ and $\cos \theta$ shown in Figure 4.32.

By now you will have realised that the graphs of $y = \sin \theta$ and $y = \cos \theta$ have the same shape, period, but different locations with respect to the θ axis (horizontal axis).

By translating the graph of $y = \sin \theta$ through 90° to the left, the graph of $y = \cos \theta$ is obtained; $\sin \theta$ and $\cos \theta$ are said to be *out of phase* by 90°. $y = \cos \theta$ has a maximum at $\theta = 0°$, whereas $y = \sin \theta$ has the corresponding maximum at $\theta = 90°$. Also $y = \cos \theta$ has a minimum at $\theta = 180°$; the corresponding minimum on $y = \sin \theta$ is at $\theta = 270°$. Hence $y = \cos \theta$ is leading $y = \sin \theta$ by 90°. This gives the identity

$\cos\theta = \sin(\theta + 90°)$ for all θ.

You can confirm this identity for particular values of θ on your calculator, for example

$$\cos 140° = \sin(140° + 90°) = \sin 230° = -0.7660$$

Further if $y = \sin(\theta + 90°)$ is plotted the graph of $y = \cos\theta$ is obtained. In general $y = \sin(\theta + \alpha)$ has a phase angle of α and represents a graph that is leading $y = \sin\theta$ by α.

To obtain the graph of $y = \sin(\theta + \alpha)$ translate the graph of $y = \sin\theta$ by α degrees to the left.

Example 4.9.4

For the sine wave shown in Figure 4.33 find the phase angle and write down its equation.

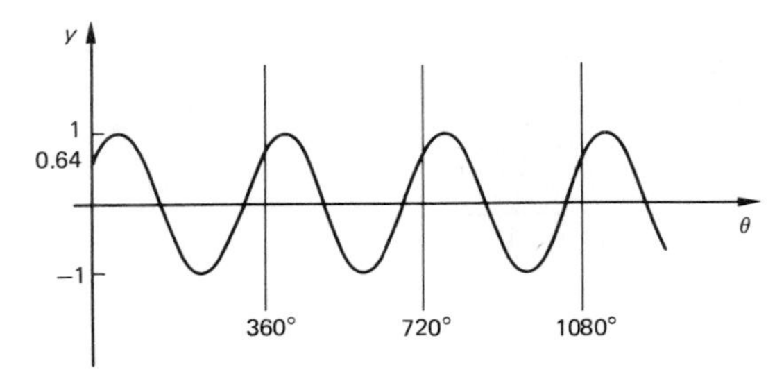

Figure 4.33

Solution

The graph cuts the y axis at $y = 0.64$ which corresponds to an angle of 40°. The phase angle is 40° so that the equation is $y = \sin(\theta + 40°)$.

Exercise 4.9.4

1. Use a computer package to plot $y = \sin\theta$ and $y = \sin(\theta + 10°)$ for θ between 0° and 720°. What simple transformation must be applied to the graph of $y = \sin\theta$ in order to obtain the graph of $y = \sin(\theta + 10°)$? Note that the graph of $\sin(\theta + 10°)$ is periodic. What is its period and amplitude?

2. Use a computer package to plot $y = \cos\theta$ and $y = \cos(\theta - 40°)$ for $-360° \leqslant \theta \leqslant 360°$. How is the graph of $y = \cos(\theta - 40°)$ obtained from the graph of $y = \cos\theta$? What is the period and amplitude of

cos(θ − 40°)? Write down cos(θ − 40°) in terms of sine.

3. Without using a computer package or calculator sketch the graphs of $y = \sin\theta$ and $y = \sin(\theta + 25°)$ for $0° \leq \theta \leq 360°$. Mark where the graphs cut the θ axis. For what values of θ is $\sin(\theta + 25°) = 1$? For what values of θ is $\sin(\theta + 25°) = -1$? Check your answers using a computer package.
4. Without using a computer package or calculator, sketch the graphs of $y = \tan\theta$ and $y = \tan(\theta + 20°)$ for $-180° \leq \theta \leq 180°$. Clearly mark all the asymptotes and the points where the graphs cut the θ axis. Check your sketch by using a computer package.
5. For each of the sine waves shown in Figure 4.34, find the period, amplitude and phase angle and write down its equation.

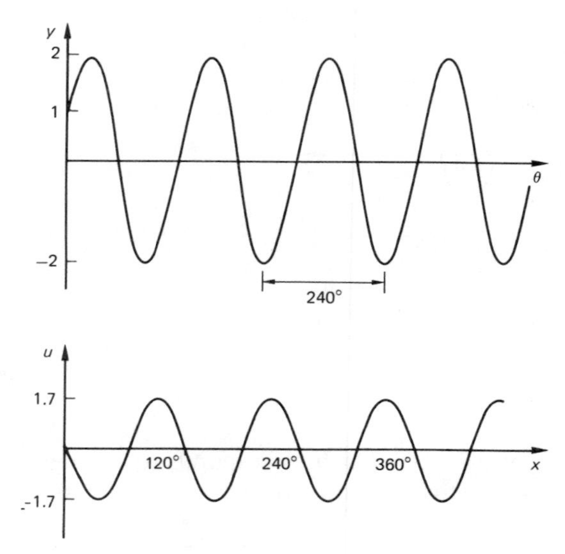

Figure 4.34

4.10 MORE TRIGONOMETRIC EQUATIONS

It is often necessary to solve trigonometric equations such as

$$\sin^2\theta = 0.16 \tag{4.10.1}$$

where $\sin^2 \theta$ means $(\sin \theta)^2$, i.e. first take the sine of θ and then square the result. For example if $\theta = 17°$, then $\sin^2 \theta = (\sin 17°)^2 = (0.2924)^2 = 0.0855$.

To solve an equation like (4.10.1), take the square root of each side giving

$$\sin \theta = \pm 0.4$$

(Do not forget the $\pm$ when taking the square root of the right hand side.)

This gives two equations to solve

$$\sin \theta = +0.4 \tag{4.10.2}$$

$$\sin \theta = -0.4 \tag{4.10.3}$$

Solve these equations graphically (Figure 4.35).

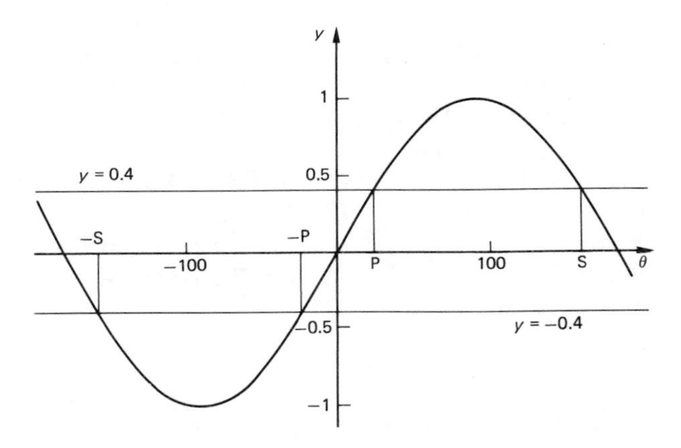

Figure 4.35

So Equation (4.10.1) has four solutions for each 360° cycle of $\sin \theta$. Using a calculator, Equation (4.10.2) has solutions $\theta = P = 23.6°$ and $\theta = S = 156.4°$, Equation (4.10.3) has solutions $\theta = -P = -23.6°$ and $\theta = -S = -156.4°$.

Combining these, the general solution to Equation (4.10.1) is

$$\theta = \pm 23.6° + 360°n$$

and

$$\theta = \pm 156.4° + 360°n$$

but these can be combined to give a more compact form for the general solution to $\sin^2 \theta = 0.16$ as

$$\theta = \pm 23.6° + 180°n$$

Exercise 4.10.1

1. Solve the following equations for angles between 0° and 360°, giving answers correct to one decimal place. In each case illustrate your solutions with a sketch graph.
   (a) $\cos^2 t = 0.082$ (b) $\sin^2 A = 0.649$
   (c) $3 \sin^2 t = 1$ (d) $\tan^2 x - 5 = 0$
   (e) $2 \sin^2 t - 3 = 0$ (f) $5 - 7 \cos^2 t = 1$

2. Solve the following for angles between 0° and 360° without using a calculator. In each case illustrate with a sketch.
   (a) $2 \sin^2 x = 1$ (b) $\tan^2 A = 3$
   (c) $\cos^2 P + 4 = 5$ (d) $4 \sin^2 t - 3 = 0$

3. Give the general solutions to the following equations in degrees, correct to one decimal place.
   For questions (a)–(c), include a sketch graph to illustrate your solutions
   (a) $\sin^2 A = 1/4$ (b) $\cos^2 A = 4/9$
   (c) $\tan^2 A = 16/9$ (d) $2 - \tan^2 x = 0$
   (e) $3 \sin^2 x = 2$ (f) $2 \cos^2 t - \sqrt{3} = 0$
   (g) $\tan^2 p + 1/2 = 2$ (h) $1 = 2 \cos^2 t$
   (i) $\sqrt{1 + \cos^2 B} = 1.1$

More complicated problems occur involving multiple angles and phase angles. The following examples show the method.

Example 4.10.1

Obtain all the solutions to the equation $\cos 2\theta = 0.6$ between 0° and 360°.

Solution

From the sketch of $y = \cos 2\theta$ and $y = 0.6$ (Figure 4.36) we can see that there are four solutions in the required range.

To obtain their values we proceed algebraically. Consider the substitution $A = 2\theta$ then the equation to be solved becomes $\cos A = 0.6$. So $A = \pm 53.13° + 360°n$. Note that it is important to find the *general solution* at *this* stage. Hence $2\theta = \pm 53.13° + 360°n$ and $\theta = \pm 26.6° + 180°n$.

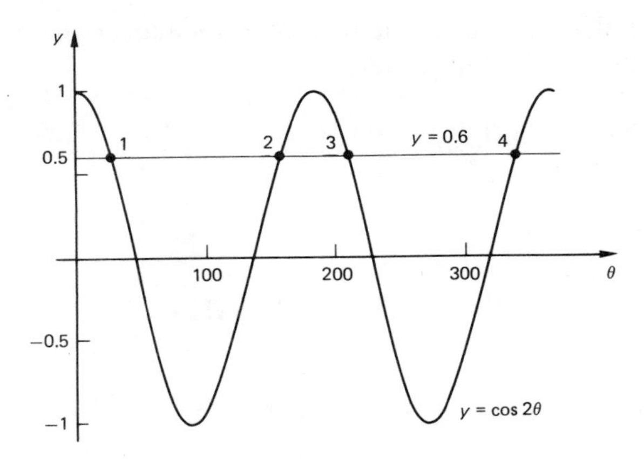

Figure 4.36

Note that the whole equation must be divided by 2, so we obtain $180°n$. This is correct since we know from the previous exercise that the period of $\cos 2\theta$ is $180°$.

To obtain the solutions in the required range $0°$ to $360°$ from the general solution give n suitable values. So $\cos 2\theta = 0.6$ has solutions $\theta = 26.6°, 153.4°, 206.6°, 333.4°$.

Example 4.10.2

Solve $\sin(\theta + 15°) = 0.4$

Solution

The easiest way is to make a simple substitution. Let $A = \theta + 15°$. So $\sin A = 0.4$. This equation can now be solved by the familiar method used earlier in the chapter.

Therefore $A = 23.6° + 360°n$ or $A = 156.4° + 360°n$

So $\theta + 15° = 23.6° + 360°n$ or $\theta + 15° = 156.4° + 360°n$

Therefore $\theta = 8.6° + 360°n$ or $\theta = 141.4° + 360°n$

Exercise 4.10.2

1. Find the general solutions of the following equations:
   (a) $\sin 2B = 0.75$ (b) $2\tan(A/2) = -3$
   (c) $3 + 4\cos 3t = 0$

2. Find all the solutions to the following equations which lie between $0°$ and $360°$.
   (a) $\sin(2/3t) = 1/3$ (b) $\cos 4x = 2$
   (c) $1 - \sin 2A = 0$

3. Find all the solutions to the following equations which lie in the range $-180°$ to $180°$.
   (a) $\cos(t - 11°) = -0.437$ (b) $3\tan(P + 38°) = 5$
   (c) $2\sin(A + 70°) - 1 = 0$ (d) $\tan(10° - B) = 0.728$

4.11 SECANT, COSECANT AND COTANGENT FUNCTIONS

In addition to the three trigonometric functions we have used so far, there are three more. These are secant, cosecant and cotangent, which are defined as follows:

$$\sec A = 1/\cos A, \quad \text{cosec } A = 1/\sin A, \quad \cot A = 1/\tan A$$

Most calculators do not have keys for these three functions but these definitions can be used to transform an equation involving sec, cosec or cot into one involving cos, sin or tan.

Example 4.11

Solve $\sec\theta = -1.35$.

Solution

Since $\sec\theta = 1/\cos\theta$, we get

$$1/\cos\theta = -1.35$$

Therefore $\cos\theta = -1/1.35 = -0.74074$.
So

$$\theta = \pm 137.8° + 360°n.$$

Exercise 4.11.1

Use your calculator to obtain the primary solution to the following equations.
   (a) $\text{cosec } P = 8$ (b) $\sec(t) = -9$
   (c) $7\cot x = 1$ (d) $\sec y = -1.527$
   (e) $\text{cosec } A + 2.493 = 0$ (f) $\cot(t) = 1.465$

2. Find the solutions in the range $0°$ to $360°$ to the following equations, giving your answers in degrees and minutes.
   (a) $\cot x = 0.102$ (b) $\sec A = -\sqrt{5}$

(c) $\cot y = -1.314$ (d) $\operatorname{cosec} P - \sqrt{3} = 0$
(e) $\operatorname{cosec} t + 1.937 = 0$ (f) $2 \sec Q = 5$

3. Find the solutions in the range $-180°$ to $180°$ to the following equations, giving your answers correct to one decimal place.
   (a) $\sec A = 1.5$ (b) $\operatorname{cosec} B = -2.3$
   (c) $3 \cot y = -4$ (d) $3 + 2 \sec x = 0$
   (e) $2 \operatorname{cosec} y - 3 = 4$ (f) $1 - \cot C = 0$

4. Use a computer to plot the following for x between $0°$ and $360°$. State the period of each function and give the equations of any asymptotes.
   (a) $y = \cos x$ and $y = \sec x$ on the same axes. (On most computers you will have to write $\sec x$ as $1/\cos x$.)
   (b) $y = \sin x$ and $y = \operatorname{cosec} x$ on the same axes.
   (c) $y = \tan x$ and $y = \cot x$ on the same axes.
   (d) Explain how the graph of $y = \cos x$ can be used to help in sketching $y = \sec x$.

4.12 CIRCULAR MEASURE OR RADIANS

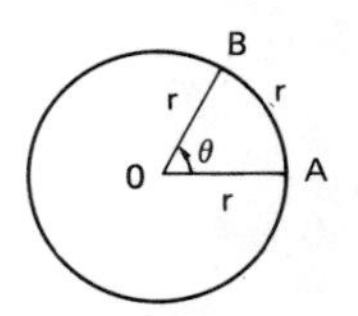

Figure 4.37

Until now we have used the degree for measuring angles. The degree is 1/360 of a full turn. An alternative measurement for angles is the radian. Consider a circle of radius r and centre O. Let angle AOB $= \theta$.

If the length of the arc AB is r then the angle θ is defined to measure *1 radian*. If the length of the arc AB is $2r$ then the angle is 2 radians, and so on. More generally, in radians, the angle θ is given by

$$\theta = \frac{\text{arc length AB}}{\text{radius OA}}$$

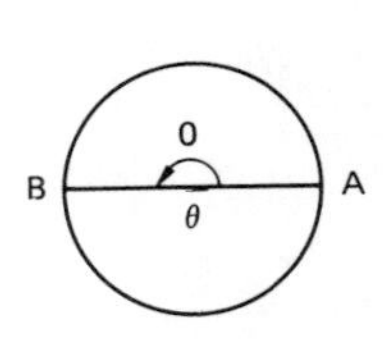

Figure 4.38

The radian is an alternative unit for measuring angles. To see its relationship to degrees consider a semicircle (Figure 4.38).

Then

$$\theta = \frac{\text{arc length AB}}{\text{radius OA}} = \frac{\pi r}{r} = \pi$$

Therefore π radians $\equiv 180°$, or 1 radian $\equiv 180/\pi$ degrees ($= 57.3°$). You should learn the following conversion table:

Table 4.1

Radians	$\frac{\pi}{6}$	$\frac{\pi}{4}$	$\frac{\pi}{3}$	$\frac{\pi}{2}$	$\frac{3\pi}{4}$	π	$\frac{3\pi}{2}$	2π
Degrees	30°	45°	60°	90°	135°	180°	270°	360°

As the radian has many theoretical advantages, in advanced mathematics it is used in preference to the degree.

If the angle θ is measured in radians, then $\sin \theta$ and $\cos \theta$ have period 2π and $\tan \theta$ has period π. Amplitudes are not affected by using radians and graphs have the same shapes as before.

All calculators and computer packages have a radian mode. The notation for radians is a small c, so that 3 radians is written as 3^c.

Example 4.12.1

Convert 50° to radians. Convert 4.21^c to degrees.

Solution

We start with the conversion rule

$$\pi^c \equiv 180°.$$

Hence

$$1° \equiv \frac{\pi^c}{180}$$

$$50° \equiv \frac{50\pi^c}{180} = 0.873^c$$

For radians to degrees, we have

$$1^c \equiv \frac{180°}{\pi} = 57.3°$$

$$4.21^c \equiv 4.21 \times \frac{180°}{\pi} = 241.2°$$

To solve trigonometric equations giving the angles in radians we follow the same procedure as for degrees but ensure that your calculator or computer is in radian mode.

Example 4.12.2

Find the general solution of the equation $\sin 4x = 0.71$.

Solution

If we let $A = 4x$ then we solve

$$\sin A = 0.71$$

A calculator gives the primary solution $P = 0.7895^c$. To see the other solutions for A we sketch a graph of $\sin A$ (Figure 4.39).

The secondary solution is $S = \pi - P = 2.3521$.

The general solution for A is

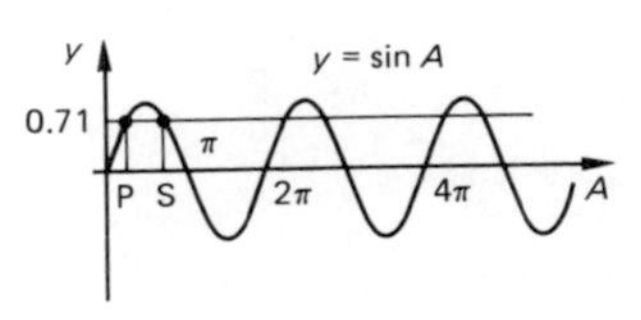

Figure 4.39

$$A = 0.7895 \pm 2\pi n, \qquad A = 2.3521 \pm 2\pi n.$$

Hence the solutions for x are found by dividing the answers by 4,

$$x = 0.1974 \pm 0.5\pi n, \; x = 0.5880 \pm 0.5\pi n.$$

Exercise 4.12.1

Use radians throughout this exercise.

1. Convert the following angles in radians into degrees.
   (a) $7\pi/3$ (b) $-\pi/6$ (c) $5\pi/4$ (d) 3π

2. Convert the following angles in degrees into radians, leaving your answers as a fraction of π (e.g. $22.5° \equiv \pi/8$ and not 0.3927 which is an approximation).
   (a) 75° (b) 300° (c) −30° (d) 450°

3. Use a computer package to graph the following functions for x between 0 and 2π radians inclusive. In each case state the period and amplitude of the function. Also give all the angles where the value of the function is zero leaving all your answers as fractions of π.
   (a) $y = 2 \sin 3x$ (b) $y = 0.5 \cos x$ (c) $y = 5 \sin 2x$

4. Solve the following equations for angles between 0 and 2π radians. Give answers as fractions of π.
   (a) $\sin A = -1$
   (b) $\cos B = 0.5$
   (c) $\sqrt{3} - 2 \sin t = 0$
   (d) $\tan P = -1$
   (e) $2 \cos 2B + 1 = 0$

5. State the period and amplitude of each of the following:
   (a) $y = \sin 3B$ (b) $y = 10 \cos 2A$

(c) $y = 0.5\cos(3x/4)$ (d) $y = \sqrt{2}\sin(2x/3)$
(e) $y = \cos(kx)$ (f) $y = a\sin(kx)$
(g) $y = b\cos(2\pi x)$ (h) $y = \sin(n\pi x)$

4.13 MODELLING WITH TRIGONOMETRIC FUNCTIONS

In the introduction we introduced various examples of physical phenomena with properties that could be described as periodic. For example, the motion of the spring-mass oscillator repeats its cycle every two seconds and has a maximum displacement of 0.05 m from its equilibrium position. Now with these two properties of the physical system we can 'work backwards' and use that trigonometric function sine as a model of the system. The period of the system is two seconds, so that the angular frequency is $2\pi/2 = \pi$ radian/s and the amplitude is 0.05. With these magnitudes we can write the displacement of the mass from equilibrium as

$$x = 0.05\sin(\pi t + \alpha)$$

The phase angle α depends on the starting position of the mass. For example, if the mass is pulled down and released from rest then $\alpha = \pi/2$.

Clearly we can use sinusoidal functions to model any system which behaves like a sine wave, all that is required is a graph describing some aspect of the physical system or data concerning the period, amplitude, angular frequency and phase angle. The following examples illustrate the method of approach.

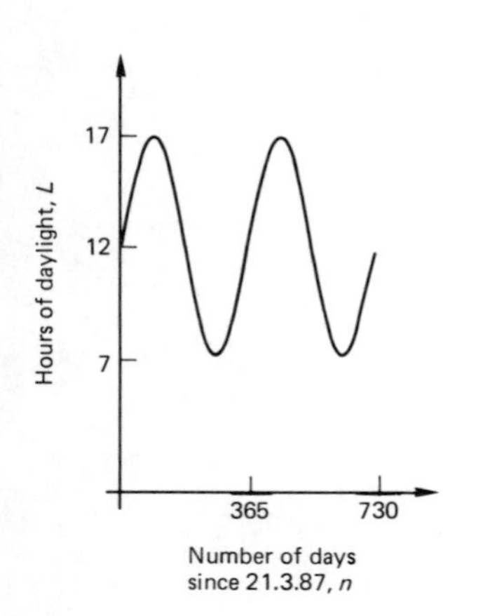

Figure 4.40

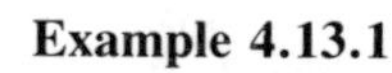

Example 4.13.1

In the south of England, the length of the day (in terms of hours of daylight) during the year follows the pattern shown in Figure 4.40.

The measurements began on 21 March 1987 and cover two years. Express the length of the day as a function of the number of days since 21 March 1987.

Solution

The first observation to make is that the length L varies between $L = 7$ and 17 so that the average value about which the sine wave is drawn is given by $L = 12$.

The period of the sine wave is 365 days, giving an angular frequency of $2\pi/365 = 0.0172$, and the phase angle is 0. The sine wave can therefore be modelled by the function $5\sin(0.0172n)$, where n is the number of days since 21 March 1987. Adding 12 to this function gives the rule for L:

$$L = 12 + 5 \sin(0.0172n).$$

Often the physical system being modelled has time as the independent variable. For example in the spring-mass oscillator the displacement of the mass is described by

$$x = a \sin(\omega t)$$

where t is measured in seconds. We have called ω the angular frequency and it is related to the period, T, by $\omega = 2\pi/T$. Now if the mass has period two seconds, for instance, it completes half a cycle per second. Similarly if the period is $\frac{1}{4}$ second then the mass completes four cycles per second. The number of cycles of an oscillation per second is called the *frequency* of the oscillation. It is related to the period by

$$f = \frac{1}{T}$$

where f is the frequency and T is the period. The unit of frequency is the *hertz* (Hz) and

1 Hz = 1 cycle per second

Thus an oscillation with a period of five seconds has a frequency of 0.2 Hz.

Frequency and angular frequency are related by

$$f = \frac{\omega}{2\pi}$$

Be careful not to confuse the two.

Example 4.13.2

In the UK, electricity is supplied at a frequency of 50 Hz. If the wave form (i.e. graph) of current across a resistance is displayed by an oscilloscope it has the shape of a sine wave. Assuming the current I has an amplitude of 100 milliamperes and that t is the time then

$$i = 100 \sin \omega t$$

What is the value of ω and the period of the sine wave?

Solution

The frequency $f = 50$ Hz.
Hence $\omega = 2\pi . 50 = 314.16$ and the period $T = 1/50 = 0.02$ seconds.

Exercise 4.13.1

1. Assuming that t represents time in seconds, find the period and frequency of the following.
   (a) $\sin(2t)$ (b) $\sin(\pi t)$ (c) $\cos(kt)$ (d) $\cos(k\pi t)$

2. A mass attached to the end of a spring oscillates so that its displacement y (in cm) from a central position is given by

   $$y = 4\cos(2\pi t)$$

   where t is the time in seconds.
   What is the maximum displacement from the central position? What is the frequency of the oscillation?

3. A mass attached to the end of a spring oscillates so that its displacement x from a central position is given by

   $$x = a\sin(\omega t)$$

   where a and ω are constants and t is the time in seconds. Assuming that the maximum displacement from the central position is 30 mm and that the mass completes two oscillations every second, determine the frequency of oscillation and the values of the constants a and ω.

4. In the USA, electricity is supplied at a frequency of 60 Hz. What equation would model the current across a resistor if the current's amplitude was 75 mA (milliamperes).

5. The horizontal range in metres, x, of a projectile is given by

   $$x = (V^2 \sin 2A)/g$$

   where A is the angle of projection, V is the velocity of projection, and g is the acceleration due to gravity. Show that the range is greatest when $A = 45°$.

6. The acceleration, a, of a pendulum in its direction of motion is given by

   $$a = -g\sin x$$

where x is the angle which the pendulum makes with the vertical and g is a constant, the acceleration due to gravity, whose value is 9.8 ms^{-2}. Sketch a graph of a against x, taking x in radians from 0 to 2π.

7. The wave forms of two notes of music, A and B, are shown in Figure 4.41.
   Which of the following statements are true?
   (a) A and B have the same frequency
   (b) A is louder than B
   (c) The period of A is twice as long as that of B
   (d) The wave forms are in phase
   (e) The amplitude of B is larger than that of A
   (f) The frequency of A is twice that of B

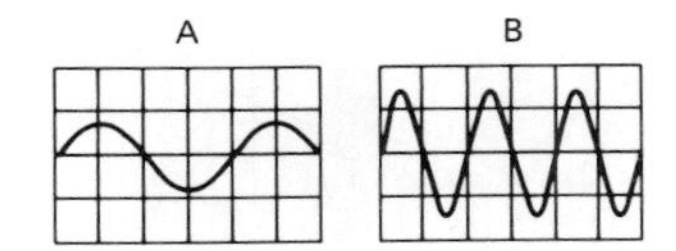

Figure 4.41

8. The graph in Figure 4.42 shows the angle to the vertical of the string of a pendulum bob as it varies with time.

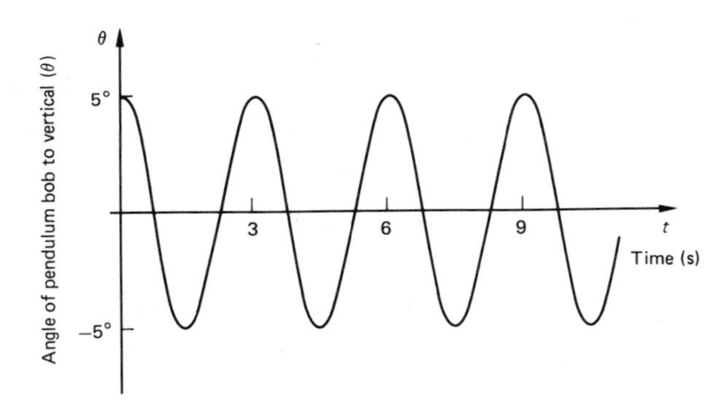

Figure 4.42

From the graph determine the amplitude, period and phase angle of the oscillations. Write down an equation describing θ as a function of time t and from your equation find the times when $\theta = 5°$.

Draw a graph which represents a pendulum swinging with half the amplitude and twice the frequency.

OBJECTIVE TEST 4

1. Reduce 500° to an angle between 0° and 360°.
2. Obtain the general solution (in degrees) to $7 \sin 2A + 3 = 0$.
3. Obtain the solutions (in radians) to $\cos(3x + \pi/4) = 0.54$ in the range 0 to 2π, giving your answers to three significant figures.

4. Solve the equation $6 \cot B = 5$ giving your answers in degrees.
5. What is the period (in degrees) of $\sin(B/3 + 30°)$.
6. What is the period (in radians) and the amplitude of $2 \cos(5x/4 + \pi/3)$.
7. How do you obtain the graph of $y = \sin(x - 30°)$ from the graph of $y = \sin x$?
8. What are the equations of the asymptotes of $y = \tan x$?
9. What is the frequency of vibration of a mass whose displacement x cm from a fixed point is given by $x = 3 \cos(5t + 0.1)$, where t = time in seconds?

5 TRIGONOMETRY

OBJECTIVES

When you have finished studying this chapter you should be able to

1. use the appropriate trigonometric identity to simplify expressions involving the trigonometric functions sine, cosine and tangent
2. quote and use the double angle formulas for $\sin(2\theta)$ and $\cos(2\theta)$
3. simplify the expression a $\sin x$ + b $\cos x$ and solve equations of the form $a \sin x + b \cos x = c$
4. define and use the inverse trigonometric functions
5. quote and use the sine rule and the cosine rule for finding angles and side lengths in oblique triangles

5.1 INTRODUCTION

In Chapter 4 we extended the idea of sine, cosine and tangent from their ratio definitions in a right angled triangle to periodic functions which can be used as models to describe oscillating systems. Often an oscillation is a combination of several basic vibrations and mathematically this involves combining sines and cosines together. In this chapter we introduce various identities for simplifying trigonometric expressions.

5.1.1 Interference of Waves

Suppose that two waves pass simultaneously through the same points in a medium, these waves may be ripples on the surface of a lake or sound waves travelling through the air from a source of sound to your ear. Each wave is unaffected as it passes through the medium, but the medium itself is affected by the displacement produced by both waves.

Consider the effect on a particle or point in the region. Each wave causes the particle to move and the total displacement of the particle is the sum of the displacements that the individual waves would produce. Thus if one wave causes a displacement x_1 and the second wave a displacement x_2 then when they occur simultaneously the displacement of the particle is given by

$$x = x_1 + x_2$$

This is called the *principle of superposition* and the effect of the two waves on the medium is called *interference*.

The effect on the medium will depend on many factors, for example whether the waves arrive in phase or out of phase or the difference in the frequency between the two waves. Figures 5.1–5.3 show various simple examples of interference.

Figure 5.1 shows two waves, A and B, of equal frequency which are in phase. The amplitude is the sum of the separate amplitudes. The resultant wave is in phase with the original waves, this is called *constructive interference*.

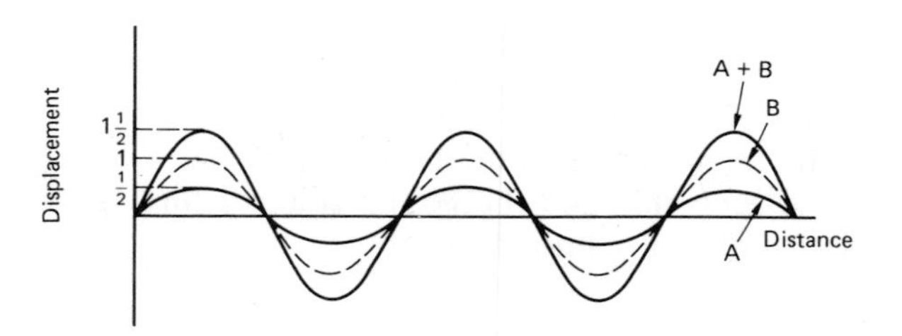

Figure 5.1

In Figure 5.2 the two waves are 180° out of phase. Now the amplitude of the resultant disturbance is reduced. This is called *destructive interference*.

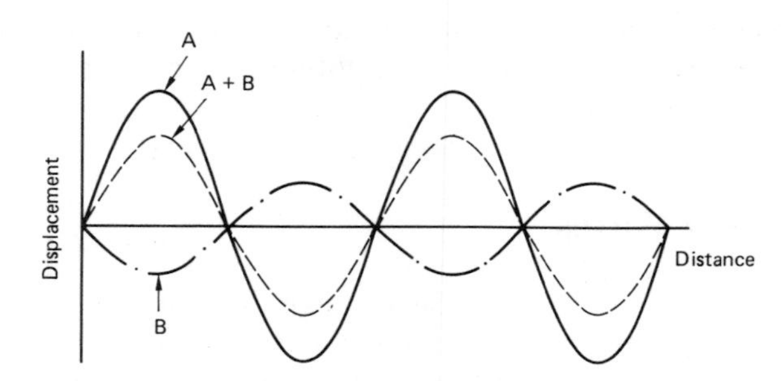

Figure 5.2

If the original two waves in Figure 5.2 have equal amplitude then the waves cancel and there is no resultant displacement of the medium. In contrast to these two situations, if the two waves have slightly different frequencies then a mixture of constructive and destructive interference occurs producing a phenomenon called *beats*. Constructive interference occurs at the times that the waves are in phase and destructive interference occurs when they are 180° out of phase.

The pattern of the resultant displacement is shown in Figure 5.3.

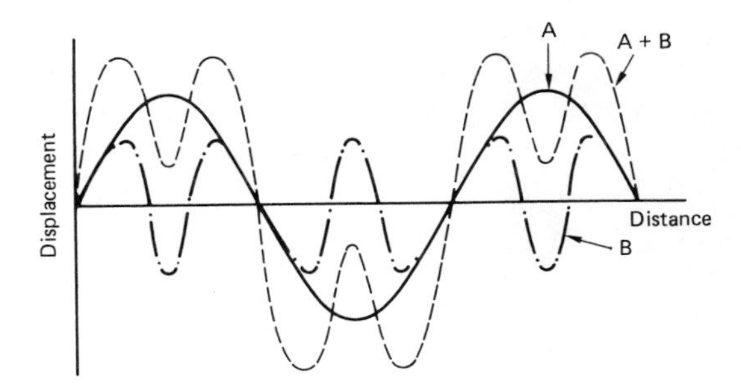

Figure 5.3

Algebraically the analysis involves trigonometric functions. The displacement of a point in a medium produced by a wave is described algebraically by

$$x = a \sin(\omega t + \varphi)$$

where a is the amplitude, ω is the frequency and φ is the phase angle. So the resultant displacement of two waves is

$$x = a_1 \sin(\omega_1 t + \varphi_1) + a_2 \sin(\omega_2 t + \varphi_2)$$

At this stage you can add two sine functions if $\omega_1 = \omega_2$ and $\varphi_1 = \varphi_2$. In this chapter we extend the ideas so that we can expand expressions such as $\sin(A + B)$ in terms of $\sin A$, $\sin B$, $\cos A$ and $\cos B$ and so that we can add sine and cosine functions together. Clearly one important application of these ideas is to the interference of waves.

5.1.2 Surveying

Suppose that you are asked to find the height of the tower in the photograph in Figure 5.4.

In theory it should be an easy task using the triangle and the tangent rule. A surveyor could measure the angle α using a theodolite and find distance AB using a tape measure. The height of the tower BC is then AB $\tan \alpha$. However it is not really that easy. Finding the angle α would cause no problem but finding the distance AB from a point perpendicularly below the top of the tower to the theodolite

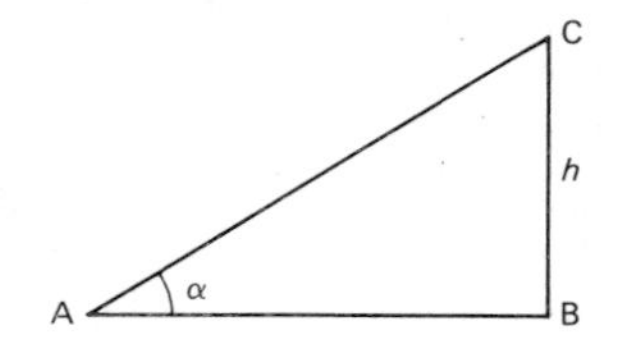

Figure 5.4

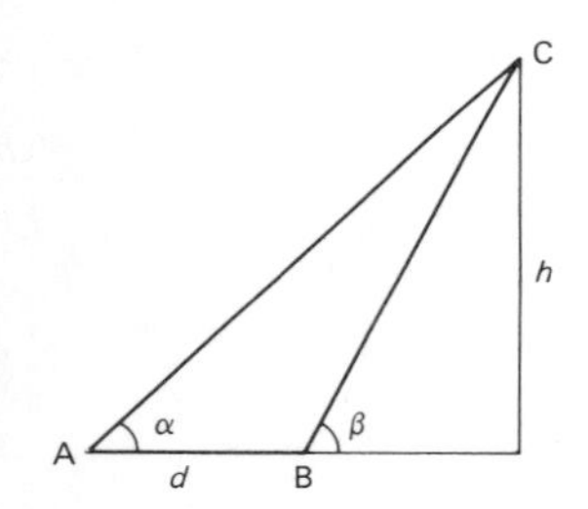

Figure 5.5

would be quite a challenge. In practice a surveyor would measure the angle of elevation at two points a convenient distance d apart (Figure 5.5).

With angles α and β and distance d the height h could be found geometrically by drawing the triangles. For an algebraic approach we would need to use the trigonometric functions in a triangle ABC (which does not contain a right-angle) to find BC and then $h = \text{BC} \sin \beta$. In Sections 5.5 and 5.6, we introduce two important rules for oblique triangles, such as ABC, so that we can find the unknown side lengths or angles.

5.2 TRIGONOMETRIC IDENTITIES

Often in problem solving, in calculus and in mathematical equations, expressions arise involving trigonometric functions and progress cannot be made until come manipulation is carried out. This manipulation requires the use of trigonometric identities, some of which occur often enough to need to be committed to memory. They are identified by the symbol [M]. In this section we explore a range of trigonometric identities.

5.2.1 Basic Forms

In Chapter 4, we reviewed the basic trigonometric functions $\sin \theta$, $\cos \theta$ and $\tan \theta$ and used them in modelling physical situations which involve periodic behaviour. We also introduced the reciprocal of these functions.

secant: $\sec\theta = \dfrac{1}{\cos\theta}$

cosecant: $\operatorname{cosec}\theta = \dfrac{1}{\sin\theta}$

cotangent: $\cot\theta = \dfrac{1}{\tan\theta}$

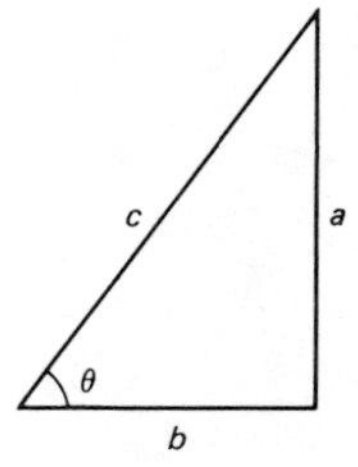

An *identity* is an equation that is true for all values of the variable involved. Perhaps one of the most commonly used trigonometric identities is derived from Pythagorus' rule. Consider a right-angled triangle with angle θ and sides a, b, c.

Figure 5.7

Using Pythagorus' rule we have

$$a^2 + b^2 = c^2$$

and dividing each term by c^2

$$\left(\frac{a}{c}\right)^2 + \left(\frac{b}{c}\right)^2 = 1$$

Now from the definitions $\sin\theta = a/c$ and $\cos\theta = b/c$ we have

$$\boxed{\sin^2\theta + \cos^2\theta = 1} \qquad (5.1)\ \boxed{\text{M}}$$

This trigonometric identity is one of the most commonly used in simplifying mathematical expressions and should be committed to memory. Two other forms are deduced from this basic one,

(a) dividing each term by $\cos^2\theta$ gives

$$\frac{\sin^2\theta}{\cos^2\theta} + 1 = \frac{1}{\cos^2\theta}$$

or

$$\tan^2\theta + 1 = \sec^2\theta$$

(b) dividing each term by $\sin^2\theta$ gives

$$1 + \frac{\cos^2\theta}{\sin^2\theta} = \frac{1}{\sin^2\theta}$$

or

$$1 + \cot^2 \theta = \operatorname{cosec}^2 \theta$$

Example 5.2.1

In mechanics the position of a projectile (x, y) fired with speed v at an angle θ to the horizontal is given by

$$x = vt \cos \theta$$

$$y = vt \sin \theta - \tfrac{1}{2}gt^2$$

where t is the time and g is a constant. By eliminating t find the equation of the path of the projectile.

Solution

From $x = vt \cos \theta$ we have

$$t = \frac{x}{v \cos \theta}$$

Substituting for t into the expression for y we have

$$y = v \sin \theta \left(\frac{x}{v\cos \theta}\right) - \tfrac{1}{2}g\left(\frac{x}{v \cos \theta}\right)^2$$

$$y = x \tan \theta - \frac{gx^2}{2v^2 \cos^2 \theta}$$

Now $1/\cos^2 \theta = \sec^2 \theta$ and $\sec^2 \theta = \tan^2 \theta + 1$ so we can write the equation of the path as

$$y = x \tan \theta - \frac{gx^2}{2v^2}(\tan^2 \theta + 1)$$

This is an illustration of using a trigonometric identify to simplify an expression so that only one trigonometric function occurs, in this case $\tan \theta$. The following example and exercises are designed to illustrate the working knowledge of the basic identities that you should try to develop.

Example 5.2.2

Prove that

$$\frac{\tan\theta + \cot\theta}{\tan\theta} = \frac{1}{\sin^2\theta}$$

Solution

We start by expressing each term on the left hand side in terms of $\sin\theta$ and $\cos\theta$. We have

$$\frac{\left(\dfrac{\sin\theta}{\cos\theta} + \dfrac{\cos\theta}{\sin\theta}\right)}{\dfrac{\sin\theta}{\cos\theta}} = \frac{(\sin^2\theta + \cos^2\theta)}{\sin\theta\cos\theta} \times \frac{\cos\theta}{\sin\theta}$$

$$= \frac{1}{\sin\theta} \times \frac{1}{\sin\theta} = \frac{1}{\sin^2\theta}$$

Exercise 5.2.1

1. Write each of the following expressions in terms of $\sin\theta$ and $\cos\theta$
   (a) $\cot\theta$
   (b) $\cot\theta\sec\theta$
   (c) $\sec^2\theta$
   (d) $\cot\theta - \cot\theta\cos^2\theta$
   (e) $\sec\theta - \sin\theta\tan\theta$
   (f) $\tan\theta + \cot\theta$
   (g) $\operatorname{cosec}\theta(\sec\theta - \cos\theta)$
   (h) $\dfrac{\tan^2\theta - 1}{\tan^2\theta + 1}$

2. Prove the following identities.
   (a) $\tan\theta\cos\theta\operatorname{cosec}\theta = 1$
   (b) $\cos^2\theta - \sin^2\theta = 2\cos^2\theta = 1$
   (c) $\cos^2\theta - \sin^2\theta = 1 - 2\sin^2\theta$
   (d) $\sin\theta\cos\theta(\cot\theta + \tan\theta) = 1$
   (e) $\cos^4\theta - \sin^4\theta = 1 - 2\sin^2\theta$
   (f) $\dfrac{\sin\theta}{1 - \cos\theta} = \dfrac{1 + \cos\theta}{\sin\theta}$

3. Show that the parametric equations $x = a\cos t$ and $y = b\sin t$ give the equation of an ellipse

$$\frac{x^2}{a^2} + \frac{y^2}{b^2} = 1$$

4. By using various angles show that the following are not trigonometric identities.
   (a) $\cos 2\theta \neq 2 \cos \theta$ (b) $\sin 3\theta \neq 3 \sin \theta$
   (c) $\tan(\theta/2) \neq \frac{1}{2} \tan \theta$

This exercise shows how very careful you must be in simplifying expressions.

5.2.2 Sum and Difference Formulas

We now introduce six important identities which involve the sine, cosine and tangent of sums and differences of two angles. They are

$$\sin(A + B) = \sin A \cos B + \cos A \sin B \quad (5.2) \quad \boxed{M}$$

$$\sin(A - B) = \sin A \cos B - \cos A \sin B \quad (5.3) \quad \boxed{M}$$

$$\cos(A + B) = \cos A \cos B - \sin A \sin B \quad (5.4) \quad \boxed{M}$$

$$\cos(A - B) = \cos A \cos B + \sin A \sin B \quad (5.5) \quad \boxed{M}$$

$$\tan(A + B) = \frac{\tan A + \tan B}{1 - \tan A \tan B} \quad (5.6)$$

$$\tan(A - B) = \frac{\tan A - \tan B}{1 + \tan A \tan B} \quad (5.7)$$

We will not prove these formulas. (If you are interested in a proof then turn to a Pure Mathematics text.)

Example 5.2.3

Consider two acute angles A and B such that $\sin A = 12/13$ and $\cos B = 3/5$. Find $\sin(A + B)$ and $\cos(A - B)$.

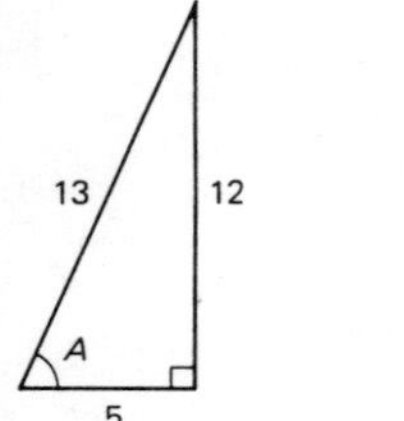

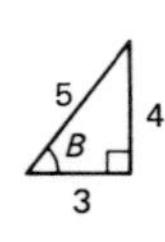

Figure 5.8

Solution

In the formulas for $\sin(A + B)$ and $\cos(A - B)$ we need values for $\cos A$ and $\sin B$. Drawing two right-angled triangles gives Figure 5.8.
Then $\cos A = 5/13$ and $\sin B = 4/5$, and

$$\sin(A + B) = \sin A \cos B + \cos A \sin B$$

$$= \left(\frac{12}{13} \times \frac{3}{5}\right) + \left(\frac{5}{13} \times \frac{4}{5}\right) = \frac{56}{65}$$

$$\cos(A - B) = \cos A \cos B + \sin A \sin B$$

$$= \left(\frac{5}{13} \times \frac{3}{5}\right) + \left(\frac{12}{13} \times \frac{4}{5}\right) = \frac{63}{65}$$

Example 5.2.4

Prove the identity

$$\sin(A + \pi/2) = \cos A$$

Solution

Using identity (5.2)

$$\sin\left(A + \frac{\pi}{2}\right) = \sin A \cos\frac{\pi}{2} + \cos A \sin\frac{\pi}{2}$$

since

$$\cos\frac{\pi}{2} = 0 \quad \text{and} \quad \sin\frac{\pi}{2} = 1$$

then

$$\sin\left(A + \frac{\pi}{2}\right) = \cos A$$

Exercise 5.2.2

1. Given that $\sin A = 15/17$, $\cos B = 3/5$ and A and B are acute, find the values of
   (a) $\sin(A + B)$ (b) $\sin(A - B)$ (c) $\cos(A + B)$
   (d) $\cos(A - B)$ (e) $\tan(A + B)$ (f) $\tan(A - B)$
   Which quadrant are the angles $(A + B)$ and $(A - B)$ in?

2. Repeat Problem 1 for $\cos A = 0.3$ and $\sin B = 0.7$.

3. Prove the following identities:
   (a) $\sin(\theta + \pi) = -\sin\theta$ (b) $\sin(\theta - \pi/2) = -\cos\theta$
   (c) $\cos(\theta - \pi) = -\cos\theta$ (d) $\tan(\theta + \pi) = \tan\theta$
   (e) $\cos(\theta + 2\pi) = \cos\theta$ (f) $\tan(\theta + \pi/4) = \dfrac{1+\tan\theta}{1-\tan\theta}$

4. By using various angles show that the following are not trigonometric identities:
   (a) $\sin(A + B) \neq \sin A + \sin B$
   (b) $\cos(A - B) \neq \cos A - \cos B$

5. Show that if $y = A \cos(\omega t + \pi)$ and $x = B \cos(\omega t - \pi)$ then $x + y = -(A + B) \cos \omega t$.
6. Starting with the identities for sine and cosine (5.2)–(5.5) prove the identites for $\tan(A + B)$ and $\tan(A - B)$.

5.2.3 Double-angle Identities

The double angle identities are among the most useful of the trigonometric identities that we develop. They are obtained from Equations (5.2)–(5.5) by setting $B = A$.

$$\boxed{\sin 2A = 2 \sin A \cos A} \qquad (5.8) \quad \boxed{M}$$

$$\boxed{\begin{aligned} \cos 2A &= \cos^2 A - \sin^2 A \\ &= 2 \cos^2 A - 1 \\ &= 1 - 2 \sin^2 A \end{aligned}} \qquad (5.9) \quad \boxed{M}$$

$$\tan 2A = \frac{2 \tan A}{1 - \tan^2 A} \qquad (5.10)$$

As an illustration of the proof of these identities, consider Equation (5.2)

$$\sin(A + B) = \sin A \cos B + \cos A \sin B$$

Setting $B = A$ we have

$$\sin(A + A) = \sin A \cos A + \cos A \sin A$$

i.e.

$$\sin(2A) = 2 \sin A \cos A$$

For two of the formulas for $\cos 2A$ we also need Identity (5.1). Consider Identity (5.4)

$$\cos(A + B) = \cos A \cos B - \sin A \sin B$$

Setting $B = A$ we have

$$\cos 2A = \cos^2 A - \sin^2 A$$

which is the first of Identity (5.9). Now writing $\cos^2 A = 1 - \sin^2 A$ from (5.1) we have

$$\cos 2A = (1 - \sin^2 A) - \sin^2 A = 1 - 2\sin^2 A$$

Exercise 5.2.3

1. Prove the Identity (5.10) for $\tan 2A$.

2. Prove the following identities:
   (a) $\sin 3x = 3\sin x - 4\sin^3 x$
   (b) $\cos 3x = 4\cos^3 x - 3\cos x$
   (c) $\sin 4x = 4\sin x\cos^3 x - 4\sin^3 x\cos x$
   (d) $\cos 4x = 1 - 8\sin^2 x\cos^2 x$
   (e) $\tan 2x = \dfrac{2\sin x}{2\cos x - \sec x}$

3. By choosing various angles θ show that the following are not trigonometric identities:
   (a) $\sin 3\theta \neq 3\sin\theta$ (b) $\cos 4\theta \neq 4\cos\theta$

4. Show that the parametric equations $x = \sin t$, $y = \cos 2t$ define the parabola $y = 1 - 2x^2$.

5. Show that
   (a) $\sin^2\theta = \dfrac{1 - \cos 2\theta}{2}$ (b) $\cos^2\theta = \dfrac{1 + \cos 2\theta}{2}$

5.2.4 Simplifying $a\sin x + b\cos x$

In the study of vibrations and waves, the displacement of an object is often given by $a\sin\omega t + b\cos\omega t$ where a, b and ω are constants. To be able to describe the properties of the wave, we need to rewrite this expression in one of the two forms $A\cos(\omega t - \varphi)$ or $A\sin(\omega t + \varphi)$. The constant A is called the amplitude and φ is the phase of the vibration. Given two waves with the same period we can compare them by comparing their amplitudes and their phases.

Consider the expression $A\sin(x + \varphi)$ and expand it using Identity (5.2).

$$A\sin(x + \varphi) = A\sin x\cos\varphi + A\sin\varphi\cos x$$

Compare the right-hand side of this expression with $a\sin x + b\cos x$. If we let

$$A\cos\varphi = a \quad \text{and} \quad A\sin\varphi = b \tag{5.11}$$

then

$$a\sin x + b\cos x = A\sin(x + \varphi).$$

Solving Equations (5.11) gives

$$A = \sqrt{(a^2 + b^2)} \quad \text{and} \quad \tan \varphi = b/a$$

Hence we deduce the identity

$$a \sin x + b \cos x = A \sin(x + \varphi)$$

$$\text{where} \quad A = \sqrt{(a^2 + b^2)} \quad \text{and} \quad \varphi = \arctan \frac{b}{a} \tag{5.12}$$

Although not an identity to remember it is important to know that such an identity exists and it can always be obtained using Equation (5.2).

Example 5.2.5

Write $8 \sin x - 15 \cos x$ in the form $A \sin(x + \varphi)$.

Solution

Expanding $A \sin(x + \varphi)$ using Identity (5.2) we have

$$8 \sin x - 15 \cos x = A \sin x \cos \varphi + A \cos x \sin \varphi$$

Comparing coefficients of $\sin x$ and $\cos x$ gives

$$A \cos \varphi = 8 \quad \text{and} \quad A \sin \varphi = -15$$

Solving for A, $A^2 = 8^2 + (-15)^2 = 289$ and hence $A = 17$.

For φ, we have $\cos \varphi = 8/17$ and $\sin \varphi = -15/17$. From graphs of $\sin \varphi$ and $\cos \varphi$ we see that φ lies between 270° and 360°, i.e. in the fourth quadrant. Hence

$$\varphi = 360° - \arccos(8/17) = 360° - 62° = 298°,$$

and

$$8 \sin x - 15 \cos x = 17 \sin(x + 298°).$$

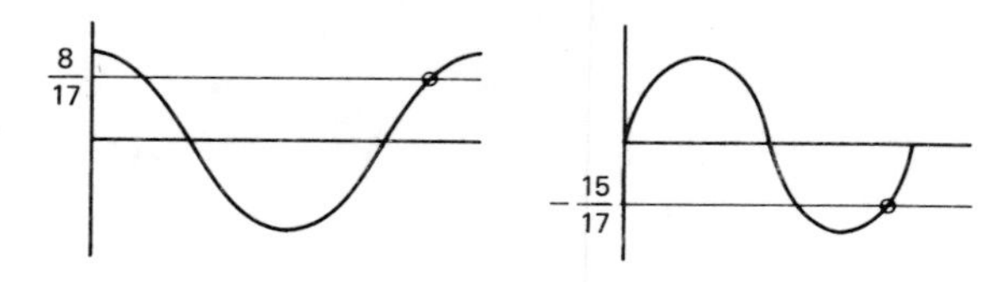

Figure 5.9

Exercise 5.2.4

1. Show that $a \sin x + b \cos x$ can be written in the form $A \cos(x - \varphi)$ where $A = \sqrt{(a^2 + b^2)}$ and $\tan \varphi = a/b$.

2. Write each of the following in the form $A \sin(x + \varphi)$ specifying the values of A and φ.
   (a) $3 \sin x + 4 \cos x$ (b) $12 \sin x + 5 \cos x$
   (c) $\sin x - \cos x$ (d) $2 \cos x - 3 \sin x$
   (e) $5 \cos x - 9 \sin x$ (f) $15 \sin x + 21 \cos x$.

   (In each case start from first principles by expanding $A \sin(x + \varphi)$ and comparing coefficients.)

5.2.5 Trigonometric Equations

We now continue the solution of trigonometric equations begun in Chapter 4. If the value of a trigonometric function is given then the angle is found using the inverse trigonometric function 'ARC' or 'INV' on your calculator. Remember that there are in theory many solutions to trigonometric equations, for example, $\theta = 30°$, $150°$, $390°$, $510°$, $750°$, $870°$ etc are all solutions to the equation $\sin 30° = 0.5$. However, in this subsection we restrict all solutions to the range $[0, 360°]$ or $[0, 2\pi]$.

Example 5.2.6

Solve the equation

$$\sin x + \cos 2x = 0$$

Solution

Using the double angle identity $\cos 2x = 1 - 2 \sin^2 x$ produces an equation containing $\sin x$. (Using a trigonometric identity is often the first step in solving a trigonometric equation.)

$$\sin x + 1 - 2 \sin^2 x = 0$$

Factorizing this quadratic in $\sin x$ gives

$$(1 + 2 \sin x)\ (1 - \sin x) = 0.$$

Thus

$$\sin x = -\tfrac{1}{2} \quad \text{or} \quad \sin x = 1$$

The three solutions for x are

$$x = 90° \quad \text{or} \quad x = 210° \quad \text{or} \quad x = 330°$$

Example 5.2.7

Solve the equation

$$8 \sin 2x - 15 \cos 2x = 10$$

Solution

For problems of this type involving $\sin \theta$ and $\cos \theta$ for the same angle θ (here $\theta = 2x$) we write the expression as $A \sin(\theta + \varphi)$. In this case the result of Example (5.2.5) gives

$$8 \sin 2x - 15 \cos 2x = 17 \sin(2x + 298°)$$

So we need to solve

$$17 \sin(2x + 298°) = 10$$

Thus

$$\sin(2x + 298°) = \frac{10}{17}$$

and then

$$2x + 298° = 396° \text{ (i.e. } 360° + 36°) \quad \text{or} \quad 504° \text{ (i.e. } 540° - 36°)$$

Solving for x we have $x = 49°$ or $x = 103°$.

(The reason that we have chosen 396° and 504° as the solutions of $\sin \theta = 10/17$ is because we require x in the range 0° to 360°. Using the result from the calculator would have given $2x + 298° = 36°$ and then $x = -131°$ which is outside the range.)

Exercise 5.2.5

1. Solve the following equations for x for $0 \leqslant x \leqslant 360°$.
   (a) $2 \sin 2x = \cos x$ (b) $\cos 2x + 2 \cos x = 0$
   (c) $\sin^2 x + \cos 2x + \cos x = 0$
   (d) $2 \sin^2 x = \sin 2x$
   (e) $2 \sec^2 x - 3 = 2 \tan x$
   (f) $4 \sin^2 x + 8 \cos x = 3$
   (g) $1 - \cos x = 2 \sin x$ (hint: use half angle $x/2$).

2. Solve the following equations for θ for $0 \leqslant \theta \leqslant 360°$.
   (a) $3 \sin \theta + 4 \cos \theta = 2$
   (b) $5 \sin \theta - 12 \cos \theta = 13$
   (c) $\sin(3\theta) + \cos(3\theta) = 1$
   (d) $2 \sin(2\theta) + 5 \cos(2\theta) = 3$

   (Hint: in each problem write the left-hand side as $a \sin(A + \varphi)$.)

5.3 INVERSE TRIGONOMETRIC FUNCTIONS

In Chapter 4 and in the previous subsection we have seen that there are many solutions of a trigonometric equation such as $\sin x = 0.5$. To solve this equation using a calculator we would press the [INV] [SIN] or [SIN⁻¹] key to obtain 30° (in degrees or 0.5236 in radians) and then deduce the other possible solutions from a graph (Figure 5.10).

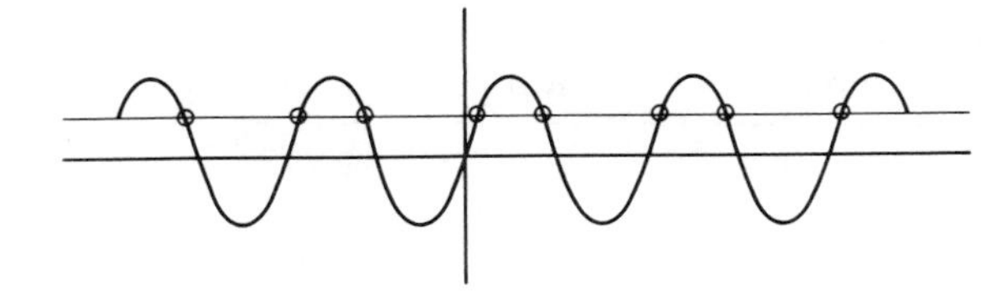

Figure 5.10 Solutions of $\sin x = 0.5$

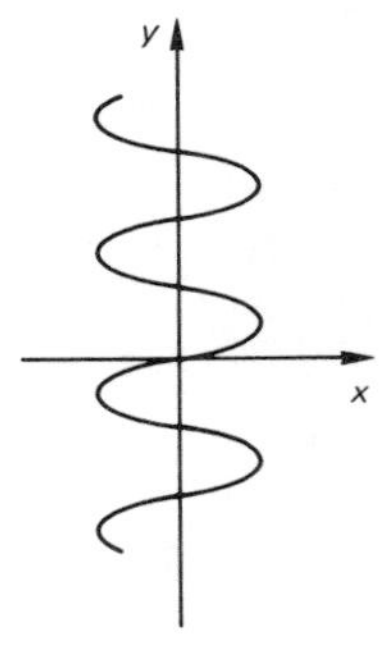

Figure 5.11 Graph of $y = \sin^{-1} x$

The rule $y = \sin^{-1} x$ (or $y = \arcsin x$) has a graph looking like a vertical sine wave (Figure 5.11). (Remember that $\sin^{-1} x$ does not mean $1/\sin x$.) To read the notation $\sin^{-1} x$ we say 'the inverse of sine x' or 'the angle whose sine is x'.

Exercise 5.3.1

1. Draw graphs showing the rules $y = \cos^{-1} x$ and $y = \tan^{-1} x$.
2. Use your calculator and the graph of Figure 5.11 to find six solutions of the equation

   $$\sin x = 0.4$$

The inverses of the trigonometric functions are not themselves functions because given a value of x there are many values of y. However, if we restrict the range of these relations, we can obtain

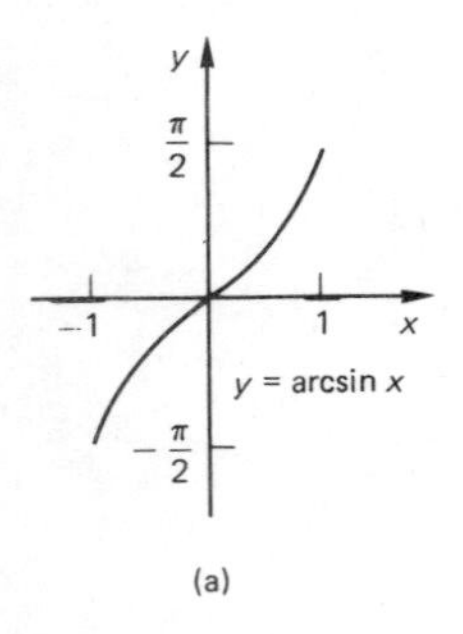

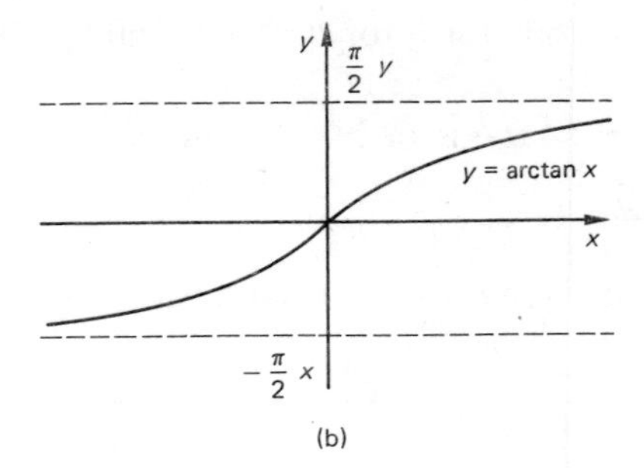

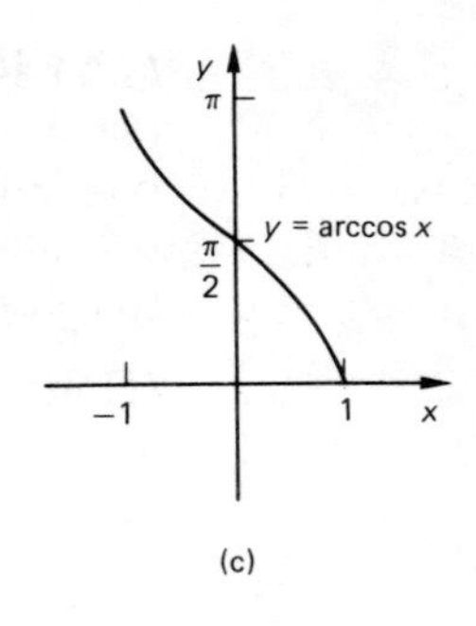

Figure 5.12 Graphs of the inverse trigonometric functions

functions. Figure 5.12 shows how these restrictions are made, which lead to *the inverse trigonometric functions* (Table 5.1).

The restricted ranges of y which turn the relations into functions are called *principal values*.

Table 5.1

Function	Domain	Range
$f(x) = \sin^{-1} x$	$-1 \leqslant x \leqslant 1$	$-\pi/2 \leqslant f \leqslant \pi/2$
$f(x) = \cos^{-1} x$	$-1 \leqslant x \leqslant 1$	$0 \leqslant f \leqslant \pi$
$f(x) = \tan^{-1} x$	$-\infty < x < \infty$	$-\pi/2 < f < \pi/2$

The values of the inverse trigonometric functions obtained from a scientific calculator are the principal values. Some calculators give the values in degrees and radians. Be sure to interpret the display on your calculator correctly.

Exercise 5.3.2

1. Find the values of the functions $f(x) = \sin^{-1} x$ and $g(x) = \cos^{-1} x$ for the following values of x.
   (a) 0.5 (b) −0.5 (c) 1 (d) $\sqrt{2}/2$
   (e) −0.62 (f) −0.95 (g) 1.37 (h) 0.19
   Give your answers in degrees and in radians.

2. Find the values of each of the following expressions.
   (a) $\tan(\cos^{-1} 0.5)$ (b) $\cos(\sin^{-1} 0.7)$
   (b) $\tan(\sin^{-1} -0.4)$ (d) $\sin(2 \sin^{-1} 0.5)$
   (c) $\cos(\tan^{-1} 4.2)$ (f) $\cos(\sin^{-1} 1.4)$

5.4 APPLICATIONS

We pause at this point to look at applications of the trigonometric formulas introduced so far. Mostly applications involve choosing and applying the appropriate formula to simplify a complicated expression involving sines and cosines.

Example 5.4

An object suspended from an elastic spring oscillates so that its displacement is given by

$$x = 3 + 5\sin(3t + \pi/2) - 5\cos(3t + \pi/4)$$

Find the amplitude and phase angle of the oscillation.

Solution

To be able to deduce the amplitude of the oscillation we need to write x in the form

$$x = 3 + a\sin(3t + \varphi)$$

then a is the amplitude and φ is the phase angle.

The first step is to expand $\sin(3t + \pi/2)$ and $\cos(3t + \pi/4)$ using Formulas (5.2) and (5.4) respectively. We have

$$\begin{aligned}\sin(3t + \pi/2) &= \sin 3t \cos \pi/2 + \cos 3t \sin \pi/2 \\ &= \cos 3t\end{aligned}$$

since $\sin \pi/2 = 1$ and $\cos \pi/2 = 0$

$$\begin{aligned}\cos(3t + \pi/4) &= \cos 3t \cos \pi/4 - \sin 3t \sin \pi/4 \\ &= \frac{\sqrt{2}}{2}(\cos 3t - \sin 3t)\end{aligned}$$

since $\sin\dfrac{\pi}{4} = \cos\dfrac{\pi}{4} = \dfrac{\sqrt{2}}{2}$.

Hence

$$x = 3 + 5\cos 3t - \frac{5\sqrt{2}}{2}(\cos 3t - \sin 3t)$$

$$= 3 + 1.46\cos 3t + 3.54\sin 3t$$

Now using Formula (5.12) we have

$$x = 3 + A \sin(3t + \varphi)$$

where $A = \sqrt{1.46^2 + 3.54^2} = 3.83$ and $\tan \varphi = \dfrac{1.46}{3.54}$

so that $\varphi = 0.39$ (in radians)

$$x = 3 + 3.83 \sin(3t + 0.39)$$

The oscillation has magnitude 3.83 and phase 0.39.

In this example we have used three formulas to solve the problem. It illustrates that you have to be fairly alert and aware of the relations that do exist.

The power of the trigonometric identities is important in the theory of waves and vibrations in physics and engineering, and we now look at the phenomenon of *beats* or *interference* of two waves. In the introduction to this chapter we saw from graphs that if two waves with equal frequency arrive in phase and have the same amplitude then the resultant amplitude is twice as large as the amplitude of the two original waves. However if the two waves are 180° out of phase and have the same amplitude then the waves cancel each other out.

What happens for other phase and frequency differences is more complicated, but we now have the mathematical techniques to investigate the phenomenon of wave interference.

The displacement of a point in a region through which a wave travels (e.g. a sound wave or a water wave) is given mathematically by

$$x = a \sin(\omega t + \varphi)$$

Clearly if two waves have the same frequency w and phase φ but different amplitudes, a and b, then the resultant wave will have frequency ω, phase φ and amplitude $a + b$.

Now consider two waves which are in phase, have the same amplitude but different frequencies

$$x_1 = a \sin(\omega_1 t) \quad \text{and} \quad x_2 = a \sin(\omega_2 t)$$

(note that for convenience we have put $\varphi = 0$ here). Then the resultant wave has equation

$$\begin{aligned} x = x_1 + x_2 &= a \sin(\omega_1 t) + a \sin(\omega_2 t) \\ &= 2a \sin[\tfrac{1}{2}(\omega_1 + \omega_2)t] \cos[\tfrac{1}{2}(\omega_1 - \omega_2)t] \end{aligned}$$

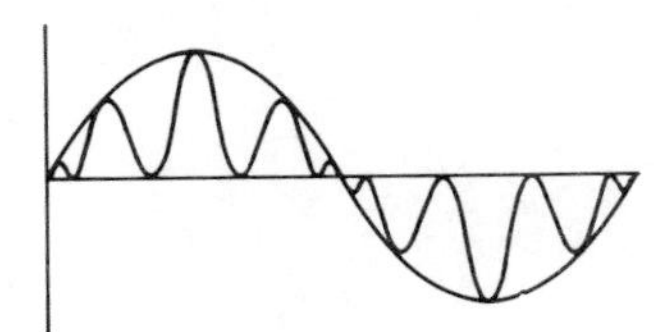

Figure 5.13

Suppose that the frequency of the two waves is close together so that $\omega_1 - \omega_2 = \alpha$ is very small, then a graph of the displacement x with time is shown in Figure 5.13.

The amplitude of the wave grows to double the original amplitude a and then decays to zero; this process is then repeated. If the original waves were sound waves, the sound would arrive in pulses getting louder and then softer. You can experience this phenomenon by tuning two strings on a guitar so that they are almost 'in tune' (i.e. they almost play the same note). Now if you pluck the strings at the same time, instead of hearing one clear note of constant amplitude, the sound appears to arrive in pulses. This idea can be used to tune a musical instrument such as a guitar. By adjusting a note produced by the instrument to a 'standard note', for example, the note given by a turning fork, then when the 'beats' disappear and a constant sound is heard the guitar is in tune with the tuning fork. Repeating this for each string on the guitar, 'tunes' the guitar. An electronic gadget about the size of a hand calculator is often used these days instead of relying on a tuning fork and the ear, but the principle is the same. The instrument receives the note as pressure pulses from the guitar and compares it with a standard pulse producing beats shown on a display. The guitar string is then tuned until these beats disappear and constant amplitude is obtained.

Exercise 5.4.1

1. Two waves A and B of equal amplitude and in phase pass simultaneously through the same medium. Wave A has frequency 495 kHz and wave B has frequency 505 kHz. Find the equation of the resultant disturbance produced by the two waves.
   Sketch a graph of the resultant disturbance.
   What is the period of the 'beats'?

2. Two waves A and B of equal amplitude and in phase pass simultaneously through the same medium. Wave A has frequency f and wave B has frequency $3f$. Find the equation of the resultant disturbance produced by the two waves. Sketch a graph of the resultant disturbance.

3. Two waves of equal amplitude and frequency differ by phase angle φ, so that their displacements are described by

   $$x_1 = a \sin \omega t \quad \text{and} \quad x_2 = a \sin(\omega t + \varphi)$$

   Find the equation of the resultant wave.

Sketch graphs of the two original waves and the resultant wave using the same scales.

Describe how the period, amplitude and phase of the resultant compare with the original waves.

Describe the resultant wave for the three cases
(a) $\varphi = 0$ (b) $\varphi = \pi/2$ (c) $\varphi = \pi$

Exercise 5.4.2 General Applications

1. The current in an electric circuit varies with decreasing amplitude, given by the equation

$$i = \sqrt{e^{-2t}(1 + \sin 2t)}$$

Show that $i = e^{-t}(\sin t + \cos t)$

2. In a simple electric circuit the instantaneous power is given by

$$P = V_m I_m \cos wt \sin wt$$

Show that $P = \dfrac{V_m I_m}{2} \sin 2wt$.

3. Show that if a water wave described by $x_1 = \sin(t - \pi/2)$ meets another water wave $x_2 = \sin(t + \pi/2)$, they will cancel each other.

4. The vibration of a cable has displacement given by

$$x = A \sin(\omega t + \pi/3) - A \cos(\omega t + \pi/6)$$

where A is constant. By expanding the sine and cosine terms write x in terms of the angle ωt.

5. In a complicated electric circuit the power is given by

$$P = (E \cos 2t + RI)^2 + (E \sin 2t + XI)^2$$

where E, R, I and X are constants. Simplify P into the form

$$P = A + B \sin(2t + \alpha)$$

Hence show that the maximum and minimum values of P are

$$(E \pm I \sqrt{R^2 + X^2})^2$$

6. The displacement from equilibrium (in centimetres) of an object oscillating on the end of a spring is given by

$$x = 3.1 \sin 4t + 4.7 \cos 4t$$

where t is the time in seconds. Write x in the form $A \sin(4t + \alpha)$ and state the amplitude and period of the oscillation.
   Find the first time at which the object is at a distance 3 cm from its equilibrium position.

7. Find the sum of the currents $i_1 = 40 \sin 5\pi t$ and $i_2 = 30 \cos 5\pi t$ in the form $a \sin(5\pi t + \alpha)$. What are the amplitude, period and phase angle of the resultant current? When is the first time that the value of the current is 50 amperes?

8. In geometry the angle between two lines whose slopes are $m_1 = \tan \theta_1$ and $m_2 = \tan \theta_2$ is $\theta_2 - \theta_1$. Show that

$$\tan(\theta_2 - \theta_1) = \frac{m_2 - m_1}{1 + m_1 m_2}$$

   Find the angle between the lines with equations $y = 2x + 1$ and $y = -3x + 4$.
   Deduce that if two lines are perpendicular then $m_1 m_2 = -1$.

9. Using the formula in Problem (8), find the equation of the line through the point (0, 2) which is perpendicular to the line $y = 2x - 3$.

5.5 THE SINE RULE

You will be familiar with the use of sine, cosine and Pythagorus' rule in solving problems involving right-angled triangles. Unfortunately not all physical situations involve right angled triangles and two important laws can be used in such problems; these are *the sine rule* and *the cosine rule*. In this subsection we prove the first of these.

Consider the triangle ABC shown in Figure 5.14. Such a triangle which does not include a right angle is called *an oblique triangle*. Note that we label the length of the side opposite angle A as a, the length of the side opposite angle B as b and similarly for C.

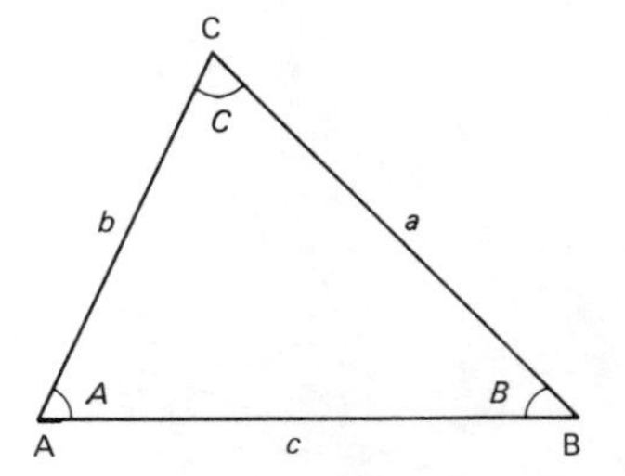

Figure 5.14

The sine rule:

$$\frac{a}{\sin A} = \frac{b}{\sin B} = \frac{c}{\sin C} \qquad (5.13) \quad \boxed{M}$$

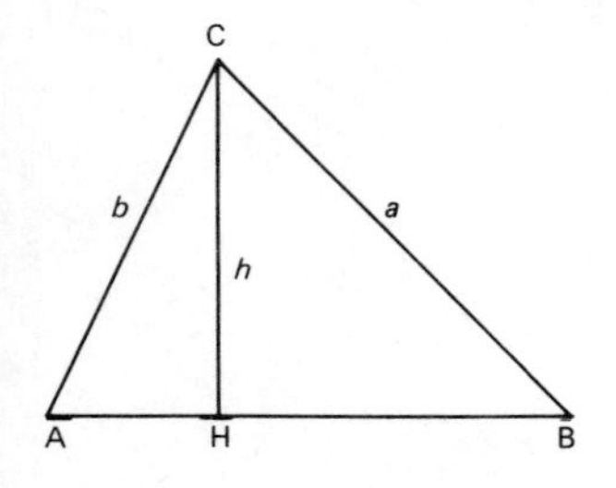

Figure 5.15

To prove the sine rule we draw in a line from C which is perpendicular to AB; suppose that this line has length h (Figure 5.15).

Now we have two right-angled triangles AHC, BHC and we can write

$$\sin A = \frac{h}{b} \quad \text{and} \quad \sin B = \frac{h}{a}$$

Thus

$$h = b \sin A \quad \text{and} \quad h = a \sin B$$

giving

$$b \sin A = a \sin B$$

or

$$\frac{a}{\sin A} = \frac{b}{\sin B}$$

Similarly, by drawing a perpendicular from A to BC we can show that

$$\frac{b}{\sin B} = \frac{c}{\sin C}$$

and hence the sine rule follows.

Example 5.5

A yacht heads on the bearing 60° from harbour for 20 km and then heads east. A customs boat intercepts the yacht having travelled 25 km. Find the bearing set by the customs boat and the distance travelled by the yacht.

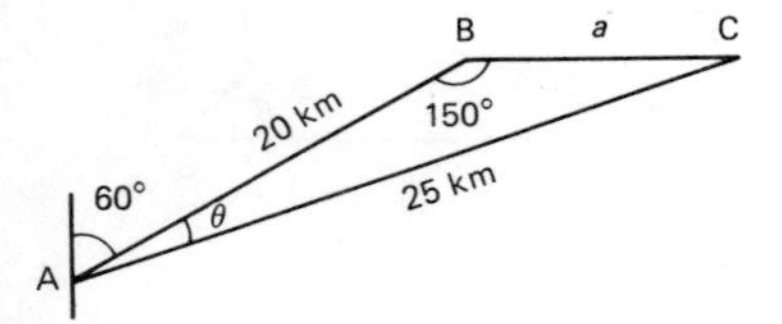

Figure 5.16

Solution

Figure 5.16 shows the path of the yacht and customs boat. We need to find angle θ and length a. The sine rule gives

$$\frac{25}{\sin 150°} = \frac{20}{\sin C} = \frac{a}{\sin \theta}$$

and also

$\theta = 180° - (150° + C) = 30° - C.$

Thus

$$\sin C = \frac{20 \sin 150°}{25} = 0.4$$

hence

$$C = 23.6°, \text{ and then } \theta = 30° - C = 6.4°.$$

(A solution to $\sin C = 0.4$ could also be $C = 156.4°$, however this can be rejected since the triangle already contains an angle of 150°.)

Solving for a,

$$a = \frac{25}{\sin 150°} \sin \theta = 5.57$$

The bearing set by the customs boat is 66.4° and the total distance travelled by the yacht is 25.57 km.

Exercise 5.5.1

1. Find the three missing parts (either angles and/or side lengths) in each of the following triangles:
   (a) $A = 60°$, $B = 45°$, $a = 6$
   (b) $A = 110°$, $B = 25°$, $a = 4.5$
   (c) $B = 78.5°$, $C = 31.6°$, $a = 11.4$
   (d) $a = 7.61$, $b = 6.23$, $A = 55°$
   (e) $b = 0.21$, $c = 0.12$, $B = 41°$
   (f) $A = 45°$, $a = 16$, $c = 7.5$
   (g) $C = 116°$, $c = 4.1$, $a = 3.4$
   (h) $a = 263$, $b = 216$, $A = 60°$

2. Two forces F_1 and F_2 act on an object and the angle between the forces is 80° (Figure 5.17). If the magnitude of the resultant R is 16 newtons when $F_1 = 11$ newtons, find the magnitude of F_2 and the angle between R and F_2.

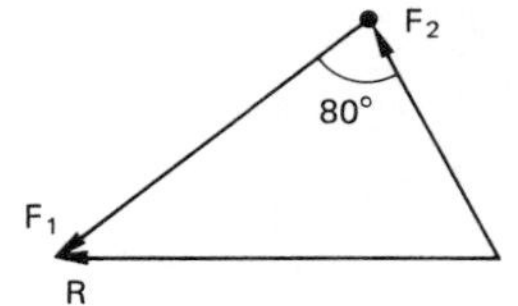

Figure 5.17

3. An aerial of height 40 m is on top of a building. From a point on the ground the top and bottom of the aerial have angles of elevation 56° and 47° respectively. Find the height of the building.

4. An aircraft has a speed of 250 knots and the pilot wishes to fly on a bearing 030° in a wind of 60 knots from due south. Find the bearing that the pilot should set in order to fly on the required bearing.

5. A surveyor standing on level ground intends to find the height of a flagpole. She measures the angle of inclination from a point A to the top of the flagpole to be 32°. She then walks 20 m to a point B which is directly in line between point A and the base of the flagpole. The angle of inclination from point B is 62°. Calculate the height of the flagpole.

5.6 THE COSINE RULE

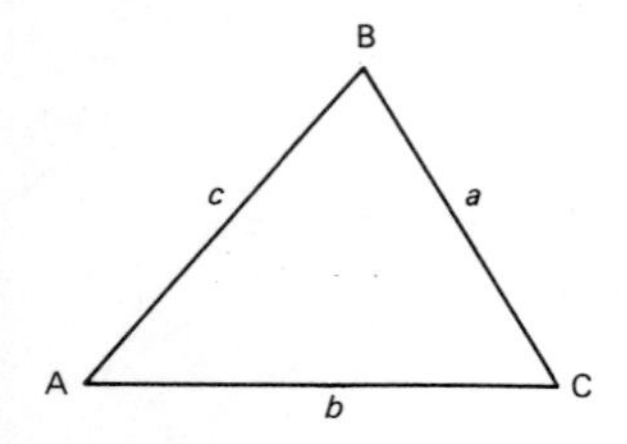

Figure 5.18

Consider the oblique triangle ABC in Figure 5.18. Another rule relating the sides and angles of the triangle is *the cosine rule*.

> The cosine rule
>
> $$c^2 = a^2 + b^2 - 2\,ab\cos C$$

(5.14) [M]

Although in Equation (5.14) we have stated the law with side length c on the left-hand side, we could equally have written

$$b^2 = a^2 + c^2 - 2ac\cos B$$

$$a^2 = b^2 + c^2 - 2bc\cos A$$

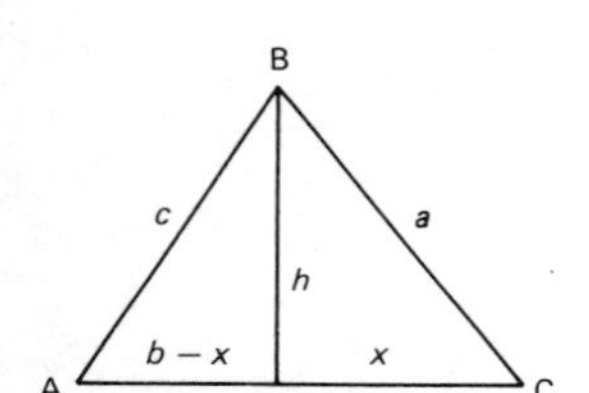

Figure 5.19

The cosine rule is essentially a generalization of Pythagorus' theorem. To prove the cosine rule, we construct a perpendicular from B to side AC and let BH = h and HC = x (Figure 5.19).

Apply Pythagorus' theorem to the two right-angled triangles ABH and CBH

$$c^2 = (b - x)^2 + h^2 \quad \text{and} \quad a^2 = x^2 + h^2$$

Subtract these two equations to give

$$c^2 - a^2 = (b - x)^2 - x^2 = b^2 - 2bx + x^2 - x^2$$

thus

$$c^2 = a^2 + b^2 - 2bx$$

But from triangle CBH, $x = a\cos C$ and hence

$$c^2 = a^2 + b^2 - 2ba\cos C$$

which is one form of the cosine rule. It is only necessary to remember one form, since by interchanging the symbols or relabelling the triangle the other forms can be obtained.

Example 5.6.1

An object experiences two forces F_1 and F_2 of magnitudes 9 and 13 newtons with an angle 100° between their directions (Figure 5.20). Find the magnitude of the resultant R.

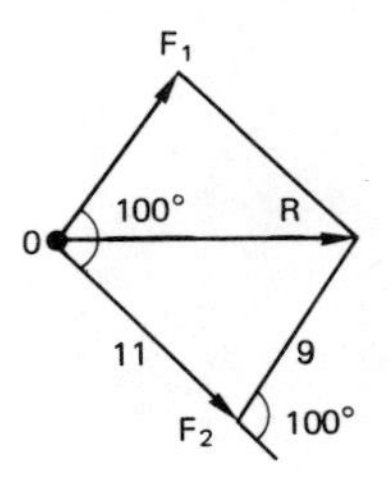

Figure 5.20

Solution

Applying the cosine rule (5.14) we have

$$R^2 = 11^2 + 9^2 - 2 \times 11 \times 9 \times \cos 80°$$
$$= 167.62$$
$$R = \sqrt{167.62} = 12.95 \text{ newtons}$$

Example 5.6.2

Given that angle B = 135° and side lengths $a = 3.15$ and $c = 2.72$ find the other angles and side length b (Figure 5.21).

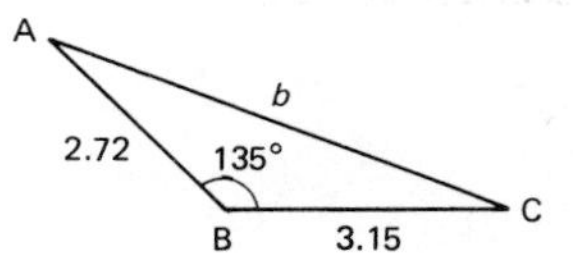

Figure 5.21

Solution

The cosine rule for side length b is

$$b^2 = a^2 + c^2 - 2ac \cos B$$
$$= 3.15^2 + 2.72^2 - 2 \times 3.15 \times 2.72 \times \cos 135°$$
$$= 29.44$$
$$b = \sqrt{29.44} = 5.43$$

To find angles A or C we use the sine rule

$$\frac{a}{\sin A} = \frac{b}{\sin B} = \frac{c}{\sin C}$$

Thus

$$\sin A = \frac{a \sin B}{b} = \frac{3.15 \sin 135°}{5.43} = 0.41$$
$$A = 24.2°$$

and then

$$C = 180° - (\text{A} + \text{B}) = 20.8°.$$

(Notice in this example, cos 135° is negative which makes the term $2ac \cos B$ negative resulting in the final term being added to $a^2 + c^2$.)

Exercise 5.6.1

1. Find the three missing parts (either angles and/or side lengths) in each of the following triangles:
   (a) $C = 75°$, $a = 4.0$, $b = 6.0$
   (b) $C = 43.1°$, $a = 2.42$, $b = 4.73$
   (c) $C = 127°$, $a = 7.3$, $b = 1.9$
   (d) $B = 110°$, $a = 180$, $c = 230$
   (e) $a = 2.1$, $b = 3.6$, $c = 4.9$
   (f) $a = 0.12$, $b = 0.63$, $c = 0.52$
   (g) $A = 73°$, $b = 12$, $c = 17$
   (h) $B = 105°$, $c = 9.41$, $a = 7.31$
2. Two forces $F_1 = 5$ N and $F_2 = 12$ N act on an object. The angle between the directions of these forces is 75°. Find the magnitude of the resultant force and the angle between the resultant and force F_2.
3. The supersonic aircraft Concorde, travelling due west at 1200 knots, encounters a wind of speed 100 knots from a bearing of 305°. Find the resultant speed of Concorde and its bearing.
4. A yacht travels 50 miles on a bearing 050° from harbour and then travels 30 miles on a bearing 130°. Find the distance and bearing of the yacht from the harbour.

OBJECTIVE TEST 5

1. Solve the equation

$$2 \tan \theta - 4 \cot \theta = \operatorname{cosec} \theta$$

giving the answers in the range $0 < \theta < \pi$.

2. If $\sin(x + \alpha) = a$ and $\cos(x + \beta) = b$ express $\cos x$ and $\sin x$ in terms of α, β, a and b.
3. If $x = 3 \sin(25t)$ and $y = 3 \sin(15t)$ find a formula for $x + y$ and sketch its graph.
4. Show that

$$\frac{\sin 4\theta(1 - \cos 2\theta)}{\cos 2\theta(1 - \cos 4\theta)} = \tan\theta$$

5. Write $4 \sin 2x + 3 \cos 2x$ in the form $R \sin(2x + \varphi)$ where R is positive and φ is an acute angle measured in degrees.

Hence find the values of x between 0 and 180° for which

$$4 \sin 2x + 3 \cos 2x = 3.2$$

6. Use the sine rule or the cosine rule (as appropriate) to find the unknown angle and side lengths in the triangles shown in Figure 5.22.

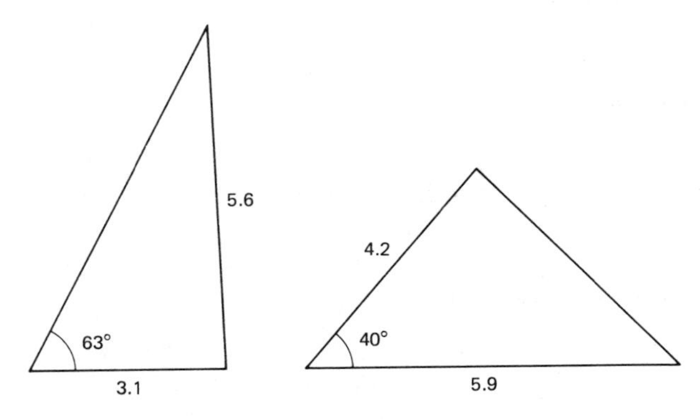

Figure 5.22

LIST OF TRIGONOMETRIC IDENTITIES

Basic Identities

$$\sin^2 \theta + \cos^2 \theta = 1 \quad \boxed{\text{M}}$$

$$\sec^2 \theta = 1 + \tan^2 \theta \quad \boxed{\text{M}}$$

$$\text{cosec}^2 \theta = 1 + \cot^2 \theta$$

Addition Formulas

$$\sin(A + B) = \sin A \cos B + \cos A \sin B \quad \boxed{\text{M}}$$

$$\sin(A - B) = \sin A \cos B - \cos A \sin B$$

$$\cos(A + B) = \cos A \cos B - \sin A \sin B \quad \boxed{\text{M}}$$

$$\cos(A - B) = \cos A \cos B + \sin A \sin B$$

$$\tan(A + B) = \frac{\tan A + \tan B}{1 - \tan A \tan B}$$

$$\tan(A - B) = \frac{\tan A - \tan B}{1 + \tan A \tan B}$$

Double-angle formulas

$$\sin 2A = 2 \sin A \cos A \qquad \boxed{M}$$

$$\cos 2A = \cos^2 A - \sin^2 A \qquad \boxed{M}$$

$$\text{or} \quad \cos 2A = 2 \cos^2 A - 1 \qquad \boxed{M}$$

$$\text{or} \quad \cos 2A = 1 - 2 \sin^2 A \qquad \boxed{M}$$

The factor formulas

$$\sin X + \sin Y = 2 \sin \left(\frac{X + Y}{2}\right) \cos \left(\frac{X - Y}{2}\right)$$

$$\sin X - \sin Y = 2 \cos \left(\frac{X + Y}{2}\right) \sin \left(\frac{X - Y}{2}\right)$$

$$\cos X + \cos Y = 2 \cos \left(\frac{X + Y}{2}\right) \cos \left(\frac{X - Y}{2}\right)$$

$$\cos X - \cos Y = - 2 \sin \left(\frac{X + Y}{2}\right) \sin \left(\frac{X - Y}{2}\right)$$

Because of their power and how often they are used you are recommended to memorize the identities labelled $\boxed{M}$.

6 FURTHER ALGEBRAIC SKILLS

OBJECTIVES

When you have completed this chapter you should be able to

1. define a sequence
2. state the limit of a sequence when it exists
3. use formulae to compute the sums of powers of positive integers
4. define arithmetic progressions and sum their terms
5. define geometric progressions and sum their terms
6. state the condition for a geometric progression to have a sum to infinity and compute such a sum
7. carry out polynomial divisions, computing quotients and remainders
8. use the factor theorem to test a polynomial for factors
9. state when a quadratic polynomial is irreducible over the real numbers
10. identify proper and improper partial fractions
11. define the three types of reduction for partial fractions
12. reduce an algebraic fraction to partial fractions
13. expand $(a + b)^n$ where n is a positive integer using Pascal's triangle and the Binomial Theorem
14. expand $(1 + x)^n$ using the Binomial Theorem and state when the expansion is valid

6.1 SEQUENCES, SERIES AND LIMITS

6.1.1 Sequences

A company has a £16 000 lathe. Its value at the end of any given year is 75% of its value at the beginning of that year.

Let v_0 = initial value of the lathe, and v_n = the value at the end of year n. So

$v_0 = £16\,000$
$v_1 = £16\,000 \times 0.75 = £12\,000$
$v_2 = £12\,000 \times 0.75 = £9000$

$v_3 = £9000 \times 0.75 = £6750$
$v_4 = £6750 \times 0.75 = £5062.50$

The values $v_0, v_1, v_2, \ldots$ are an example of a sequence.

Definition: A sequence is a list of numbers which occur in a given order.

Consider the sequence 2, 4, 6, 8,... Call the terms of this sequence $t_1, t_2, t_3, \ldots, t_n, \ldots$, where $t_n = n$th term. t_1 is the first term.

The rule for generating this sequence is 'Add 2 to the previous term'. So the sequence is given by

$$t_1 = 2$$

$$t_n = t_{n-1} + 2, \quad \text{for } n = 2, 3, 4, \ldots$$

An expression of this type is often called a *recurrence relation*. An explicit formula for the nth term is

$$t_n = 2n \quad \text{for } n = 1, 2, 3, \ldots$$

Consider another sequence, namely 2, 4, 8, 16,... Call the terms of this sequence $a_1, a_2, a_3, \ldots, a_n, \ldots$ The rule for generating this sequence is 'Double the previous term'. So

$$a_1 = 2$$

$$a_n = 2a_{n-1}, \quad \text{for } n = 2, 3, 4, \ldots$$

An explicit formula for the nth term is

$$a_n = 2^n \quad \text{for } n = 1, 2, 3, \ldots$$

Finally suppose that a mill is set to machine 0.1 mm off a block of metal on each pass. Initially the block of metal is 20 mm thick. We can find a formula for its thickness t_n on the nth pass as follows:

Initial thickness, $t_0 = 20\text{m}$
Thickness after first pass, $t_1 = 20 - 0.1 = 19.9$ mm
Thickness after second pass, $t_2 = 19.9 - 0.1 = 19.8$ mm
Thickness after nth pass, $t_n = (20 - 0.1n)$ mm

Exercise 6.1.1

In Problems 1–10, you are given the nth term of a sequence, t_n. Write down its first three terms t_1, t_2 and t_3.

1. $t_n = \left(\frac{1}{10}\right)^n$

2. $t_n = \left(\frac{1}{10}\right)^{n-1}$

3. $t_n = 2 + \left(\frac{1}{10}\right)^n$

4. $t_n = 3 + \left(\frac{1}{10}\right)^{n-1}$

5. $t_n = 3\left(\frac{1}{4}\right)^n$

6. $t_n = (-1)^n$

7. $t_n = \left(-\frac{1}{10}\right)^{n-1}$

8. $t_n = 3 + \left(-\frac{1}{10}\right)^{n-1}$

9. $t_n = 1 + 2n$

10. $t_n = 3 - \left(\frac{n}{2}\right)$

In Problems 11–18, you are given t_1, t_2 and t_3, the first three terms of a sequence. In each case write down a formula for t_n, the nth term of the sequence.

11. $\frac{1}{4}, \frac{1}{16}, \frac{1}{64}$ (each term is $\frac{1}{4}$ of its predecessor)

12. $1, \frac{1}{4}, \frac{1}{16}$ (each term is $\frac{1}{4}$ of its predecessor)

13. $1, \frac{1}{3}, \frac{1}{9}$

14. $-2, +2, -2$ (Hint: number 6 may help)

15. $3, -3, 3$

16. 0.1, 0.01, 0.001

17. 5.1, 5.01, 5.001 (Hint: use 5 plus your answer to 16)

18. 0.9, 0.99, 0.999 (Subtract a suitable sequence from 1, 1, 1,...)

19. Each year a car depreciates, so that its value at the end of the year is 70% of its value at the beginning. If the car cost £9500, what is its value at the end of the first three years? What is its value at the end of year n?

6.1.2 Limits of Sequences

The sequence 2, 4, 8, 16,... has terms which increase without bound. We say that this sequence has no limit.

The sequence 0.9, 0.99, 0.999,... also has terms which increase. However in this case there is a definite bound on the growth. The terms appear to be approaching a limiting value of 1. If the terms of the sequence are t_1, t_2, t_3,..., t_n,... then the nth term is given by

$$t_n = 1 - \left(\frac{1}{10}\right)^n$$

The larger n becomes, the nearer $(1/10)^n$ approaches zero, and hence the nearer t_n approaches 1. We say that 1 is the limiting value of this sequence and we write

$$\lim_{n\to\infty} t_n = 1$$

By this statement we do not mean that any term t_n is actually equal to 1, but that by choosing a sufficiently large value for n, we can make all subsequent terms of the series as close to 1 as we like. For example, to be within one millionth of the limit 1, we would need to take a term beyond the sixth term. Each term beyond the sixth term will be within the required tolerance of the limit.

Example 6.1.5

Fnd the limit of the sequence $1\frac{1}{3}$, $1\frac{1}{9}$, $1\ \frac{1}{27}$,...

What is the first term of the sequence which is within $\frac{1}{100}$ of the limit?

Solution

Let the sequence be t_1, t_2, t_3,... Then

$$t_n = 1 + \left(\frac{1}{3}\right)^n$$

So

$$\lim_{n\to\infty} t_n = 1 \quad \text{as} \quad \left(\frac{1}{3}\right)^n \to 0 \quad \text{as} \quad n \to \infty.$$

Let

$$t = \lim_{n\to\infty} t_n$$

We require the difference between t_n and t to be less than $\frac{1}{100}$.
So

$$|t_n - t| < \frac{1}{100}$$

$$\left(\frac{1}{3}\right)^n < \frac{1}{100}$$

Take logs,

$$n \ln\left(\frac{1}{3}\right) < \ln\left(\frac{1}{100}\right)$$

So

$$n > \frac{\left(\ln\frac{1}{100}\right)}{\ln\left(\frac{1}{3}\right)}$$

(note $\ln \frac{1}{3} < 0$ so the direction of the inequality changes).

Therefore $n > 4.19$.

Now n must be an integer, so the first term which is within $\frac{1}{100}$ of the limit is the fifth.

Exercise 6.1.2

In Problems 1–6 you are given the first four terms, t_1, t_2, t_3 and t_4 of a sequence. Write down a formula for the nth term t_n, and the limit of the sequence if it exists.

1. $1, \frac{1}{4}, \frac{1}{16}, \frac{1}{64}$
2. $3\frac{1}{2}, 3\frac{1}{4}, 3\frac{1}{8}, 3\frac{1}{16}$
3. $1, -1, 1, -1$
4. $5, 10, 15, 20$
5. $2, 1\frac{1}{2}, 1, \frac{1}{2}$
6. 11, 10.1, 10.01, 10.001
7. Each hour a battery loses 20% of its charge, i.e. if C is the charge at the start of any hour, the charge at the end is $0.8C$. Write down a formula for the charge at the end of the nth hour and show that as n increases the battery's charge approaches zero.

8. A square has sides of length $2l$. The midpoints of each side are joined to form a new square. What is the length of its sides? From the second square, a third square is formed in the same manner. If the process is continued, what is the length of the sides of the nth square? What happens if the process is continued indefinitely?

6.1.3 Series

When the terms of a sequence are added together a series is obtained.

Example 6.1.6

A company aims to increase its turnover by 25% per annum. At the beginning of the present financial year its turnover was £0.5 million. What is the total turnover over the next four years, if its goal is achieved?

Solution

Let t_n = turnover at end of nth year.
Then

$$t_1 = 0.5 \times 1.25 \times £10^6, \quad t_2 = t_1 \times 1.25$$

$$t_3 = t_2 \times 1.25 = t_1 \times 1.25^2 \quad \text{and} \quad t_4 = t_3 \times 1.25 = t_1 \times 1.25^3$$

So total turnover is $t_1 + t_2 + t_3 + t_4$

$$= t_1 + t_1 \times 1.25 + t_1 \times 1.25^2 + t_1 \times 1.25^3$$
$$= t_1(1 + 1.25 + 1.25^2 + 1.25^3)$$
$$= 0.5 \times 1.25\,(1 + 1.25 + 1.25^2 + 1.25^3) \times £10^6$$
$$\simeq £3\,603\,516$$

6.1.4 The Sigma Notation

A series is obtained by summing (or adding together) the terms of a sequence. If we want to add together the first ten terms of the sequence $t_1, t_2, t_3, \ldots$ we could write

$$S_{10} = t_1 = t_2 + \ldots + t_{10}$$

or more simply

$$S_{10} = \sum_{n=1}^{10} t_n$$

The symbol Σ is sigma, the Greek letter 's' for sum. The numbers below and above the Σ indicate where the summation is to start and finish. So for example

$$\sum_{n=1}^{50} t_n = t_1 + t_2 + \ldots + t_{50}$$

$$\sum_{n=10}^{20} t_n = t_{10} + t_{11} + \ldots + t_{20}$$

$$\sum_{i=10}^{20} t_i = t_{10} + t_{11} + \ldots + t_{20}$$

$$\sum_{i=5}^{80} a_i = a_5 + a_6 + \ldots + a_{80}$$

If a formula is available for the terms of the sequence, this can be used in the notation for the series:

$$\sum_{n=1}^{5} \left(\frac{1}{2}\right)^n = \frac{1}{2} + \frac{1}{4} + \ldots + \frac{1}{32}$$

$$\sum_{i=10}^{15} 3^i = 3^{10} + 3^{11} + \ldots + 3^{15}$$

Exercise 6.1.3

Write the following series using the sigma notation. Do not attempt to find the sum of the series

1. $1 + \frac{1}{10} + \frac{1}{100} + \ldots + \frac{1}{10^6}$
2. $\frac{1}{2} + \frac{1}{4} + \frac{1}{8} + \ldots + \frac{1}{256}$
3. $1 + 3 + 5 + \ldots + 9$
4. $2 - 4 + 6 - 8 + \ldots - 20$
5. $1 + x + x^2 + \ldots + x^{30}$

6.1.5 Some Useful Series

The following results are quoted without proof:

Summing series of powers of integers

$$1 + 2 + 3 + \dots + n = \sum_{i=1}^{n} i = \frac{1}{2} n(n + 1)$$

$$1^2 + 2^2 + 3^2 + \dots + n^2 = \sum_{i=1}^{n} i^2 = \frac{1}{6} n(n + 1)(2n + 1)$$

$$1^3 + 2^3 + 3^3 + \dots + n^3 = \sum_{i=1}^{n} i^3 = \frac{1}{4} n^2(n + 1)^2$$

Example 6.1.7

Find $5^2 + 6^2 + 7^2 + \dots + 13^2$.

Solution

Let $S_n = \sum_{i=1}^{n} i^2$, then we require $S_{13} - S_4$. From the formula

$$S_{13} = \frac{1}{6} \times 13 \times 14 \times 27 = 819$$

$$S_4 = \frac{1}{6} \times 4 \times 5 \times 9 = 30$$

So

$$5^2 + 6^2 + 7^2 + \dots + 13^2 = 819 - 30 = 789$$

Exercise 6.1.4

Use the formulae given above to find the following sums:

1. $1 + 2 + \dots + 100$
2. $1^2 + 2^2 + \dots + 20^2$
3. $1^3 + 2^3 + \dots + 16^3$
4. $100 + 101 + \dots + 200$
5. $10^2 + 11^2 + \dots + 19^2$
6. $20^3 + 21^3 + \dots + 30^3$

6.2 ARITHMETIC PROGRESSIONS

A sequence in which you go from term-to-term by adding or subtracting the same number each time is called an *arithmetic sequence* or *arithmetic progression* (AP).

Example 6.2.1

position n:	1	2	3	4	5	6
term t_n:	7	11	15	19	23	27

add 4

Example 6.2.2

position n:	1	2	3	4	5	6
term t_n:	10	7	4	1	−2	−5

subtract 3

For arithmetic sequences we can work out the general term by working out how many times you need to apply the term-to-term rule.

For example in the sequence in Example 6.2.1 above

1st term, $t_1 = 7$
2nd term, $t_2 = 7 + 4$
3rd term, $t_3 = 7 + 4 \times 2$
4th term, $t_4 = 7 + 4 \times 3$
5th term, $t_5 = 7 + 4 \times 4$
6th term, $t_6 = 7 + 4 \times 5$
nth term, $t_n = 7 + 4 \times (n - 1) = 4n + 3$

This is the most useful way of describing a sequence. We can find any term using the formula for the general term t_n. For example

10th term, $t_{10} = 4 \times 10 + 3 = 43$
50th term, $t_{50} = 4 \times 50 + 3 = 203$
1000th term, $t_{1000} = 4 \times 1000 + 3 = 4003$

In general, an arithmetic progression will have the form:

$$a, a + d, a + 2d, a + 3d, \ldots$$

where a = first term and d = common difference.

If we call the terms of the AP $t_1, t_2, t_3, \ldots$ then,

$$t_1 = a$$
$$t_2 = a + d$$
$$t_3 = a + 2d$$
$$\vdots$$

and $t_n = a + (n - 1)d$ is the nth term.

Example 6.2.3

Find the common difference d, and the twelfth term of the arithmetic progression

$$7, 10, 13, 16, \ldots$$

Solution

The common difference $d = 10 - 7 = 3$. The first term is $a = 7$. The nth term is $t_n = a + (n - 1)d$. So $t_{12} = 7 + (12 - 1)3 = 40$.

Example 6.2.4

The fourth term of an AP is 7.5 and its seventh term is 5.1. Find its first and twentieth term.

Solution

Let the AP have first term a and common difference d. Then the nth term, $t_n = a + (n - 1)d$.

So

$$\text{fourth term, } t_4 = a + 3d = 7.5 \qquad (1)$$

and

$$\text{seventh term, } t_7 = a + 6d = 5.1 \qquad (2)$$

Subtracting

$$-3d = 2.4$$

So

$$d = -0.8$$

So from (1)

$$a = 7.5 - 3d = 7.5 + 2.4 = 9.9$$

The twentieth term is

$$t_{20} = a + 19d = 9.9 + 19(-0.8) = -5.3$$

So

$$t_1 = 9.9 \quad \text{and} \quad t_{20} = -5.3$$

Exercise 6.2.1

1. Find the tenth term of the AP 2, 5, 8,...
2. Find the hundredth term of the AP 0.5, 0.75, 1.0,...
3. An AP has eighth term equal to 7 and twelfth term equal to 15. Find its first term.
4. An AP has its fifth term equal to 7 and its first term is three times as large as its sixth term. Find its first term and common difference.
5. How many terms are there in the AP 1, 2, 3,..., 10?
6. How many terms are there in the AP 100, 101,..., 200?
7. How many terms are there in the AP 200, 202,..., 300?

6.2.1 Summing the terms of an Arithmetic Progression

Before considering the general case, take a particular example.

Example 6.2.3

Find the sum of the even numbers between 50 and 250 inclusive.

Solution

We need to find $S = 50 + 52 + 54 + \ldots + 250$. This is the sum of an AP with first term $a = 50$ and common difference $d = 2$. If n is the number of terms, $250 = 50 + (n - 1)2$, so $n = 101$. Now write S in ascending and descending orders.

$$S = 50 + 52 + \ldots + 248 + 250$$

$$S = 250 + 248 + \ldots + 52 + 50$$

Add

$$2S = \underbrace{300 + 300 + \ldots + 300 + 300}_{n\,=\,101 \text{ terms}}$$

So

$$2S = 101 \times 300$$

Hence $S = 15150$

Now consider the general case. Let S_n be the sum of n terms of an AP with first term a and common difference d. The last (nth) term in the sum is $a + (n - 1)d$. Write S_n with its terms in ascending and descending order.

$$S_n = a + (a + d) + \ldots + [a + (n - 2)d] + [a + (n - 1)d]$$
$$S_n = [a + (n - 1)d] + [a + (n - 2)d] + \ldots + (a + d) + a$$

Add

$$2S_n = \underbrace{[2a + (n - 1)d] + [2a + (n - 1)d] + \ldots + [2a + (n - 1)d] + [2a + (n - 1)d]}_{n \text{ terms}}$$

So

$$2S_n = n[2a + (n - 1)d]$$

Therefore

> The sum of n terms of an AP with first term a and common difference d,
>
> $$S_n = \frac{n}{2}[2a + (n - 1)d]$$

Example 6.2.4

Find the sum $1 + 1\frac{1}{2} + 2 + \ldots + 60$.

Solution

This is an AP, with first term $a = 1$ and common difference $d = 1\frac{1}{2} - 1 = \frac{1}{2}$.

The last term $t_n = 1 + \frac{1}{2}(n - 1) = 60$, so $n = 119$. So the sum is $S_{119} = \frac{119}{2}(2 \times 1 + 118(\frac{1}{2})) = 3629\frac{1}{2}$

Exercise 6.2.2

1. Find the sum $1 + 2 + 3 + \ldots + 100$.
2. Find the sum $1 + 1.9 + 2.8 + \ldots + 23.5$.
3. Find the sum of the first ten terms of the AP 3, 2.5,...
4. Find the sum of the first 20 and the first 30 terms of the AP 5, 9, 13,... Hence find the sum of terms 21 to 30 inclusive.

5. Show that the sum of n terms of an AP is given by $S_n = \frac{1}{2}n(a + l)$ where a = first term and l = last term. Hence find the sum $50 + 55 + \ldots + 100$.

6.3 GEOMETRIC PROGRESSIONS

A sequence in which you go from term-to-term by multiplying by the same number each time is called a *geometric sequence* or *geometric progression* (GP).

Example 6.3.1

position n:	1	2	3	4	5
term t_n:	4	12	36	108	324

$\times 3$

Example 6.3.2

position n:	1	2	3	4	5
term t_n:	80	40	20	10	5

$\times \frac{1}{2}$

Example 6.3.3

Find the general term for the geometric sequence of Example 6.3.1.

Solution

1st term, $t_1 = 4 = 4$
2nd term, $t_2 = 4 \times 3 = 4 \times 3^1$
3rd term, $t_3 = 4 \times 3 \times 3 = 4 \times 3^2$
4th term, $t_4 = 4 \times 3 \times 3 \times 3 = 4 \times 3^3$
5th term, $t_5 = 4 \times 3 \times 3 \times 3 \times 3 = 4 \times 3^4$

.
.
.

nth term, $t_n = 4 \underbrace{\times 3 \times 3 \times \ldots \times 3}_{(n-1) \text{ lots of } 3}$

nth term, $t_n = 4 \times 3^{n-1}$

If the first term of a GP is a and the common ratio is ρ, then the GP is:

$$a,\ a\rho,\ a\rho^2,\ a\rho^3,\ \ldots$$

If the terms of the AP are $t_1, t_2, t_3, \ldots$, then

$$t_1 = a$$
$$t_2 = a\rho$$
$$t_3 = a\rho^2$$
$$\vdots$$

so $t_n = a\rho^{n-1}$ is the nth term.

Example 6.3.4

Find the common ratio and the fifth and tenth terms of the GP

$$4,\ 3.2,\ 2.56,\ \ldots$$

Solution

First term $\quad a = 4$

Common ratio $\rho = \dfrac{3.2}{4} = \dfrac{2.56}{3.2} = 0.8$

Fifth term $\quad t_5 = a\rho^{5-1} = 4(0.8)^4 = 1.6384$

Tenth term $\quad t_{10} = a\rho^{10-1} = 4(0.8)^9 = 0.5369$ (4 d.p.)

Example 6.3.5

A ball is thrown vertically upwards. It reaches a height of 2.2 m. On falling back to earth it bounces back, reaching three quarters of the original height. This pattern continues on subsequent bounces: each bounce height is three quarters of its predecessor. What height does it reach on its tenth bounce?

Solution

On the first bounce, height reached $= b_1 = 2.2 \times (\frac{3}{4})$
On the second bounce, height reached $= b_2 = b_1 \times (\frac{3}{4}) = 2.2 \times (\frac{3}{4})^2$. The bounce heights clearly form a GP.

The height of the n bounce is $b_n = 2.2 \times (\frac{3}{4})^n$. Therefore on the tenth bounce, height is

$$b_{10} = 2.2 \times (\tfrac{3}{4})^{10} = 0.124 \text{ m (to three decimal places)}$$

Example 6.3.6

A sum of £500 is invested at an interest rate of 7.2% p.a. The interest is compounded annually. What is the total investment at the end of six years? What would the total investment at the end of six years be if the interest was compounded twice a year at a rate of 3.6% per six months?

Solution

(i) Interest compounded annually
Let C_n = the total capital at the end of year n.
So

$$C_1 = 500 \times 107.2\% \equiv 500 \times 1.072$$

$$C_2 = C_1 \times 107.2\% \equiv (500 \times 1.072) \times 1.072$$
$$= 500 \times (1.072)^2$$

So this situation is modelled by a GP with first term 500×1.072 and common ratio 1.072. Hence capital at the end of six years is

$$C_6 = 500 \times (1.072)^6 = £758.82$$

(ii) Interest compounded every six months
Let C_n = the total capital at the end of the nth six month period. 7.2% per annum is equivalent to 3.6% every six months.
So

$$C_1 = 500 \times 103.6\% \equiv 500 \times 1.036$$

$$C_2 = C_1 \times 103.6\% \equiv 500 \times (1.036)^2$$

Note C_2 = capital at end of first year. Again the situation is modelled by a GP.

The capital at the end of six years (= twelve six month periods) is $C_{12} = 500 \times (1.036)^{12} = £764.34$.

This is equivalent to a simple interest rate of 8.81% p.a., compared with 7.33% p.a. if the interest was compounded annually.

Exercise 6.3.1

1. Find the common ratio and seventh term of the GP: 4.5, 5.4, 6.48, ...

2. Find the common ratio and the ninth term of the GP: 80, −64, 51.2, ...
3. A GP has a first term of 9 and a common ratio of 0.85. Which is the first term in this GP which is smaller than 1?
4. The charge of a capacitor halves every second. How long does it take to reach 1% of its original charge? Give your answer to the nearest second.
5. A company accountant depreciates the company's machine tools at a rate of 20% p.a. (compound), so at the end of one year they are only worth 80% of their original cost. What is the value of a £20 000 machine tool at the end of four years?
6. On each stroke a vacuum pump draws out 17.5% of the air in a container present at the beginning of the stroke. What percentage of the original air remains after eight strokes?

6.3.1 Summing the terms of a Geometric Progression

Before considering the general case we take a particular example.

Example 6.3.7

A ball bounces so that the length of each bounce is a half of the previous bounce. The first bounce covers a total distance of 2 m (i.e. 1 m up and 1 m down). What distance does it cover in ten bounces?

Solution

First bounce, $b_1 = 2$
Second bounce, $b_2 = b_1(\frac{1}{2}) = 2(\frac{1}{2})$
The third bounce, $b_3 = b_2(\frac{1}{2}) = 2(\frac{1}{2})^2$

So this situation is modelled by a GP with first term 2 and common ratio $\frac{1}{2}$.

The total distance covered in ten bounces is

$$S_{10} = 2 + 2(\tfrac{1}{2}) + 2(\tfrac{1}{2})^2 + \dots + 2(\tfrac{1}{2})^9$$

This could be found by direct summation but it is much quicker to proceed as follows:

$$S_{10} = 2 + 2(\tfrac{1}{2}) + 2(\tfrac{1}{2})^2 + \dots + 2(\tfrac{1}{2})^9$$

So

$$\tfrac{1}{2}S_{10} = \qquad 2(\tfrac{1}{2}) + 2(\tfrac{1}{2})^2 + \ldots + 2(\tfrac{1}{2})^9 + 2(\tfrac{1}{2})^{10}$$

Subtract

$$\tfrac{1}{2}S_{10} = 2 \qquad\qquad - 2(\tfrac{1}{2})^{10}$$

So

$$S_{10} = 2(2 - 2(\tfrac{1}{2})^{10}) = 4 - (\tfrac{1}{2})^8$$

Therefore

Distance covered is 3.996 m (to three decimal places)

In the general case consider summing n terms of a GP with first term a and common ratio ρ. The nth term is $a\rho^{n-1}$, so the sum of n terms is

$$S_n = a + a\rho + a\rho^2 + \ldots + a\rho^{n-1} \qquad (1)$$

Multiplying (1) by the common ratio ρ

$$\rho S_n = a\rho + a\rho^2 + \ldots + a\rho^{n-1} + a\rho^n \qquad (2)$$

Subtract (2) from (1)

$$(1 - \rho)S_n = a - a\rho^n$$

So that provided $\rho \neq 1$

$$S_n = \frac{a(1 - \rho^n)}{1 - \rho}$$

If $\rho = 1$, the GP is $a, a, a, \ldots$ and $S_n = na$.

The sum of n terms of a GP with first term a and common ratio ρ is given by

$$S_n = \frac{a(1 - \rho^n)}{1 - \rho}$$

provided $\rho \neq 1$

Example 6.3.8

Find the sum of the first 15 terms of the GP: 2, 1.8, 1.62, ...

Solution

This GP has first term $a = 2$ and common ratio $\rho = 1.8/2 = 0.9$. $n = 15$.

The sum of the first fifteen terms is

$$S_{15} = \frac{2(1-0.9^{15})}{1-0.9} = 20(1 - 0.2059)$$

$$= 15.882 \quad \text{(to three decimal places)}$$

Exercise 6.3.2

1. Find the sum of the first ten terms of the GP: 4, 6, 9, ...
2. Find the sum of the first eight terms of the GP: 16, -14, 12.25, ...
3. A GP has first term 18 and common ratio 0.9. Find the sum of terms 10 to 15 inclusive.
4. A GP has a common ratio of 2. The sum of its first six terms is 15.75. Find its first term.
5. The sum of the first three terms of a GP is 26 and the sum of the first six terms is 1690. Find the first term and the common ratio of the GP.

6.4 SUMMING GP'S TO INFINITY AND REPEATING DECIMALS

Everyone will be familiar with the representation of $\frac{1}{3}$ in decimal (fraction) form as $0.\dot{3}$ where $0.\dot{3}$ means 0.333..., that is a decimal point followed by an infinitely long chain of three's. However, if decimal notation was being used we would have to round the representation and obtain an approximation, say 0.333 (three decimal places) or 0.333 333 (six decimal places).

It is possible to form a sequence of decimal approximations, A_1, A_2,... to the fraction $\frac{1}{3}$.

Let

$$A_1 = 0.3 = \frac{3}{10}$$

$$A_2 = 0.33 = \frac{3}{10} + \frac{3}{100}$$

$$A_3 = 0.333 = \frac{3}{10} + \frac{3}{100} + \frac{3}{1000}$$

and so on, giving the general term

$$A_n = \underbrace{0.33 \ldots 3}_{n \text{ three's}} = \frac{3}{10} + \frac{3}{100} + \ldots + \frac{3}{10^n}$$

So each of the approximations $A_1, A_2, A_3, \ldots, A_n, \ldots$ is actually a sum of terms of a sequence. The sequence concerned is:

$$\frac{3}{10}, \frac{3}{100}, \frac{3}{1000}, \ldots, \frac{3}{10^n}, \ldots$$

Now these terms are the terms of a GP with first term 3/10 and common ratio 1/10.

Using the formula for the sum of n terms of a GP (see Section 6.3), the nth approximation is,

$$A_n = \frac{3}{10}\left[1 - \left(\frac{1}{10}\right)^n\right] \Big/ \left[1 - \frac{1}{10}\right]$$

So

$$A_n = \frac{1}{3}\left[1 - \left(\frac{1}{10}\right)^n\right]$$

Now as $n \to \infty, (1/10)^n \to 0$. So as $n \to \infty$, $A_n \to 1/3$, which is what we would expect since $A_1, A_2, \ldots, A_n, \ldots$ are approximations to 1/3.

In fact, given any repeating decimal it is always possible to find its fractional representation, as shown in the next example.

Example 6.4.1

Express $0.\dot{8}\dot{3}$ as a common fraction.

Solution

$$0.\dot{8}\dot{3} = 0.838383\ldots = \frac{83}{100} + \frac{83}{10000} + \frac{83}{1000000} + \ldots$$

Let the nth approximation be

$$A_n = \frac{83}{100} + \frac{83}{(100)^2} + \frac{83}{(100)^3} + \ldots + \frac{83}{(100)^n} \tag{1}$$

So A_n consists of a decimal point followed by a string of n 83's. It is the sum of the terms of a GP with first term 83/100 and common ratio 1/100.

Multiply Equation (1) by the common ratio, 1/100, and subtract this from Equation (1). This gives

$$A_n = \frac{83}{100} + \frac{83}{(100)^2} + \frac{83}{(100)^3} + \ldots + \frac{83}{(100)^n}$$

$$\frac{1}{100} A_n = \frac{83}{(100)^2} + \frac{83}{(100)^3} + \ldots + \frac{83}{(100)^n} + \frac{83}{(100)^{n+1}}$$

Subtract

$$\left(1 - \frac{1}{100}\right) A_n = \frac{83}{100} \qquad - \frac{83}{(100)^{n+1}}$$

So

$$A_n = \frac{\frac{83}{100}\left[1 - \frac{1}{(100)^{n+1}}\right]}{1 - \frac{1}{100}} = \frac{83}{99}\left[1 - \left(\frac{1}{100}\right)^{n+1}\right]$$

As $n \to \infty$

$$\left(\frac{1}{100}\right)^{n+1} \to 0, \text{ so } A_n \to \frac{83}{99}.$$

So

$$0.\dot{8}\dot{3} = \frac{83}{99}$$

Exercise 6.4.1

Use the method of Example 6.4.1 to express the following repeating decimals as common fractions:

1. $0.\dot{5}$ 2. $0.\dot{3}\dot{7}$ 3. $0.\dot{4}1\dot{7} = 0.417417\ldots$

4. $5.\dot{1}\dot{8}$ (Hint. write as $5 + 0.\dot{1}\dot{8}$)
5. $5.1\dot{8}$ (Hint. write as $5.1 + 0.0\dot{8}$)
6. $3.8\dot{9}\dot{2}$

6.4.1 Summing GP's to Infinity – The General Case

From Section 6.3.1 we know that the sum of n terms of a GP with first term a and common ratio ρ is given by

$$S_n = \frac{a(1 - \rho^n)}{1 - \rho}$$

Now if $-1 < \rho < 1$, then as $n \to \infty$, $\rho^n \to 0$ and so $S_n \to a/(1 - \rho)$ which has a definite value.

If $\rho = 1$, then $S_n = \underbrace{a + a + \ldots + a}_{n\ a\text{'s}}$, so as $n \to \infty$, $S_n \to \infty$.

If $\rho = -1$, then $S_n = a - a + a - a + \ldots$, so

$$S_n = \begin{cases} a & \text{if } n \text{ is odd} \\ 0 & \text{if } n \text{ is even} \end{cases}$$

and hence S_n has no definite value.

If $\rho > 1$, then $\rho^n \to \infty$ as $n \to \infty$ and $S_n \to \infty$.

If $\rho < -1$, then ρ^n oscillates between positive and negative numbers whose size is increasing without bound.

Hence only in the case $-1 < \rho < 1$ does the limit of S_n as $n \to \infty$ have a definite value.

Before giving the general result it is useful to consider a particular example.

Example 6.4.2

A tower is built from blocks. The first block is 1 m high, the second 0.5 m high, and thereafter every block is half the height of its predecessor. How tall is the tower when ten blocks have been used? How tall is it when twenty blocks have been used? How tall is it when n blocks have been used? What happens to the height of the tower as n increases?

Solution

The heights of the blocks form a GP: $1, \frac{1}{2}, \frac{1}{4}, \ldots$ The height of the tower is given by the sum of these terms.

Let S_n = height of tower of n blocks.
For ten blocks,

$$S_{10} = \frac{1(1 - (\frac{1}{2})^{10})}{1 - \frac{1}{2}} = 2(1 - (\tfrac{1}{2})^{10})$$

$$= 1.998047 \text{ m} \quad \text{(six decimal places)}$$

For twenty blocks,

$$S_{20} = \frac{1(1 - (\frac{1}{2})^{20})}{1 - \frac{1}{2}} = 2(1 - (\tfrac{1}{2})^{20})$$

$$= 1.999998 \text{ m} \quad \text{(six decimal places)}$$

For n blocks,

$$S_n = \frac{1(1 - (\frac{1}{2})^{n})}{1 - \frac{1}{2}} = 2(1 - (\tfrac{1}{2})^{n})$$

Now as n increases, $(\frac{1}{2})^n \to 0$, and S_n gets closer and closer to 2. Clearly S_n will never actually be equal to 2, but by taking enough blocks we can make the height of the tower as close to 2 as we like. We call the limiting value of S_n as $n \to \infty$, the sum to infinity of the GP. We will write,

$$\text{Sum to infinity } S = \lim_{n\to\infty} S_n = \lim_{n\to\infty} 2(1 - (\tfrac{1}{2})^n) = 2$$

This is another example of a limiting process (see Section 6.1). For the sum to infinity of a GP the general result is

> *The sum to infinity of a geometric progression*
>
> Let a GP have first term a and common ratio ρ. Then provided $-1 < \rho < 1$, the sum to infinity is given by:
>
> $$S = \frac{a}{1 - \rho}$$

Example 6.4.3

Find the sum to infinity of the GP whose first term is 10 and whose common ratio is $-\frac{3}{4}$.

Solution

In this example $a = 10$ and $\rho = -\frac{3}{4}$. Since $-1 < \rho < 1$, the sum to infinity exists and is

$$S = \frac{10}{1 - (-\frac{3}{4})} = \frac{40}{7}$$

Example 6.4.4

A ball is dropped from a window 3 m above ground level. It rebounds to 1 m on its first bounce, and on each bounce after this it reaches one third of its previous height. What distance does it cover before coming to rest?

Solution

Let d_n = distance covered on first bounce (up and down). Then

$$d_1 = 1 + 1 = 2 \text{ m},\ d_2 = 2 \times (\tfrac{1}{3}),\ \ldots,\ d_n = 2 \times (\tfrac{1}{3})^{n-1}.$$

So we require the sum to infinity of a GP with first term $a = 2$ and common ratio $\rho = \frac{1}{3}$. This is

$$S = \frac{a}{1 - \rho} = \frac{2}{1 - \frac{1}{3}} = 3$$

But the ball initially dropped 3 m from the window before starting bouncing. So total distance before coming to rest is 6 m

Note that we have modelled the process by a GP and taken a sum to infinity. Taken literally this would mean that the ball bounced forever and never came to rest. Although this clearly does not happen it seems that the model is a good approximation to reality.

Exercise 6.4.2

For Questions 1–5 find the sum to infinity of the given GP if the sum exists.

1. First term = 30, common ratio = 3/5
2. First term = $\frac{1}{2}$, common ratio = 0.9
3. First term = $\frac{1}{2}$, common ratio = 1.1
4. First term = −5, common ratio = $-\frac{1}{2}$
5. First term = 5, common ratio = −3/2
6. A GP has common ratio 1/5 and sum to infinity 10. What is its first term?

7. A particle P moves along a straight line AB, where AB = 1 m. The particle travels from A to B. It then reverses direction, but only travels 0.7 m along BA before coming to rest. It then reverses direction, travelling 0.49 m towards B. This process continues indefinitely. The distance covered at any stage is 0.7 of the distance at the previous stage. What distance does the particle cover? What can you say about the particle's ultimate position?

8. A particle P moves in the x–y plane. It starts at the origin and moves 10 units along the positive x axis. It then turns left, travelling a further 5 units along a straight line. This process is continued indefinitely, that is: P travels a distance d_n, then turns left, where

   $$d_n = \tfrac{1}{2}d_{n-1} \text{ for } n = 2, 3, \ldots \text{ and } d_1 = 10.$$

   What distance does the particle cover and what are the coordinates of its ultimate position?

9. A pole is driven into the ground by a pile driver. On the first strike, the pole sinks 0.5 m into the ground. On each subsequent strike, it sinks a distance which is 62% of the depth sunk on the previous strike. What is the maximum depth which the pole can reach? What is the depth reached after n strikes? How many strikes are necessary to reach a depth within 1cm of the theoretical maximum depth?

10. A wire is used to make a sequence of circular rings. The radius of the first ring is 10 cm. The radius of each subsequent ring is two-thirds of its predecessor's radius. If this process were continued indefinitely how much wire would be used?

6.5 POLYNOMIAL DIVISION

Multiplication of polynomials has been covered in Section 2.3. In this section division of polynomials is considered.

This process has much in common with long division of integers. Recall in this process we require to compute quotients and remainders.

Example 6.5.1

```
                    153   <—— Quotient
Divisor ——>       7|1073  <—— Dividend
                    7
                   ---
                    373
                    35
                   ---
                     23
                     21
                    ---
                      2   <—— Remainder
```

Therefore $1073 \div 7 = 153$, remainder 2 which is equivalent to saying $1073 = 7 \times 153 + 2$. In the next example the process is examined in more detail.

Example 6.5.2

```
                      37
                  83|3105
(1) ——>             249
                    ---
(2) ——>              615
(3) ——>              581
                     ---
(4) ——>               34
```

Solution

$3105 \div 83 = 37$, remainder 34 which is equivalent to $3105 = 83 \times 37 + 34$.

For the process to work, each digit must be correctly weighted. Division cannot commence until stage (1), which is $310 \div 83 = 3$ remainder 61.

This division is actually $3100 \div 83 = 30$, remainder 610, but the weighting of digits takes care of this. At stage (2), the next digit 5 is brought down, and stage (3) gives $615 \div 83 = 7$, remainder 34. There are no more digits to bring down. 34 is smaller than the divisor 83, so the process stops and 34 is the remainder.

Polynomial division is similar to this process. There will be a quotient and a remainder as a result of carrying out the division.

The first question to be answered is: when is the process halted? In integer division the process terminates when the divisor is larger than the number to be divided (see Examples 6.5.1 and 6.5.2, last stage).

The value of the polynomial depends on the value substituted for its variable, so it is pointless to ask: Is $x + 7$ bigger than $2x - 1$? If $x = 0$, $x + 7 = 7$ and $2x - 1 = -1$, but if $x = 10$, $x + 7 = 17$ and

$2x - 1 = 19$. So the idea of basing the 'size' of a polynomial on its value will not work.

Instead, the degree of the polynomial is used. Recall (from Section 2.7) the degree of a polynomial is the highest power of the variable which appears, so

$$x^2 + 10x - 1 \quad \text{is of degree 2}$$
$$1 - 6t^5 \quad \text{is of degree 5}$$

The process of polynomial division terminates when the degree of the divisor is greater than the degree of the remaining polynomial at that stage.

The layout of polynomial long division is similar to that for integer long division.

The polynomials must be written in descending powers of the variable. The following examples illustrate the process.

Example 6.5.3

Find the quotient and remainder when $6x^3 + 27x^2 + 22x - 10$ is divided by $2x + 5$.

Solution

First arrange each polynomial in order of descending powers of x.

Then proceed as follows:

$$\begin{array}{rll}
 & 3x^2 \longleftarrow & (1) \\
2x + 5 & \overline{)6x^3 + 27x^2 + 22x - 10} \longleftarrow & (2) \\
 & \underline{6x^3 + 15x^2} \longleftarrow & (3) \\
 & 12x^2 + 22x \longleftarrow & (4)
\end{array}$$

The $3x^2$ in line 1 is the first term of the quotient and is obtained by dividing $6x^3$ by $2x$, the term of highest degree in the divisor. The divisor, $2x + 5$, is then multiplied by $3x^2$ and the result is written down at line 3. Line 3 is subtracted from the first two terms of line 2. The result is written down at line 4 together with the next term from line 2.

The process of division is now repeated on line 4. Dividing $12x^2$ by $2x$ we get $6x$.

The whole process is

$$\begin{array}{r l}
 & 3x^2 + 6x - 4 \longleftarrow \text{quotient} \\
2x + 5 & \overline{)6x^3 + 27x^2 + 22x - 10} \\
 & \underline{6x^3 + 15x^2} \\
 & \quad 12x^2 + 22x \\
 & \quad \underline{12x^2 + 30x} \\
 & \qquad -8x - 10 \\
 & \qquad \underline{-8x - 20} \\
 & \qquad\qquad 10 \longleftarrow \text{remainder}
\end{array}$$

The process stops at this stage, since 10 is not divisible by $2x$. Thus when $6x^3 + 27x^2 + 22x - 10$ is divided by $2x + 5$, the quotient is $3x^2 + 6x - 4$ and the remainder is 10.

It is easy to check the answer by multiplication

$$6x^3 + 27x^2 + 22x - 10$$
$$= (2x + 5)\underset{\text{(quotient)}}{(3x^2 + 6x - 4)} + \underset{\text{(remainder)}}{10}$$

Example 6.5.4

Divide $t^4 + 6t^2 - 5t + 4$ by $t^2 - 2t + 3$.

Solution

The first polynomial has no t^3 term. In setting out the long division, a t^3 term is required, so it must be recorded as $0t^3$.

$$\begin{array}{r l}
 & t^2 + 2t + 7 \\
t^2 - 2t + 3 & \overline{)t^4 + 0t^3 + 6t^2 - 5t + 4} \\
 & \underline{t^4 - 2t^3 + 3t^2} \\
 & \quad 2t^3 + 3t^2 - 5t \\
 & \quad \underline{2t^3 - 4t^2 + 6t} \\
 & \qquad 7t^2 - 11t + 4 \\
 & \qquad \underline{7t^2 - 14t + 21} \\
 & \qquad\qquad 3t - 17
\end{array}$$

The process stops here since the degree of $(3t - 17)$ is 1 which is less than the degree of $(t^2 - 2t + 3)$ which is 2. So $(t^4 + 6t^2 - 5t + 4) \div (t^2 - 2t + 3) = t^2 + 2t + 7$, remainder $3t - 17$.

Again this may be checked by multiplication

$$t^4 + 6t^2 - 5t + 4 = (t^2 - 2t + 3)\underset{\text{(quotient)}}{(t^2 + 2t + 7)} + \underset{\text{(remainder)}}{(3t - 17)}$$

Although polynomial division will never involve fractional or negative powers of the variable there is no reason why the coefficients cannot be fractions, as shown in Example 6.5.5.

Example 6.5.5

Divide $1 - y^2$ by $2y + 1$.

Solution

$$
\begin{array}{r|l}
 & -\tfrac{1}{2}y + \tfrac{1}{4} \\
\hline
2y + 1 & -y^2 + 0y + 1 \\
 & \underline{-y^2 - \tfrac{1}{2}y} \\
 & \qquad\quad \tfrac{1}{2}y + 1 \\
 & \qquad\quad \underline{\tfrac{1}{2}y + 4} \\
 & \qquad\qquad\quad \tfrac{3}{4}
\end{array}
$$

Answer: Quotient $-\frac{1}{2}y + \frac{1}{4}$, remainder $\frac{3}{4}$.

Exercise 6.5.1

Find the quotient and remainder by long division in the following questions. Check your answers by multiplying out: dividend = divisor × quotient + remainder.

1. $(x^3 - 2x^2 + 3) \div (x - 1)$
2. $(15 - 7x^2 + 6x^3 - 14x) \div (3x - 5)$
3. $(12x^3 - 22x^2 + 16x - 7) \div (2x - 3)$
4. $(6x^3 + x^2 - 10x + 3) \div (2x^2 + 3x - 6)$
5. $(2 - t^3) \div (t - 1)$
6. $(v^2 - 2v + 3) \div (2v - 1)$

6.6 THE REMAINDER AND FACTOR THEOREMS

Long division of polynomials is a time-consuming process.

Consider what happens when we divide a polynomial $p(x)$ by a polynomial $(x - a)$ of degree one. We will obtain a quotient $q(x)$ and a remainder r. r will be an integer, since we know that degree r is less than degree $(x - a)$ which is 1.

We can write

$$p(x) = (x - a)q(x) + r \qquad (1)$$

(e.g. $x^3 - x^2 + 1 = (x - 2)(x^2 + x + 2) + 5$)

Equation (1) is true for all values of x, in particular when $x = a$. So we get

$$p(a) = (a - a)q(a) + r = r$$

This leads to

The remainder theorem

If a polynomial $p(x)$ is divided by $x - a$, then the remainder $r = p(a)$

Example 6.6.1

Find the remainder when $p(x) = x^{10} - 7x^3 + 5$ is divided by $x - 1$.

Solution

By the remainder theorem

$$r = p(1) = 1^{10} - 7 \times 1^3 + 5 = -1$$

Example 6.6.2

Find the remainder when $3x^4 + x^3 - 4x + 5$ is divided by $x + 2$.

Solution

In this case $x - a \equiv x + 2$ so $a = -2$, and $p(x) \equiv 3x^4 + x^3 - 4x + 5$, so by the remainder theorem,

$$\begin{aligned} r = p(-2) &= 3(-2)^4 + (-2)^3 - 4(-2) + 5 \\ &= 53 \end{aligned}$$

6.6.1 The Factor Theorem

7 is a factor of 56, because if 56 is divided by 7, the remainder is zero.

In a similar way $x - 3$ is a factor of $x^2 - x - 6$ since if $x^2 - x - 6$ is divided by $x - 3$, the remainder is zero. This fact is easy to check using the remainder theorem. When $p(x) = x^2 - x - 6$ is divided by $x - 3$ the remainder is

$$p(3) = 3^2 - 3 - 6 = 0$$

These ideas lead to

> *The factor theorem*
> $(x{-}a)$ is a factor of a polynomial $p(x)$ if and only if $p(a) = 0$

Example 6.6.3

Is $x - 2$ a factor of $x^4 - 16$?

Solution

$x - a \equiv x - 2$ so $a = 2$, and $p(x) \equiv x^4 - 16$. Therefore

$$p(2) = 2^4 - 16 = 0,$$

so $x - 2$ is a factor of $x^4 - 16$ by the factor theorem.

Example 6.6.4

Is $x + 3$ a factor of $2x^3 + 5x^2 + x + 6$?

Solution

$x - a \equiv x + 3$ so $a = -3$, and $p(x) \equiv 2x^3 + 5x^2 + x + 6$.

Therefore

$$p(-3) = 2(-3)^3 + 5(-3)^2 + (-3) + 6 = -6 \neq 0.$$

So $x + 3$ is not a factor of $2x^3 + 5x^2 + x + 6$.

6.6.2 Using the Factor Theorem to Factorize Polynomials

The factor theorem's main application is in the factorization of polynomials. The method involves writing down all the possible linear factors of the given polynomial, and then testing each using the factor theorem until a factor is found. The following examples illustrate the method.

Example 6.6.5

Factorize $x^3 + 4x^2 + x - 6$.

Solution

Possible linear factors are required. To find these, the possible factors of the highest power of x and the possible factors of the constant are required.

Possible factors of x^3: x only

Possible factors of -6: ± 1, ± 2, ± 3, ± 6

Therefore the possible linear factors are

$(x + 1)$, $(x - 1)$, $(x + 2)$, $(x - 2)$,
$(x + 3)$, $(x - 3)$, $(x + 6)$, $(x - 6)$.

$$p(x) = x^3 + 4x^2 + x - 6$$

Is $x + 1$ a factor? $p(-1) = (-1)^3 + 4(-1)^2 + (-1) - 6 = -4 \neq 0$. So $x + 1$ is not a factor.

Is $x - 1$ a factor? $p(1) = (1)^3 + 4(1) + (1) - 6 = 0$. So $x - 1$ is a factor.

The other factors could be tested by the factor theorem, but it is probably just as quick to divide $p(x)$ by $(x - 1)$ using long division

$$\begin{aligned} p(x) &= x^3 + 4x^2 + x - 6 \\ &= (x - 1)(x^2 + 5x + 6) \end{aligned}$$

So

$$p(x) = (x - 1)(x + 2)(x + 3) \quad \text{by inspection}$$

Example 6.6.6

Factorize $x^4 + 4x^3 - 5x^2 - 36x - 36$.

Solution

The possible linear factors are

$(x \pm 1)$, $(x \pm 2)$, $(x \pm 3)$, $(x \pm 4)$, $(x \pm 6)$,
$(x \pm 9)$, $(x \pm 12)$, $(x \pm 18)$, $(x \pm 36)$

$$p(x) = x^4 + 4x^3 - 5x^2 - 36x - 36$$

Is $(x + 1)$ a factor? $p(-1) = -8 \neq 0$. So $(x + 1)$ is not a factor.
Is $(x - 1)$ a factor? $p(1) = -72 \neq 0$. So $(x - 1)$ is not a factor.
Is $(x + 2)$ a factor? $p(-2) = 0$. So $(x + 2)$ is a factor.
Divide $p(x)$ by $(x + 2)$

$$p(x) = x^4 + 4x^3 - 5x^2 - 36x - 36 = (x + 2)(x^3 + 2x^2 - 9x - 18)$$
$$= (x + 2)q(x)$$

where $q(x) = x^3 + 2x^2 - 9x - 18$.

Now factorize $q(x)$. Its possible factors are

$$(x \pm 1),\ (x \pm 2),\ (x \pm 3),\ (x \pm 6),\ (x \pm 9).$$

By the factor theorem, it is found that $(x + 1)$ and $(x - 1)$ are not factors. However,

$$q(-2) = (-2)^3 + 2(-2)^2 - 9(-2) - 18 = 0$$

so $(x + 2)$ is a factor. Dividing $q(x)$ by $(x + 2)$

$$q(x) = (x + 2)(x^2 - 9)$$
$$= (x + 2)(x + 3)(x - 3) \quad \text{by inspection.}$$

So

$$p(x) = (x + 2)^2(x + 3)(x - 3)$$

Exercise 6.6.1

For Problems 1 to 5 use the remainder theorem to find the remainder when the given division is carried out.

1. $(x^3 - x^2 + 3x + 2) \div (x - 1)$
2. $(2x^3 + x^2 - 7x + 2) \div (x + 1)$
3. $(x^3 - 8) \div (x - 2)$
4. $(3t^3 - 2t^2 - 5t + 3) \div (t - 1)$
5. $(4y^3 + 5y^2 - 3y + 6) \div (y + 3)$

For Problems 6 to 8 use the factor theorem to find whether $f(x)$ is a factor of $p(x)$.

6. $p(x) = x^3 - 3x^2 - 4x + 12$, $f(x) = x + 2$
7. $p(x) = 3x^3 - 4x^2 - 5x + 2$, $f(x) = x + 1$
8. $p(x) = 2x^4 - x^3 + 3x^2 - 5x + 1$, $f(x) = x - 2$
9. Is $t - 3$ a factor of $2t^3 - 5t^2 - 5t + 6$?
10. Is $y + 1$ a factor of $y^4 + 2y^2 + 1$?
11. Factorize $x^3 - 3x^2 - 10x + 24$
12. Factorize $t^3 + 9t^2 + 20t + 12$

6.6.3 Generalization of the Remainder and Factor Theorems

In the examples considered so far, the factors tested have all been of the form $x - a$. Both theorems can be extended for testing more general factors of the form $ax - b$.

> *Generalization of the remainder theorem*
>
> When the polynomial $p(x)$ is divided by $(ax - b)$ the remainder
>
> $$r = p\left(\frac{b}{a}\right)$$
>
> *Generalization of the factor theorem*
>
> $(ax - b)$ is a factor of the polynomial $p(x)$ if $p\left(\frac{b}{a}\right) = 0$

Example 6.6.7

Find the remainder when $x^3 - 2x^2 + x - 3$ is divided by $2x - 1$.

Solution

Let $p(x) \equiv x^3 - 2x^2 + x - 3$ and $ax - b \equiv 2x - 1$ so $a = 2, b = 1$.
By the remainder theorem

$$r = p\left(\frac{b}{a}\right) = p\left(\frac{1}{2}\right)$$

$$= \left(\frac{1}{2}\right)^3 - 2\left(\frac{1}{2}\right)^2 + \frac{1}{2} - 3$$

$$= \frac{1}{8} - \frac{1}{2} + \frac{1}{2} - 3 = -2\frac{7}{8}$$

Example 6.6.8

Is $4x + 1$ a factor of $4x^3 - 3x^2 + 3x + 1$?

Solution

Let $p(x) \equiv 4x^3 - 3x^2 + 3x + 1$.

$$ax - b \equiv 4x + 1, \quad \text{so } a = 4, \quad b = -1$$

$$p\left(\frac{b}{a}\right) = p\left(-\frac{1}{4}\right) = 4\left(-\frac{1}{4}\right)^3 - 3\left(-\frac{1}{4}\right)^2 + 3\left(-\frac{1}{4}\right) + 1 = 0$$

So $4x + 1$ is a factor of $4x^3 - 3x^2 + 3x + 1$.

Exercise 6.6.2

1. Find the remainder when $x^3 + x + 2$ is divided by $3x - 1$.
2. Find the remainder when $2x^3 - 3x^2 - 4x + 1$ is divided by $2x + 1$.
3. Find the remainder when $1 - t + t^2 - t^3$ is divided by $2t - 3$.
4. Is $2x - 1$ a factor of $3x^3 - 2x^2 - 5x + 3$?
5. Is $3x + 1$ a factor of $6x^3 - x^2 - 10x - 3$?
6. Is $3t - 2$ a factor of $3t^3 - 5t^2 + 8t - 4$?

For Problems 7 to 10 factorize the expressions:

7. $4x^3 - 4x^2 - 7x - 2$
8. $2x^3 - 5x^2 - 4x + 3$
9. $3t^4 + 7t^3 - 24t^2 - 12t + 16$
10. $6y^4 - 19y^3 + 13y^2 + 4y - 4$

6.7 IRREDUCIBLE QUADRATIC POLYNOMIALS

Consider the polynomial $p(x) \equiv x^2 + 1$. Its only possible linear factors are $(x + 1)$ and $(x - 1)$,
but

$$p(-1) = (-1)^2 + 1 = 2 \neq 0$$

and

$$p(1) = (1)^2 + 1 = 2 \neq 0$$

so neither $(x + 1)$ nor $(x - 1)$ are factors.

Hence $x^2 + 1$ cannot be factorized using factors with real number coefficients. Such a polynomial is called an *irreducible quadratic polynomial* (over the real numbers).

In practice the easiest way to see if a quadratic polynomial is irreducible is to use its discriminant (see Section 2.6).

Example 6.7.1

Factorize $x^3 + 3x^2 + 4x + 4$.

Solution

Let

$$p(x) \equiv x^3 + 3x^2 + 4x + 4$$

Possible factors are $(x \pm 1)$, $(x \pm 2)$, $(x \pm 4)$. Using the factor theorem $(x + 1)$ and $(x - 1)$ are not factors, but

$$p(-2) = (-2)^3 + 3(-2)^2 + 4(-2) + 4 = 0$$

so $(x + 2)$ is a factor.
Dividing out

$$p(x) \equiv (x + 2)(x^2 + x + 2)$$

Now $x^2 + x + 2$ has a discriminant $(b^2 - 4ac)$ equal to

$$1^2 - 4(1)(2) = -7 < 0$$

so it is irreducible.
So $x^3 + 3x^2 + 4x + 4 = (x + 2)(x^2 + x + 2)$
and this is the complete factorization.

Exercise 6.7.1

Factorize each polynomial as far as possible.

1. $x^2 + 9$
2. $x^2 - 9$
3. $x^3 - 1$
4. $x^3 + 1$
5. $x^3 - 3x^2 + 2x - 6$
6. $t^4 + t^3 - 3t^2 - t + 2$
7. $2y^3 + 7y^2 + 6y + 9$
8. $3z^3 + 5z^2 + 7z - 3$

6.8 PARTIAL FRACTIONS

Until now problems which involve algebraic fractions have usually been concerned with combining two or more such fractions into one single fraction, for example

$$\frac{4}{2t+1} + \frac{3}{t-1} = \frac{4(t-1)+3(2t+1)}{(2t+1)(t-1)} = \frac{10t-1}{(2t+1)(t-1)}$$

With partial fractions we wish to reverse this process, that is split

$$\frac{10t-1}{(2t+1)(t-1)}$$

into the sum of two fractions.

Before going into the details, it is useful to recall some terminology.

The top of a fraction is called its *numerator* and the bottom of a fraction is called its *denominator*. Thus,

$$\frac{3x+1}{x^2-2} \quad \begin{array}{l} \longleftarrow \text{Numerator} \\ \longleftarrow \text{Denominator} \end{array}$$

An arithmetic fraction is called a *proper fraction* if its numerator is smaller than its denominator. So 1/2, 3/7 and 11/12 are proper fractions. If the numerator is bigger than or equal to the denominator the fraction is *improper*, for example 10/9, 12/3 or 4/4 are improper fractions.

Algebraic fractions are classified as proper or improper according to the following rule based on the degree of the numerator and denominator.

> *Rule*
>
> An algebraic fraction is proper if degree (numerator) $<$ degree (denominator).
>
> An algebraic fraction is improper if degree (numerator) $\geqslant$ degree (denominator).

Example 6.8.1

$$\frac{3-t^2}{t^3+1}$$

is a proper fraction, since degree (numerator) = 2 and degree (denominator) = 3.

$$\frac{3 - t^2}{t^2 + 1}$$

is an improper fraction, since degree (numerator) = 2 and degree (denominator) = 2.

$$\frac{3 - t^2}{t + 1}$$

is an improper fraction, since degree (numerator) = 2 and degree (denominator) = 1.

Any improper arithmetic fraction can be converted into an integer plus a proper arithmetic fraction, for example

$$\frac{9}{7} = 1\frac{2}{7} \quad \text{and} \quad \frac{44}{9} = 4\frac{8}{9}.$$

In a similar way, any improper algebraic fraction may be converted into a polynomial plus a proper algebraic fraction. This can be done by using long division.

Example 6.8.2

Convert

$$\frac{2 + t - 3t^2}{t + 2}$$

into a polynomial plus a proper algebraic fraction.

Solution

$$\begin{array}{r l l}
 & \;\; -3t + 7 & \longleftarrow \text{quotient} \\
t + 2 & \overline{)\,-3t^2 + t + 2} & \\
 & \;\; \underline{-3t^2 - 6t} & \\
 & \qquad\quad 7t + 2 & \\
 & \qquad\quad \underline{7t + 14} & \\
 & \qquad\qquad - 12 & \longleftarrow \text{remainder}
\end{array}$$

So

$$\frac{2 + t - 3t^2}{t + 2} = -3t + 7 - \frac{12}{t + 2} \quad \begin{array}{l} \longleftarrow \text{remainder} \\ \longleftarrow \text{denominator} \end{array}$$

quotient

Example 6.8.3

Convert

$$\frac{2x^4 + 3x^3 - 4x - 5}{x^2 + 2x - 3}$$

into a polynomial plus a proper algebraic fraction.

Solution

$$\begin{array}{r l}
 & \underline{2x^2 - x + 8} \longleftarrow \text{quotient} \\
x^2 + 2x - 3 \, \big) & 2x^4 + 3x^3 + 0x^2 - 4x - 5 \\
 & \underline{2x^4 + 4x^3 - 6x^2} \\
 & \quad - x^3 + 6x^2 - 4x - 5 \\
 & \quad \underline{- x^3 - 2x^2 + 3x} \\
 & \qquad\qquad 8x^2 - 7x - 5 \\
 & \qquad\qquad \underline{8x^2 + 16x - 24} \\
 & \qquad\qquad\quad - 23x + 19 \longleftarrow \text{remainder}
\end{array}$$

So

$$\frac{2x^4 + 3x^3 - 4x - 5}{x^2 + 2x - 3} = 2x^2 - x + 8 + \frac{19 - 23x}{x^2 + 2x - 3}$$

Exercise 6.8.1

In Problems 1–6 state whether the fraction is proper or improper.

1. $\dfrac{x^3 + 1}{1 - x^3}$

2. $\dfrac{y^2 + 2y + 11}{7y^2 - 2y + 1}$

3. $\dfrac{7y^2 - 2y + 1}{10y + 13}$

4. $\dfrac{2t^3 + 3t^2 - 4t - 3}{(t^2 + 1)(t^2 - 3)}$

5. $\dfrac{2 - z + z^2}{3z^3 + 1}$

6. $\dfrac{(x + 1)(x^2 + x + 1)}{(3x - 2)(x + 7)}$

In Problems 7–10 express the improper fraction as the sum of a polynomial and a proper fraction.

7. $\dfrac{4x - 2}{x - 1}$

8. $\dfrac{3t^2 + 6t + 6}{t^2 + 3t + 1}$

9. $\dfrac{s^2 + s + 1}{s - 1}$ 10. $\dfrac{-y(2y^2 - 5y + 1)}{1 + 2y - y^2}$

6.8.1 Finding Partial Fractions

Rule 1
Only proper algebraic fractions may be split into partial fractions. If an improper fraction is given, this must first be reduced to a polynomial plus a proper algebraic fraction.

Rule 2
Given a proper algebraic fraction it may be split into partial fractions by identifying it as belonging to one of three basic types (see below).

Type 1 Denominator Contains Unrepeated Linear Factors

Example 6.8.4

Split

$$\frac{7v - 8}{(v + 1)(v - 2)}$$

into partial fractions.

Solution

This is a proper fraction. Its denominator consists of two linear factors. Let

$$\frac{7v - 8}{(v + 1)(v - 2)} \equiv \frac{A}{v + 1} + \frac{B}{v - 2}$$

where A and B are constants whose values must be found. If the two fractions on the right hand side are combined, we have

$$\frac{7v - 8}{(v + 1)(v - 2)} \equiv \frac{A(v - 2) + B(v + 1)}{(v + 1)(v - 2)}$$

$$\equiv \frac{(A + B)v + (B - 2A)}{(v + 1)(v - 2)}$$

Equating the numerators

$$7v - 8 \equiv (A + B)v + (B - 2A)$$

The coefficients of v^1 and v^0 (constants) must be the same on each side so

$$A + \mathrm{B} = 7$$
$$-2A + \mathrm{B} = -8$$

Solving these gives $A = 5$ and $B = 2$, so

$$\frac{7v - 8}{(v + 1)(v - 2)} \equiv \frac{5}{v + 1} + \frac{2}{v - 2}$$

The numerators of the partial fractions may also be found by substitution as the following example shows.

Example 6.8.5

Split

$$\frac{5t^2 - 7t - 7}{(t - 1)(t + 2)(2t + 1)}$$

into partial fractions.

Solution

This is a proper fraction since degree (numerator) = 2 and degree (denominator) = 3.
Let

$$\frac{5t^2 - 7t - 7}{(t - 1)(t + 2)(2t + 1)} \equiv \frac{A}{t - 1} + \frac{B}{t + 2} + \frac{C}{2t + 1}$$

where A, B and C are constants whose values are to be found.
Combining the fractions on the right hand side

$$\frac{5t^2 - 7t - 7}{(t - 1)(t + 2)(2t + 1)} \equiv$$

$$\frac{A(t + 2)(2t + 1) + B(t - 1)(2t + 1) + C(t - 1)(t + 2)}{(t - 1)(t + 2)(2t + 1)}$$

Equating numerators

$$5t^2 - 7t - 7 \equiv$$

$$A(t + 2)(2t + 1) + B(t - 1)(2t + 1) + C(t - 1)(t + 2) \qquad (*)$$

At this stage, do not multiply the brackets out on the right hand side. Instead substitute the values of t that make the brackets zero.

Put $t = 1$ in (*): $-9 = A(3)(3) + 0 + 0$

$\therefore A = -1$

Put $t = -2$ in (*): $27 = 0 + B(-3)(-3) + 0$

$\therefore B = 3$

Put $t = -\frac{1}{2}$ in (*): $-\dfrac{9}{4} = 0 + 0 + C\left(-\dfrac{3}{2}\right)\left(\dfrac{3}{2}\right)$

$\therefore C = 1$

So

$$\frac{5t^2 - 7t - 7}{(t - 1)(t + 2)(2t + 1)} \equiv \frac{-1}{t - 1} + \frac{3}{t + 2} + \frac{1}{2t + 1}$$

This method of finding the numerators is usually faster than the method of equating coefficients.

Exercise 6.8.2

Split the following into partial fractions:

1. $\dfrac{2x}{(x + 1)(x - 1)}$

2. $\dfrac{5 - 4t}{(1 - 2t)(1 + t)}$

3. $\dfrac{t + 11}{t^2 + t - 2}$ (factorize the denominator first)

4. $\dfrac{11 - 2x - x^2}{(x^2 - 1)(x - 2)}$ (factorize the denominator first)

5. $\dfrac{1}{(x + 1)(x - 1)}$

Type 2 Denominator Contains Repeated Linear Factors

Example 6.8.6

Split

$$\frac{5v - 1}{(v + 1)^2(v - 2)}$$

into partial fractions.

Solution

The fraction is a proper fraction. One factor, $v + 1$, is squared, and is termed a repeated factor. It appears twice in the partial fraction denominators; once squared and once to power one.

Let

$$\frac{5v - 1}{(v + 1)^2(v - 2)} \equiv \frac{A}{(v + 1)^2} + \frac{B}{v + 1} + \frac{C}{v - 2}$$

repeated factor is squared – so it appears twice

Equating denominators

$$5v - 1 \equiv A(v - 2) + B(v + 1)(v - 2) + C(v + 1)^2 \qquad (*)$$

Do not multiply the right hand side out.

Put $v = -1$ in (*): $-6 = A(-3) + 0 + 0$ giving $A = 2$

Put $v = 2$ in (*): $9 = 0 + 0 + C(3)^2$ Giving $C = 1$

It is not possible to find B by making the other two terms zero, so we choose any convenient value for v.

Put $v = 0$ in (*): $-1 = A(-2) + B(1)(-2) + C(1)^2$:

But $A = 2$ and $C = 1$, so $2B = 1 - 4 + 1$ hence $B = -1$

So

$$\frac{5v - 1}{(v + 1)^2(v - 2)} \equiv \frac{2}{(v + 1)^2} - \frac{1}{v + 1} + \frac{1}{v - 2}$$

Example 6.8.7

Split

$$\frac{x^3 - 7x^2 + 11x + 3}{(x - 1)^3(x + 1)}$$

into partial fractions.

Solution

Degree numerator = 3; degree denominator = 3 + 1 = 4. So this is a proper fraction. The denominator contains a cubed factor, so let

$$\frac{x^3 - 7x^2 + 11x + 3}{(x-1)^3(x+1)} \equiv \frac{A}{(x-1)^3} + \frac{B}{(x-1)^2} + \frac{C}{(x-1)} + \frac{D}{x+1}$$

repeated factor is cubed – so it appears three times

So

$$\frac{x^3 - 7x^2 + 11x + 3}{(x-1)^3(x+1)} \equiv$$

$$\frac{A(x+1) + B(x-1)(x+1) + C(x-1)^2(x+1) + D(x-1)^3}{(x-1)^3(x+1)}$$

Equate numerators,

$$x^3 - 7x^2 + 11x + 3 \equiv$$
$$A(x+1) + B(x-1)(x+1) + C(x-1)^2(x+1) + D(x-1)^3$$

Do not multiply out the right hand side. Choose values for x which make factors zero.

Put $x = 1$ in (*): $8 = A(2) + 0 + 0 + 0 \quad \therefore A = 4.$
Put $x = -1$ in (*): $-16 = 0 + 0 + 0 + D(-2)^3 \therefore D = 2.$

There are no other values which make factors zero, so choose convenient values for x. Use the values for A and D.

Put $x = 0$ in (*): $3 = 4 + B(-1)(1) + C(-1)^2(1) - 2.$
Put $x = 2$ in (*): $5 = 12 + B(1)(3) + C(1)^2(3) + 2.$

These give $-B + C = 1$, and $B + C = -3$ which may be solved to give $B = -2$ and $C = -1$. So

$$\frac{x^3 - 7x^2 + 11x + 3}{(x-1)^3(x+1)} \equiv \frac{4}{(x-1)^3} - \frac{2}{(x-1)^2} - \frac{1}{x-1} + \frac{2}{x+1}$$

Exercise 6.8.3

Split the following into partial fractions.

1. $\dfrac{2t^2 + 2t + 1}{(t+2)^2(t-3)}$

2. $\dfrac{2x^2 + 3x + 2}{(2x+1)^2(2x+3)}$

3. $\dfrac{3x^2 + 3x - 5}{x^3 - x^2}$ (factorize the denominator)

4. $\dfrac{3t^3 + 2t^2 + t + 1}{t^3(t + 1)}$

5. $\dfrac{3 - 7x + 7x^2 - 2x^3}{x^2(x - 1)^2}$

Type 3 Denominator Contains Irreducible Quadratic Factors

An *irreducible quadratic factor* is a factor which will not factorize over the real numbers.

If such a factor is present, its numerator must take the form $Av + B$ or $Ax + B$ or $At + B$, etc. This is unlike the linear or repeated linear factors whose numerators are just constants.

Example 6.8.8

Split

$$\frac{v^2 - v + 1}{(v^2 + 1)(v - 2)}$$

into partial fractions.

Solution

The expression is a proper fraction since degree (numerator) = 2 and degree (denominator) = 3. The factor $v^2 + 1$ is an irreducible quadratic factor.

Let

$$\frac{v^2 - v + 1}{(v^2 + 1)(v - 2)} \equiv \frac{Av + B}{v^2 + 1} + \frac{C}{v - 2}$$

irreducible quadratic

So

$$\frac{v^2 - v + 1}{(v^2 + 1)(v - 2)} \equiv \frac{(Av + B)(v - 2) + C(v^2 + 1)}{(v^2 + 1)(v - 2)}$$

Equating numerators

$$v^2 - v + 1 \equiv (Av + B)(v - 2) + C(v^2 + 1) \qquad (*)$$

Put $v = 2$ in (*): $3 = 0 + C(2^2 + 1)$ giving $C = 3/5$.
Put $v = 0$ in (*): $1 = B(-2) + C(1)$ giving $-2B = 1 - C$, thus $B = -1/5$
Put $v = 1$ in (*): $1 = (A + B)(1 - 2) + C(1^2 + 1)$
$1 = -A - B + 2C$

So

$$A = 2C - B - 1 = \frac{6}{5} + \frac{1}{5} - 1 = \frac{2}{5}$$

So

$$\frac{v^2 - v + 1}{(v^2 + 1)(v - 2)} \equiv \frac{2v - 1}{5(v^2 + 1)} + \frac{3}{5(v - 2)}$$

Exercise 6.8.4

Split the following into partial fractions:

1. $\dfrac{5x^2 + x + 8}{(x^2 + 2)(x + 1)}$

2. $\dfrac{t^2 - t + 9}{t(t^2 + 3)}$

3. $\dfrac{2s^3 + 3s^2 + 4s + 7}{(s + 1)(s + 2)(s^2 + 1)}$

4. $\dfrac{3v^2 - 3v + 1}{v^3 + v}$ (factorize the denominator)

5. $\dfrac{t^3 - t^2 + t - 4}{t^2(t^2 + 1)}$

Improper Partial Fractions

With improper partial fractions, the improper fraction must first be reduced to a polynomial plus a proper fraction. The proper fraction is then split into partial fractions, taking care to identify its type carefully.

Example 6.8.9

Split

$$\frac{x^3 - x - 1}{(x - 1)(x + 1)}$$

into partial fractions.

Solution

Since degree (numerator) = 3 and degree (denominator) = 2 this is an improper fraction. Divide out, noting that denominator = $x^2 - 1$.

$$\begin{array}{r|l} & x \\ \hline x^2 - 1 & x^3 + 0x^2 - x - 1 \\ & \underline{x^3 + 0x^2 - x} \\ & \qquad\qquad\quad -1 \end{array}$$

So

$$\frac{x^3 - x - 1}{(x - 1)(x + 1)} = x - \frac{1}{(x - 1)(x + 1)}$$

The proper fraction

$$\frac{-1}{(x - 1)(x + 1)}$$

may then be split into partial fractions. Let

$$\frac{-1}{(x - 1)(x + 1)} \equiv \frac{A}{x - 1} + \frac{B}{x + 1}$$

So

$-1 \equiv A(x + 1) + B(x - 1).$
Put $x = 1$ in (*): $-1 = 2A + 0$ giving $A = -\frac{1}{2}$.
Put $x = -1$ in (*): $-1 = 0 + B(-2)$ giving $B = \frac{1}{2}$.
So

$$\frac{x^3 - x - 1}{(x - 1)(x + 1)} \equiv x - \frac{1}{2(x - 1)} + \frac{1}{2(x + 1)}$$

Exercise 6.8.5

Split the following into a polynomial plus partial fractions:

1. $\dfrac{2v^3 + v^2 + v + 3}{v(v + 1)}$

2. $\dfrac{2t^2 + 5t}{(t + 1)^2}$

3. $\dfrac{x^4 + 4x^2 + x + 6}{x^3 + 2x}$ (factorize the denominator)

4. $\dfrac{5x^2 - 14x + 10}{x^2 - 3x + 2}$ (factorize the denominator)

6.8.2 Partial Fractions – Summary

Given a fraction to be put into partial fractions obey the following steps:

1. If the fraction is improper, divide out to obtain a polynomial plus a proper fraction.
2. Identify the correct form for the partial fraction from the following three types.

   Type 1
   Denominator has linear factors only, e.g.

$$\frac{px + q}{(x + a)(x + b)} \equiv \frac{A}{x + a} + \frac{B}{x + b}$$

   Type 2
   Denominator has a repeated factor, e.g.

$$\frac{px + q}{(x + a)^2(x + b)} \equiv \frac{A}{(x + a)^2} + \frac{B}{x + a} + \frac{C}{x + b}$$

   Type 3
   Denominator has an irreducible quadratic factor, e.g.

$$\frac{px + q}{(x^2 + a)(x + c)} \equiv \frac{Ax + B}{x^2 + a} + \frac{C}{x + c}$$

3. Find the values of the constants A, B, C..., by equating numerators.

Exercise 6.8.6

Put the following into partial fractions. Examples of all three types and improper fractions are included. It may be necessary to factorize denominators.

1. $\dfrac{1}{x^2 - 9}$
2. $\dfrac{1}{v^2 + 4v + 3}$
3. $\dfrac{4(x + 1)}{x^2 - 4}$
4. $\dfrac{3v - 1}{(v + 3)(2v + 1)}$
5. $\dfrac{x^2 + 3x + 3}{3x(x^2 + 3x + 2)}$
6. $\dfrac{t^2 + 2}{t^2 - t}$
7. $\dfrac{t^2 + 2}{t^3 + t}$
8. $\dfrac{-2x(2x^2 + 2x + 1)}{(x^2 + 1)(x + 1)^2}$

6.9 EXPANDING $(a + b)^n$ WHERE n IS A POSITIVE INTEGER

Two techniques are available for this task. Pascal's triangle is useful for small values of n. The other method involves using the binomial theorem.

6.9.1 Pascal's triangle and the expansion $(a + b)^n$

If we want to calculate $(a + b)^2$, $(a + b)^3$ and $(a + b)^4$ we could proceed as follows:

$$
\begin{aligned}
(a + b)^2 &= (a + b)(a + b) = a^2 + 2ab + b^2 \\
(a + b)^3 &= (a + b)(a + b)^2 \\
&= (a + b)(a^2 + 2ab + b^2) \text{ using the last result} \\
&= a^3 + 2a^2b + ab^2 + a^2b + 2ab^2 + b^3 \\
&= a^3 + 3a^2b + 3ab^2 + b^3 \\
(a + b)^4 &= (a + b)(a + b)^3 \\
&= (a + b)(a^3 + 3a^2b + 3ab^2 + b^3) \text{ using the last result} \\
&= a^4 + 4a^3b + 6a^2b^2 + 4ab^3 + b^4
\end{aligned}
$$

Now $(a + b)^5$ and $(a + b)^6$ could be calculated by the same method, but it would be very tedious to do so. Instead, examine the pattern which is emerging in the expansions.

Each term in the expansion consists of a product of an integer and a power of a and b. The integers are called *coefficients*. For example,

Powers of a and b add up to 3

$$(a + b)^3 = 1a^3b^0 + 3a^2b^1 + 3a^1b^2 + 1a^0b^3$$

Coefficients

Here a^3 has been written as $1a^3b^0$ to make the pattern clearer.

So the powers of a and b in any term are easy to predict. In $(a + b)^n$, each term will contain a product of a power of a and a power of b. These powers will add up to n.

The coefficients are more difficult to determine, but may be obtained from *Pascal's Triangle* (Figure 6.1).

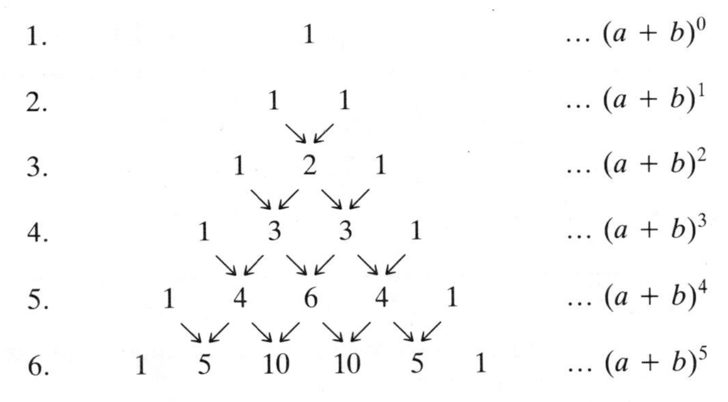

Figure 6.1 Pascal's triangle

Pascal's triangle can be extended indefinitely. It is bordered by 1s and each interior number is the sum of the two closest numbers in the line above (indicated by arrows).

The third row: '1 2 1' gives the coefficients for

$$(a + b)^2 = 1a^2b^0 + 2a^1b^1 + 1a^0b^2 = a^2 + 2ab + b^2$$

The sixth row: '1 5 10 10 5 1' gives

$$\begin{aligned}(a + b)^5 &= 1a^5b^0 + 5a^4b^1 + 10a^3b^2 + 10a^2b^3 + 5a^1b^4 + 1a^0b^5 \\ &= a^5 + 5a^4b + 10a^3b^2 + 10a^2b^3 + 5ab^4 + b^5\end{aligned}$$

Check that the powers of a and b in each term add up to 5.

Pascal's triangle can be used to compute the expansions of more difficult expansions as the next two examples show.

Example 6.9.1

Use Pascal's triangle to expand $(x + 2y)^4$.

Solution

The coefficients come from row five of Pascal's triangle.

Substitute $a = x$ and $b = 2y$. Note carefully with this substitution the use of brackets and that since $b = 2y$, the 2 as well as y is raised to the given power.

$$\begin{aligned}(x + 2y)^4 &= 1x^4 + 4x^3(2y) + 6x^2(2y)^2 + 4x(2y)^3 + 1(2y)^4 \\ &= x^4 + 8x^3y + 24x^2y^2 + 32xy^3 + 16y^4\end{aligned}$$

Example 6.9.2

Use Pascal's triangle to expand $(2p - q)^3$.

Solution

$(2p - q)^3 = (2p + (-q))^3$, so let $a = 2p$ and $b = -q$. The coefficients come from row four of Pascal's triangle. Again note the use of brackets.

$$\begin{aligned}(2p - q)^3 &= 1(2p)^3 + 3(2p)^2(-q) + 3(2p)(-q)^2 + 1(-q)^3 \\ &= 8p^3 - 12p^2q + 6pq^2 - q^3\end{aligned}$$

Exercise 6.9.1

Write out the first eight rows of Pascal's triangle and use this to expand the following expressions:

1. $(x - y)^6$
2. $(a + b)^7$
3. $(3p + q)^5$
4. $(3x - 2y)^4$
5. $(1 - x^2)^5$
6. $(3v - \frac{1}{2})^4$
7. $(1 - 1/x)^3$
8. $(t + 1/t)^6$

6.9.2 Use of the Binomial Theorem in Expanding $(a + b)^n$ where n is a non-negative Integer

Pascal's triangle becomes cumbersome for large values of n, so it is preferable to use the binomial theorem. Before stating the theorem it is necessary to consider factorials and combinations.

Factorials

The notation $n!$ (n factorial) is used to represent the product $n(n-1)(n-2)\ldots3.2.1$. We define $0! = 1$.

Most calculators have a factorial key. Make sure you can use it.

Sometimes we want to leave factorials in product form, as the next example shows.

Example 6.9.3

Without using a calculator evaluate $7!/5!$ and $12!/8!$ leaving your answer in product form.

Solution

Noting that $7! = 7.6.5.4.3.2.1 = 7.6.5!$
then

$$\frac{7!}{5!} = \frac{7.6.5!}{5!} = 7.6$$

Similarly

$$\frac{12!}{8!} = \frac{12.11.10.9.8!}{8!} = 12.11.10.9$$

Exercise 6.9.2

Use your calculator to evaluate the factorials in Problems 1–5.

1. $4!$ 2. $0!$ 3. $10!$ 4. $\frac{7!}{3!}$ 5. $\frac{10!}{5!}$

Without using a calculator, evaluate the factorials in Problems 6–10 leaving your answer in product form.

6. $\frac{9!}{8!}$ 7. $\frac{11!}{9!}$ 8. $\frac{11!}{9!2!}$ 9. $\frac{n!}{(n-1)!}$

10. $\frac{n!}{(n-3)!}$

Combinations

The notation ${}^nC_r$ or ${}_nC_r$ or $\binom{n}{r}$ is used to represent $n!/r!(n-r)!$. It is sometimes read as 'n choose r'. n and r are both non negative integers and $0 \leqslant r \leqslant n$.

Some calculators have a key dedicated to this function; on others it is necessary to use the factoral key for each of the three factorials.

Example 6.9.4

Evaluate ${}^3C_2$, ${}^6C_3$ and ${}^{10}C_4$.

Solution

$${}^3C_2 = \frac{3!}{2!1!} = 3$$

$${}^6C_3 = \frac{6!}{3!3!} = 20$$

$${}^{10}C_4 = \frac{10!}{4!6!} = 210$$

Example 6.9.5

Evaluate ${}^5C_r$ for $r = 0, 1, \ldots, 5$ and confirm that these are the coefficients which appear in row six of Pascal's triangle (Figure 6.1).

Solution

$${}^5C_0 = \frac{5!}{0!5!} = \frac{1}{0!} = 1$$

$${}^5C_1 = \frac{5!}{1!4!} = \frac{5}{1} = 5$$

$${}^5C_2 = \frac{5!}{2!3!} = \frac{5.4}{2!} = 10$$

$${}^5C_3 = \frac{5!}{3!2!} = \frac{5.4}{2!} = 10$$

$${}^5C_4 = \frac{5!}{4!1!} = \frac{5}{1} = 5$$

$$^5C_5 = \frac{5!}{5!0!} = \frac{1}{0!} = 1$$

Referring to Figure 6.1 confirms that these numbers form row six of Pascal's triangle.

Exercise 6.9.3

Use your calculator to evaluate the expressions in Problems 1–6.

1. $^6C_2$
2. $^8C_3$
3. $_{10}C_5$
4. $_{12}C_{11}$
5. $\binom{3}{1}$
6. $\binom{7}{2}$
7. Prove that $^nC_r = {}^nC_{n-r}$ for any r such that $0 \leqslant r \leqslant n$. Why does this mean that Pascal's triangle is symmetrical?
8. Prove that $^nC_1 = n$ for any value of n.
9. Prove $^nC_{r-1} + {}^nC_r = {}^{n+1}C_r$. How does this relate to Pascal's triangle?

6.9.3 The Binomial Theorem for $(a + b)^n$ where n is a non-negative Integer

If n is a non-negative integer and a and b are any real numbers, then

$$(a + b)^n = a^n + {}^nC_1a^{n-1}b + {}^nC_2a^{n-2}b^2 + \dots + {}^nC_ra^{n-r}b^r + \dots + b^n$$

The coefficient pattern 1, $^nC_1$, $^nC_2$,...,1 is simply row $(n + 1)$ of Pascal's triangle.

Note that the expansion for $(a + b)^n$ has $(n + 1)$ terms. The powers of a start of n and descend to 0. The powers of b start at 0 and ascend to n. The sum of the powers of a and b in each term is n. The coefficient of the term $a^{n-r}b^r$ is ^nC_r.

Example 6.9.6

Find the first three terms in the expansion of $(2x - y)^{10}$ in ascending powers of y.

Solution

$(2x - y)^{10} \equiv [2x + (-y)]^{10}$, so take $a = 2x$ and $b = (-y)$ in the binomial theorem. The first three terms are:

$$(2x)^{10} + {}^{10}C_1(2x)^9(-y) + {}^{10}C_2(2x)^8(-y)^2$$
$$= 1024x^{10} + 10 \times 512x^9(-y) + 45 \times 256x^8(-y)^2$$
$$= 1024x^{10} - 5120x^9y + 11520x^8y^2$$

Example 6.9.7

Find the coefficient of t^4 in the expansion of $(2 + 3t)^8$.

Solution

Using the binomial theorem with $a = 2$ and $b = 3t$, the general term is

$${}^8C_r(2)^{8-r}(3t)^r$$

The required term has $r = 4$, and is

$${}^8C_4(2)^{8-4}(3t)^4 = 70 \times 16 \times 81t^4$$

So the coefficient of t^4 is 90 720.

Exercise 6.9.4

1. Use the binomial theorem to expand $(u + 2v)^3$.
2. Use the binomial theorem to expand $(x - \frac{1}{2})^4$.
3. Use the binomial theorem to expand $(2t^2 + 1/t)^5$.

In Problems 4–9 use the binomial theorem to find

4. the third term in the expansion of $(2s - t)^{10}$
5. the term containing x^6 in the expansion of $(x + y)^{12}$
6. the coefficient of u^3 in the expansion of $(3u - 2v)^8$
7. the coefficient of q^2 in the expansion of $(\frac{1}{2}p + q)^7$
8. the coefficient of t^2 in the expansion of $(t + 1/(2t))^6$
9. the coefficient of x^3 in the expansion of $(3 - x)^2 (1 + 2x)^5$

6.10 THE BINOMIAL THEOREM FOR THE EXPANSION OF $(1 + x)^n$

The binomial theorem and Pascal's triangle in Section 6.9 restricted expansions to non-negative integer powers. However there is another version of the binomial theorem which allows negative and fractional powers, but it is only valid for a limited range of values of the variable x.

$$(1 + x)^n = 1 + nx + \frac{n(n - 1)}{2!}x^2 + \frac{n(n - 1)(n - 2)}{3!}x^3 + \ldots$$

$$+ \frac{n(n - 1)\ldots(n - r + 1)}{r!}x^r + \ldots \text{ provided } |x| < 1$$

In this version, there is no restriction on the power n, but x may only take values between -1 and $+1$, exclusive. Also note that the expansion has an infinite number of terms in it. There is a pattern to each term:

1. The coefficient of x^r has a denominator $r!$.
2. The coefficient of x^r has a numerator which consists of a product of r numbers, starting at n and descending by 1 at each step.

Example 6.10.1

Write down the first four terms in the expansion of $\sqrt{1 + x}$ and $1/(1 + x)$.

Solution

$$\sqrt{1 + x} = (1 + x)^{1/2}, \text{ so } n = 1/2 \text{ in the binomial theorem}$$

$$\sqrt{1 + x} = 1 + \tfrac{1}{2}x + \frac{\frac{1}{2}(-\frac{1}{2})}{2!}x^2 + \frac{\frac{1}{2}(-\frac{1}{2})(-\frac{3}{2})}{3!}x^3 + \ldots$$

$$= 1 + \frac{1}{2}x - \frac{1}{8}x^3 + \frac{1}{16}x^4 \ldots \text{ provided } |x| < 1$$

$$\frac{1}{1 + x} = (1 + x)^{-1}, \text{ so } n = -1 \text{ in the binomial theorem}$$

$$\frac{1}{1 + x} = 1 + (-1)x + \frac{(-1)(-2)}{2!}x^2 + \frac{(-1)(-2)(-3)}{3!}x^3 + \ldots$$

$$= 1 - x + x^2 - x^3 + \ldots \text{ provided } |x| < 1$$

In this version of the binomial theorem, the expansion must be of the form $(1 + \text{term})^n$, i.e. 1 must be present. It is possible to deal with more complex expressions as shown in the following examples.

Example 6.10.2

Write down the first four terms in the expansion of $(1 - 2y)^{-1/2}$ and state the range of values of y for which the expansion is valid.

Solution

$(1 - 2y)^{-1/2} \equiv [1 + (-2y)]^{-1/2}$ so take $n = -\frac{1}{2}$ and $x = -2y$ in the binomial theorem.

$$(1 - 2y)^{-1/2} = 1 + (-\tfrac{1}{2})(-2y) + \frac{(-\frac{1}{2})(-\frac{3}{2})(-2y)^2}{2!}$$

$$+ \frac{(-\frac{1}{2})(-\frac{3}{2})(-\frac{5}{2})(-2y)^3}{3!} + \ldots$$

$$= 1 + y + \frac{3}{2}y^2 + \frac{5}{2}y^3 + \ldots$$

$(1 + x)^n$ is valid for $|x| < 1$ so $(1 - 2y)$ is valid for $|-2y| < 1$, that is $|y| < 1/2$.

Example 6.10.3

Write down the first four terms in the expansion of $(2 + t)^{-3}$ and state its range of validity.

Solution

The binomial theorem applies to the expansion of expressions such as $(1 + x)^n$.

$$(2 + t)^{-3} \equiv \left[2\left(1 + \frac{t}{2}\right)\right]^{-3} \equiv 2^{-3}\left(1 + \frac{t}{2}\right)^{-3} \equiv \frac{1}{8}\left(1 + \frac{t}{2}\right)^{-3}$$

Use the binomial theorem on $(1 + t/2)^{-3}$, i.e. take $x = t/2$ and $n = -3$.

$$(2 + t)^{-3} \equiv \frac{1}{8}\left(1 + \frac{t}{2}\right)^{-3}$$

$$= \frac{1}{8}\left[1 + (-3)\left(\frac{t}{2}\right) + \frac{(-3)(-4)}{2!}\left(\frac{t}{2}\right)^2 + \frac{(-3)(-4)(-5)}{3!}\left(\frac{t}{2}\right)^3 + \ldots\right]$$

$$= \frac{1}{8} - \frac{3t}{16} + \frac{3t^2}{16} - \frac{5t^3}{32} + \ldots$$

The expansion is valid for $|t/2| < 1$, that is $|t| < 2$.

Example 6.10.4

Given that $|x| < 1/3$, expand $(1 + 3x)^{1/3}$ as far as the term in x^3. Hence find $^3\sqrt{1.03}$ correct to five decimal places.

Solution

Using the binomial theorem

$$(1 + 3x)^{1/3} = 1 + \tfrac{1}{3}(3x) + \frac{\frac{1}{3}(-\frac{2}{3})(3x)^2}{2!} + \frac{\frac{1}{3}(-\frac{2}{3})(-\frac{5}{3})(3x)^3}{3!} + \ldots$$

$$= 1 + x - x^2 + \frac{5}{3}x^3 + \ldots$$

Now this expansion is valid for $|3x| < 1$, that is $|x| < 1/3$ as given. Put $x = 0.01$

$$(1 + 0.03)^{1/3} = {}^3\sqrt{1.03}$$

$$= 1 + (0.01) - (0.01)^2 + \frac{5}{3}(0.01)^3$$

$$= 1.00990$$

Exercise 6.10.1

In Problems 1–4, use the binomial theorem to write down the first four terms in the given expansion. State when the expansion is valid.

1. $(1 - 2x)^{-2}$
2. $(2 + y)^{1/5}$
3. $(3 - 2t)^{2/5}$
4. $(x - y)^{1/2}$

5. Write down the first three terms in the expansion of $(100 + x)^{1/2}$. Hence find $\sqrt{103}$ correct to the three decimal places.
6. By writing $\sqrt[3]{61}$ as $\sqrt[3]{(64 - 3)}$ and using a suitable binomial expansion, find $\sqrt[3]{61}$ correct to three decimal places.
7. Expand $(1 + x)/(2 + x)$ as far as the term in x^3.

OBJECTIVE TEST 6

1. Give a suitable formula for the nth term of the sequence with first four terms

$$t_1 = -3, \quad t_2 = \frac{3}{2}, \quad t_3 = -\frac{3}{4}, \quad t_4 = \frac{3}{8}$$

2. What is the limit of the sequence whose nth term is

$$t_n = \frac{5^n + 4^n}{5^n}\ ?$$

3. Two hundred oil drums are stacked on their sides. There are twenty drums in the bottom row, nineteen drums in the next row, eighteen in the row after that, and so on. How many rows are there in the stack and how many drums are on the top row?

4. A sheet of paper 0.12 mm thick is folded in two. This is then folded in two again. The process is repeated a further two times. How thick is the folded paper?

5. What is the remainder when $t^5 + 4t - 1$ is divided by $2t + 1$?

6. Factorize $2x^3 + x^2 - 13x + 6$.

7. Put

$$\frac{3x^2 + 3x + 2}{2x^3 + 3x^2 + 4x + 6}$$

into partial fractions.

8. Put

$$\frac{4x^3 + 2x^2 - x + 1}{2x^2 + x}$$

into partial fractions.

9. Expand $(2x - \frac{1}{2}y)^4$.

10. Find the first three terms in the expansion of $(8 - t)^{1/3}$.

7 DIFFERENTIATION

OBJECTIVES

When you have completed this chapter you should be able to

1. distinguish between average and instantaneous rates of change
2. find the derivative of a function by a limiting process
3. use rules to differentiate polynomial, power, trigonometric, exponential and logarithmic functions
4. use rules to differentiate composite functions, products of functions and quotients of functions
5. find and identify the nature of stationary points
6. use differentiation as a tool in curve sketching

7.1 AVERAGE RATES OF CHANGE VERSUS INSTANTANEOUS RATES OF CHANGE

Speed is the most familiar rate of change. Speed measures the rate of change of distance with respect to time. We may speak of the average speed for a journey and the speed at any instant.

For example, consider a typical train journey between Newcastle and Carlisle. The train from Newcastle takes 1 hour 37 minutes to cover its $61\frac{3}{4}$ mile journey.

$$\text{Average speed} = \frac{\text{distance covered}}{\text{time taken}}$$

$$= \frac{61.75}{1.62} = 38.2 \text{ mph}$$

Clearly the train does not cover its journey at a constant speed of 38.2 mph. It has seven stops to make en route and must brake and accelerate to take account of these. The train's speed would be changing all the time, and this would be obvious from observing the driver's speedometer. The speedometer would indicate how fast the train is travelling at any particular instant, i.e. the instantaneous speed.

As a second example, consider Table 7.1 which shows the results of a test on a car. The car accelerated from rest along a straight test track. Its distance, s metres, from the starting point was recorded each second. Each second the speedometer was read to give the instantaneous speed v in ms^{-1}. The results are recorded against time in seconds.

Table 7.1

Time, t(s)	0.0	1.0	2.0	3.0	4.0	5.0	6.0	7.0	8.0
Distance, s(m)	0.0	0.7	2.8	6.6	12.3	20.4	31.7	46.8	66.9
Instantaneous Speed, v (ms^{-1})	0.0	1.4	2.9	4.7	6.8	9.6	13.0	17.4	23.0

From the information given it is not possible to say what the instantaneous speed was at a time not recorded in the table. However if the speed v is plotted against time t on a graph, the data points from Table 7.1 could be joined up by a curve which would give a good approximation for the instantaneous speed v at any time, say $t =$ 2.3 seconds. Figure 7.1 shows this graph, and this shows that when $t = 2.3$s, the instantaneous speed is 3.4 ms^{-1}

Alternatively, from the distance data in Table 7.1 it is possible to compute average speeds.

The formula for calculating average speed is

$$v_{av} = \frac{\Delta s}{\Delta t}$$

where Δs = change in distance travelled in the time interval Δt.

So for the whole eight-second period,

Distance travelled, $\Delta s = 66.9 - 0.0 = 66.9$ m

Change in time, $\Delta t = 8.0 - 0.0 = 8.0$ s

Hence

$$\text{Average speed } v_{av} = \frac{\Delta s}{\Delta t} = \frac{66.9}{8.0} = 8.4 \text{ ms}^{-1}.$$

Alternatively, the journey could be divided into two four second intervals, and the average speed in each half period could be calculated.

For the first 4s period, $\Delta s = 12.3 - 0.0 = 12.3$ m and $\Delta t = 4.0 - 0.0 = 4.0$ s. So $v_{av} = \Delta s/\Delta t = 12.3/4 = 3.1$ ms^{-1}

For the second 4s period, $\Delta s = 66.9 - 12.3 = 54.6$ m and $\Delta t = 8.0 - 4.0 = 4.0$ s. So $v_{av} = \Delta s/\Delta t = 54.6/4.0 = 13.7$ ms^{-1}

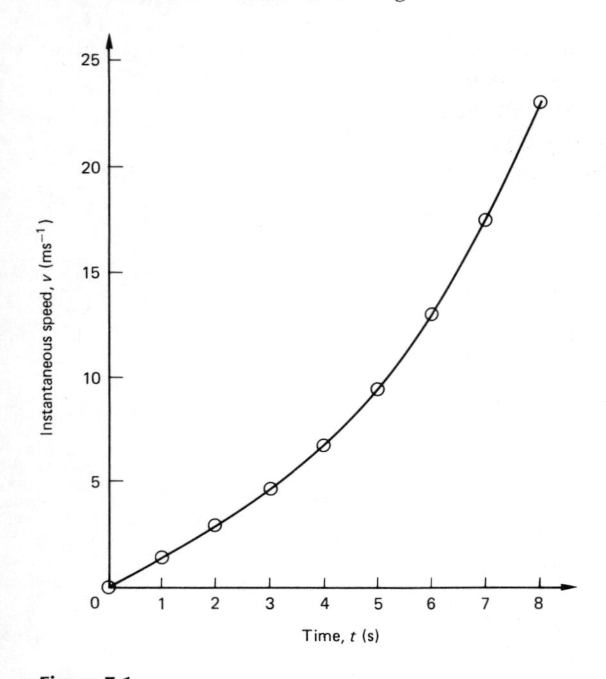

Figure 7.1

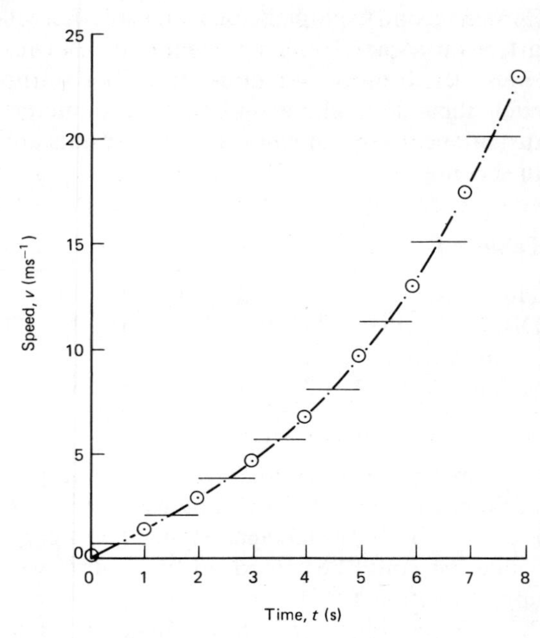

Figure 7.2 ⊙ instantaneous speed; — · — · — interpolation curve for instantaneous speed; ——— average speed over 1 s period.

Exercise 7.1.1

Table 7.2 shows the results of computing average speeds over successive one second periods. For example, between $t = 3.0$ s and $t = 4.0$ s, distance travelled $\Delta s = 12.3 - 6.6 = 5.7$ m and change in time $\Delta t = 4.0 - 3.0 = 1.0$ s. So average speed $v_{av} = \Delta s / \Delta t = 5.7$ ms^{-1}.
Copy and complete Table 7.2.

Table 7.2

Time, t(s)	0.0	1.0	2.0	3.0	4.0	5.0	6.0	7.0	8.0
Distance, travelled, s(m)	0.0	0.7	2.8	6.6	12.3	20.4	31.7	46.8	66.9
Average speed, v_{av} (ms^{-1})	*	0.7			5.7				20.1 *

Note: Average speeds are recorded over time intervals of one second. Hence the offset in the columns of the last row of the table. For example, 0.7 ms^{-1} is the average speed over the interval 0.0 s to 1.0 s.

7.1.1 Comparison of Average and Instantaneous Speeds

The instantaneous speeds from Table 7.1 are plotted on a speed time graph (Figure 7.2). The data points have been joined by a smooth curve, since it is assumed that the speed is changing smoothly. Also plotted in Figure 7.2 are the eight average speeds over successive one-second periods from the complete Table 7.2. Note that the average speeds do not change smoothly, but give a stepped pattern. Also the instantaneous speed curve cuts each average speed 'step'. This is because the car must have an instantaneous speed equal to the average speed at some time.

7.1.2 A Note on the Terms 'Speed' and 'Velocity'

In this section we have used the terms 'distance', 'speed' and 'time' to illustrate rates of change. In kinematics (the area of applied mathematics which deals with speed and acceleration), it is the displacement and velocity which are of interest. They are vector quantities with magnitude and direction. The importance and properties of vectors are discussed in Chapter 11.

Displacement of an object is its distance from a fixed point measured in a given direction. For example town A may be three miles due north of town B. If we take our direction as being 'due north' then the displacement of A relative to B is +3 miles. However the displacement of B relative to A is −3 miles, the negative sign indicating 'due south'.

The velocity of an object is its speed in a given direction. Velocities may be positive or negative, indicating the direction of travel. Speeds are never negative since they only indicate how fast an object is moving and take no account of the direction.

Acceleration is also a vector quantity. It is the rate of change of velocity.

If displacement is measured in metres (m) then velocity is measured in metres per second (ms^{-1}) and acceleration is measured in metres per second per second (ms^{-2}).

For the rest of this chapter we shall use displacement to describe the position of an object and velocity to describe the rate of change of displacement. Do not worry about the subtle differences at this stage.

Exercise 7.1.2

1. Suppose that the distances and speeds of Table 7.1 had been recorded every half second, and a speed time graph was plotted. Average speeds over half-second intervals could be computed.

   Would the 'steps' of the average speed graph follow the instantaneous speed interpolation curve more accurately than those in Figure 7.2? Give reasons for your answer.
2. The radioactive element radon decays naturally to the element polonium. So over time, the mass of radon will

decrease. Table 7.3 gives the results of recording the mass of radon (m grams) over a two-minute period. From this table, the average rate of decay over successive ten-second intervals may be computed. For example, the interval from $t = 20.0$ to $t = 30.0$, the average rate of increase of mass is:

$$\frac{\text{Change in mass}}{\text{Change in time}} = \frac{\Delta m}{\Delta t} = \frac{6.861\text{–}7.779}{30.0\text{–}20.0} = \frac{-0.918}{10.0}$$

$$\text{Average decay rate} = 0.092 \text{ gs}^{-1}.$$

Note: This is negative since mass is decreasing.

Compute the average rates in gs^{-1} over each ten second interval. Plot them on a graph of decay rate against time.

Does this graph provide enough information to sketch in a curve which models the instantaneous decay rate?

Table 7.3 Radioactive decay of radon

Time t(s)	0.0	10.0	20.0	30.0	40.0	50.0	60.0
Mass, m(g)	10.000	8.820	7.779	6.861	6.051	5.337	4.707
Time t(s)	70.0	80.0	90.0	100.0	110.0	120.0	
Mass, m(g)	4.151	3.661	3.229	2.848	2.512	2.215	

7.2 AVERAGE RATES OF CHANGE

The ideas of Section 7.1 which were used to calculate average speeds and decay rates, can be extended to calculate the average rate of change for other quantities.

Let Q be any quantity which varies with time. The average rate of change of Q between times $t = t_1$ and $t = t_2$ is defined as follows:

Average rate of change of a quantity Q

Average rate of change of Q over the time interval from $t = t_1$ to $t = t_2$ is

$$\frac{\Delta Q}{\Delta t}$$

where ΔQ = change in $Q = Q_{t_2} - Q_{t_1}$

and Δt = change in $t = t_2 - t_1$

and Q_{t_1} = value of Q at time t_1

and Q_{t_2} = value of Q at time t_2

Example 7.2.1

The temperature T of a metal changes from 1500 K to 1410 K over a five-minute period. Calculate the average rate of change of T.

Solution

The change in temperature, $\Delta T = 1410 - 1500 = -90$ K
and the change in time, $\Delta t = 5$ minutes.
So the average rate of change of temperature

$$= \frac{\Delta T}{\Delta t} = \frac{-90}{5} = -18 \text{ Kmin}^{-1}$$

Example 7.2.2

Table 7.4 records the velocity of a car in metres per second over 2 minutes. Acceleration is the rate of change of velocity.

Calculate the average accelerations over the four thirty-second intervals.

Table 7.4

Time, t(s)	0	30	60	90	120
Velocity, v(ms^{-1})	7	31	32	15	6

Solution

$$\text{Average acceleration} = \frac{\Delta v}{\Delta t} \text{ ms}^{-2}$$

The results are given in Table 7.5.

Table 7.5

Time, t(s)	0		30		60		90		120
Change in velocity, Δv		24		1		-17		-9	
Change in time, Δt		30		30		30		30	
Average acceleration		0.80		0.03		-0.57		-0.30	

7.3 INSTANTANEOUS RATES OF CHANGE

If we have a quantity Q which varies with time, we may calculate the average rate of change $\Delta Q/\Delta t$ over smaller and smaller time intervals Δt. As Δt becomes smaller the average rate of change will approximate the instantaneous rate of change with an increasing degree of accuracy.

For example, suppose that the velocity v (in ms^{-1}) of a body at time $t(s)$ is given by the formula:

$$v = t^3$$

The average acceleration between times $t = t_1$ and $t = t_2$ seconds is given by

$$\text{Average acceleration} = \frac{\Delta v}{\Delta t} = \frac{v_2 - v_1}{t_2 - t_1}$$

where v_1 is the velocity at time $t = t_1$ and v_2 is the velocity at time $t = t_2$.

To find the instantaneous acceleration at time $t = 1.5$ s, calculate $\Delta v/\Delta t$ for a sequence of decreasing time intervals centred on $t = 1.5$ s.

The first time interval could run from $t_1 = 1.25$ s to $t_2 = 1.75$ s, giving a time difference of $\Delta t = 0.5$ s.

In this case

$$v_1 = (t_1)^3 = (1.25)^3 = 1.953125 \text{ ms}^{-1}$$

and

$$v_2 = (t_2)^3 = (1.75)^3 = 5.359375 \text{ ms}^{-1}$$

Hence

$$\Delta v = v_2 - v_1 = 3.406250 \quad \text{(six decimal places)}$$

The second time interval could run from $t_1 = 1.45$ s to $t = 1.55$ s, giving $\Delta t = 0.10$.

Further shorter time intervals can be taken. The results are tabulated (correct to six decimal places) in Table 7.6.

From this table it seems that the average accelerations $\Delta v/\Delta t$ have settled down to 6.750 000 ms^{-2}.

We say that the average accelerations are approaching a limiting value of 6.75 ms^{-2}. From this we may conclude that, at time $t = 1.5$ seconds, the instantaneous acceleration is 6.75 ms^{-2}.

Table 7.6

Δt	t_1	t_2	$v_1 = t_1^3$	$v_2 = t_2^3$	Δv	$\Delta v/\Delta t$
0.50	1.25	1.75	1.953 125	5.359375	3.406 25	6.812 5
0.10	1.45	1.55	3.048 625	3.723875	0.675 25	6.752 5
0.01	1.495	1.505	3.341 362	3.408623	0.067 500	6.750 025
0.001	1.499 5	1.500 5	3.371 626	3.378376	0.006 750	6.750 000
0.0001	1.499 95	1.500 05	3.374 663	3.375338	0.000 675	6.750 000

An alternative approach to finding the instantaneous acceleration at time $t = 1.5$ s is as follows.

Fix t_1 at 1.5 s. Take $t_2 = t_1 + \Delta t$ for decreasing values of Δt. The results are shown in Table 7.7. The average accelerations approach a constant value of 6.75 ms^{-2}, which agrees with the result of Table 7.6.

Although the processes used in Tables 7.6 and 7.7 are different, they lead to the same final result. Note that in each case we have a sequence of decreasing time intervals that eventually 'home-in' on $t = 1.5$ s.

Table 7.7

Δt	t_1	$t_2 = t_1 + \Delta t$	$v_1 = t_1^3$	$v_2 = t_2^3$	Δv	$\Delta v/\Delta t$
0.50	1.5	2.0	3.375	8.000 000	4.625 000	9.250 000
0.10	1.5	1.6	3.375	4.096 000	0.721 000	7.210 000
0.01	1.5	1.51	3.375	3.442 951	0.067 951	6.795 000
0.001	1.5	1.501	3.375	3.381 755	0.006 755	6.754 501
0.0001	1.5	1.500 1	3.375	3.375 675	0.000 675	6.750 450

Exercise 7.3.1

1. A particle moves so that its displacement s (in metres) from a fixed point P at time t (seconds) is given by

$$s = t^2 + t$$

Estimate the instantaneous velocity of the particle at time $t = 3$s by drawing up a table similar to Table 7.6. Take the first time interval from $t_1 = 2.75$ s to $t_2 = 3.25$ s, giving a value of $\Delta t = 0.5$ s. Consider other time intervals, centred on $t = 3$ s, and show that as the interval width decreases, the average velocities approach a constant value.

2. An electric current i is given by

$$i = 1 - t^2$$

where t is the time and $0 \leqslant t \leqslant 1$. Estimate the instantaneous rate of change of the current at time $t = 0.2$ s by drawing up a table similar to Table 7.7. Fix $t_1 = 0.2$ and let $t_2 = t_1 + \Delta t$ where $\Delta t = 0.1, 0.01, 0.001, \ldots$

7.3.1 An Algebraic Approach to the Problem

The numerical processes used in Tables 7.6 and 7.7 to estimate the instantaneous acceleration may be subject to numerical errors as Δt gets smaller. So instead of giving Δt values, an algebraic approach will be used.

Fix $t_1 = 1.5$ (as in Table 7.7) then $t_2 = 1.5 + \Delta t$. But $v = t^3$ so when $t = t_1$, $v_1 = 1.5^3$, and when

$$t = t_2,\ v_2 = (1.5 + \Delta t)^3$$
$$= 1.5^3 + 3(1.5)^2\Delta t + 3(1.5)(\Delta t)^2 + (\Delta t)^3$$

Hence

$$\Delta v = v_2 - v_1 = 3(1.5)^2\Delta t + 3(1.5)(\Delta t)^2 + (\Delta t)^3$$

So the average acceleration is

$$\frac{\Delta v}{\Delta t} = \frac{3(1.5)^2\Delta t + 3(1.5)(\Delta t)^2 + (\Delta t)^3}{\Delta t}$$

Simplifying

$$\frac{\Delta v}{\Delta t} = 3(1.5)^2 + 3(1.5)\Delta t + (\Delta t)^2$$

To find the instantaneous acceleration at $t = 1.5$, let Δt approach zero. As Δt approaches zero, $\Delta v/\Delta t$ approaches $3(1.5)^2 = 6.75$ which is the result obtained numerically in Tables 7.6 and 7.7.

As $\Delta t \to 0$

$$\frac{\Delta v}{\Delta t} \to 6.75$$

This is an example of a limiting process, and it may be used to obtain instantaneous rates of change. The following definition is used.

> *Instantaneous rate of change of a quantity Q with respect to time t*
>
> Let Q be a function of time. Then the instantaneous rate of change of Q with respect to time t is
>
> $$\frac{\mathrm{d}Q}{\mathrm{d}t} = \lim_{\Delta t \to 0} \left(\frac{\Delta Q}{\Delta t} \right)$$

The expression

$$\lim_{\Delta t \to 0} \left(\frac{\Delta Q}{\Delta t} \right)$$

means

1. Work out the average rate of change of Q in a time span Δt. This is $\Delta Q/\Delta t$.
2. Simplify $\Delta Q/\Delta t$.
3. Let Δt approach zero.

To return to the example above where $v = t^3$. The quantity Q is velocity v. Let the velocity at time t be $v = t^3$. Let the velocity at time $t + \Delta t$ be $v + \Delta v = (t + \Delta t)^3$. So the change in the velocity $\Delta v = (t + \Delta t)^3 - t^3$.

But

$$(t + \Delta t)^3 = t^3 + 3t^2\Delta t + 3t(\Delta t)^2 + (\Delta t)^3$$

so

$$\Delta v = 3t^2\Delta t + 3t(\Delta t)^2 + (\Delta t)^3$$

Hence

$$\frac{\Delta v}{\Delta t} = 3t^2 + 3t\Delta t + (\Delta t)^2$$

$$\frac{\mathrm{d}v}{\mathrm{d}t} = \lim_{\Delta t \to 0} \left(\frac{\Delta v}{\Delta t} \right) = \lim_{\Delta t \to 0} (3t^2 + 3t\Delta t + (\Delta t)^2)$$

$$= 3t^2$$

This result shows the power of the algebraic/symbolic approach over the numerical approach of Tables 7.6 and 7.7, because it gives the instantaneous acceleration at any time t, not just when $t = 1.5$ s, i.e. if velocity $v = t^3$, then acceleration $a = 3t^2$.

Example 7.3.1

If displacement $s = t^4$, show that the velocity $v = 4t^3$.

Solution

At time $t + \Delta t$, displacement $s + \Delta s$ is given by

$$(t + \Delta t)^4 = t^4 + 4t^3\Delta t + 6t^2(\Delta t)^2 + 4t(\Delta t)^3 + (\Delta t)^4$$

so

$$\Delta s = (s + \Delta s) - s$$

$$\Delta s = 4t^3\Delta t + 6t^2(\Delta t)^2 + 4t(\Delta t)^3 + (\Delta t)^4$$

$$\frac{\Delta s}{\Delta t} = 4t^3 + 6t^2\Delta t + 4t(\Delta t)^2 + (\Delta t)^3$$

So the instantaneous velocity

$$v = \lim_{\Delta t \to 0}\left(\frac{\Delta s}{\Delta t}\right) = 4t^3$$

Exercise 7.3.2

1. Write a program which will enable tables such as 7.6 and 7.7 to be constructed. The program must be able to:

   (a) Accept $f(t)$, which is a function of t. (If the language you are using does not allow you to input $f(t)$, define it within your program.)
   (b) Ask the user to input a value for t_1, the time at which the rate of change is to be calculated.
   (c) Calculate the average rate of change of $f(t)$ for $\Delta t = 0.1, 0.01, 0.001, 0.0001$ and print these results out.

   Check your program on the data of Tables 7.6 and 7.7.

2. Use your program to investigate the rates of change of $f(t) = t^2$ and $f(t) = t^5$ at time $t_1 = 2.0$ seconds.

3. By taking limits as in Example 7.3.1, show if $s = t^2$ then $v = 2t$ and if $s = t^5$, then $v = 5t^4$.

7.4 GRADIENTS OF GRAPHS

Consider the problem of finding the gradient of $y = x^3$ when $x = 1.5$. The gradient of the curve at $x = 1.5$ is given by the slope of the tangent to $y = x^3$ at P where $x = 1.5$, see Figure 7.3..

This could be determined graphically by drawing Figure 7.3 to scale, constructing the tangent at P(1.5, 3.375) and measuring its slope. However this would be subject to numerical errors as in any graphical process. Instead a numerical approximation could be used.

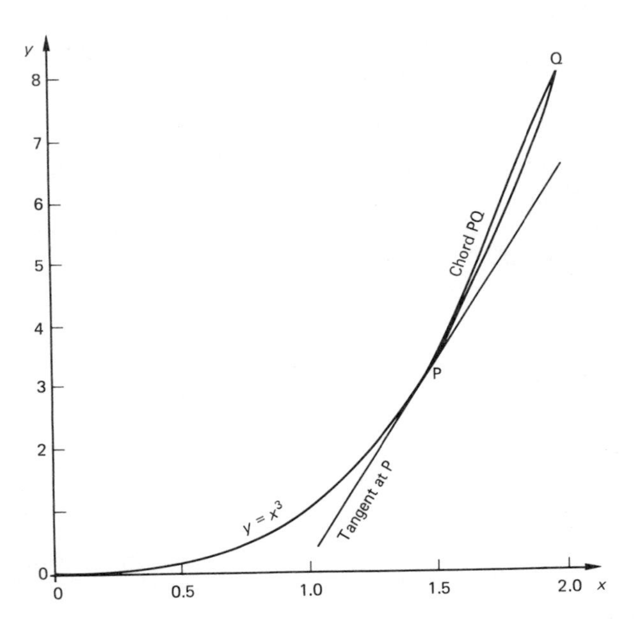

Figure 7.3

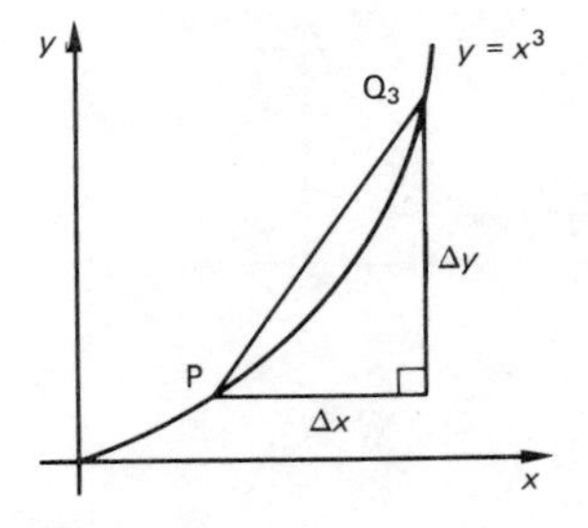

Figure 7.4

Take a point Q_n on $y = x^3$ (Figure 7.4). Then the gradient of the chord PQ_n is approximately equal to the gradient of the tangent at P. In Figure 7.3, Q is (2, 8), so the gradient of the chord PQ_3

$$PQ = \frac{2^3 - 1.5^3}{2 - 1.5} = \frac{8 - 3.375}{0.5} = 9.25$$

Therefore gradient of tangent $\simeq$ gradient of chord PQ = 9.25.

Obviously this is a poor approximation to the gradient of the tangent. To get a better approximation let Q slide down $y = x^3$ towards P, to give a sequence of improving approximations to the gradient of the tangent at P(1.5, 3.375), see Table 7.8 using the data from the graphs in Figures 7.4 and 7.5.

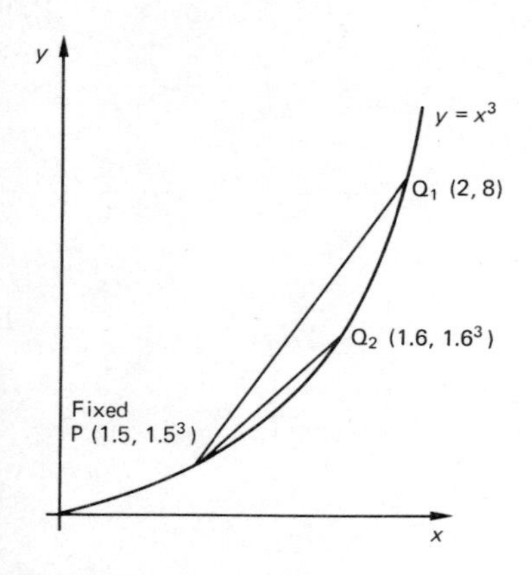

Figure 7.5
Part of the sequence of chords PQ_n

Table 7.8 Gradients of chords PQ_n to $y = x^3$ with P(1.5, 3.375)

P	Q_n	Δx	Δy	$\frac{\Delta y}{\Delta x}$ Gradient of chord PQ
(1.5, 3.375)	(2, 8)	0.5	4.625	9.250000
(1.5, 3.375)	(1.6, 4.096)	0.1	0.721	7.210000
(1.5, 3.375)	(1.51, 3.442951)	0.01	0.067951	6.795100
(1.5, 3.375)	(1.501, 3.381755)	0.001	0.006755	6.754501
(1.5, 3.375)	(1.5001, 3.375675)	0.0001	0.000675	6.750450

From Table 7.8 we can predict that as $\Delta x \to 0$ the gradient of the chord PQ $\to$ 6.75, that is the gradient of the tangent at P is 6.75.

This is precisely the same process which was carried out in Table 7.7 for finding the instantaneous velocity $v = \mathrm{d}s/\mathrm{d}t$ for $s = t^3$.

The correspondence between rates of change and gradients is

$$\text{Average rate of change } \frac{\Delta s}{\Delta t} \longleftrightarrow \text{Gradient of chord } \frac{\Delta y}{\Delta x}$$

$$\text{Instantaneous rate of change } \frac{\mathrm{d}s}{\mathrm{d}t} \longleftrightarrow \text{Gradient of tangent}$$

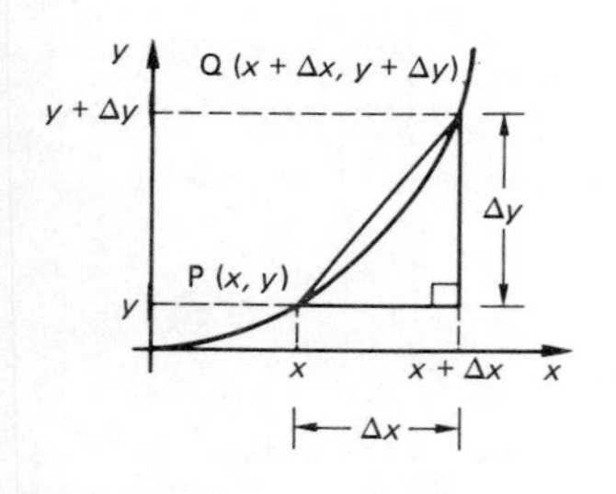

Figure 7.6

It is best to find the gradient at P by a limiting process, see Figure 7.6.

$$y + \Delta y = (x + \Delta x)^3$$
$$= x^3 + 3x^2\Delta x + 3x(\Delta x)^2 + (\Delta x)^3$$
$$y = x^3$$

Thus

$$\Delta y = 3x^2\Delta x + 3x(\Delta x)^2 + (\Delta x)^3$$

So the slope of the chord PQ is

$$\frac{\Delta y}{\Delta x} = 3x^2 + 3x\Delta x + (\Delta x)^2$$

The slope of the tangent at P is

$$\frac{dy}{dx} = \lim_{\Delta x \to 0} (3x^2 + 3x\Delta x + (\Delta x)^2) = 3x^2$$

Again the process is identical to that for finding ds/dt from $\Delta s/\Delta t$.

Example 7.4.1

If $y = x^2 - 3x + 1$, find dy/dx.

Solution

$$y + \Delta y = (x + \Delta x)^2 - 3(x + \Delta x) + 1$$

$$\therefore \quad y + \Delta y = x^2 + 2x\Delta x + (\Delta x)^2 - 3x - 3\Delta x + 1$$

$$y \quad = x^2 \quad - 3x \quad + 1$$

Subtract

$$\Delta y = 2x\Delta x + (\Delta x)^2 - 3\Delta x$$

So

$$\frac{\Delta y}{\Delta x} = 2x + \Delta x - 3$$

and

$$\frac{dy}{dx} = \lim_{\Delta x \to 0} \left(\frac{\Delta y}{\Delta x}\right) = \lim_{\Delta x \to 0} (2x + \Delta x - 3) = 2x - 3$$

The limiting process can be used for any variable.

Exercise 7.4.1

By taking appropriate limits as in Examples 7.3.1 and 7.4.1 find the following:

1. If $s = 1 - t^2$, find ds/dt.
2. If $v = 3t^3 - 4t^2$, find dv/dt.
3. If $y = \frac{1}{2}x^2 + x$, find dy/dx.

7.5 DIFFERENTIATION – SUMMARY OF CONCEPTS

Given a function $f(x)$ where x is the independent variable.

1. The derivative of $f(x)$ is the *(instantaneous) rate of change* of f with respect to x.
2. Algebraically, the derivative of $f(x)$ with respect to x is

$$\lim_{\Delta x \to 0} \frac{f(x + \Delta x) - f(x)}{\Delta x} = \frac{\mathrm{d}f}{\mathrm{d}x}$$

3. The derivative of $f(x)$ with respect to x is the *gradient* (or slope) of the graph $y = f(x)$.

Given a function $f(t)$ where t is the independent variable.

1. The derivative of $f(t)$ is the *(instantaneous) rate of change* of f with respect to t.
2. Algebraically, the derivative of $f(t)$ with respect to t is

$$\lim_{\Delta t \to 0} \frac{f(t + \Delta t) - f(t)}{\Delta t} = \frac{\mathrm{d}f}{\mathrm{d}t}$$

3. The derivative of $f(t)$ with respect to t is the *gradient* (or slope) of the graph $y = f(t)$.

The derivative of a function is a new function. For example, if $f(x) = x^3$ then $\mathrm{d}f/\mathrm{d}x = 3x^2$.

These are different functions as we can see from the graphs in Figure 7.7.

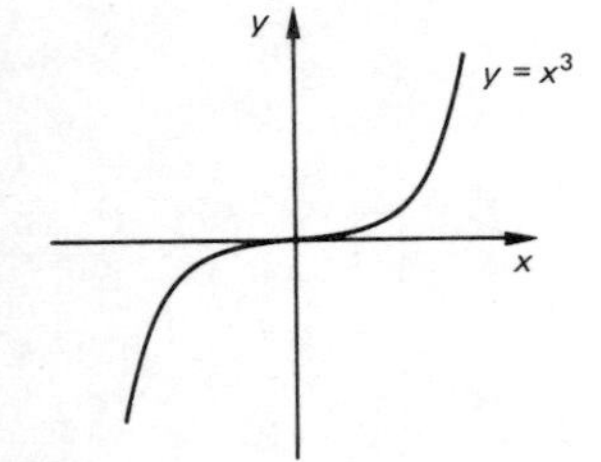

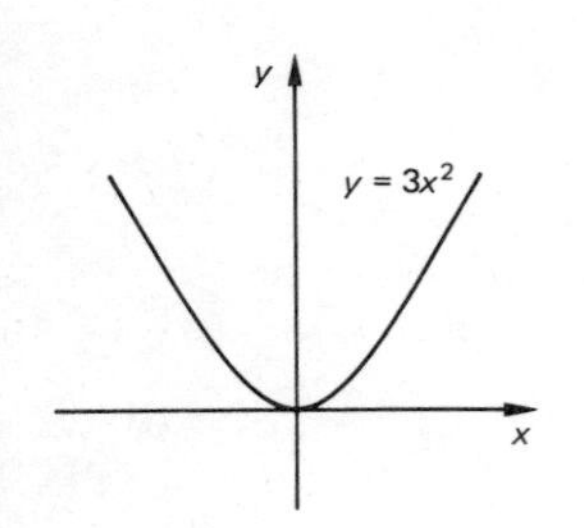

Figure 7.7

7.6 RULES FOR DIFFERENTIATION

Differentiation from first principles by taking limits as in Section 7.5 is very time consuming. You have probably noticed a pattern emerging, namely

$$y = x^n \Rightarrow \frac{\mathrm{d}y}{\mathrm{d}x} = nx^{n-1} \tag{7.6.1}$$

This rule is valid for all values of n, positive integers, negative integers or fractions.

If a is a constant, then $y = a \Rightarrow \dfrac{\mathrm{d}y}{\mathrm{d}x} = 0$ (7.6.2)

Example 7.6.1

If $y = x^{5/2} + 7$, find $\mathrm{d}y/\mathrm{d}x$.

Solution

By rule (7.6.1) with $n = 5/2$ and rule (7.6.2), $dy/dx = (5/2)x^{3/2} + 0 = (5/2)x^{3/2}$

Example 7.6.2

If $s = 1/\sqrt{t}$, find ds/dt.

Solution

Rule (7.6.1) is easily adapted for the variables s and t, to give

$$s = t^n \Rightarrow \frac{ds}{dt} = nt^{n-1}$$

To use the rule, we require powers of s.
Now $s = 1/\sqrt{t} = t^{-1/2}$, so using rule (7.6.1) with $n = -\frac{1}{2}$

$$\frac{ds}{dt} = -\tfrac{1}{2}t^{-3/2}$$

Example 7.6.3

If $p = 1/v$, find dp/dv.

Solution

By rule (7.6.1),

$$p = v^n \Rightarrow \frac{dp}{dv} = nv^{n-1}$$

Again, powers are required. $p = 1/v = v^{-1}$
So

$$\frac{dp}{dv} = -1v^{-2} = -\frac{1}{v^2}$$

Exercise 7.6.1

Use rule (7.6.1) to find the following derivatives:

1. $y = x^5, \frac{dy}{dx}$
2. $y = \frac{1}{x^3}, \frac{dy}{dx}$

3. $s = \dfrac{1}{t} + 1, \dfrac{ds}{dt}$
4. $v = \sqrt{t}, \dfrac{dv}{dt}$
5. $x = t\sqrt{t}, \dfrac{dx}{dt}$
6. $p = v^{1.1} + \frac{1}{2}, \dfrac{dp}{dv}$
7. $p = v^{-1.1} - 7, \dfrac{dp}{dv}$
8. $y = \sqrt[3]{u}, \dfrac{dy}{du}$
9. $y = \dfrac{1}{\sqrt[4]{u}}, \dfrac{dy}{du}$
10. $y = x, \dfrac{dy}{dx}$

7.6.1 Differentiating Sums, Differences and Multiples

The following rules are stated without proof.

Let $f(x)$ and $g(x)$ be functions of x, then:

$$y = f(x) \pm g(x) \Rightarrow \frac{dy}{dx} = \frac{df(x)}{dx} \pm \frac{dg(x)}{dx} \tag{7.6.3}$$

This rule says 'differentiate the functions of f and g separately and add (or subtract) the answers'.

Example 7.6.4

If $y = x^2 + x^3$, then $dy/dx = 2x + 3x^2$.

Let $f(x)$ be a function of x, and c be a constant, then

$$y = cf(x) \Rightarrow \frac{dy}{dx} = c\frac{df(x)}{dx} \tag{7.6.4}$$

This rule says 'differentiate the function f and multiply the result by the constant c'.

Example 7.6.5

If $y = 10x^2$, then $dy/dx = 10 \times (2x) = 20x$.

Rules (7.6.3) and (7.6.4) are usually used together, and extend to other variables easily.

Example 7.6.6

If $y = 5x^3 + 4x^2 - 3x + 1$, find dy/dx.

Solution

$$\frac{dy}{dx} = 5 \times (3x^2) + 4 \times (2x) - 3 \times 1 + 0$$

$$= 15x^2 + 8x - 3$$

Example 7.6.7

If $v = 7t^2 - 3/t$, find dv/dt.

Solution

First write every term in with powers of t

$$v = 7t^2 - 3t^{-1}$$

so

$$\frac{dv}{dt} = 7 \times (2t) - 3 \times (-t^{-2}) = 14t + 3t^{-2} = 14t + \frac{3}{t^2}$$

Exercise 7.6.2

Use rules (7.6.1)–(7.6.4) to find the following derivatives:

1. $y = x^4 - x^3 + 4x^2 - 8x + 9; \frac{dy}{dx}$
2. $y = 1 - \sqrt{x}; \frac{dy}{dx}$
3. $y = \frac{5}{x} - \frac{3}{x^2} + \frac{4}{x^4}; \frac{dy}{dx}$
4. $s = \sqrt{t} + 3t; \frac{ds}{dt}$
5. $v = 1 - \frac{8}{t^3}; \frac{dv}{dt}$
6. $p = v^{-1.2} + 0.5; \frac{dp}{dv}$

7. $y = 4\sqrt[3]{u} + 7\sqrt{u}$; $\dfrac{dy}{du}$

8. $u = 1 - t^{2/3} - \dfrac{2}{t^{1/2}}$; $\dfrac{du}{dt}$

9. $T = 3t^2 - \frac{1}{4}t + \frac{1}{2}$; $\dfrac{dT}{dt}$

10. $\theta = \pi t^2 + \dfrac{\pi}{2}$; $\dfrac{d\theta}{dt}$

7.7 THE FUNCTIONAL NOTATION AND SECOND DERIVATIVES

7.7.1 The Functional Notation

There is an alternative notation for the derived function. If $y = x^3$ then we have written $dy/dx = 3x^2$. We could instead say $y = f(x)$ where $f(x) = x^3$. Then the derived function of f is given by $f'(x) = 3x^2$. This is just a different notation — there are no new concepts involved.

Example 7.7.1

If $f(x) = 1 - x^2 - 7x^3$, $f'(x) = -2x - 21x^2$. f' is the result of differentiating f with respect to x.

Example 7.7.2

If $f(v) = v^{0.8}$, then $f'(v) = 0.8v^{-0.2}$. f' is the result of differentiating f with respect to v.

Exercise 7.7.1

Find $f'(x)$ if

1. $f(x) = x^{3/2}$ 2. $f(x) = \dfrac{1}{\sqrt{x}}$ 3. $f(x) = x - \sqrt{x}$

4. $f(x) = 10 + 3x + \dfrac{7}{x}$ 5. $f(x) = x^{2.1} - x^{1.1}$

7.7.2 Second Derivatives

If $s = t^4 - t^2 + 1$ represents displacement, differentiating once gives velocity

$$v = \frac{ds}{dt} = 4t^3 - 2t$$

Differentiating v gives acceleration

$$a = \frac{dv}{dt} = 12t^2 - 2$$

So acceleration a is obtained from the displacement s by differentiating twice. We write

$$a = \frac{d^2s}{dt^2}$$

which is the second derivative of s with respect to t. In the functional notation the second derivative of $f(x)$ is written $f''(x)$.

Example 7.7.3

If $f(x) = \sqrt{x} + x^{10}$ find $f''(x)$.

Solution

$$f(x) = x^{1/2} + x^{10} \Rightarrow f'(x) = \tfrac{1}{2}x^{-1/2} + 10x^9$$
$$\Rightarrow f''(x) = -\tfrac{1}{4}x^{-3/2} + 90x^8$$

Exercise 7.7.2

In Problems 1–5, find d^2s/dt^2.

1. $s = t^3$ 2. $s = 1 - t^2$ 3. $s = t$ 4. $s = \dfrac{1}{t}$ 5. $s = \dfrac{1}{t^2}$

In Problems 6–10, find $f''(x)$.

6. $f(x) = \sqrt[3]{x}$
7. $f(x) = 1 + x + x^2$
8. $f(x) = \dfrac{1}{6}x^3 + \dfrac{1}{2}x^2$
9. $f(x) = \dfrac{1}{3\sqrt{x}}$
10. $f(x) = 6 + 3x$

7.8 APPLICATIONS

Examples 7.8.1

Find the gradient of the curve whose equation is $y = 2x^3 + 3.5x^2 - 3x + 1$ when $x = 0$, 1 and 2. For which values of x is the gradient of the curve zero? Find the coordinates of the points where the gradient is zero.

Solution

The gradient is given by dy/dx

$$y = 2x^3 + 3.5x^2 - 3x + 1 \quad \text{so} \quad \frac{dy}{dx} = 6x^2 + 7x - 3$$

So when $x = 0$,

$$\frac{dy}{dx} = 6 \times 0 + 7 \times 0 - 3 = -3$$

when $x = 1$,

$$\frac{dy}{dx} = 6 \times 1^2 + 7 \times 1 - 3 = 10$$

when $x = 2$,

$$\frac{dy}{dx} = 6 \times 2^2 + 7 \times 2 - 3 = 35$$

So the gradients at $x = 0$, 1 and 2 are -3, 10, and 35.

When the gradient is zero, $dy/dx = 0$ so $6x^2 + 7x - 3 = 0$. Factorizing $(2x + 3)(3x - 1) = 0$

$$x = -\frac{3}{2} \quad \text{and} \quad \frac{1}{3}$$

When $x = -\frac{3}{2}$,

$$y = 2\left(-\frac{3}{2}\right)^3 + 3.5\left(-\frac{3}{2}\right)^2 - 3\left(-\frac{3}{2}\right) + 1 = 6.625$$

When $x = \frac{1}{3}$,

$$y = 2\left(\frac{1}{3}\right)^3 + 3.5\left(\frac{1}{3}\right) - 3\left(\frac{1}{3}\right) + 1 = 0.463 \text{ (three decimal places)}$$

So the gradient is zero at $(-1.5, 6.625)$ and $(0.333, 0.463)$.

Example 7.8.2

The temperature T(°C) of a liquid at time t (seconds) is given by $T = 100 + 0.5t - 0.01t^2$. When does it begin to cool?

Solution

The rate of change of temperature is given by $\mathrm{d}T/\mathrm{d}t$. When $\mathrm{d}T/\mathrm{d}t > 0$, it is heating up and when $\mathrm{d}T/\mathrm{d}t < 0$, it is cooling. So we need to find when $\mathrm{d}T/\mathrm{d}t = 0$.

$$T = 100 + 0.5t - 0.01t^2 \Rightarrow \frac{\mathrm{d}T}{\mathrm{d}t} = 0.5 - 0.02t$$

$$\frac{\mathrm{d}T}{\mathrm{d}t} = 0 \Rightarrow 0.5 - 0.02t = 0 \Rightarrow t = \frac{0.5}{0.02} = 25 \text{ s}$$

Now

$$\frac{\mathrm{d}T}{\mathrm{d}t} = 0.5 - 0.02t = 0.02\,(25 - t)$$

so if

$$t < 25, \frac{\mathrm{d}T}{\mathrm{d}t} > 0 \quad \text{and if} \quad t > 25, \frac{\mathrm{d}T}{\mathrm{d}t} < 0$$

So the liquid starts to cool after $t = 25$ s.

Example 7.8.3

The displacement, s (metres) of a particle at time t (seconds) is given by

$$s = 0.3t + 0.6t^2 - 0.02t^{5/2}$$

Find its velocity and acceleration at time $t = 3$ s.

Solution

Velocity,

$$v = \frac{\mathrm{d}s}{\mathrm{d}t} = 0.3 + 1.2t - 0.05t^{3/2}$$

Acceleration,

$$a = \frac{dv}{dt} = 1.2 - 0.075t^{1/2}$$

So when $t = 3$ s, $v = 3.640\ \text{ms}^{-1}$, $a = 1.070\ \text{ms}^{-2}$ (both answers: three decimal places)

Example 7.8.4

Find the equation of the tangent and normal to the curve whose equation is $y = 2 + 5x - x^2$ at the point (1,6).

Solution

At any given point, the tangent to the curve has the same gradient as the curve. The gradient of the curve with equation $y = 2 + 5x - x^2$ is given by

$$\frac{dy}{dx} = 5 - 2x$$

So at (1,6) the gradient is $5 - 2 \times 1 = 3$.

The tangent is a straight line. From Section 1.4 the equation of the straight line through (x_1, y_1) with gradient m is

$$y - y_1 = m(x - x_1)$$

so with $(x_1, y_1) = (1, 6)$ and $m = 3$

$$y - 6 = 3(x - 1)$$

$$\therefore y = 3x + 3$$

The normal is a straight line at right angles to the tangent. Now, straight lines at right angles have the product to their gradients equal to -1.

So

$$\text{Gradient of normal} = \frac{-1}{\text{gradient of tangent}} = -\frac{1}{3}$$

Hence the normal is a straight line through $(x_1, y_1) = (1, 6)$ with gradient $-1/3$. Its equation is

$$y - 6 = -\frac{1}{3}(x - 1)$$

$3y + x = 19$

So the equations of the tangent and normal at $(1, 6)$ are $y = 3x + 3$ and $3y + x = 19$.

Exercise 7.8.1

1. Find the gradient of the curve $y = \sqrt{x}$ at the point where $x = 9$.
2. Find the gradient of the curve $y = 1/x$ when $x = 3$.
3. Show that there is no point on the curve $y = 1/x^2$ where the gradient is zero.
4. Find the coordinates of the two points on the curve $y = 5x^3 - 7x^2 + 3x + 2$ where the gradient is zero.
5. Show that if $x > 1$, then the gradient of the curve $y = 1/x + x$ is positive.
6. A beam AB has a bending moment M (kNm) given by $M = 6x^2 - 12x$ where x is the distance from A in metres. Find the rate of change of M with respect to x. When is this rate zero?
7. The period of oscillation, p(s), of a simple pendulum is given by

   $$p = 2\pi \sqrt{\frac{l}{g}}$$

   where l is the length of the pendulum (m) and g is the acceleration due to gravity (ms^{-2}). If the pendulum's length changes (due to a change in temperature say), its period will change. Find the rate of change of period with respect to length when $l = 0.3$ m. Take $g = 9.8\ ms^{-2}$ and give your answer to three decimal places.
8. The kinetic energy K (joules) of a body mass m (kg) travelling at v (ms^{-1}) is given by

   $$K = \tfrac{1}{2}mv^2$$

   Find the rate of change of K with respect to velocity of a body of mass 10 kg at the instant that its velocity is 25 ms^{-1}.
9. The power P (watts) dissipated in a resistor of R (ohms) when a current of I (amperes) flows is given by $P = I^2R$. Find the rate of change of power with respect to current when $I = 3$ mA and $R = 1$ kΩ.

10. A body moves along a straight line AB so that its displacement s (m) from a point A at time t (s) is given by $s = t^3 + 3$. Show that the body is initially (at $t = 0$ s) 3 m from A and that it always moves away from A.

11. A particle moves so that its displacement s from a point P at time t is given by $s = 1 + 2t - 0.1t^2$. Find when its velocity is zero and show that its acceleration is constant.

12. Find the equation of the tangent to the curve with equation

$$y = \frac{3}{x} + x^2$$

at the point where $x = 3$.

13. Find the coordinates of the two points on the graph of $y = 12/x$ where the gradient is -3. Find the equations of the tangents at these points.

14. Find the equation of the normal to the curve $y = 2x^2 + 3x + 3$ when $x = -1$. Where does this normal cut the curve again?

15. Show that $y = 8x + 11$ is tangent to the cubic $y = x^3 - 3x^2 - x + 6$. Find the equation of the other tangent to the cubic with gradient 8.

7.9 FURTHER RULES FOR DIFFERENTIATION

7.9.1 The Chain Rule or Differentiation by Substitution

$$\frac{dy}{dx} = \frac{dy}{du} \cdot \frac{du}{dx}$$

This is used for differentiating composite functions, that is functions formed from two or more 'simple' functions, for example

$$y = \sqrt{1 + x^2} \quad \text{or } y = \frac{2}{1 - t^2}$$

Example 7.9.1

If $y = (1 + x^3)^{1/2}$, find dy/dx.

Solution

Use the substitution $u = 1 + x^3$, so $y = u^{1/2}$.
Hence $du/dx = 3x^2$ and $dy/du = \frac{1}{2}u^{-1/2} = \frac{1}{2}(1 + x^3)^{-1/2}$.
By the chain rule

$$\frac{dy}{dx} = \frac{dy}{du} \cdot \frac{du}{dx} = \tfrac{1}{2}(1 + x^3)^{-1/2} \times 3x^2$$

$$\frac{dy}{dx} = \frac{3}{2}x^2(1 + x^3)^{-1/2}$$

Example 7.9.2

If $s = 1/(3 + t^2)$, find ds/dt.

Solution

The chain rule works for any variables, so

$$\frac{ds}{dt} = \frac{ds}{du} \cdot \frac{du}{dt}$$

$$s = \frac{1}{3 + t^2}$$

Let $u = 3 + t^2$, hence $s = 1/u = u^{-1}$
So

$$\frac{du}{dt} = 2t \quad \text{and} \quad \frac{ds}{du} = -u^{-2} = \frac{-1}{u^2} = \frac{-1}{(3 + t^2)^2}$$

By the chain rule

$$\frac{ds}{dt} = \frac{ds}{du} \cdot \frac{du}{dt} = -\frac{2t}{(3 + t^2)^2}$$

Exercise 7.9.1

Find dy/dx in Problems 1–4

1. $y = (2 - x^4)^{1/2}$
2. $y = \dfrac{1}{(4 - x)^2}$
3. $y = (x^2 + x + 1)^3$
4. $y = (x^3 - 1)^{0.1}$

Find ds/dt in Problems 5–8

5. $s = \sqrt{t^3 - 3t^2 + t - 1}$ 6. $s = \dfrac{1}{\sqrt{1 + 3t}}$

7. $s = \dfrac{1}{1 + t} + t$ 8. $s = t^2 + \sqrt{1 - t}$

9. If $v = (1 - t^2)^{1.2}$, find dv/dt.

10. If $p = 1 + \dfrac{1}{v} + \sqrt[3]{v^2 - 2}$, find dp/dv.

7.9.2 Applications of the Chain Rule

The chain rule is very useful in relating rates of change.

Example 7.9.3

The rate of increase of the radius of a sphere is 0.5 mm per second. Find the rate of increase of the volume of the sphere when the radius is 30 cm.

Solution

Let V be the volume of the sphere and r be its radius. Then $V = (4/3)\pi r^3$. We need to find the rate of increase of volume, i.e. dV/dt. We are given that the rate of increase of radius $dr/dt = 0.5\ \text{mms}^{-1}$

By the chain rule

$$\frac{dV}{dt} = \frac{dV}{dr} \cdot \frac{dr}{dt}$$

so

$$\frac{dV}{dt} = 0.5\frac{dV}{dr}$$

Now

$$V = \frac{4}{3}\pi r^3 \quad \text{so} \quad \frac{dV}{dr} = 4\pi r^2$$

so

$$\frac{dV}{dt} = 0.5 \times 4\pi r^2 = 2\pi r^2$$

So when $r = 10$ cm $\equiv 100$ mm (note units must be consistent),

$$\frac{dV}{dt} = 2\pi(100)^2 \simeq 6.283 \times 10^4 \text{ mm}^3\text{s}^{-1}.$$

Example 7.9.4

The pressure, p, and volume, v, of a gas are known to obey Boyle's Law, $pv = c$ where c is a constant. Suppose the volume is given by $v = 0.04t^2 + 0.32t + 0.6$ at time t, find the rate of change of pressure.

Solution

We need to find dp/dt, the rate of change of pressure. Now

$$v = 0.04t^2 + 0.32t + 0.6$$

so the rate of change of volume is

$$\frac{dv}{dt} = 0.08t + 0.32.$$

The derivatives dp/dt and dv/dt can be related by the chain rule

$$\frac{dp}{dt} = \frac{dp}{dv} \cdot \frac{dv}{dt}$$

so dp/dv must be found.
But $pv = c \Rightarrow p = cv^{-1}$, so $dp/dv = -cv^{-2} = -c/v^2$.

$$\frac{dp}{dt} = \frac{-c(0.08t + 0.32)}{v^2}$$

Example 7.9.5

The velocity v (ms^{-1}) of a particle is given by

$$v = \frac{1}{1 + s^2}$$

where s(m) is its displacement from a fixed point. Find its velocity and acceleration when $s = 2$ m.

Solution

When $s = 2$ m, $v = \dfrac{1}{1 + 2^2} = \dfrac{1}{5}\,\text{ms}^{-1}$

Acceleration $a = \mathrm{d}v/\mathrm{d}t$. Since v is not given in terms of t, use the chain rule to give

$$a = \frac{\mathrm{d}v}{\mathrm{d}t} = \frac{\mathrm{d}v}{\mathrm{d}s} \cdot \frac{\mathrm{d}s}{\mathrm{d}t} = \frac{\mathrm{d}v}{\mathrm{d}s}\, v = v\,\frac{\mathrm{d}v}{\mathrm{d}s}$$

But

$$v = \frac{1}{1 + s^2} = (1 + s^2)^{-1}$$

so

$$\frac{\mathrm{d}v}{\mathrm{d}s} = \frac{-2s}{(1 + s^2)^2} \quad \text{(using a substitution and the chain rule)}$$

Therefore

$$a = v\,\frac{\mathrm{d}v}{\mathrm{d}s} = \frac{-2s}{(1 + s^2)^3}$$

So when $s = 2$ m,

$$a = \frac{-4}{(1 + 4)^3} = -\frac{4}{125}\,\text{ms}^{-2}$$

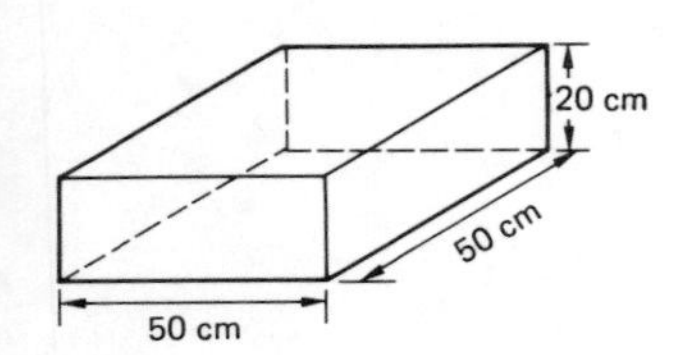

Figure 7.8

Example 7.9.6

A car has a petrol tank whose faces are rectangles as shown in Figure 7.8. It holds petrol to a depth of h cm. A petrol pump is delivering fuel at a rate of 0.5 litre per second. How fast is h changing?

Solution

We need to find $\mathrm{d}h/\mathrm{d}t$. Let V = volume of petrol in the tank. So $V = 2500h$ cm^3.

$$\frac{\mathrm{d}V}{\mathrm{d}t} = 0.5 \text{ litre s}^{-1} \equiv 500 \text{ cm}^3\text{s}^{-1}$$

Now $V = 2500h \Rightarrow dV/dh = 2500$.

By the chain rule

$$\frac{dh}{dt} = \frac{dh}{dV} \cdot \frac{dV}{dt}$$

but $dV/dh = 2500$ so $dh/dV = \dfrac{1}{2500}$

$$\therefore \frac{dh}{dt} = \frac{1}{2500} \times 500 = 0.2 \text{ cms}^{-1}$$

Exercise 7.9.2

1. The surface area, S, of a sphere of radius r is given by $S = 4\pi r^2$. If r is increasing by 1.5 mm per second, find the rate of increase of S when $r = 25$ cm.
2. The surface area S of a sphere is increasing at a rate of 2.4 cm^2s^{-1}. Find the rate of increase of the radius when the radius is 12 cm.
3. A current I flows through a resistance R. The power P developed is given by $P = I^2R$. Find the rate of change of power if

   $$I = 1 + \frac{1}{t + 1}$$

   where t is the time. Assume that R is constant.
4. A conveyor belt delivers coal onto a stockpile at the rate of 0.25 m^3 per second. The coal forms a conical heap whose height is equal to its radius. How fast is the height of the cone increasing when it consists of 20m^3 of coal? (Volume of cone, height h, base radius r is $\frac{1}{3}\pi r^2 h$.)
5. A cone has a vertical height h which is five times its base radius. Show that its volume is

   $$\frac{1}{75}\pi h^3$$

   The cone is supported with its vertex lowermost and its axis of symmetry vertical. It is being filled with water at the rate of 0.02 litres per second. The depth of water in the cone is w. Find an expression for the rate of change of w in cms^{-1}.

6. The velocity v (ms^{-1}) of a particle is given by

$$v = \frac{1}{2s + 1}$$

where s(m) is its displacement from a fixed point. Find its acceleration in terms of s. What is the acceleration when $s = 1$ m, 10 m and 100 m. What happens to the acceleration as s increases?

7.10 MAXIMA AND MINIMA

7.10.1 Stationary Points

Consider the graph of $y = f(x)$ where f is some function of x. The gradient of the graph is given by $f'(x)$. If $f'(x) > 0$ the function is increasing, if $f'(x) < 0$ it is decreasing. However if $f'(x) = 0$, the function is neither increasing nor decreasing. Points where $f'(x) = 0$ are called stationary points. At a stationary point, the tangent to $y = f(x)$ will be parallel to the x axis. In Figure 7.9, there are stationary points at P, Q, R and S.

Stationary point

The function f has a stationary point at $x = a$ if $f'(a) = 0$. Alternatively the graph of $y = f(x)$ has a stationary point at $x = a$ if

$$\frac{dy}{dx} = f'(a) = 0$$

at $x = a$

Example 7.10.1

Find the stationary points of $f(x) = 6x^3 + 3x^2 - 12x + 1$.

Solution

$f'(x) = 18x^2 + 6x - 12$

For stationary points $f'(x) = 0$, so

$18x^2 + 6x - 12 = 0,$

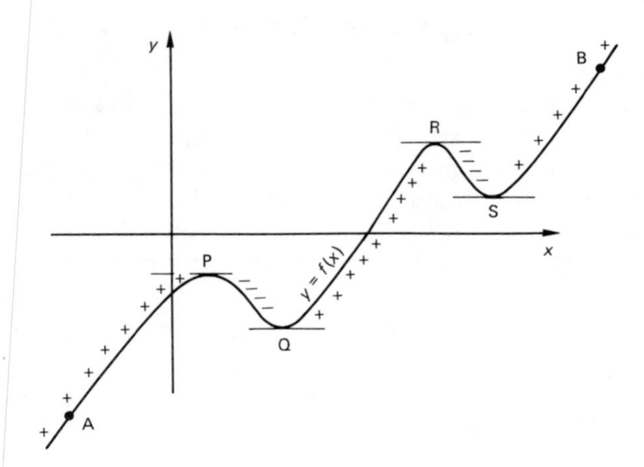

Figure 7.9 + means positive gradient, $f'(x) > 0$, $f(x)$ increasing;
− means negative gradient, $f'(x) < 0$, $f(x)$ decreasing

$$6(3x^2 + x - 2) = 0$$

$$(3x - 2)(x + 1) = 0$$

Hence, $x = 2/3$ and $x = -1$ are the stationary points.

Example 7.10.2

Find the stationary points of the curve whose equation is $s = t + 1/t$.

Solution

Instead of the variables x and y, we have t and s. We must find where $\mathrm{d}s/\mathrm{d}t = 0$.

$$s = t + t^{-1} \quad \text{so} \quad \frac{\mathrm{d}s}{\mathrm{d}t} = 1 - t^2$$

$$\frac{\mathrm{d}s}{\mathrm{d}t} = 0 \Rightarrow 1 - t^2 = 0 \Rightarrow t = \pm 1$$

So there are stationary points at $t = 1$ and -1.

Example 7.10.3

Find the stationary points of the curve whose equation is $v = t^3 + 2t + 3$.

Solution

For stationary points, $\mathrm{d}v/\mathrm{d}t = 0$.

Now $\mathrm{d}v/\mathrm{d}t = 3t^2 + 2$, so $3t^2 + 2 = 0$ or $t^2 = -2/3$. This equation has no real solutions, so there are no stationary points. In fact $\mathrm{d}v/\mathrm{d}t = 3t^2 + 2 > 0$ for all values of t, so the gradient of the graph is always positive and the function is always increasing.

Exercise 7.10.1

For Problems 1–5, find $f'(x)$ and hence state whether the function is increasing or decreasing at the given value of x.

1. $f(x) = 1 - x^2$; $x = 2$
2. $f(x) = 1/x$; $x = 1$
3. $f(x) = x^3 + 2x^2 - 1$; $x = 3$
4. $f(x) = \dfrac{3}{1 + x^2}$; $x = 0$
5. $f(x) = \sqrt{x^2 + 1}$; $x = -1$

For Problems 6–12 find the coordinates of the stationary points if they exist.

6. $y = x - 3x^2$
7. $y = 4x^2 - 1/x$
8. $v = t^3 + 1$
9. $s = \sqrt{1 + t^2}$
10. $p = v^{1.2}$
11. $p = v^{0.2}$
12. $y = u^5 - 5u + 1$

7.10.2 Maxima, minima and points of inflexion

Stationary points occur when $f'(x) = 0$, and the tangent to the curve $y = f(x)$ is parallel to the x axis. It is possible to classify stationary points into three categories.

Local Maximum

In Figure 7.10 the tangent to P is parallel to the x axis and so

$$\frac{\mathrm{d}y}{\mathrm{d}x} = f'(a) = 0$$

and there is a *local maximum* at P.

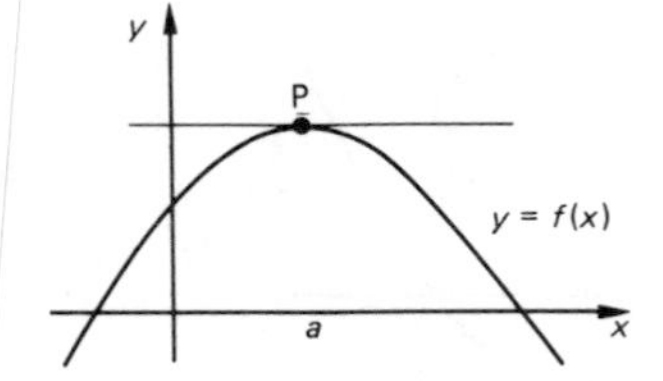

Figure 7.10

The term 'local' maximum is used, because for all points in the neighbourhood of $x = a$, $f(a) > f(x)$. This does not guarantee an absolute maximum. In Figure 7.9, both P and R are local maxima, but they are not absolute maxima. For example the value of the function at B exceeds the values at P and R.

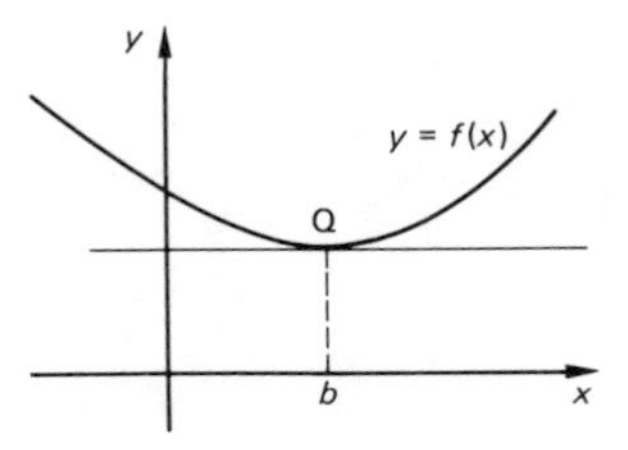

Figure 7.11

Local Minimum

In Figure 7.11 at Q the tangent is parallel to the x axis, and

$$\frac{dy}{dx} = f'(b) = 0$$

For all points in the neighbourhood of $x = b, f(b) < f(x)$, so there is a local minimum at $x = b$. The term 'local' applies for similar reasons to the last section. In Figure 7.9 both Q and S are local minima, but A is a point on the curve giving an even lower value.

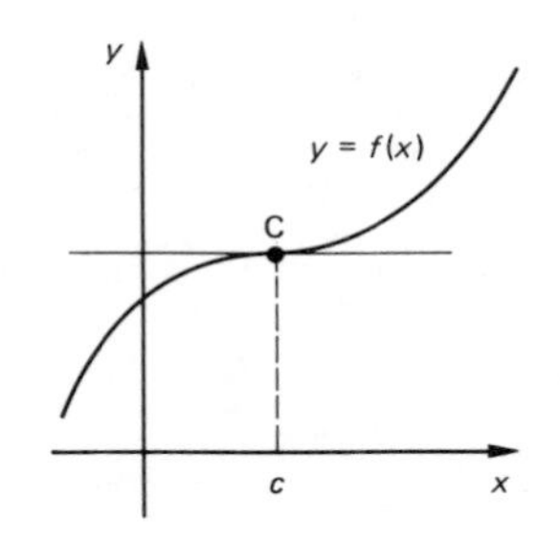

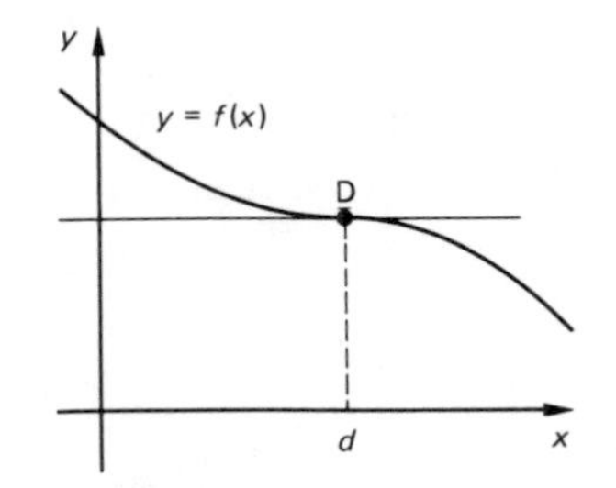

Figure 7.12

Point of Inflexion

In Figure 7.12, at C, $f'(c) = 0$ and at D, $f'(d) = 0$, but in neither case is there a local maximum or minimum. Points such as these are called points of inflexion. At a point of inflexion the curve crosses its tangent. (Note the gradient at a point of inflexion does not have to be zero, but in this section we are not interested in anything other than stationary points.)

7.10.3 Classification of Stationary Points

Method 1: Gradient sign diagrams

Passing through a maximum the gradient changes from positive to negative (see Figure 7.13).

Passing through a minimum the gradient changes from negative to positive.

Passing through a point of inflexion the sign of the gradient does not change.

These facts can be used to classify any stationary point.

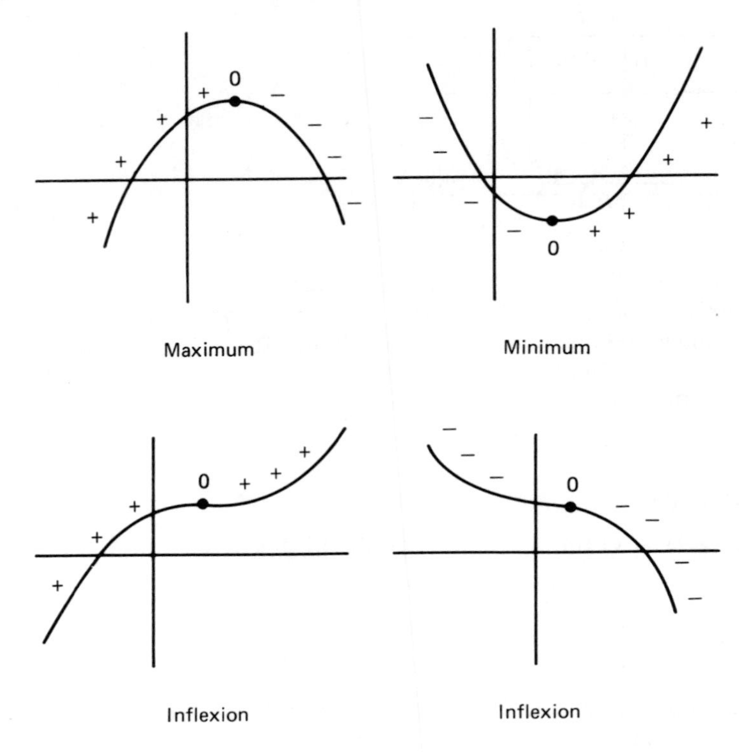

Figure 7.13

Example 7.10.4

Find and classify the stationary points of $y = x^3 + 3/x$.

Solution

$$\frac{dy}{dx} = 3x^2 - \frac{3}{x^2}$$

For stationary points $dy/dx = 0$, so $x^2 = 1/x^2$ or $x^4 = 1$. $\therefore x = \pm 1$. So there are stationary points when $x = \pm 1$.

When $x = 1$

$$y = 1 + \frac{3}{1} = 4$$

Examine the sign of the gradient in the neighbourhood of $x = 1$. $x = 0.9$ and $x = 1.1$ are sufficiently near, since the other stationary point is at $x = -1$.

x	0.9	1.0	1.1
$\frac{dy}{dx} = 3x^2 - \frac{3}{x^2}$	$-$	0	$+$
Slope	╲	–	╱

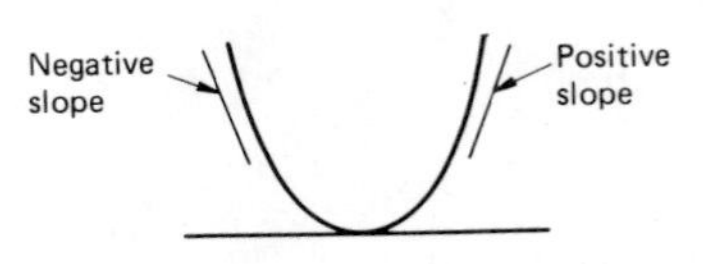

Figure 7.14

The shape of the slope indicates a minimum (see Figure 7.15). So (1, 4) is a minimum.

When $x = -1$

$$y = (-1)^3 + \frac{3}{(-1)} = -4$$

x	-1.1	-1.0	-0.9
$\frac{dy}{dx} = 3x^2 - \frac{3}{x^2}$	$+$	0	$-$
Slope	╱	–	╲

Note $-1.1 < -1.0 < -0.9$; the x values must be in ascending order or the slopes will give the wrong picture. So $(-1, -4)$ is a maximum.

Exercise 7.10.2

Find the coordinates of the stationary points of the following equations and classify them.

1. $y = x^2 + 2x + 3$
2. $y = x^3 + 1$
3. $s = \frac{1}{t + 1}$
4. $v = t^2 + \frac{1}{t^2}$

Method 2: Use of the Second Derivative

$dy/dx = f(x)$ measures the rate of increase of $y = f(x)$, so $d^2y/dx^2 = f''(x)$ measures the rate of increase of $dy/dx = f'(x)$.

Now if $f'(x) > 0$, the gradient of $y = f(x)$ is positive so f is increasing. Similarly if $f''(x) > 0$, then f' is increasing, that is the gradient is increasing.

Now suppose we have $f'(a) = 0$ and $f''(x) > 0$ in the neighbourhood of $x = a$. This means that the gradient of $y = f(x)$ is increasing in the neighbourhood of $x = a$, and is zero at $x = a$. This can only mean there is a minimum at $x = a$ (Figure 7.15.)

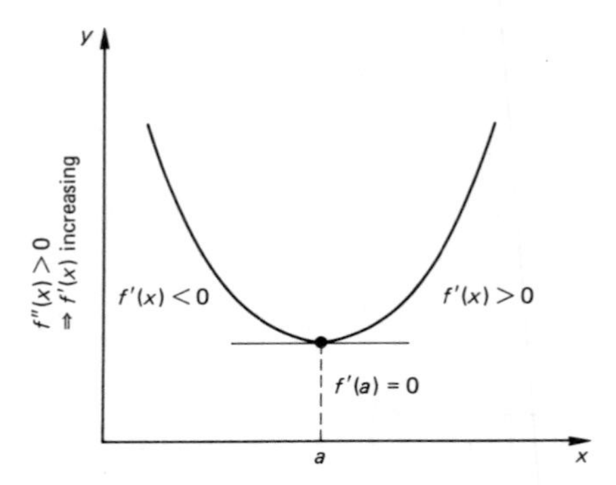

Figure 7.15

By similar reasoning $f'(a) = 0$ and $f''(x) < 0$ in the neighbourhood of $x = a$ implies $x = a$ is a maximum.

These implications cannot be reversed; $x = a$ is a maximum does *not* imply $f''(a) < 0$. For example if $f(x) = 1 - x^4$, then $f'(x) = -4x^4$. So there is a turning point at $x = 0$. By doing a sketch of $y = f(x)$ confirm that $(0, 1)$ is a maximum. But $f''(x) = -12x^2$, so $f''(0) = 0$.

In fact, $f''(a) = 0$ can imply a maximum, a minimum or a point of inflexion at $x = a$.

Example 7.10.5

Give functions for which $f'(a) = 0$ and $f''(a) = 0$ to illustrate the three types of stationary points.

Solution

$f(x) = x^3 \Rightarrow f'(x) = 3x^2 \therefore$ Stationary point at $x = 0$. $f''(x) = 6x$ so $f''(0) = 0$. This point is a point of inflexion (from a gradient sign diagram).

$f(x) = x^4 \Rightarrow f'(x) = 4x^3 \therefore$ Stationary point at $x = 0$. $f''(x) = 12x^2$ so $f''(0) = 0$. This is a minimum (from a gradient sign diagram.)

$f(x) = 1 - x^4 \Rightarrow f'(x) = -4x^3 \therefore$ Stationary point at $x = 0$. $f''(x) = -12x^2$ so $f''(0) = 0$. This is a maximum (from a gradient sign diagram).

Despite the fact that the second derivative method does not always identify the stationary point, it is quicker than using gradient sign diagrams. So the following method of finding and identifying stationary points should be used.

> *Finding and identifying the stationary points of $y = f(x)$*
>
> 1. Find the values of x where
>
> $$\frac{dy}{dx} = f'(x) = 0$$
>
> Call these points $x = a, b, c, \ldots$ for each point.
>
> 2. Find the sign of $dy^2/dx^2 = f''(a)$
>
> $f''(a) < 0 \Rightarrow x = a$ is a maximum
>
> $f''(a) > 0 \Rightarrow x = a$ is a minimum
>
> $f''(a) = 0$ gives *no* information
>
> 3. If $f''(a) = 0$ use a gradient sign diagram to identify the point.

Example 7.10.6

Find and classify the turning points of $y = f(x) \equiv 4x^4 + 2x + 1$.

Solution

1. $$\frac{dy}{dx} = f'(x) = 16x^3 + 2$$

For stationary points $dy/dx = 0$ so $16x^3 + 2 = 0$

$$\therefore x^3 = -\frac{1}{8} \quad \therefore x = -\frac{1}{2} \quad \therefore y = 4(-\tfrac{1}{2})^4 + 2(-\tfrac{1}{2}) + 1 = \frac{1}{4}$$

So there is one stationary point at $(-\frac{1}{2}, \frac{1}{4})$.

2. $$\frac{d^2y}{dx^2} = f''(x) = 48x^2$$

When $x = -\frac{1}{2}$, $d^2y/dx^2 = f''(-\frac{1}{2}) = 48(-\frac{1}{2})^2 > 0$
So the point $(-\frac{1}{2}, \frac{1}{4})$ is a minimum and is the only stationary point.

Example 7.10.7

Find and classify the turning points of

$$y = f(x) \equiv x^3 + 2x^2 + x - 1.$$

Solution

1. $\dfrac{dy}{dx} = f'(x) = 3x^2 + 4x + 1$

   For stationary values $dy/dx = f'(x) = 0$.
   So $3x^2 + 4x + 1 = 0$. Hence $(3x + 1)(x + 1) = 0$, giving $x = -1/3$ or $x = -1$.

   If $x = -\dfrac{1}{3}$, $y = \left(-\dfrac{1}{3}\right)^3 + 2\left(-\dfrac{1}{3}\right)^2 + \left(-\dfrac{1}{3}\right) - 1 = -\dfrac{31}{27}$.

   If $x = -1$, $y = (-1)^3 + 2(-1)^2 + (-1) - 1 = -1$.

2. $\dfrac{d^2y}{dx^2} = f''(x) = 6x + 4$

   When $x = -1/3$, $d^2y/dx^2 = f''(-1/3) = 6(-1/3) + 4 > 0$

   $\left(-\dfrac{1}{3}, -\dfrac{31}{27}\right)$ is a minimum.

   When $x = -1$, $d^2y/dx^2 = f''(-1) = 6(-1) + 4 < 0$

   $(-1, -1)$ is a maximum.

Exercise 7.10.3

Find and classify the turning points for the following equations:

1. $y = 4 + x - x^2$
2. $y = 12x - x^3$
3. $f(x) = 9x + \dfrac{1}{x}$
4. $s = (3t^2 + 1)^2$
5. $v = 6t^2 - 3t^4$

7.11 FURTHER RULES FOR DIFFERENTIATION

Unfortunately if $y = (3x + 2)(x + 1)^{1/2}$ or $y = x^2/(1 + 2x)$ have to be differentiated, it is not possible to differentiate each part separately and then multiply or divide the results. The following rules must be used.

7.11.1 The Product Rule for Differentiation

Let u and v be functions of x, and suppose that $y = uv$. Then

$$\frac{dy}{dx} = u\frac{dv}{dx} + v\frac{du}{dx} \tag{7.11.1}$$

7.11.2 The Quotient Rule for Differentiation

Let u and v be functions of x, and suppose that $y = u/v$. Then

$$\frac{dy}{dx} = \frac{v\dfrac{du}{dx} - u\dfrac{dv}{dx}}{v^2} \tag{7.11.2}$$

Example 7.11.1

If $y = (3x + 2)(x + 1)^{1/2}$, find dy/dx.

Solution

Let $u = 3x + 2$ and $v = (x + 1)^{1/2}$. So $y = uv$.

$$\frac{du}{dx} = 3 \quad \text{and} \quad \frac{dv}{dx} = \tfrac{1}{2}(x + 1)^{-1/2} \quad \text{(by the chain rule)}$$

So

$$\frac{dy}{dx} = u\frac{dv}{dx} + v\frac{du}{dx}$$

$$= (x + 2)\tfrac{1}{2}(x + 1)^{-1/2} + (x + 1)^{1/2} \times 3$$

$$= \tfrac{1}{2}(x + 1)^{-1/2}\,[(x + 2) + 6(x + 1)]$$

$$\frac{dy}{dx} = \tfrac{1}{2}(x + 1)^{-1/2}\,(9x + 8)$$

Example 7.11.2

If $y = x^2/(1 + 2x)$ find dy/dx.

Solution

Let $u = x^2$ and $v = 1 + 2x$, then $y = (u/v) \cdot (du/dx) = 2x$ and $dv/dx = 2$.

So

$$\frac{dy}{dx} = \frac{v\dfrac{du}{dx} - u\dfrac{dv}{dx}}{v^2}$$

$$= \frac{(1 + 2x)2x - x^2 2}{(1 + 2x)^2}$$

$$= \frac{2x^2 + 2x}{(1 + 2x)^2} = \frac{2x(x + 1)}{(1 + 2x)^2}$$

Example 7.11.3

The displacement, s, by a particle at time, t, is given by

$$s = \frac{8t}{1 + t^2}$$

Find an expression for the velocity at time t.

Solution

Velocity $v = ds/dt$. Let $u = 8t$ and $w = 1 + t^2$, then $s = u/w$. w has been chosen to avoid confusion between the velocity v in this problem, and v in Formula (7.11.2).

Replace v by w, y by s and x by t in (7.11.2) to obtain

$$\frac{ds}{dt} = \frac{w\dfrac{du}{dt} - u\dfrac{dw}{dt}}{w^2}$$

Now $u = 8t \Rightarrow du/dt = 8$ and $v = 1 + t^2 \Rightarrow dw/dt = 2t$

Hence

$$v = \frac{ds}{dt} = \frac{(1 + t^2)8 - 8t(2t)}{(1 + t^2)^2}$$

$$v = \frac{8(1 - t^2)}{(1 + t^2)^2}$$

Exercise 7.11.1

Find dy/dx in Problems 1–6.

1. $y = x^2(2x + 1)^3$
2. $y = x\sqrt{x + 1} + x^2$
3. $y = (x^2 + 1)(1 - x^2)^{3/2}$
4. $y = \dfrac{x^3}{4 + x}$
5. $y = \dfrac{2 + 3x}{10 - 7x}$
6. $y = \dfrac{x^2}{\sqrt{1 + 2x^2}} + 2x$

Find ds/dt in Problems 7–10.

7. $s = (1 - t)^{1/2}\,(3 + 2t)$
8. $s = t^2(6 - t)^{1/3}$
9. $s = t + \dfrac{t}{1 + t^2}$
10. $s = t^3 + \dfrac{t^{3/2}}{3 + 2t}$

In Problems 11–12 find and identify the stationary points.

11. $y = \dfrac{x}{x^2 + 1}$
12. $y = \dfrac{x}{(x + 1)^2}$

7.12 FURTHER APPLICATIONS OF MAXIMA AND MINIMA

Maxima and minima have many applications in practical problems. Fortunately it is rarely necessary to identify the nature of the stationary values formally as in Section 7.10, as the interpretation of the problem usually makes this clear.

Example 7.12.1

An open rectangular box is to be made from a 60 cm × 80 cm sheet of metal. What should the dimensions be to contain the maximum possible volume?

Solution

A plan view of the sheet of metal is shown in Figure 7.16. The box will be made by cutting an x by x cm square from each corner.

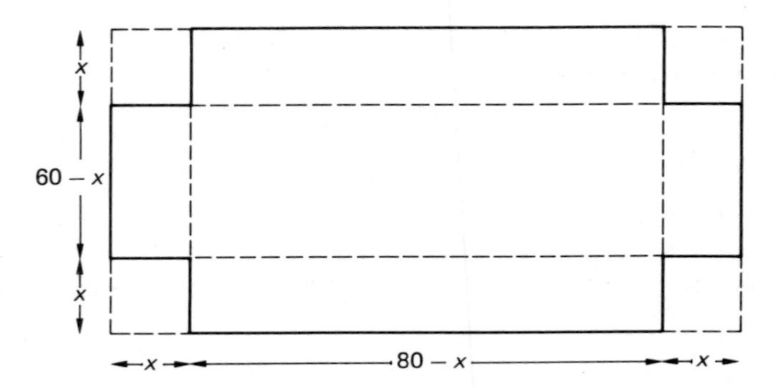

Figure 7.16

So volume of the box is

$$V = (80 - x)(60 - x)x$$
$$V = 4800x - 140x^2 + x^3$$

The problem requires us to maximize V.

$$\frac{dV}{dx} = 4800 - 280x + 3x^2$$

For maximums and minimums $dV/dx = 0$ so

$$4800 - 280x + 3x^2 = 0$$
$$x = 22.63 \text{ cm} \quad \text{or} \quad 70.70 \text{ cm}.$$

Since it is physically impossible to take two 70.70 cm cuts from a sheet of width 60 cm, the required answer is 22.63 cm.

So for the box to have maximum volume its dimensions should be 14.7 cm × 34.7 cm × 22.6 cm.

Example 7.12.2

A cylindrical fuel storage tank must hold 12000 litres. Find its dimensions if its surface area is to be minimized. What is the area of metal used in constructing the tank?

Solution

Work throughout in centimetres. So volume,
$V = 12000\ \text{l} \equiv 1.2 \times 10^7 \text{cm}^3$.

Let the tank have radius r and length l. Then

$$V = \pi r^2 l = 1.2 \times 10^7 \qquad (7.12.1)$$

and the surface area of the tank S consists of its two circular ends and its curved cylindrical surface. So

$$s = 2\pi r^2 + 2\pi r l = 2\pi r(r + l) \qquad (7.12.2)$$

We have to minimize S, but as it stands we cannot differentiate since S is a function of two variables, r and l. However Equation (7.12.1) is available, and this allows r to be expressed in terms of l or conversely. It is easiest to take

$$l = \frac{1.2 \times 10^7}{\pi r^2} \qquad (7.12.3)$$

from (7.12.1) and substitute into (7.12.2) to give

$$S = 2\pi r \left(r + \frac{1.2 \times 10^7}{\pi r^2}\right)$$

Now S is a function of one variable, r. Multiply out the brackets and differentiate with respect to r

$$S = 2\pi r^2 + \frac{2.4 \times 10^7}{r}$$

$$\frac{\mathrm{d}S}{\mathrm{d}r} = 4\pi r - \frac{2.4 \times 10^7}{r^2}$$

For maximums and minimums, $\mathrm{d}S/\mathrm{d}r = 0$. So

$$r^3 = \frac{2.4 \times 10^7}{4\pi}$$

and $r = 124$ cm (to nearest centimetre)

From (7.12.3),

$$l = \frac{1.2 \times 10^7}{\pi(124)^2} = 248 \text{ cm}$$

Therefore to minimize the amount of metal used in constructing the tank, the dimensions should be 248 cm long with a radius of 124 cm. The surface area of metal used is $S = 2\pi(124)(124 + 248) \simeq 2.90 \times 10^5 \text{ cm}^2$ or 29.0 m^2.

Example 7.12.2 illustrates a technique which is very useful in this type of problem. In modelling the situation two equations are obtained. One of these (7.12.2) involves the quantity to be maximized/minimized, as a function of two variables.

The other (7.12.1) is a constraint equation which enables a relation to be obtained between the variables. Using this (7.12.2) is reduced to a function of one variable which can be differentiated.

Exercise 7.12.1

1. A 3 m beam AB carries a point load 2 m from A. At a distance of x m from A, the deflection in mm is

$$y = 0.25 \times (8 - x^2).$$

Find the maximum deflection of the beam and its distance from A.

2. The power dissipated in an electric current is given by $P = I^2R$ watts where I is the current flowing (amperes) and R is the resistance (ohms). A circuit consists of two resistors R_1 and R_2 in parallel. The sum of their resistances is 100 kΩ and a constant current of 2 mA flows in the circuit. Find the values of R_1 and R_2 which maximize the power in the circuit. What is the maximum power? You may use the fact that the total resistance R in the parallel circuit is given by

$$\frac{1}{R} = \frac{1}{R_1} + \frac{1}{R_2}$$

3. The fuel economy E of a car in miles per gallon is

$$E = 35 + 2.07 \times 10^{-2}v^2 - 3.85 \times 10^{-6}v^4$$

where v is the speed in miles per hour ($5 \leqslant v \leqslant 70$). What is the most economical fuel consumption, and at what speed is it achieved?

4. A contractor has 150 m of security fencing and wishes to enclose a rectangular area (Figure 7.17.) The area is divided into two parts with fencing. The areas, A_1 and A_2, are such that $A_1 = 2A_2$. What are the dimensions of the enclosure so that the total enclosed area is a maximum?

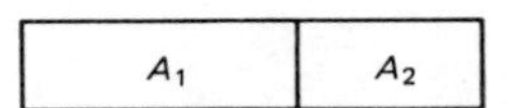

Figure 7.17

5. A window is in the shape of a rectangle with a semicircle on top of it. Lead beading is applied around its edges, and also as framing (Figure 7.18). If 700 cm of lead beading is available, what are the dimensions of the window such that its area is a maximum. (You may assume the beading is of negligible width.)

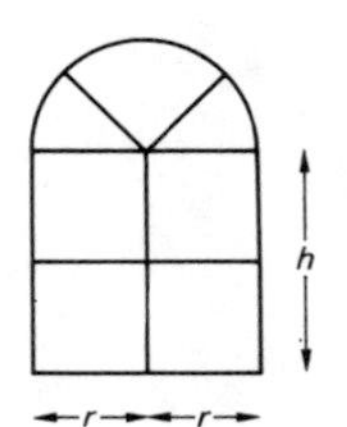

Figure 7.18

6. Two oil pipes with circular cross-sections are such that the sum of their diameters must not exceed 1 m. What should their diameters be to give the maximum possible combined cross sectional area? (When you have differentiated, formally identify your answer as a maximum or minimum. If you cannot obtain the answer by differentiation, try a sketch graph.)

7.13 DIFFERENTIATION OF TRIGONOMETRIC FUNCTIONS

Before deriving the results for differentiating $\sin x$ and $\cos x$, you are asked to carry out the following numerical investigation.

Exercise 7.13.1

In this investigation use the program you developed for numerical differentiation in Exercise 7.3.

1. Most computers are programmed to work with angles in radians for trigonometric functions. Make sure that your computer is working in radians. For a range of angles for x from 0 to 2π ($\simeq 6.3$), plot the slope of $\sin x$ against x. What graph do you obtain?
2. Repeat Question 1 with $\cos x$.

7.13.1 Differentiation of sin x

In order to differentiate $\sin x$, the value of the limit of

$$\frac{\sin x}{x}$$

as x approaches zero is required. The value of this limit is 1. This result will not be proved, but it may be confirmed by a numerical investigation.

Exercise 7.13.2

Set your calculator to radian mode. Make a table with headings x and $\sin x/x$, and work out $\sin x/x$ for $x = 1$. 0.1, 0.01, 0.001, 0.0001 and 0.00001. Confirm

$$\lim_{x\to 0}\left(\frac{\sin x}{x}\right) = 1$$

provided x is in radians.

7.13.2 Differentiation of sin x (continued)

Let $y = \sin x$

then $y + \Delta y = \sin(x + \Delta x)$

Subtracting

$$\Delta y = \sin(x + \Delta x) - \sin x$$

$$= 2\cos\left(x + \frac{\Delta x}{2}\right)\sin\left(\frac{\Delta x}{2}\right)$$

from the factor formula in Chapter 5.

So

$$\frac{\Delta y}{\Delta x} = \frac{2\cos\left(x + \frac{\Delta x}{2}\right)\sin\left(\frac{\Delta x}{2}\right)}{\Delta x}$$

$$= \frac{\cos\left(x + \frac{\Delta x}{2}\right)\sin\left(\frac{\Delta x}{2}\right)}{\frac{\Delta x}{2}}$$

$$= \cos\left(x + \frac{\Delta x}{2}\right)\left(\frac{\sin\left(\frac{\Delta x}{2}\right)}{\frac{\Delta x}{2}}\right)$$

Now as $\Delta x \to 0$, $\cos\left(x + \frac{\Delta x}{2}\right) \to \cos x$ and $\left(\frac{\sin\left(\frac{\Delta x}{2}\right)}{\frac{\Delta x}{2}}\right) \to 1$

provided Δx is in radians. Hence

$$\frac{\mathrm{d}y}{\mathrm{d}x} = \lim_{\Delta x \to 0} \left[\cos\left(x + \frac{\Delta x}{2}\right)\left(\frac{\sin\left(\frac{\Delta x}{2}\right)}{\frac{\Delta x}{2}}\right)\right] = \cos x$$

So,

$$y = \sin x \Rightarrow \frac{\mathrm{d}y}{\mathrm{d}x} = \cos x \quad \text{provided } x \text{ is in radians} \tag{7.13.1}$$

7.13.3 Differentiation of cos x

The identity $\cos A - \cos B \equiv -2 \sin\left(\frac{A + B}{2}\right)\sin\left(\frac{A - B}{2}\right)$ (from the factor formulas in Chapter 5) is required, as well as the limit

$\lim_{x \to 0}\left(\frac{\sin x}{x}\right) = 1$ provided x is in radians.

Let $y = \cos x$
so

$$y + \Delta y = \cos(x + \Delta x)$$

So,

$$\Delta y = \cos(x + \Delta x) - \cos x$$

$$= -2 \sin\left(x + \frac{\Delta x}{2}\right)\sin\left(\frac{\Delta x}{2}\right)$$

and

$$\frac{\Delta y}{\Delta x} = \frac{-2\sin\left(x + \frac{\Delta x}{2}\right)\sin\left(\frac{\Delta x}{2}\right)}{\Delta x}$$

$$= -\sin\left(x + \frac{\Delta x}{2}\right)\left(\frac{\sin\left(\frac{\Delta x}{2}\right)}{\frac{\Delta x}{2}}\right)$$

$$\frac{dy}{dx} = \lim_{\Delta x \to 0}\left[-\sin\left(x + \frac{\Delta x}{2}\right)\left(\frac{\sin\left(\frac{\Delta x}{2}\right)}{\frac{\Delta x}{2}}\right)\right]$$

Now as $\Delta x \to 0$, $\sin\left(x + \frac{\Delta x}{2}\right) \to \sin x$ and $\frac{\sin\left(\frac{\Delta x}{2}\right)}{\frac{\Delta x}{2}} \to 1$

provided x is in radians. Hence,

$$y = \cos x \Rightarrow \frac{dy}{dx} = -\sin x \tag{7.13.2}$$

provided x is in radians

7.13.4 Differentiation of tan x

Since results (7.13.1) and (7.13.2) are available, it is not necessary to differentiate tan x from first principles.

$$y = \tan x = \frac{\sin x}{\cos x}$$

Let $u = \sin x$ and $v = \cos x$, then $y = u/v$, $du/dx = \cos x$ and $dv/dx = -\sin x$. So by the quotient rule

$$\frac{dy}{dx} = \frac{v\dfrac{du}{dx} - u\dfrac{dv}{dx}}{v^2}$$

$$= \frac{\cos x \cos x - \sin x\,(-\sin x)}{\cos^2 x}$$

$$= \frac{\cos^2 x + \sin^2 x}{\cos^2 x}$$

$$= \frac{1}{\cos^2 x} \text{ by identity 5.1 in Chapter 5}$$

$$= \sec^2 x$$

$y = \tan x \Rightarrow \dfrac{dy}{dx} = \sec^2 x$
provided x is in radians

Example 7.13.1

If $y = \cos(\frac{1}{2}x)$, find dy/dx.

Solution

$y = \cos(\frac{1}{2}x)$. Let $u = \frac{1}{2}x$ so $y = \cos u$.

$$\frac{du}{dx} = \frac{1}{2} \quad \text{and} \quad \frac{dy}{du} = -\sin u = -\sin(\tfrac{1}{2}x)$$

$$\frac{dy}{dx} = \frac{dy}{du} \cdot \frac{du}{dx} = -\tfrac{1}{2}\sin(\tfrac{1}{2}x)$$

Example 7.13.2

If $y = \sin^3 x$, find dy/dx.

Solution

$y = \sin^3 x = (\sin x)^3$. Let $u = \sin x$, so $y = u^3$.

$$\frac{du}{dx} = \cos x \quad \text{and} \quad \frac{dy}{du} = 3u^2 = 3(\sin x)^2 = 3\sin^2 x$$

$$\frac{dy}{dx} = \frac{dy}{du} \cdot \frac{du}{dx} = 3 \sin^2 x \cos x$$

Example 7.13.3

If $s = t^4 \cos \omega t$, where ω is a constant, find ds/dt.

Solution

$$s = t^4 \cos \omega t$$

Let $u = t^4$ and $v = \cos \omega t$, so $s = uv$.

$$\frac{du}{dt} = 4t^3 \quad \text{and} \quad \frac{dv}{dt} = -\omega \sin \omega t \quad \text{(by the chain rule)}$$

$$\frac{ds}{dt} = u\frac{dv}{dt} + v\frac{du}{dt} = -\omega t^4 \sin \omega t + 4t^3 \cos \omega t$$

$$\frac{ds}{dt} = t^3(4 \cos \omega t - \omega t \sin \omega t)$$

Exercise 7.13.3

1. By writing $\cot x$ as $\cos x/\sin x$ find $d/dx\ (\cot x)$.
2. By writing $\sec x$ as $(\cos x)^{-1}$ and using the chain rule, find $d/dx\ (\sec x)$.
3. Find $d/dx\ (\operatorname{cosec} x)$.

In Problems 4–10, differentiate the expression with respect to x.

4. $\cos 5x$
5. $\tan(\frac{1}{2}x)$
6. $\cos x + 2 \sin x$
7. $\sin^2 x (\equiv (\sin x)^2)$
8. $3 \tan^2 x$
9. $\sin(x^2)$
10. $\cos^2 2x$
11. $\sin \alpha \cos x$ (α = constant)
12. $\sin x \cos x$

In Problems 13–18 you are given the displacement, s, of a particle. Find its velocity and acceleration. Throughout ω and α are constants.

13. $s = 2 \sin \omega t$
14. $s = \operatorname{cosec} 2\omega t$
15. $s = t \cos \omega t$
16. $s = \cos(2\omega t + \alpha)$
17. $s = \sin^2(\omega t - \alpha)$
18. $s = \cos \omega t \sin \omega t$

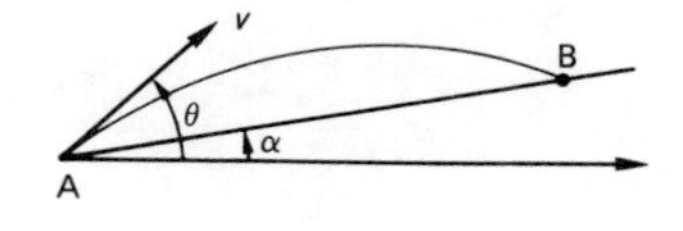

Figure 7.19

19. A particle is projected from A at an angle of θ to the horizontal and hits a plane inclined at an angle of α to the horizontal at B (Figure 7.19.).
    The range up the plane is

$$AB = \frac{2v^2\cos\theta\,\sin(\theta - \alpha)}{g\cos^2\alpha}$$

    Assuming that v, g and α are constants, find the value of θ which maximizes AB.

20. Figure 7.20 shows a cone with slant height l and vertical angle 2θ. Show its volume is $\frac{1}{3}\pi l^3 \sin^2\theta\cos\theta$. Assuming that l is constant, find the value of θ which gives the maximum volume. What is the maximum volume?

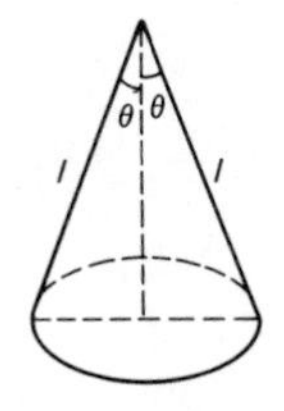

Figure 7.20

7.14 DIFFERENTIATION OF EXPONENTIAL FUNCTIONS

The exponential functions whose graphs are given by $y = 2^x$, $y = e^x$ and $y = 3^x$ are shown in Figure 7.21.

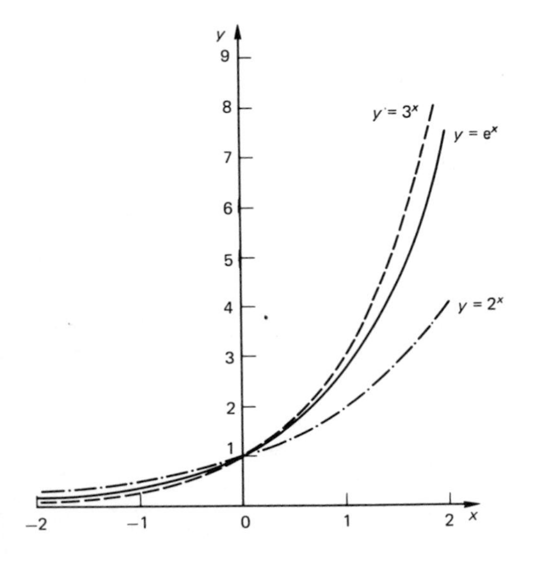

Figure 7.21

From the graph, if $x > 0$ the slope of 3^x is greater than the slope of e^x which in turn is greater than the slope of 2^x.

Exercise 7.14.1

1. Make a table with the following headings:

x	2^x	Slope	e^x	Slope	3^x	Slope

Use the program you wrote for Exercise 7.3.2 or a calculator to obtain numerical approximations for the slopes of these three functions for values of x between -2 and 2. Take an increment for x of 0.25. Graph your results.

2. When $x = 1.5$ your results for 2^x may be $2^x = 2.83$ and slope $= 1.96$. The ratio

$$\frac{\text{slope}}{2^x} = \frac{1.96}{2.83} = 0.69 \quad \text{when} \quad x = 1.5$$

Calculate this ratio for other values of x and confirm it is constant.

3. Calculate the ratio

$$\frac{\text{slope}}{e^x}$$

for various values of x. Confirm that this ratio is constant. What is its value?

4. Repeat Problem (3) for

$$\frac{\text{slope}}{3^x}$$

5. Let $y = e^x$. Then if x increases to $x + \Delta x$, y will change to $y + \Delta y$ where

$$y + \Delta y = e^{x+\Delta x}$$

$$\text{So } \Delta y = e^{x+\Delta x} - e^x = e^x(e^{\Delta x}-1)$$

and

$$\frac{\Delta y}{\Delta x} = \frac{e^x(e^{\Delta x}-1)}{\Delta x}$$

Use your calculator to show that as $\Delta x \to 0$, $\Delta y/\Delta x \to e^x$.

7.14.1 Differentiation of e^x

From the numerical investigations of differentiating exponential functions in Exercise 7.14.1, we obtain the result

$$\frac{\text{slope}}{e^x} = 1 \quad \text{for all values of } x$$

and the result

$$\text{if } y = e^x \text{ then } \lim_{\Delta x \to 0} \frac{\Delta y}{\Delta x} = e^x$$

$$\boxed{y = e^x \Rightarrow \frac{dy}{dx} = e^x}$$

This result can be rigorously proved, but we shall not do so here. The result has been confirmed by the numerical investigations and will be used in the following examples and exercises.

Example 7.14.1

If $y = e^{-x}$ find dy/dx.

Solution

$y = e^{-x}$. Let $u = -x$, so $y = e^u$. Then $du/dx = -1$ and $dy/du = e^u = e^{-x}$.

$$\therefore \frac{dy}{dx} = \frac{dy}{du} \cdot \frac{du}{dx} = -e^{-x}$$

Example 7.14.2

If $f(x) = 5e^{x^2}$ find $f'(x)$.

Solution

Let $y = f(x) = 5e^{x^2}$. Then $f'(x) = dy/dx$.
Let $u = x^2$ so $y = 5e^u$. Then $du/dx = 2x$ and $dy/du = 5e^u = 5e^{x^2}$
So

$$f'(x) = \frac{dy}{dx} = \frac{dy}{du}\frac{du}{dx} = 5e^{x^2} \times 2x = 10xe^{x^2}$$

Example 7.14.3

If $s = e^{3t}/(1 + t^2)$ find ds/dt.

Solution

Let $u = e^{3t}$ and $v = 1 + t^2$ so $s = u/v$.
So

$$\frac{du}{dt} = 3e^{3t} \text{ (by the chain rule) and } \frac{dv}{dt} = 2t$$

$$\frac{ds}{dt} = \frac{v\dfrac{du}{dt} - u\dfrac{dv}{dt}}{v^2} = \frac{(1 + t^2)3e^{3t} - e^{3t}(2t)}{(1 + t^2)^2}$$

$$\frac{ds}{dt} = \frac{e^{3t}(3t^2 - 2t + 1)}{(1 + t^2)^2}$$

Exercise 7.14.2

Throughout this exercise a, α and ω are constants. In Problems 1–6 find $f'(x)$.

1. $f(x) = e^{-6x}$
2. $f(x) = \frac{1}{2}(e^x + e^{-x})$
3. $f(x) = e^{-x^2}$
4. $f(x) = xe^{2x}$
5. $f(x) = e^{1-x}$
6. $f(x) = x^2e^{-x}$

In Problems 7–12 find ds/dt.

7. $s = e^{-2t} \sin t$
8. $s = e^{at} \cos \omega t$
9. $s = \dfrac{e^t - e^{-t}}{e^t + e^{-t}}$
10. $s = \dfrac{\sin(\omega t + \alpha)}{e^t}$
11. $s = t^3e^{t^2}$
12. $s = \dfrac{e^{at}}{t}$
13. Find the equation of the tangent and normal to $y = e^x$ at the point P(1, e). This tangent and normal cut the x axis at A and B. Find the coordinates of A and B and the area of the triangle ABP.
14. Identify the turning point on the graph of $y = xe^x$.
15. A particle moves along the x axis such that its displacement x from the origin at time $t \geqslant 0$ is

$$x = 5e^{-3t} \sin 2t$$

Find its velocity and acceleration at time t, and show that

$$\frac{d^2x}{dt^2} + 6\frac{dx}{dt} + 13x = 0$$

Find when the velocity is first zero, and show that the particle reverses its motion once every $\pi/2$ seconds.

7.14.2 Differentiation of a^x, a = constant $\neq$ e, $a > 0$

To differentiate 2^x, 3^x,... etc, the method is to change the function to a power of e and then differentiate.

Example 7.14.4

If $y = 5^x$, find dy/dx.

Solution

$$y = 5^x \Rightarrow \ln y = x \ln 5 \Rightarrow e^{\ln y} = e^{x\ln 5} \Rightarrow y = e^{x \ln 5}$$

Now ln 5 is a constant so

$$\frac{dy}{dx} = \ln 5 \, e^{x \ln 5}$$

But

$$e^{x \ln 5} = 5^x, \quad \text{so} \quad \frac{dy}{dx} = 5^x \ln 5$$

Example 7.14.5

If $y = 2^{-x}$, find dy/dx.

Solution

$$y = 2^{-x} \Rightarrow \ln y = -x \ln 2 \Rightarrow y = e^{-x \ln 2}$$

$$\frac{dy}{dx} = e^{-x \ln 2} \times (-\ln 2) = -2^{-x} \ln 2$$

(since ln 2 is a constant.)

Exercise 7.14.3

1. Use the method of Examples 7.14.4 and 7.14.5 to find dy/dx if $y = a^x$ where $a =$ constant $\neq 0 > 0$.
2. Use your answer to Question (1) to find dy/dx if $y = 2^x$ and $y = 3^{2x}$ (i.e. $-(3^2)^x$).

7.15 DIFFERENTIATION OF LOGARITHMIC FUNCTIONS

7.15.1 Differentiation of ln *x*

$$\text{If } y = \ln x, \quad \text{then } \frac{dy}{dx} = \frac{1}{x} \tag{7.15.1}$$

A formal proof of this result will not be given. The next exercise confirms the result by a numerical investigation.

Exercise 7.15.1

Use your program from Exercise 7.3.2 to complete a table with the following headings.

x	$\ln x$	Gradient	$\frac{1}{x}$

Take values of x from -0.5 to 2.5 in increments of 0.5. Confirm that the gradient is given by $1/x$.

Result (7.15.1) can be used to differentiate other logarithmic functions.

Example 7.15.1

If $y = \ln(x^2 + 3)$, find dy/dx.

Solution

Let $u = x^2 + 3$, so $y = \ln u$.

$$\frac{du}{dx} = 2x \quad \text{and} \quad \frac{dy}{du} = \frac{1}{u} = \frac{1}{x^2 + 3}$$

So

$$\frac{dy}{dx} = \frac{dy}{du}\frac{du}{dx} = \frac{2x}{x^2 + 3}$$

Example 7.15.2

If $f(x) = \ln(9x)$, find $f'(x)$.

Solution

A substitution could be used, but it is easier to use a property of logarithmic functions, namely

$$\ln(AB) = \ln A + \ln B$$

(from the property of logarithms, see Section 3.5.2).

$$\therefore f(x) = \ln(9x) = \ln 9 + \ln x$$

and

$$f'(x) = 0 + \frac{1}{x} = \frac{1}{x}$$

since $\ln 9$ is a constant.

Example 7.15.3

If $s = 6t^3 \ln(1 - t)$, find ds/dt.

Solution

s is a product of two functions, so let $u = 6t^3$ and $v = \ln(1 - t)$. Then $s = uv$, and

$$\frac{du}{dt} = 18t^2 \quad \text{and} \quad \frac{dv}{dt} = \frac{-1}{1 - t} \quad \text{(by the chain rule)}$$

$$\therefore \frac{ds}{dt} = u\frac{dv}{dt} + v\frac{du}{dt} = \frac{-6t^3}{1 - t} + \ln(1 - t) \times 18t^2$$

$$\frac{ds}{dt} = 18t^2 \ln(1 - t) + \frac{6t^3}{t - 1}$$

Exercise 7.15.2

Throughout this exercise a is a constant. Find $f'(x)$ in Problems 1–6.

1. $f(x) = \ln(5x)$
2. $f(x) = \ln x^2$
3. $f(x) = \ln x^2 + \ln 6$
4. $f(x) = \ln(6x^2)$
5. $f(x) = \ln x^3 + \ln a$
6. $f(x) = \ln(ax^{1/2})$

Find ds/dt in Problems 7–16.

7. $s = t + \ln(3.02t)$
8. $s = \ln(1 + t)$
9. $s = \dfrac{1}{t} + \ln\left(\dfrac{5}{t}\right)$
10. $s = \ln(1 + at^2)$
11. $s = a^2 \ln(2t)$
12. $s = t^2 \ln(2t)$
13. $s = \dfrac{\ln t^2}{t}$
14. $s = \dfrac{t^2}{\ln t}$
15. $s = \ln(e^t)$
16. $s = (\ln t)(\ln t^3)$
17. Find the equation of the tangent to the curve $y = \ln x$ at the point $(2, \ln 2)$.
18. Find the coordinates of all the stationary points of $y = \ln(1 + x^2)$ and classify them.

7.16 APPLICATIONS OF DIFFERENTIATION – CURVE SKETCHING

Curve sketching is a useful skill, even if you have a computer with a graph drawing package. Curve sketching is as much an art as a science, but some useful points to consider are:

1. Is the function you are sketching defined for all x? For example $1/x$ is not defined for $x = 0$, $\sqrt{x + 1}$ is not defined for $x < -1$.
2. Where does the graph cut the coordinate axes?
3. Does the graph have stationary values?
4. When is the graph concave up or concave down?

$$\frac{d^2y}{dx^2} > 0 \Rightarrow \text{concave up}$$

$$\frac{d^2y}{dx^2} < 0 \Rightarrow \text{concave down.}$$

5. What happens as $x \to \infty$? As $x \to -\infty$?
6. What happens as x approaches values where the function is undefined?

Example 7.16.1

Sketch

$$y = \frac{2x + 1}{x - 1}$$

Solution

The reference letters are marked on Figure 7.22.

When $x = 1$, $x - 1 = 0$, so y is not defined when $x = 1$, since the denominator is zero.

Now as $x \to 1$ from above (e.g. $x = 1.1, 1.01, 1.001,...$) then $(x - 1) \to 0$ from above $(x - 1 = 0.1, 0.01, 0.001,...)$ and $2x - 1 \to 2 - 1 = 1$. So $y \to \infty$ (see A, Figure 7.22).

If $x \to 1$ from below (e.g. $x = 0.9, 0.99, 0.999,...$) then $x - 1 \to 0$ from below $(x - 1 = -0.1, -0.01, -0.001,...)$ and $2x - 1 \to 1$. So $y \to \infty$ (see B, Figure 7.22).

When $x = 0$, $y = -1$ (see C, Figure 7.22).

When $y = 0$, $2x + 1 = 0$, so $x = -\frac{1}{2}$ (see D, Figure 7.22).

$$\frac{dy}{dx} = \frac{(x - 1)(2) - (2x + 1)(1)}{(x - 1)^2} = -\frac{3}{(x - 1)^2}$$

so

$$\frac{dy}{dx} \neq 0 \text{ for any value of } x.$$

Moreover $dy/dx < 0$ for all $x \neq 1$, so the function has a graph which is decreasing.

$$\frac{d^2y}{dx^2} = 6(x - 1)^3$$

So if $x > 1$, $\dfrac{d^2y}{dx^2} > 0$ (concave up) (see E, Figure 7.22) and if $x < 1$, $\dfrac{d^2y}{dx^2} < 0$ (concave down) (see F, Figure 7.22)

$$y = \frac{2x + 1}{x - 1} = \frac{2x - 2 + 3}{x - 1} = 2 + \frac{3}{x - 1}$$

Now as $x \to \infty$, $3/(x - 1) \to 0$, so $y \to 2 + 0 = 2$ and as $x \to -\infty$, $y \to 2$ (see G, Figure 7.22).

The graph appears in Figure 7.22. Note that no values are plotted, this is a sketch.

The lines $y = 2$ and $x = 1$ are asymptotes for the graph (see H, Figure 7.22).

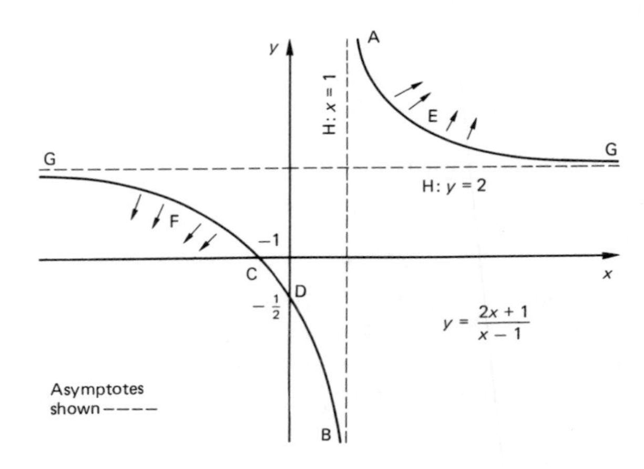

Figure 7.22

Example 7.16.2

Sketch $y = e^{-x} \sin x$.

Solution

Refer to Figure 7.23 for the sketch. y is defined for all values of x.

When $x = 0$, $y = e^{-0} \sin 0 = 0$ (see E, Figure 7.23).

When $y = 0$, $0 = e^{-x}\sin x$.

Now $e^{-x} \neq 0$, so $\sin x = 0$. $\therefore$ $x = n\pi$, for $n = 0, \pm 1, \pm 2, \ldots$ (see B, Figure 7.23).

$$\frac{dy}{dx} = e^{-x} (\cos x - \sin x)$$

$$\frac{dy}{dx} = 0 \Rightarrow e^{-x}(\cos x - \sin x) = 0$$

Now $e^{-x} \Rightarrow 0$ so $\cos x - \sin x = 0$, $\tan x = 1$
So

$$x = \frac{\pi}{4} + n\pi, \text{ for } n = 0, \pm 1, \pm 2, \ldots$$

giving

$$y = e^{(\pi/4 + n\pi)} \sin(\pi/4 + n\pi).$$

Now

$$\sin\left(\frac{\pi}{4} + n\pi\right) = \frac{1}{\sqrt{2}}(-1)^{n+1}$$

So the turning points are

$$A\left(\frac{\pi}{4}, \frac{1}{\sqrt{2}} e^{-\pi/4}\right), \qquad B\left(\frac{5\pi}{4}, -\frac{1}{\sqrt{2}} e^{-5\pi/4}\right)$$

$$C\left(-\frac{3\pi}{4}, -\frac{1}{\sqrt{2}} e^{3\pi/4}\right), \qquad D\left(-\frac{7\pi}{4}, \frac{1}{\sqrt{2}} e^{7\pi/4}\right)$$

and so on (Figure 7.23).
As $x \to \infty$, $e^{-x} \to 0$, so $y \to e^{-x} \sin x \to 0$ since $-1 \leqslant \sin x \leqslant 1$.

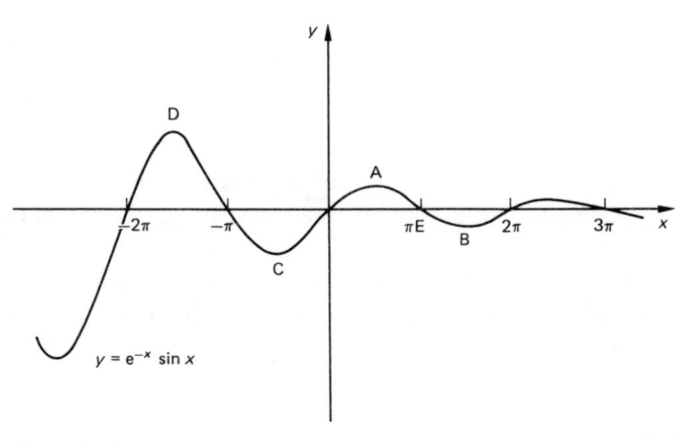

Figure 7.23

As $x \to -\infty$, $e^{-x} \to \infty$. As $x \to -\infty$, $\sin x$ oscillates between $+1$ and -1 inclusive. So $y = e^{-x} \sin x$ has unbounded oscillation as $x \to -\infty$

Example 7.16.3

Sketch

$$y = \frac{x^2 + 1}{x + 2}$$

Solution

(See Figure 7.24).

When $x = -2$, $x + 2 = 0$, so y is not defined.

Now as $x \to -2$, $x^2 + 1 \to 5$.

As $x \to -2$ from below (e.g. $x = -2.1, -2.01, -2.001, \ldots$) $(x + 2) \to 0$ from below ($x + 2 = -0.1, -0.01, -0.001, \ldots$) so $y \to -\infty$.

As $x \to -2$ from above (e.g. $x = -1.9, -1.99, -1.999, \ldots$) $(x + 2) \to 0$ from above so $y \to \infty$.

So $x = -2$ is an asymptote (see A, Figure 7.24).

When $x = 0$, $y = \frac{1}{2}$ (see B, Figure 7.24).

$$\text{When } y = 0, \frac{x^2 + 1}{x + 2} = 0, \text{ so } x^2 + 1 = 0.$$

But there is no value of x to satisfy this, so the graph does not cut the x axis.

$$y = \frac{x^2 + 1}{x + 2} \Rightarrow \frac{dy}{dx} = \frac{x^2 + 4x - 1}{(x + 2)^2}$$

$$\frac{dy}{dx} = 0 \Rightarrow x^2 + 4x - 1 = 0 \Rightarrow x = -2 \pm \sqrt{5}$$

i.e. $x \simeq -4.236$ or $x \simeq 0.236$. So there are stationary points at $(-4.236, -8.472)$ and $(0.236, 0.472)$ to an accuracy of three decimal places (see C, D, Figure 7.24).

It would be possible formally to classify the nature of these points, but d^2y/dx^2 is fairly difficult to find. The sketch should make identification easy.

What happens as $x \to \infty$?

$$y = \frac{x^2 + 1}{x + 2} = x - 2 + \frac{5}{x + 2} \quad \text{by long division.}$$

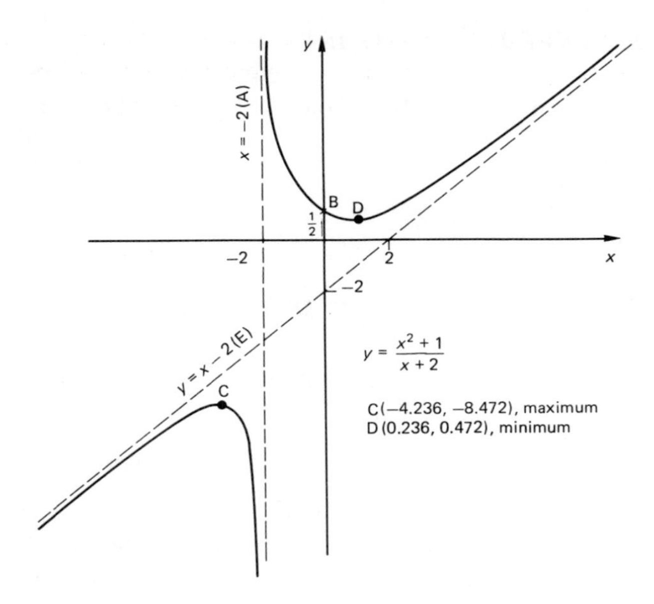

Figure 7.24

Now as

$$x \to \pm\infty, \frac{5}{x + 2} \to 0$$

so as

$$x \to \pm\infty, y \to x - 2$$

Hence $y = x - 2$ is an asymptote (see E, Figure 7.24).

Exercise 7.16.1

Sketch the following curves:

1. $y = x^3 - 3x + 1$
2. $y = x^3 + 3x + 1$
3. $y = \dfrac{3x + 1}{1 - 2x}$
4. $y = \dfrac{x^2 - 8}{x - 3}$
5. $y = x \sin x$
6. $y = e^x \cos 2x$
7. $y = x\, e^x$
8. $y = e^{-x^2}$
9. $y = x \ln x$
10. $y = \dfrac{e^{2x}}{x + 1}$

7.17 MISCELLANEOUS EXERCISES

These exercises include topics from the whole of the present chapter.

Exercise 7.17.1

In Problems 1–9 find $f'(x)$.

1. $f(x) = 3\sin(2x)$
2. $f(x) = \tan\left(\frac{\pi}{4} - \frac{x}{2}\right)$
3. $f(x) = \sin^2 x + \cos^2 x$
4. $f(x) = x^2\,\mathrm{e}^{-2x}\sin(2x)$
5. $f(x) = \frac{\cos(2x)}{x^2}$
6. $f(x) = \sin(2x)\cos(3x)$
7. $f(x) = \mathrm{e}^{3x}\ln(2x+1)$
8. $f(x) = \tan(x)\sin(2x)$
9. $f(x) = 3\cos(1-4x)$

Differentiate with respect to t in Problems 10–17.

10. $5\mathrm{e}^{4t}$
11. $7\mathrm{e}^{-2t}$
12. $t\,\mathrm{e}^{2t}$
13. e^{t^2}
14. $\mathrm{e}^{2t}\sin 4t$
15. t^2/e^{4t}
16. $\ln(t^2)$
17. $\ln(3-4t)$
18. An object moves in a straight line so that its position after t seconds is given by

    $$s = 1 + 2t + 5t^2$$

    Find the velocity of the object when (a) $t = 0$, (b) $t = 2$ and (c) $t = 4$.
19. The temperature at a point on a heated rod varies according to the rule

    $$T = 20 + 100\mathrm{e}^{-5t}$$

    Find the rate of change of temperature at (a) $t = 0$ and (b) $t = 1$.
20. The population of a yeast culture is given by the formula

    $$M(t) = 4.3\,\mathrm{e}^{-2.1t}$$

    Find the rate of change of the population.
21. A particle moves so that its displacement as a function of time is

    $$s(t) = 0.3\sin(0.7t)$$

Find the velocity of the particle when (a) $t = 0$ and (b) $t = 1$. Find the acceleration of the particle when (c) $t = 0$ and (d) $t = 1$.

22. The radius r cm of a circular blot on a piece of blotting paper t seconds after it was first viewed is given by

$$r(t) = 12 - \frac{9}{t}$$

(a) Calculate the radius of the blot after 3 seconds.
(b) Find the time when the blot has radius 3 cm.
(c) Find the rate at which the radius is changing when the radius is 3 cm. Is the radius then increasing or decreasing?
(d) What is the largest value of r?

23. A stone thrown upwards to that its height in metres is given by

$$h(t) = 5 + 21t - 5t^2$$

where t is in seconds.

(a) Find the velocity of the stone when $t = 1, 2, 3$ and 4.
(b) For what range of values of t is the stone
    (i) climbing,
    (ii) falling?

24. A particle moves along the y axis so that its position at time $t \geqslant 0$ is given by

$$y = \frac{4t^2 + t + 4}{t^2 + 1}$$

Find when its velocity is zero. What is its greatest distance from the origin? What happens to the particle as t increases?

25. A particle moves along the x axis so that its position at time $t \geqslant 0$ is given by

$$x = (2t + 3)e^{-0.5t}$$

Show that its velocity is positive for $0 \leqslant t < \frac{1}{2}$. When is it travelling at its greatest velocity? What is its velocity at this time?

26. The velocity, $f(\text{ms}^{-1})$ of a jet of water issuing from a hose at the base of a water tower is given by $v^2 = 2gh$ where $g = 9.8 \text{ ms}^{-2}$ and h is the height in metres of the water surface in the tower. If h is falling at 2 cm per second, what is the rate of change of the velocity of the water leaving the hose when $h = 30$ m?

27. Find the coordinates of the point on $y = x^2 - 3x + 1$ where the gradient of the curve is 3. Hence find the equation of the normal to the curve at that point.

28. Let $y = x^3 + 3x^2 + 12x - 3$.
    (a) Find the value of y when $x = 2$.
    (b) Find the value of dy/dx when $x = 2$.
    (c) Find the equation of the tangent at the point on the curve where $x = 2$.
    (d) Find the equation of the normal at the same point.

29. The power delivered into the load X of a class A amplifier of output resistance R is given by

$$P(X) = \frac{V^2X}{(X + R)^2}$$

where V is the output voltage.
    (a) Find the value of X such that P is a maximum.
    (b) Sketch a graph of the function $P(X)$.

30. Frequency stability in the cathode-coupled oscillator can be studied with the aid of the correction factor

$$f(\alpha) = 1 - \frac{L}{16C}(\alpha - 1/R)^2$$

where L, R and C are constants.
    (a) Show that the maximum value of f is obtained when $\alpha = 1/R$.
    (b) Sketch a graph of the function $f(\alpha)$.

31. Sketch the curves of the following functions:

(a) $f(x) = x + \dfrac{1}{x}$

(b) $f(x) = \dfrac{(x + 2)}{(x - 6)}$

(c) $f(x) = \dfrac{x^2}{(1 - x^2)}$

(d) $f(x) = \dfrac{(x - 3)}{(x + 1)}$

32. For a belt drive the formula relating the power transmitted P to the speed of the belt V is given by

$$P(v) = Tv - av^3$$

where T is the tension in the belt and a is a constant.
(a) Find the speed such that the belt delivers maximum power.
(b) Sketch the graph of the function $P(v)$.

OBJECTIVE TEST 7

1. If the displacement of a particle is $s = 3t^2 - \dfrac{1}{1 + t}$, at time t, find the average velocity between $t = 2$ and $t = 3$ and the instantaneous velocity at $t = 2.5$.
2. Find the gradient of $y = \mathrm{e}^{-0.1x} \sin 4x$ when $x = 1$.
3. If $f(x) = (1 - x^2) \ln x$, find $f'(x)$.
4. If $u = \dfrac{1}{1 - p}$, find $\dfrac{\mathrm{d}u}{\mathrm{d}p}$ and $\dfrac{\mathrm{d}^2u}{\mathrm{d}p^2}$.
5. Find the equation of the tangent to $y = x + \mathrm{e}^{-x}$ at (0, 1).
6. The radius of a sphere is increasing at the rate of 0.8 mms^{-1}. What is the rate of increase of the volume of the sphere when the radius is 60 mm?
7. The displacement s of a particle at time t is given by $s = (t^2 - 4)\,\mathrm{e}^{-t}$ for $t \geqslant 0$. Find its maximum velocity.
8. Find the coordinates of the stationary point of $y = x^2 \ln x$.
9. If $y = \dfrac{\mathrm{e}^{x^2}}{1 + x}$, find $\dfrac{\mathrm{d}y}{\mathrm{d}x}$.
10. If $y = \ln(x \sin x)$, find $\dfrac{\mathrm{d}y}{\mathrm{d}x}$.

8 INTEGRATION

OBJECTIVES

When you have completed this chapter you should be able to

1. approximate areas by summing the areas of rectangles
2. approximate areas by summing the areas of trapezia
3. express a definite integral as a limit of a sum
4. relate definite integrals to areas and state the convention for the signs attached to areas
5. state the relationship between integration and differentiation
6. integrate polynomial, exponential, power and trigonometric functions
7. apply substitutions to evaluate integrals
8. integrate by parts
9. use partial fractions to help with integration
10. calculate volumes of solids of revolution
11. locate the centroid of a body

8.1 DETERMINING DISPLACEMENTS FROM VELOCITIES

If a vehicle travels with constant velocity, v, its displacement is given by

$$s = vt$$

where t is the time which the journey takes.

For example, a vehicle travelling at a constant speed of 60 kmh^{-1} for $\frac{3}{4}$h travels a distance

$$s = 60 \times \tfrac{3}{4} = 45 \text{ km}$$

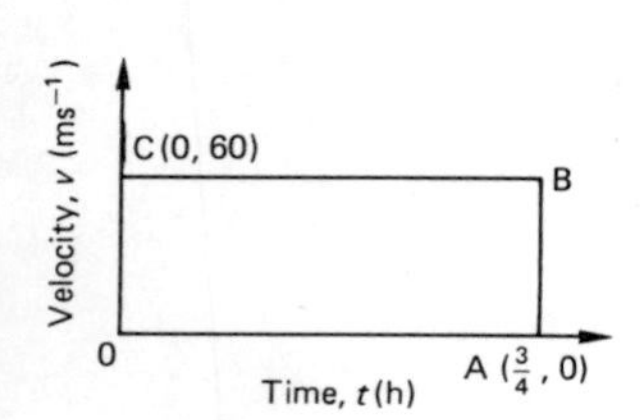

Figure 8.1

This result can be interpreted graphically. Figure 8.1 shows a graph of velocity (v) against time (t). The displacement is the area under the v–t graph, that is the area of the rectangle OABC, which is OA $\times$ OC $= 60 \times \frac{3}{4} = 45$ km.

In practice no journeys are made at constant velocity. So the problem is: How is displacement calculated if the velocity varies?

For example, suppose a particle moves so that its velocity at time t(s) is

$$v = \tfrac{1}{4}t(8 - t^3)\ (\mathrm{ms}^{-1})$$

where $0 \leqslant t \leqslant 2$.

Let

$$v(t) = \text{velocity at time } t$$

so

$$v(0) = 0 = \text{velocity at time } t = 0$$

$$v(1) = 1.75 = \text{velocity at time } t = 1$$

$$v(2) = 0 = \text{velocity at time } t = 2$$

The graph of velocity against time is shown in Figure 8.2. In order to find the displacement at $t = 2$ seconds, the area under the curve must be found, since area under the velocity–time curve gives displacement.

Several different methods for approximating this area follow. The first three are based on areas of rectangles.

Method 1: Divide the time interval into two equal parts

The total time interval is 2 seconds; dividing this in two gives increments of $\Delta t = 1.0$ seconds (Figure 8.3).

Over each half interval, the average velocity could be taken. This is perfectly acceptable, but it is easier to take the velocity in the middle of each subinterval. So

$$\text{Area of first rectangle} = v(0.5)\Delta t$$
$$= \tfrac{1}{4}(0.5)(8 - 0.5^3) \times 1 = 0.9844$$

$$\text{Area of second rectangle} = v(1.5)\Delta t$$
$$= \tfrac{1}{4}(1.5)(8 - 1.5^3) \times 1 = 1.7344$$

So the estimate of the displacement from $t = 0$ to $t = 2$ is

$$v(0.5)\Delta t + v(1.5)\Delta t = 2.72 \quad \text{(to two decimal places)}$$

Method 2: Divide the time interval into ten equal parts

Total time interval = 2.0 s. Time increment, $\Delta t = 0.2$.

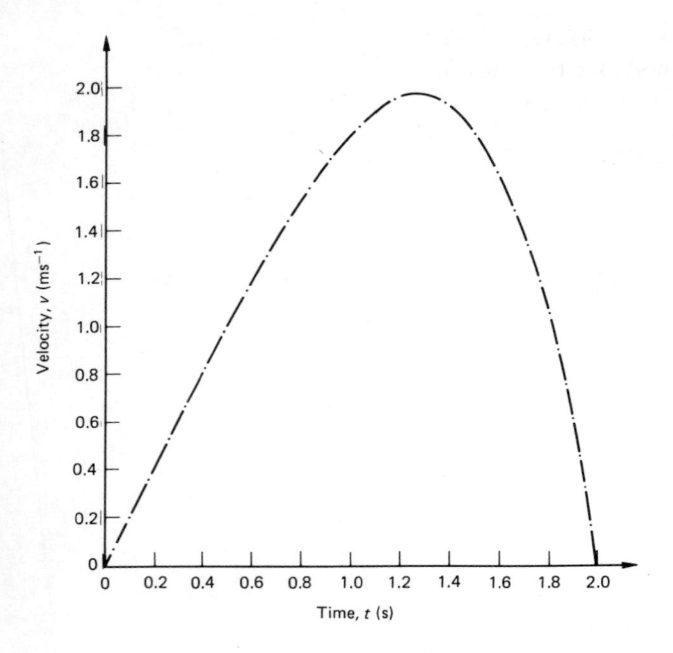

Figure 8.2

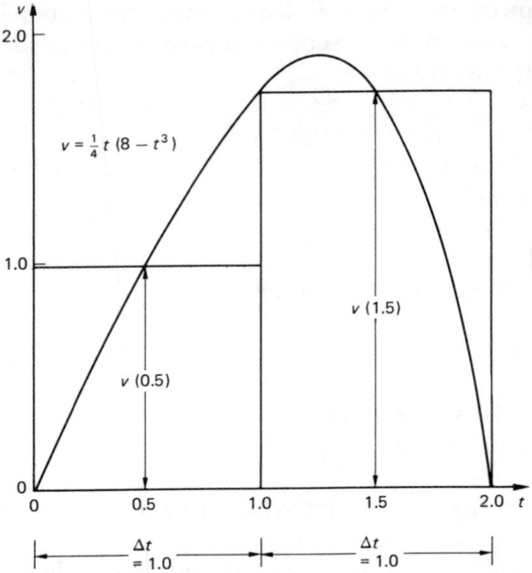

Figure 8.3

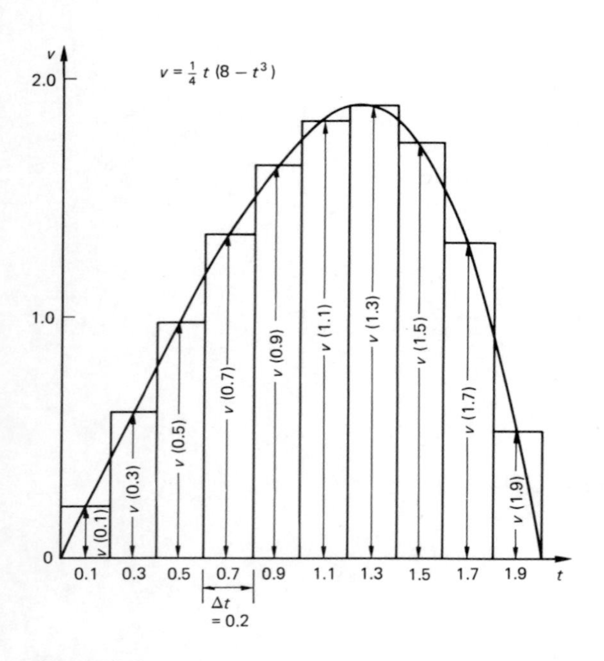

Figure 8.4

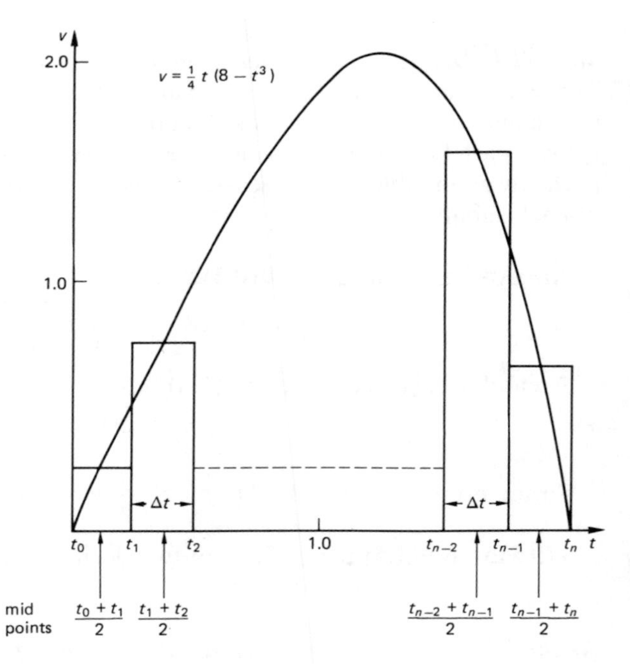

Figure 8.5

Over each subinterval, take the velocity at its mid-point. For example, for $0 \leqslant t \leqslant 0.2$, take $v(0.1) \simeq 0.200$, and for $0.2 \leqslant t \leqslant 0.4$, take $v(0.3) \simeq 0.598$, and so on.

Geometrically, the area under the v–t curve is approximated by summing the areas of ten rectangles (Figure 8.4). So

$$\begin{aligned}
\text{Displacement} &= \text{Sum of the areas of the ten rectangles} \\
&= v(0.1) \times 0.2 + v(0.3) \times 0.2 + \ldots + v(1.9) \times 0.2 \\
&= 0.2[v(0.1) + v(0.3) + \ldots + v(1.9)] \qquad (*) \\
&= 0.2(0.199975 + 0.597975 + \ldots + 0.541975) \\
&= 2.41 \quad \text{(to two decimal places)}
\end{aligned}$$

Note the factorization of 0.2 from every term at step (*); this reduces the arithmetic.

Method 3: Divide the time interval into n equal parts

This is a generalization of methods (1) and (2). Total time interval = 2 s, so time increment $\Delta t = 0.2/n$.

Let the points of subdivision be $t_0, t_1, t_2, \ldots, t_n$. Note that there are $(n + 1)$ of them (Figure 8.5). (If you find this difficult to understand think of n fence panels; $n + 1$ fence posts are needed.)

The points of subdivision are all Δt apart, so

$$t_1 = t_0 + \Delta t$$

$$t_2 = t_1 + \Delta t = t_0 + 2\Delta t$$

$$t_3 = t_2 + \Delta t = t_0 + 3\Delta t$$

$$\vdots$$

$$t_i = t_0 + i\Delta t \quad \text{for the } i\text{th division point.}$$

The height of each rectangle is given by the value of v at the mid-point of the relevant subinterval.

$$\text{Height of rectangle 1 is } v\left(\frac{t_0 + t_1}{2}\right)$$

$$\text{Height of rectangle 2 is } v\left(\frac{t_1 + t_2}{2}\right)$$

$$\vdots \qquad \vdots$$

Height of rectangle i is $v\left(\frac{t_{i-1}+t_i}{2}\right)$

The displacement $\simeq$ sum of the areas of the rectangles

$$= v\left(\frac{t_0+t_1}{2}\right)\Delta t + v\left(\frac{t_1+t_2}{2}\right)\Delta t + \ldots + v\left(\frac{t_{n-1}+t_n}{2}\right)\Delta t$$

$$= \left[v\left(\frac{t_0+t_1}{2}\right) + v\left(\frac{t_1+t_2}{2}\right) + \ldots + v\left(\frac{t_{n-1}+t_n}{2}\right)\right]\Delta t$$

Using the sigma notation

$$\text{Displacement} \simeq \left[\sum_{i=1}^{n} v\left(\frac{t_{i-1}+t_i}{2}\right)\right]\Delta t$$

where

$$\Delta t = \frac{\text{total time interval}}{n}, \; t_0 = \text{start time}$$

and

$$t_i = t_0 + i\Delta t$$

Example 8.1.1

Use the method of approximation by rectangles to estimate the increase in displacement from $t = 1$ s to $t = 5$ s for a particle whose velocity is given by $v = t^2 - 2t + 2$. Take $n = 8$ equal subintervals.

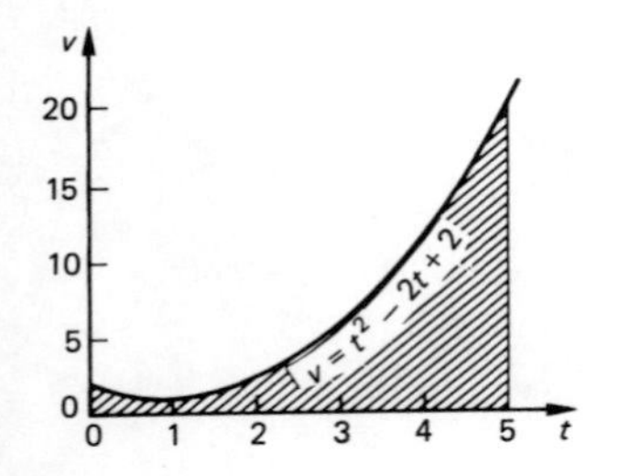

Figure 8.6

Solution

The v–t graph is shown in Figure 8.6. The displacement is the area under the graph from $t = 1$ to $t = 5$ s (shaded).

Since start time = 1 s and finish time = 5 s, and there are $n = 8$ subintervals, $\Delta t = (5 - 1)/8 = 0.5$ s. So

$$s \simeq \left[\sum_{n=1}^{8} v\left(\frac{t_{i-1}+t_i}{2}\right)\right]\Delta t$$

where $t_0 = 1$ and $t_i = t_{i-1} + \Delta t$ and $\Delta t = 0.5$. The results are set out in Table 8.1.

Table 8.1

i	t_i	Midpoint $= \dfrac{t_{i-1} + t_i}{2}$	$v\left(\dfrac{t_{i-1} + t_i}{2}\right)$
0	1.0		
1	1.5	1.25	1.0625
2	2.0	1.75	1.5625
3	2.5	2.25	2.5625
4	3.0	2.75	4.0625
5	3.5	3.25	6.0625
6	4.0	3.75	8.5625
7	4.5	4.25	11.5625
8	5.0	4.75	15.0625
		SUM	50.50000

From Table 8.1

$$\sum_{i=1}^{8} v\left(\frac{t_{i-1} + t_i}{2}\right) = 50.5$$

So

$$\text{Displacement} \simeq \left[\sum_{i=1}^{8} v\left(\frac{t_{i-1} + t_i}{2}\right)\right] \Delta t$$

$$= 50.5 \times 0.5 = 25.25 \text{ m}$$

Exercise 8.1.1

Write a program or use a spreadsheet to compute an approximation to the displacement, s, given the velocity, v, as a function of time, t. Use the midpoint rectangle rule of Section 8.1.

Your program should allow the user to imput the start and stop times, and the number of sub-intervals required. If the computer language you are using permits, also input the formula for velocity as a function of time. Otherwise you may change a statement which defines velocity as a function of time within the program.

As test data use Example 8.1.1.

What happens as you increase the number of subintervals? Does s settle down to a given value? If so, what value? If s does not settle down, explain what is happening.

8.1.4 Negative displacements and 'negative' areas

Consider a particle whose velocity v (ms^{-1}) and time t (s) is given by $v = 4 - t^2$ ($t \geq 0$).

Now for $0 \leq t < 2$, $v > 0$. When $t = 2$, $v = 0$ and when $t > 2$, $v < 0$.

Suppose that this particle starts from a point 0. For the first two seconds $v \geq 0$, so it is moving away from 0. At $t = 2$, $v = 0$, so it is at rest. After this for $t > 2$, $v < 0$ which means that the particle moves back towards 0 (Figure 8.7). In fact it will eventually reach 0 and go beyond it.

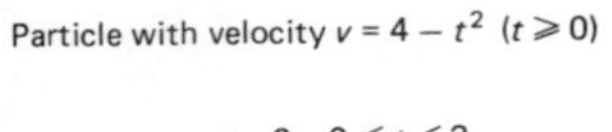

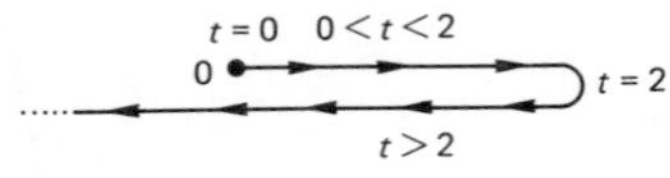

Figure 8.7

Hence the displacement s will initially be positive, and will reach a maximum positive value at $t = 2$ s. Thereafter it will decrease, become zero and then more and more negative.

Now consider the velocity–time graph (Figure 8.8). Consider the problem of finding the displacement at $t = 3$ s. The particle has obviously gone out to its maximum positive displacement at $t = 2$ s and is returning to 0.

The shaded area A_1 (Figure 8.8) represents the displacement from $t = 0$ to $t = 2$. This displacement is positive, since the particle is travelling away from 0.

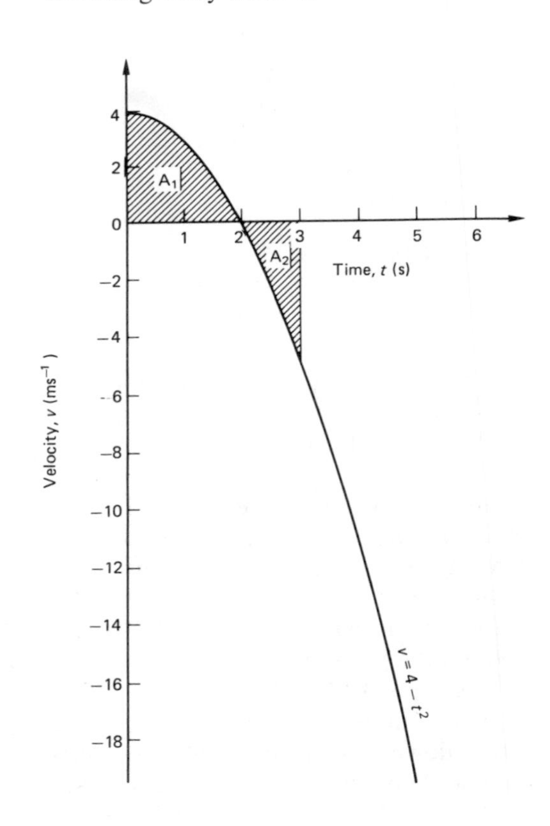

Figure 8.8

The shaded area A_2 (Figure 8.8) represents the displacement from $t = 2$ to $t = 3$. This displacement is negative as the particle is travelling towards 0.

Thus an area below the t axis has a negative sign attached to it. So the displacement from $t = 0$ to $t = 3$ is given by $A_1 + A_2$ with the convention that A_1 is positive and A_2 is negative.

Table 8.2 gives an approximation to the displacement using $n = 15$ subintervals and an increment of $\Delta t = (3 - 0)/15 = 0.2$.

Table 8.2 $v = 4 - t^2$. Table of approximations to displacement with $t_0 = 0$, $\Delta t = 0.2$

i	Midpoint $= \dfrac{t_{i-1} + t_i}{2}$	$v\left(\dfrac{t_{i-1} + t_i}{2}\right)$	
1	0.1	3.99	
2	0.3	3.91	
3	0.5	3.75	
4	0.7	3.51	
5	0.9	3.19	$A_1 = 26.7$
6	1.1	2.79	
7	1.3	2.31	
8	1.5	1.75	
9	1.7	1.11	
10	1.9	0.39	
11	2.1	−0.41	
12	2.3	−1.29	
13	2.5	−2.25	$A_2 = -11.65$
14	2.7	−3.29	
15	2.9	−4.41	
	SUMS	+15.05	15.05

So

$$\sum_{i=1}^{15} v\left(\frac{t_{i-1} + t_i}{2}\right) = 15.05$$

Hence Displacement $s \simeq 15.05\ \Delta t = 15.05 \times 0.2 = 3.01$ m.

In Table 8.2, note how the velocities for midpoints 1 to 10 are positive, while those for 11 to 15 are negative. This gives a positive value for A_1 and a negative value for A_2.

It is important to realise that the sign adjustments for A_1 and A_2 are taken care of by the algebra. There is no need for us to worry about them. However it is very important to be aware of the signs attached to areas. The rule is:

Remember that the displacement is not the same as the distance travelled. In this problem, the displacement tells us that after 3 seconds the particle is 3.01 m from 0, however the actual distance travelled is $(26.7 + 11.65) \times 0.2 = 7.67$ m.

Areas above the t axis are considered as positive
Areas below the t axis are considered as negative

Exercise 8.1.2

1. Your program or spreadsheet from Exercise 8.1.1 will automatically take account of the sign convention. Test it on the data of Example 8.1.2. By taking more subintervals in the program find the best approximation you can for the displacement from $t = 0$ to $t = 3$ s.
2. Repeat Problem 1 to estimate the displacement between $t = 0$ and $t = 2$ seconds and $t = 2$ and $t = 3$ seconds.
3. Use your program to estimate the time at which the displacement is zero for the second time (i.e. at what time does the particle return to 0?).

8.2 THE TRAPEZIUM RULE

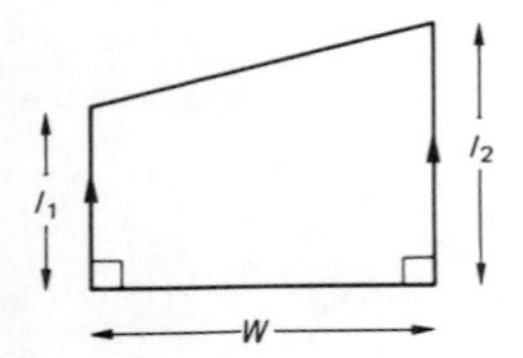

Figure 8.9

In the previous section, areas under graphs were estimated by means of summing the areas of rectangles. An alternative to this is to use trapezia.

A trapezium is a quadrilateral (four sided figure) with one pair of sides parallel. The type of trapezium required is shown in Figure 8.9. It has its base at right angles to its parallel sides. Its area is $\frac{1}{2}w(l_1 + l_2)$, where w is the width and l_1, l_2 the lengths of the parallel sides.

The idea of the trapezium rule is to approximate the area under a curve by trapezia. This involves approximating curved segments of the given graph by straight lines (Figure 8.10).

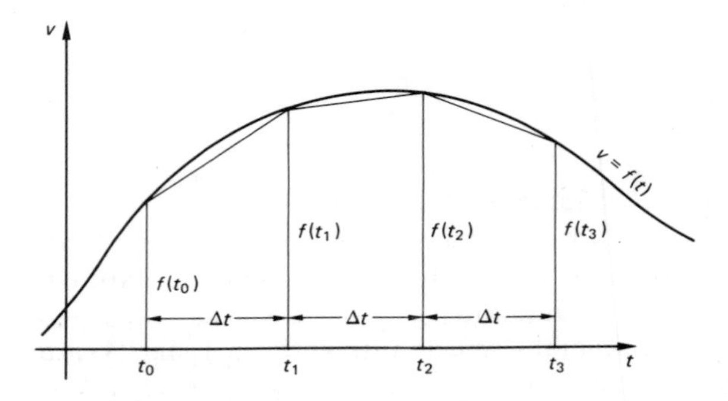

Figure 8.10

The area under $v = f(t)$ has been approximated by the sum of the areas of three trapezia with equal width

$$\Delta t = \frac{t_3 - t_0}{3}$$

So, Area under $v = f(t)$ from $t = t_0$ to $t = t_3$ is

$$s \simeq \tfrac{1}{2}\Delta t[f(t_0) + f(t_1)] + \tfrac{1}{2}\Delta t[f(t_1) + f(t_2)] + \tfrac{1}{2}\Delta t[f(t_2) + f(t_3)]$$
$$= \tfrac{1}{2}\Delta t\{f(t_0) + f(t_3) + 2[f(t_1) + f(t_2)]\}$$

In the general case we would take n trapezia, giving the following result.

$$s \simeq \tfrac{1}{2}\Delta t\{f(t_0) + f(t_n) + 2[f(t_1) + \dots + f(t_{n-1})]\}$$

where

$$\Delta t = \frac{t_n - t_0}{n} \quad \text{and} \quad t_i = t_{i-1} + i\Delta t.$$

Example 8.2.1

Use the trapezium rule with six subintervals to estimate the displacement from $t = 0$ to $t = 3$ seconds of a particle whose velocity is $v = 4 - t^2$ ($t \geqslant 0$). Compare the result with Table 8.2 and Exercise 8.1.2 question 1.

Solution

$n = 6$ subintervals, so $\Delta t = (3 - 0)/6 = 0.5$.

$t_0 = 0.0$, $t_6 = 3.0$ and $t_i = t_{i-1} + 0.5$

The results are shown in Table 8.3.

$$s \simeq \tfrac{1}{2}\Delta t\{f(t_0) + f(t_6) + 2[f(t_1) + \dots + f(t_5)]\}$$
$$= \tfrac{1}{2}\Delta t\,[-1.00 + 2(6.25)] = \tfrac{1}{2} \times 0.5 \times 11.5 = 2.875 \text{ m}$$

Exercise 8.2.1

1. Write a program or use a spreadsheet to implement the trapezium rule. Use Example 8.2.1 as test data. Investigate the effect of taking more subintervals and compare your results with Exercise 8.1.2, Question 1.

2. Use your program to estimate the displacement of the particle of Example 8.2.1 between $t = 0$ and $t = 2$ seconds and $t = 2$ and $t = 3$ seconds.

Table 8.3

n	t_n	$f(t_n) = 4 - t_n^2$	
0	0.0	4.00	
1	0.5		3.75
2	1.0		3.00
3	1.5		1.75
4	2.0		0.00
5	2.5		−2.25
6	3.0	−5.0	
		(1)	(2)
	SUMS	−1.00	6.25
	GRAND TOTAL = (1) + 2(2)	11.5	(3)

8.3 FORMULAE FOR AREAS UNDER CURVES

We now consider the problem of finding a formula for the area A under a curve $y = f(x)$ starting from $x = 0$ and finishing at $x = a$ (Figure 8.11). The reason for starting at $x = 0$ and finishing at $x = a$ will be explained later.

A set of simple 'power' functions are considered for $f(x)$.

1. $y = f(x) \equiv x^0 \equiv 1$
The graph of $y = x^0 \equiv 1$ is shown in Figure 8.12. The shaded area $A = 1 \times a = a$.

2. $y = f(x) \equiv x$
From Figure 8.13 the shaded area $A = \frac{1}{2}a \times a = a^2/2$.

3. $y = f(x) \equiv x^2$
The required area is shaded in Figure 8.14.
We shall 'sandwich' A between two other areas called L (lower sum) and U (upper sum), which represent the sum of the areas of a 'staircase' of rectangles.

A cannot be calculated directly. It is necessary to carry out a limiting process.

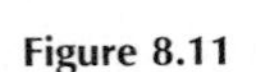

Figure 8.11

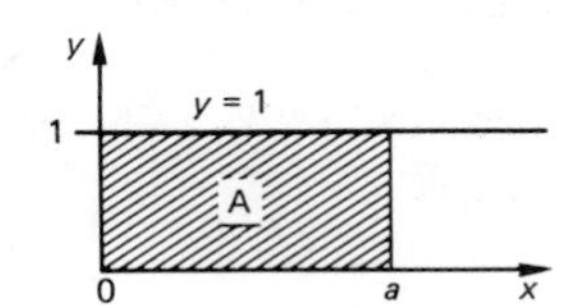

Figure 8.12

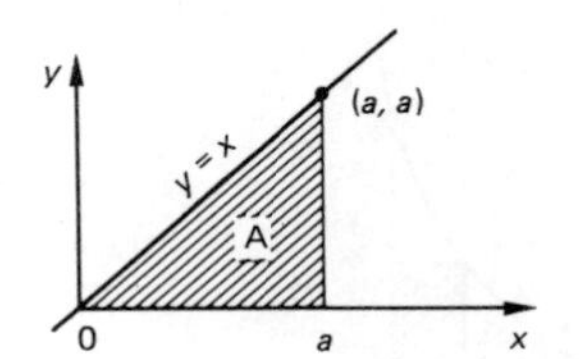

Figure 8.13

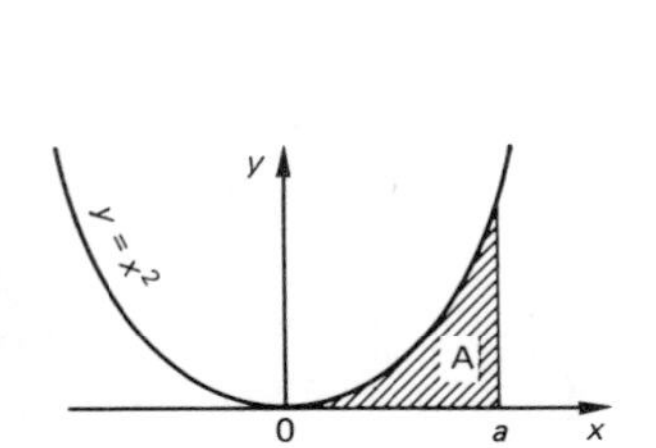

Figure 8.14

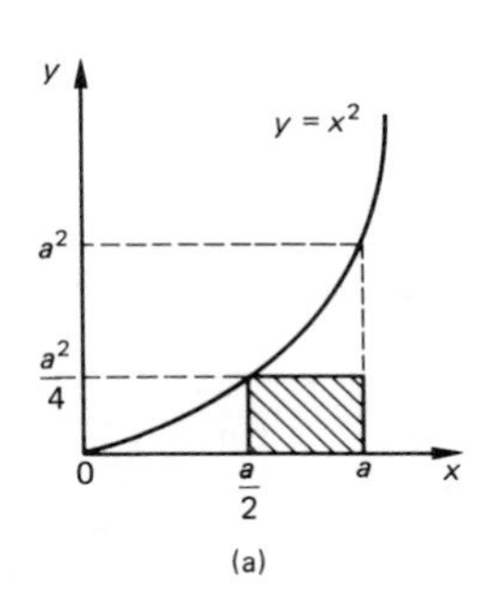

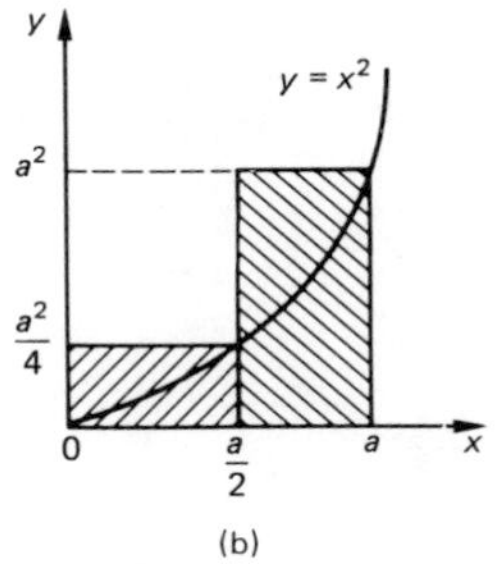

Figure 8.15

For the first approximation, divide the interval $(0, a)$ into two equal parts, and then construct the 'staircase' as shown in Figure 8.15.

Using Figure 8.15(a)	*Using Figure 8.15(b)*
Area of rectangle on $\left[0, \dfrac{a}{2}\right]$	Area of rectangle on $\left[0, \dfrac{a}{2}\right]$
is $\dfrac{a}{2} \times 0 = 0$	is $\dfrac{a}{2} \times \dfrac{a^2}{4} = \dfrac{a^3}{8}$
Area of rectangle on $\left[\dfrac{a}{2}, a\right]$	Area of rectangle on $\left[\dfrac{a}{2}, a\right]$
is $\dfrac{a}{2} \times \dfrac{a^2}{4} = \dfrac{a^3}{8}$	is $\dfrac{a}{2} \times a^2 = \dfrac{a^3}{2}$
Therefore Lower sum	Therefore Upper sum
$L_2 = 0 + \dfrac{a^3}{8} = \dfrac{a^3}{8}$	$U_2 = \dfrac{a^3}{8} + \dfrac{a^3}{2} = \dfrac{5a^3}{8}$

(The term 'sum' is used since we are adding the areas of rectangles. The subscripts on L_2 and U_2 refer to the fact that $[0, a]$ is divided into two parts.)

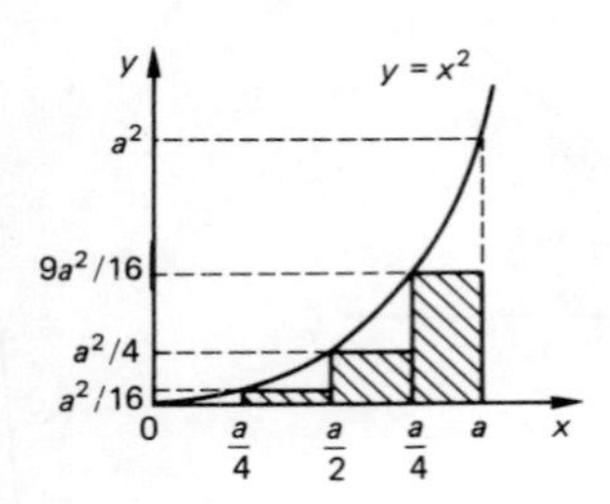

Figure 8.16

Now clearly $L_2 < A < U_2$ so

$$\frac{a^3}{8} < A < \frac{5a^3}{8}$$

and A has been 'sandwiched' between two areas.

Now take a better approximation, by dividing $[0, a]$ into four equal parts. Consider the 'lower staircase' in Figure 8.16.

$$\text{Area of rectangle on } \left[0, \frac{a}{4}\right] \text{ is } \frac{a}{4} \times 0 = 0$$

$$\text{Area of rectangle on } \left[\frac{a}{4}, \frac{a}{2}\right] \text{ is } \frac{a}{4} \times \frac{a^2}{16} = \frac{a^3}{64}$$

$$\text{Area of rectangle on } \left[\frac{a}{2}, \frac{3a}{4}\right] \text{ is } \frac{a}{4} \times \frac{a^2}{4} = \frac{a^3}{16}$$

$$\text{Area of rectangle on } \left[\frac{3a}{4}, a\right] \text{ is } \frac{a}{4} \times \frac{9a^2}{16} = \frac{9a^3}{64}$$

$$\text{Therefore, Lower sum } L_4 = 0 + \frac{a^3}{64} + \frac{a^3}{16} + \frac{9a^3}{64} = \frac{7a^3}{32}$$

Exercise 8.3.1

1. Draw a clearly labelled diagram of the corresponding 'upper staircases'.

$$\text{Show } U_4 = \frac{15a^3}{32}. \quad \text{Deduce } \frac{7a^3}{32} < A < \frac{15a^3}{32}.$$

2. Divide $[0, a]$ into eight equal parts. Draw diagrams for the lower and upper staircases. Deduce that

$$\frac{35a^3}{128} = L_8 < A < U_8 = \frac{51a^3}{128}$$

8.3.1 The *n*th approximation to A

From Problem 2, Exercise 8.3.1, it can be seen that L_8 is a better approximation to A than L_4, and in turn L_4 is better than L_2. The same applies for upper sums. This improvement in approximations is easy to explain. The more subdivisions we have, the nearer the 'staircase' comes to $y = x^2$. Rather than calculate L_{16}, L_{32}, etc explicitly it is quicker to proceed straight to L_n (Figure 8.17). In the work that follows, the result for the sum of the squares of the first n integers is needed (Section 6.1).

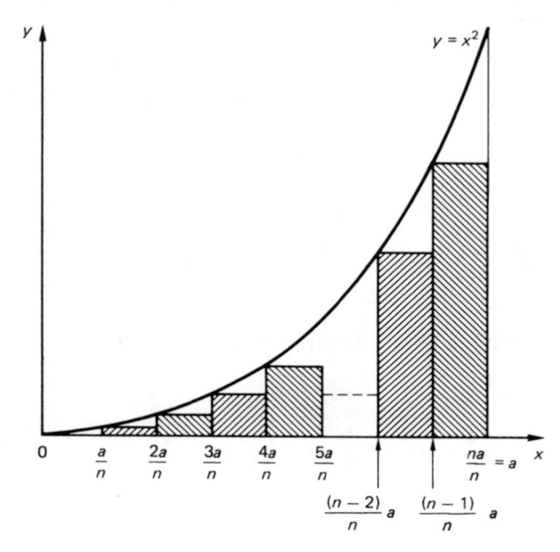

Figure 8.17

All n rectangles have base $= a/n$, so we get

$$\text{Area of first rectangle} = \frac{a}{n} \times 0$$

$$\text{Area of second rectangle} = \frac{a}{n} \times \left(\frac{a}{n}\right)^2$$

$$\text{Area of third rectangle} = \frac{a}{n} \times \left(\frac{2a}{n}\right)^2$$

$$\text{Area of fourth rectangle} = \frac{a}{n} \times \left(\frac{3a}{n}\right)^2$$

$$\text{Area of } (n-1)\text{th rectangle} = \frac{a}{n} \times \left(\frac{(n-2)a}{n}\right)^2$$

Area of nth rectangle is $\frac{a}{n} \times \left(\frac{(n-1)a}{n}\right)^2$

Adding,

$$L_n = \frac{a^3}{n}\left[0 + \left(\frac{1}{n}\right)^2 + \left(\frac{2}{n}\right)^2 + \left(\frac{3}{n}\right)^2 + \dots + \left(\frac{n-2}{n}\right)^2 + \left(\frac{n-1}{n}\right)^2\right]$$

Therefore

$$L_n = \left(\frac{a}{n}\right)^3 [0 + 1^2 + 2^2 + 3^2 + \dots + (n-2)^2 + (n-1)^2]$$

To evaluate L_n we need the sum of the squares of the first $(n - 1)$ integers. From Section 6.1.

$$1^2 + 2^2 + 3^2 + \dots + r^2 = \frac{1}{6} r(2r + 1)(r + 1)$$

Hence,

$$L_n = \left(\frac{a}{n}\right)^3 \times \frac{1}{6} \times (n-1) \times (2n-1) \times n$$

$$= \frac{a^3}{6n^2}(n-1)(2n-1)$$

Therefore

$$L_n = \frac{a^3}{6}\frac{(n-1)}{n}\frac{(2n-1)}{n} = \frac{a^3}{6}\left(1 - \frac{1}{n}\right)\left(2 - \frac{1}{n}\right)$$

Now as we take more and more rectangles (i.e. let $n \to \infty$) we get a better and better approximation to A. In fact we can say

$$A = \lim_{n\to\infty} L_n$$

Therefore

$$A = \lim_{n\to\infty}\left[\frac{a^3}{6}\left(1 - \frac{1}{n}\right)\left(2 - \frac{1}{n}\right)\right] = \lim_{n\to\infty}\left(\frac{a^3}{6} \times 1 \times 2\right) = \frac{a^3}{3}$$

Exercise 8.3.2

Draw a sketch showing U_n.

$$\text{Show } U_n = \left(\frac{a}{n}\right)^3 \times \frac{1}{6} n\,(2n + 1)\,(n + 1)$$

$$= \frac{a^3}{6}\left(1 + \frac{1}{n}\right)\left(2 + \frac{1}{n}\right).$$

Find $A = \lim_{n\to\infty} U_n$ and check that it gives the same result as $\lim_{n\to\infty} L_n$.

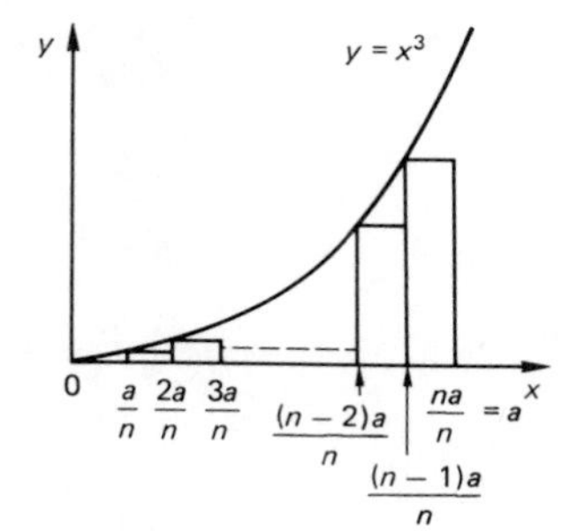

Figure 8.18

4. $y = f(x) \equiv x^3$
Consider the nth lower sum, see Figure 8.18.

$$L_n = 0 + \frac{a}{n} \times \left(\frac{a}{n}\right)^3 + \frac{a}{n} \times \left(\frac{2a}{n}\right)^3 + \ldots + \frac{a}{n}\left(\frac{(n-1)a}{n}\right)^3$$

$$= \left(\frac{a}{n}\right)^4 (0 + 1^3 + 2^3 + \ldots + (n-1)^3)$$

$$= \left(\frac{a}{n}\right)^4 \left(\frac{(n-1)^2}{4}(n)^2\right)$$

using the result $\sum_{i=1}^{n} i^3 = \frac{n^2(n+1)^2}{4}$ from Section 6.1.

So

$$L_n = \frac{a^4}{4}\frac{1}{n^4}(n-1)^2 n^2 = \frac{a^4}{4}\frac{1}{n^2}(n-1)^2$$

$$= \frac{a^4}{4}\left(1 - \frac{1}{n}\right)^2.$$

So as $n \to \infty$ $L_n \to \frac{a^4}{4}$. Hence $A = \frac{a^4}{4}$.

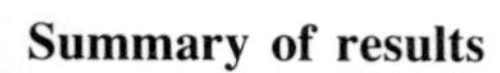

Summary of results

The area A bounded by $y = x^n$, the x axis, $x = 0$ and $x = a$ (Figure 8.19) is given in Table 8.4.

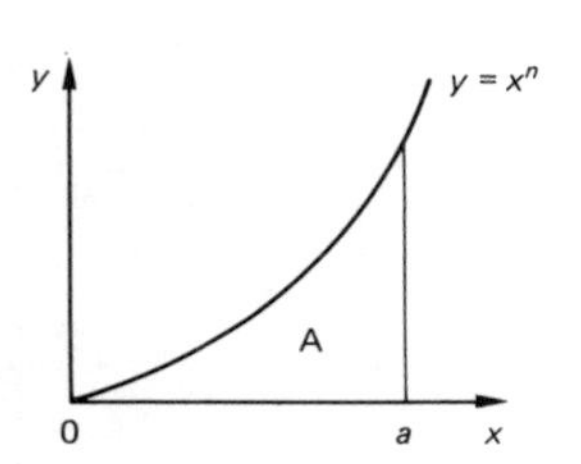

Figure 8.19

Table 8.4

Function	Area A
$y = x^0 = 1$	a
$y = x$	$\dfrac{a^2}{2}$
$y = x^2$	$\dfrac{a^3}{3}$
$y = x^3$	$\dfrac{a^4}{4}$

The pattern is easy to see. The area under $y = x^n$ between the x axis, $x = 0$ and $x = a$ should be $a^{n+1}/(n + 1)$.

This result is correct for all powers n (except -1), but we shall not prove it here.

Example 8.3.1

Find the shaded areas in Figure 8.20.

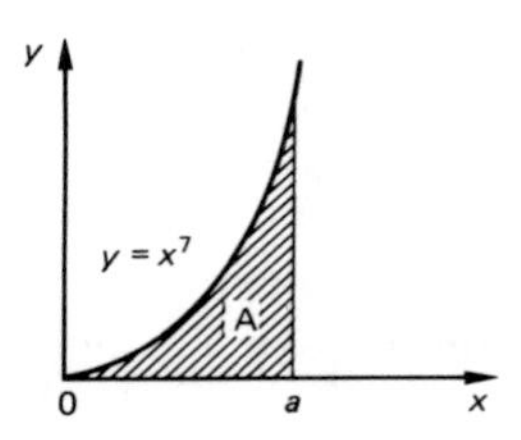

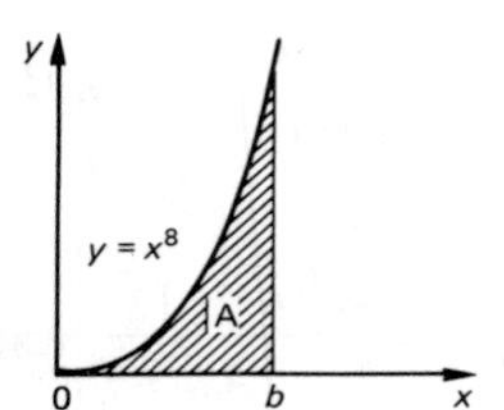

Figure 8.20

Solution

For $y = x^7$, the shaded area is $a^{7+1}/(7+1) = a^8/8$.
For $y = x^8$, up to $x = b$ the shaded area is $b^{8+1}/(8+1) + b^9/9$.

Example 8.3.2

Find the shaded areas in Figure 8.21.

Solution

These questions can be done by differences of areas using the results in Table 8.4. Take $y = x^2$ (Figure 8.22).

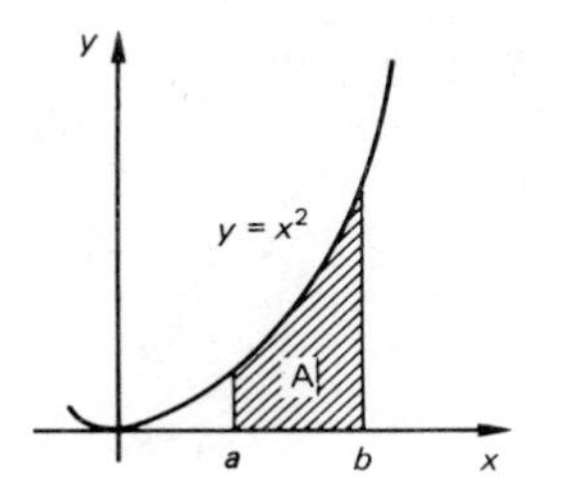

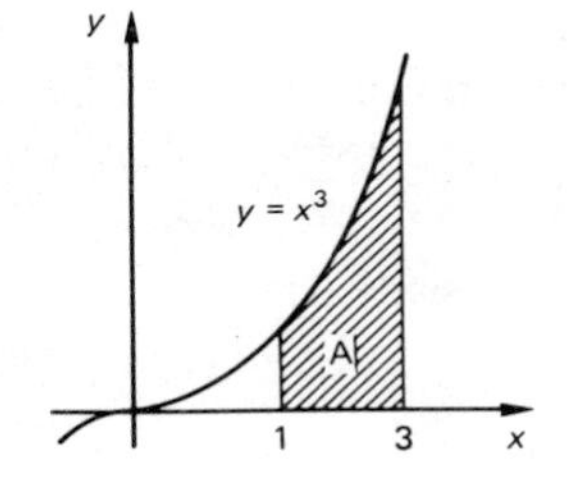

Figure 8.21

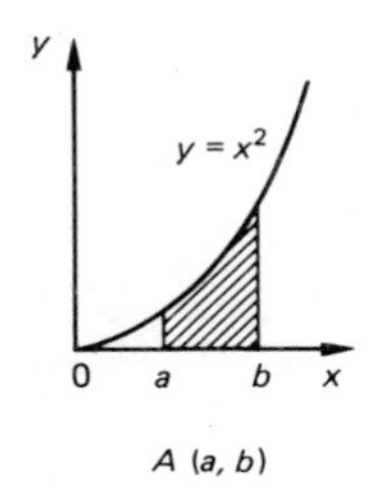

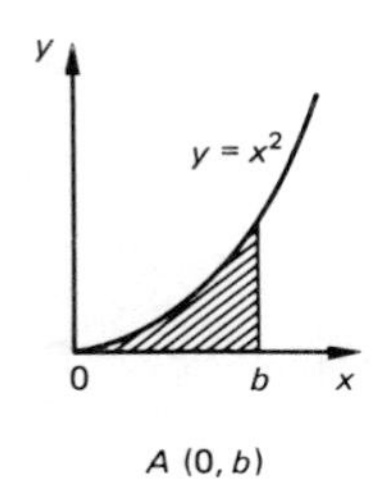

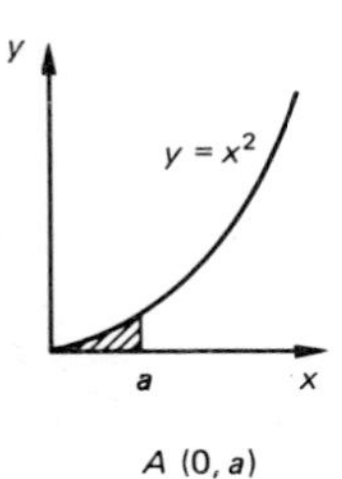

Figure 8.22

Let the area under $y = x^2$ above the x axis and bounded by $x = a$ and $x = b$ be $A = A(a, b)$. So (see Figure 8.22)

$$A(a, b) = A(0, b) - A(0, a)$$

$$= \frac{b^3}{3} - \frac{a^3}{3}$$

Similarly for $y = x^3$

$$A(1, 3) = A(0, 3) - A(0, 1)$$

$$= \frac{3^4}{4} - \frac{1^4}{4} = \frac{80}{4} = 20$$

8.3.2 Notation

The process of finding areas under curves is called *definite integration*.

Definite integration is a special summation process — in fact it involves summing the areas of n rectangles and then taking the limit as $n \to \infty$.

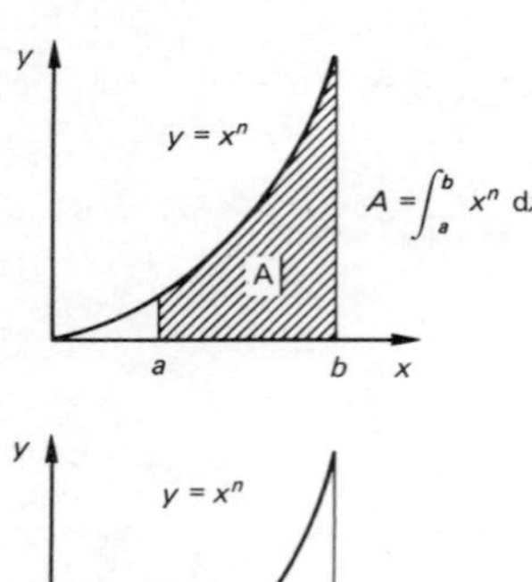

Figure 8.23

If the shaded area A in Figure 8.23 (upper) is required, it is obtained by taking the areas of many small rectangles, and adding them up. A typical rectangle is shown in Figure 8.23 (lower).

Its area is $x^n \Delta x$, so

$$A \simeq \sum_{x=a}^{x=b} x^n \Delta x$$

As we take more and more rectangles, $\Delta x \to 0$, so

$$A = \lim_{\Delta x \to 0} \sum_{x=a}^{x=b} x^n \Delta x$$

and we write

$$A = \int_a^b x^n \, dx$$

where $\int_a^b$ suggests a summation process, which starts at $x = a$ and finishes at $x = b$.

This summation process actually gives a 'signed' area as seen in Example 8.1.4. The reason is that x^n may be negative (see Figure 8.24 for $y = x^3$).

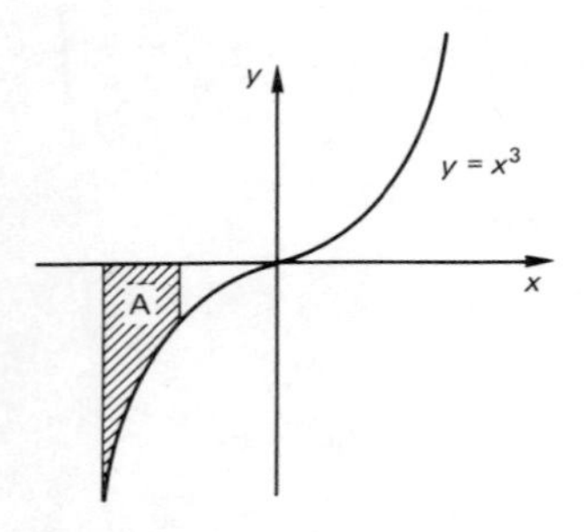

Figure 8.24

As before

Areas above the x axis are given a positive sign Areas below the x axis are given a negative sign

So integrals give a signed area. You should be aware of this because it may lead to problems (see Example 8.3.2).

8.3.3 Evaluation of $\int_a^b x^n \, dx$

Since $\int_a^b x^n \, dx$ represents the area bounded by $y = x^n$, the x axis and the lines $x = a$ and $x = b$ (Figure 8.25), it can be evaluated as the difference of two areas.

From Figure 8.25 we have $A = A_1 - A_2$ but

$$A = \int_a^b x^n \, dx, \ A_1 = \int_0^b x^n \, dx \text{ and } A_2 = \int_0^a x^n \, dx$$

so

$$\int_a^b x^n \, dx = \int_0^b x^n \, dx - \int_0^a x^n \, dx$$

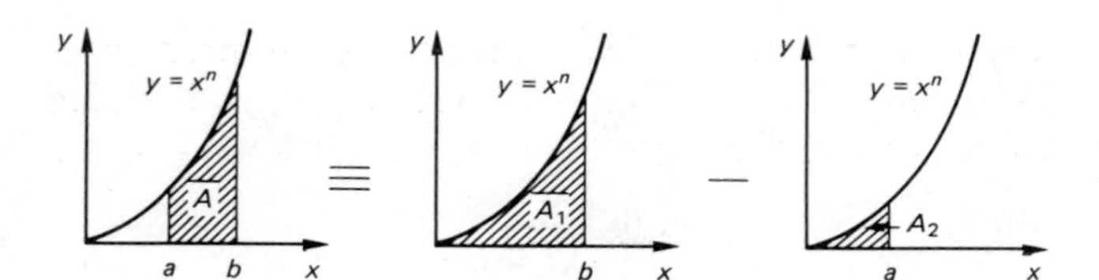

Figure 8.25

but we know

$$\int_0^b x^n \, dx = \frac{b^{n+1}}{n+1} \quad \text{and} \quad \int_0^a x^n \, dx = \frac{a^{n+1}}{n+1}.$$

So

> *Rule 1*
>
> $$\int_a^b x^n \, dx = \frac{b^{n+1}}{n+1} - \frac{a^{n+1}}{n+1}$$
>
> provided $n \neq -1$

Example 8.3.2

Find $\int_{-1}^{1} x^3 \, dx$, $\int_{-1}^{0} x^3 \, dx$ and $\int_{0}^{1} x^3 \, dx$.

Also find the shaded area in Figure 8.26.

Solution

$$\int_{-1}^{1} x^3 \, dx = \frac{(1)^4}{4} - \frac{(-1)^4}{4} = \frac{1}{4} - \frac{1}{4} = 0$$

from Rule 1 with $b = 1$, $a = -1$ and $n = 3$.

$$\int_{-1}^{0} x^3 \, dx = \frac{(0)^4}{4} - \frac{(-1)^4}{4} = 0 - \frac{1}{4} = -\frac{1}{4}$$

from Rule 1 with $b = 0$, $a = -1$ and $n = 3$.

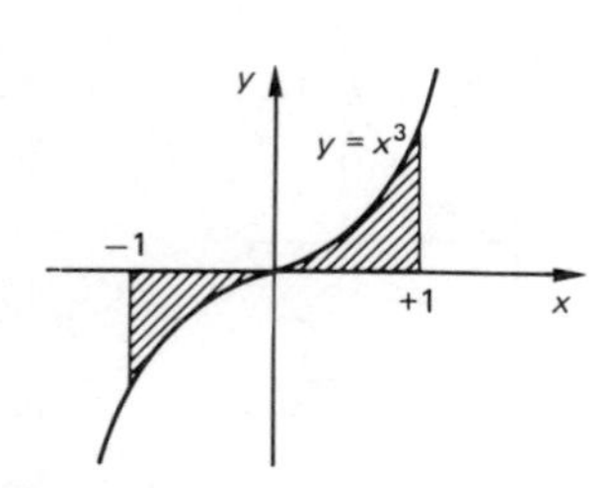

Figure 8.26

$$\int_0^1 x^3\,dx = \frac{(1)^4}{4} - \frac{(0)^4}{4} = \frac{1}{4}$$

from Rule 1 with $b = 0$, $a = 1$ and $n = 3$.

Note that the first integral was zero. At first this seems surprising, but this integral does *not* represent the shaded area in Figure 8.26. The reason for this is explained by the fact that

$$\int_{-1}^0 x^3\,dx = -\frac{1}{4} \quad \text{and} \quad \int_0^1 x^3\,dx = \frac{1}{4}$$

Integrals attach signs to areas, and when you add these two 'signed' areas together, the result is zero. The required area is

$$\left|-\frac{1}{4}\right| + \left|\frac{1}{4}\right| = \frac{1}{4} + \frac{1}{4} = \frac{1}{2}$$

So in using integrals to evaluate areas, negative signs are ignored.

Exercise 8.3.3

For Problems 1–6 evaluate the following integrals using Rule 1.

1. $\int_p^q x^5\,dx$
2. $\int_{-1}^3 x^3\,dx$
3. $\int_1^4 x\,dx$
4. $\int_1^2 x^{1/2}\,dx$
5. $\int_1^2 x^{-2}\,dx$
6. $\int_1^4 x^{3/2}\,dx$
7. Evaluate the areas shaded in Figure 8.27.

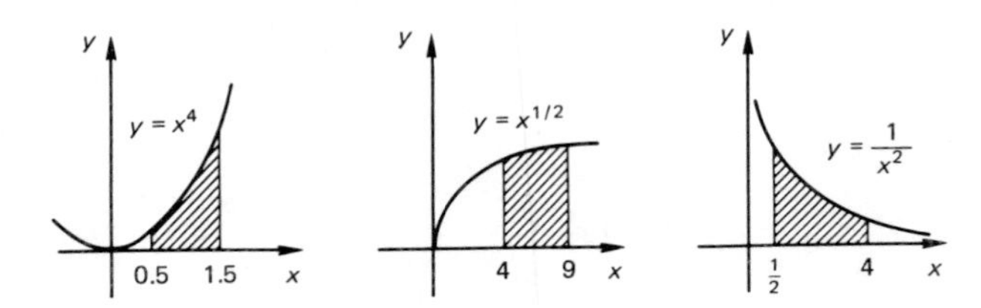

Figure 8.27

8. Evaluate $\int_{-2}^0 x^3\,dx$, $\int_0^3 x^3\,dx$ and $\int_{-2}^3 x^3\,dx$.

   Also find the area shaded in Figure 8.28.

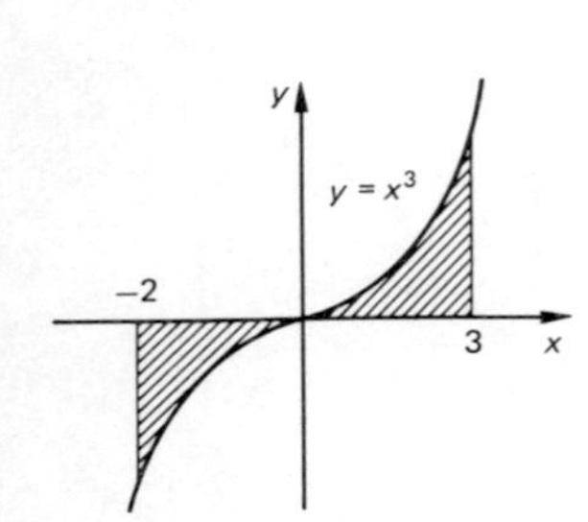

Figure 8.28

8.4 THE RELATIONSHIP BETWEEN INTEGRATION AND DIFFERENTIATION

We know that $\int_0^a x^n\,dx = \dfrac{a^{n+1}}{n+1}$. Now when $\dfrac{x^{n+1}}{n+1}$ is

differentiated the result is $\dfrac{(n+1)x^{(n+1)-1}}{(n+1)} = x^n$.

So it seems that differentiation and integration are in a sense opposite processes. This idea is made precise by the Fundamental Theorem of the Calculus which relates differentiation and integration.

Suppose that we have determined A, the area bounded by $y = f(x)$, the x axis and the vertical lines through $x = a$ and $x = x$ (Figure 8.29). The value of A depends on f, a and x. However, if we fix a and f then A depends on the position of x alone. Move x left, and A decreases. Move x right and A increases. So let x increase by a small amount Δx, resulting in the area increasing by ΔA. Now if Δx is small, the figure PQRS has an area which is approximately equal to the area of a rectangle height PS and width PQ.

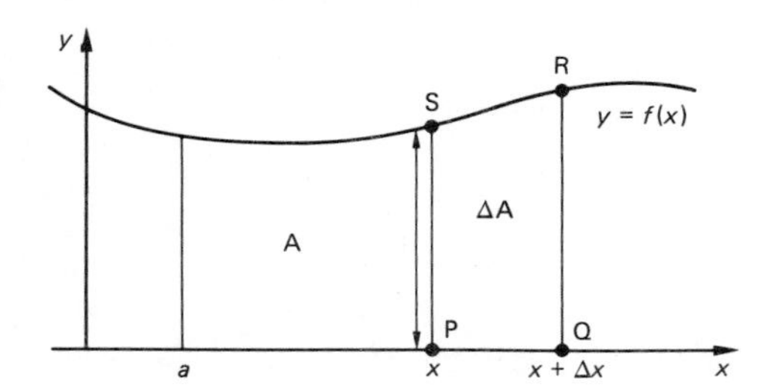

Figure 8.29

So

$$\Delta A \simeq y\Delta x$$

or

$$\frac{\Delta A}{\Delta x} \simeq y \tag{8.4.1}$$

But A depends on x alone, so $A = F(x)$ for some function F. Also $y = f(x)$. Hence from (8.4.1)

$$\frac{\Delta F(x)}{\Delta x} \simeq f(x) \tag{8.4.2}$$

From Section 7.3

$$\lim_{\Delta x \to 0} \frac{\Delta F(x)}{\Delta x} = \frac{dF(x)}{dx}$$

So

$$\frac{dF(x)}{dx} = f(x)$$

from taking the limit of (8.4.2) as $\Delta x \to 0$.

This equation means that if you differentiate the function $F(x)$ with respect to x, the result is $f(x)$.

Now $F(x)$ is the area function and $f(x)$ is the function defining the graph. In the notation we have used before

$$\text{If } A = F(x) = \int_a^x f(x)\,dx$$

$$\text{then } \frac{dF(x)}{dx} = f(x)$$

Example 8.4.1

Find a function $F(x)$ such that $dF(x)/dx = x^3$.

Solution

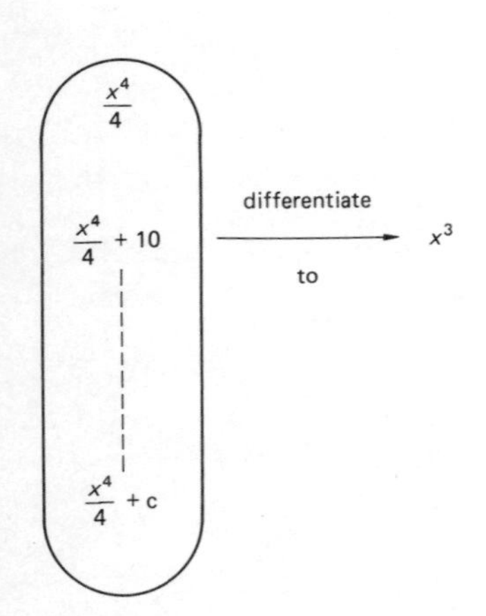

Figure 8.30

One solution is $F_1(x) \equiv x^4/4$ since $dF_1(x)/dx = 4 \times x^3/4 = x^3$. Another is $F_2(x) \equiv x^4/4 + 10$ since $dF_2(x)/dx = x^3$.

In fact any function of the form

$$F_3(x) \equiv \frac{x^4}{4} + c$$

where c is a constant will give $dF_3(x)/dx = x^3$.

This is summarized in Figure 8.30.

Example 8.4.2

Find a function $F(x)$ such that $dF(x)/dx = 5x^2$.

Solution

Differentiation reduces powers by one. So a candidate for $F(x)$ is ax^3, where a = constant. Now

$$F(x) = ax^3 \Rightarrow \frac{dF(x)}{dx} = 3ax^2$$

$$3ax^2 = 5x^2 \Rightarrow 3a = 5 \quad \text{so} \quad a = \frac{5}{3}$$

$$\therefore F(x) = \frac{5}{3}x^3 + c$$

where c is any constant. Since the constant c may take any value, it is called an arbitrary constant.

Example 8.4.3

Find a function $F(x)$ such that $dF(x)/dx = \cos x$.

Solution

If $F(x) = \sin x + c$ where c is an arbitrary constant then $dF(x)/dx = \cos x$.

Exercise 8.4.1

In Problems 1–6 find a function $F(x)$ which gives

1. $\dfrac{dF(x)}{dx} = 1$
2. $\dfrac{dF(x)}{dx} = x^2$
3. $\dfrac{dF(x)}{dx} = e^x$
4. $\dfrac{dF(x)}{dx} = \frac{3}{4}x^3$
5. $\dfrac{dF(x)}{dx} = \sin x$
6. $\dfrac{dF(x)}{dx} = x^2 + 3x$

8.5 DEFINITE AND INDEFINITE INTEGRALS

An integral which relates to a (signed) area has a definite value, since the value of the area is not arbitrary. For example,

$$\int_a^b x^2 \, dx = \frac{b^3}{3} - \frac{a^3}{3}$$

A definite integral is a function with limits — it represents a (signed) area. However if we are interested in the process which produces the function $x^3/3 + c$ from the function x^2 we have an indefinite integral

$$\int x^2 \, dx = \frac{x^3}{3} + c$$

where c is an arbitrary constant.

An indefinite integral has no limits — it represents a process which relates one function to another. To be precise

$$\text{If} \int f(x) \, dx = F(x) + c$$

$$\text{then} \frac{dF(x)}{dx} = f(x)$$

This is of great practical use. Our knowledge of differentiation can be used to help us with integration.

Example 8.5.1

Find (a) $\int \cos x \, dx$, (b) $\int e^x \, dx$ and (c) $\int (x^2 + 3x + 2) \, dx$

Solution

(a) To find $\int \cos x \, dx$ we need a function $F(x)$ such that

$$\frac{dF(x)}{dx} = \cos x$$

Now

$$\frac{d}{dx}(\sin x) = \cos x$$

Also

$$\frac{d}{dx}(\sin x + c) = \cos x$$

So

$$\int \cos x \, dx = \sin x + c$$

(b) For $\int e^x\,dx$ we need $F(x)$ such that $dF(x)/dx = e^x$.

Now

$$\frac{d}{dx}(e^x + c) = e^x$$

So

$$\int e^x\,dx = e^x + c$$

where c is an arbitrary constant.

(c) $\int (x^2 + 3x + 2)\,dx$

This can be taken stage by stage

$$\frac{d}{dx}\left(\frac{x^3}{3}\right) = x^2, \frac{d}{dx}\left(\frac{3}{2}x^2\right) = 3x \quad \text{and} \quad \frac{d}{dx}(2x) = 2$$

so

$$\frac{d}{dx}\left(\frac{x^3}{3} + \frac{3}{2}x^2 + 2x + c\right) = x^2 + 3x + 2$$

$$\int (x^2 + 3x + 2)\,dx = \frac{x^3}{3} + \frac{3}{2}x^2 + 2x + c$$

A full table of indefinite integrals is given in Appendix 2

Many problems can be done by referring to Appendix 2, but it is advisable to try some more difficult examples without referring to these results.

Example 8.5.2

Find (a) $\int \sin 2x\,dx$, (b) $\int e^{-3x}\,dx$ and (c) $\int \cos\left(\frac{1}{2}x + \frac{\pi}{2}\right)dx$

Solution

These integrals can all be done by substitution (see Section 8.7), but solutions are possible by 'trial and error' as well.

(a) $\int \sin 2x \, dx$

If $F(x) = \cos 2x$, then $dF(x)/dx = -2 \sin 2x$. The result we want is $\sin 2x$, so this is no good. Try $F(x) = -\frac{1}{2} \cos 2x$. Then $dF(x)/dx = -\frac{1}{2}(-2 \sin 2x) = \sin 2x$.

So

$$\int \sin 2x dx = -\tfrac{1}{2} \cos 2x + c$$

(b) $\int e^{-3x} \, dx$

If $f(x) = F(x) = e^{-3x}$, then $dF(x)/dx = -3e^{-3x}$, which is not good. But if $F(x) = -\frac{1}{3} e^{-3x}$, then $dF(x)/dx = -\frac{1}{3}(-3e^{-3x}) = e^{-3x}$.

So

$$\int e^{-3x} \, dx = -\tfrac{1}{3}e^{-3x} + c$$

(c) $\int \cos\left(\frac{1}{2}x + \frac{\pi}{2}\right) dx$

$$\frac{d}{dx}\left(\sin\left(\frac{1}{2}x + \frac{\pi}{2}\right)\right) = \frac{1}{2}\cos\left(\frac{1}{2}x + \frac{\pi}{2}\right)$$

so

$$\int \cos\left(\frac{1}{2}x + \frac{\pi}{2}\right) dx = 2 \sin\left(\frac{1}{2}x + \frac{\pi}{2}\right)$$

Example 8.5.3

Find (a) $\int \cos t \, dt$, (b) $\int e^u \, du$, (c) $\int (y^2 + \sin y) \, dy$

Solution

Although Appendix 2 uses the variable x, the results are easily related to other variables.

(a) $\int \cos t \, dt = -\sin t + c$

(b) $\int e^u \, du = e^u + c$

(c) $$\int (y^2 + \sin y)\,dy = \int y^2\,dy + \int \sin y\,dy = \frac{y^3}{3} - \cos y + c$$

Exercise 8.5.1

Evaluate the following integrals:

1. $\int (\sin t + e^t)\,dt$
2. $\int (u^2 + 3u + 8)\,du$
3. $\int \frac{1}{v^2}\,dv$
4. $\int p^{-1/2}\,dp$
5. $\int \frac{1}{\sqrt{x}}\,dx$
6. $\int e^{-x}\,dx$
7. $\int \sec^2 x\,dx$
8. $\int \operatorname{cosec} x \cot x\,dx$

8.6 FURTHER DEFINITE INTEGRALS

Definite integrals can be evaluated using the table of indefinite integrals in Appendix 2, but the arbitrary constant is not required as Example 8.6.1 shows.

Example 8.6.1

Evaluate

$$\int_1^2 e^x\,dx$$

Solution

$$\int e^x\,dx = e^x + c$$

so

$$\int_1^2 e^x\,dx = [e^x + c]_1^2$$

where the square brackets means substitute the limit values for x in the expression and take their difference.

$$[e^x + c]_1^2 = (e^2 + c) - (e^1 + c) = e^2 - e^1 \simeq 4.671$$

This shows that the arbitrary constant is not required in evaluating a definite integral. In future we write

$$\int_1^2 e^x \, dx = [e^x]_1^2 = e^2 - e^1 \simeq 4.671$$

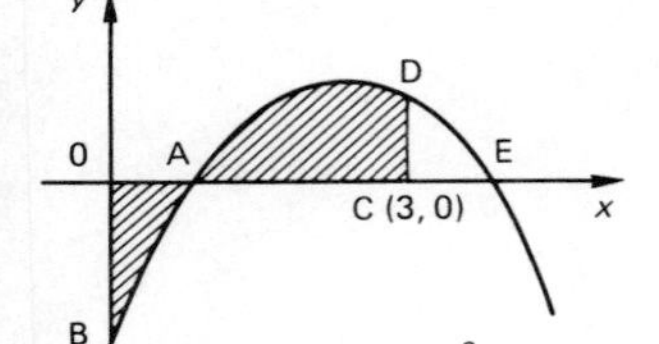

Figure 8.31

Example 8.6.2

Find the shaded area in Figure 8.31.

Solution

Integration will assign a negative value to area OAB and a positive value to area ACD, so we cannot compute the shaded area by

$$\int_0^3 (-x^2 + 5x - 4) \, dx$$

Instead evaluate each area separately. First find the x coordinate of A.

$$y = 0 \Rightarrow -x^2 + 5x - 4 = 0 \Rightarrow -(x - 1)(x - 4) = 0$$
$$\Rightarrow x = 1 \text{ (at A) or } 4 \text{ (at E)}$$

$$\int_0^1 (-x^2 + 5x - 4) \, dx = \left[-\frac{x^3}{3} + \frac{5x^2}{2} - 4x \right]_0^1 = -1\tfrac{5}{6}$$

So area OAB $= 1\frac{5}{6}$.

$$\int_1^3 (-x^2 + 5x - 4) \, dx = \left[-\frac{x^3}{3} + \frac{5x^2}{2} - 4x \right]_1^3$$
$$= 1\tfrac{1}{2} - (-1\tfrac{5}{6}) = 3\tfrac{1}{3}$$

So Area ACD $= 3\frac{1}{3}$.

Hence, the shaded area in Figure 8.3 is area OAB + area ACD $= 5\frac{1}{6}$.

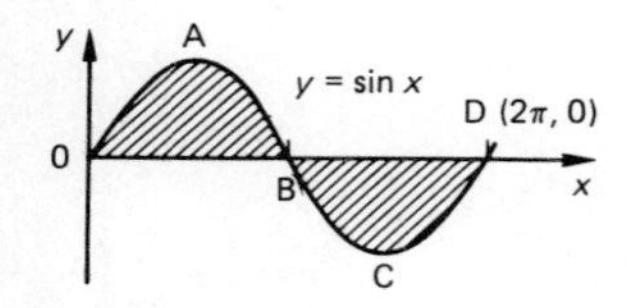

Figure 8.32

Example 8.6.3

Find the shaded area in Figure 8.32.

Solution

Integration will assign a positive value to area OAB and a negative value to area BCD. Because the sine curve is symmetrical, area OAB = area BCD.

At B, $x = \pi$, so the shaded area in Figure 8.32 is

$$2\int_0^{\pi} \sin x \, dx = 2\,[-\cos x]_0^{\pi} = 2[-\cos \pi - (-\cos 0)]$$

$$= 2[-(-1)-(-1)] = 4$$

Exercise 8.6.1

Calculate the shaded areas in Problems 1–6 (Figure 8.33).

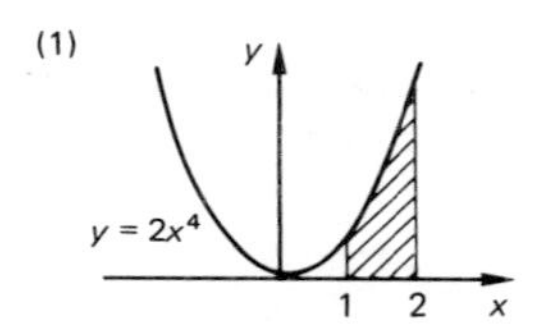

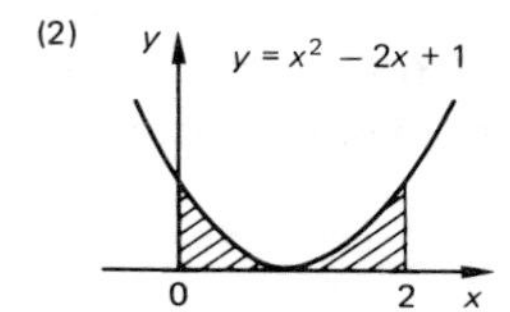

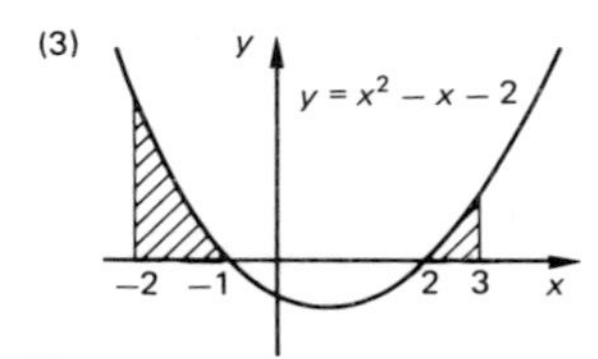

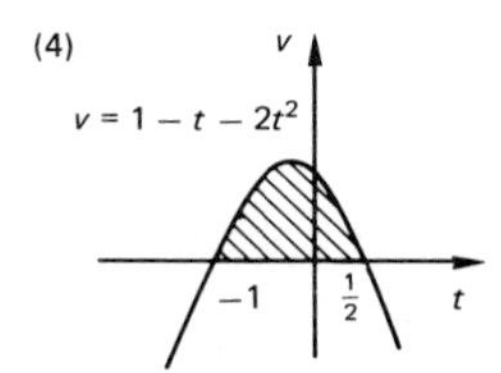

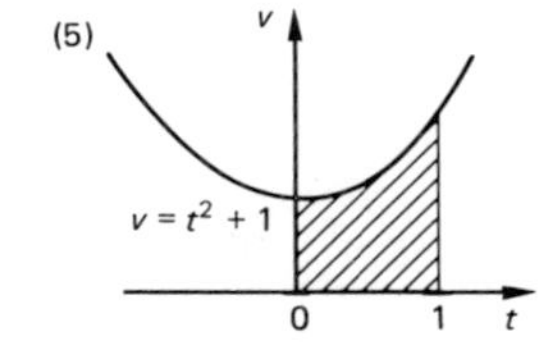

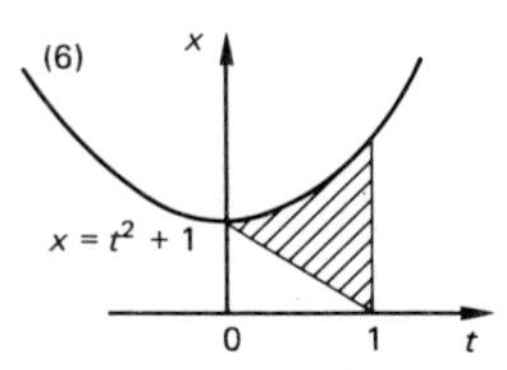

Figure 8.33

For Problems 7–12, evaluate the integrals given.

7. $\displaystyle\int_{-2}^{3} (v^2 + 3)\, dv$

8. $\displaystyle\int_{1}^{4} (y^2 + \sqrt{y})\, dy$

9. $\displaystyle\int_{0}^{1} \sqrt{x}\,(x + 2)\, dx$ (NB. multiply the brackets out *before* integration)

10. $\displaystyle\int_{1}^{2} \left(3 + \frac{1}{t^2} + \frac{1}{t^4}\right) dt$

11. $\displaystyle\int_{1}^{9} \left(\sqrt{x} + \frac{1}{\sqrt{x}}\right) dx$

12. $\int_{4}^{11} \sqrt{x+5}\, dx$

For Problems 13–15 sketch the graph, shade the required area (as was done in Problems 1–6) and then determine the area asked for.

13. Find the area enclosed by $y = x(4 - x)$, the lines $x = 1$, $x = 2$ and the x axis.

14. Find the area enclosed by $y = x^2 + 3$, the lines $x = -1$, $x = 2$ and the x axis.

15. Find the area enclosed by $y = \sqrt{x}$ (where $\sqrt{\ }$ means the positive square root), the lines $x = 4$, $x = 9$ and the x axis.

For Problems 16 and 17 of Figure 8.34, relate each shaded area to an integral. The integral should have a negative value in each case – but consider the area as positive.

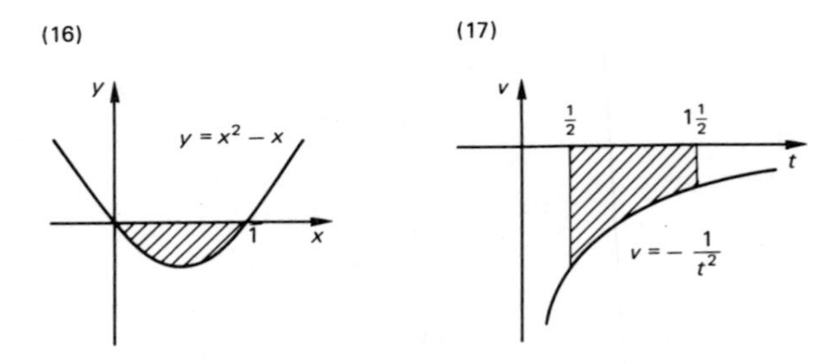

Figure 8.34

For Problems 18–20 find where each curve cuts the x axis. Sketch the curve and find the area enclosed by the curve and the x axis.

18. $y = (x - 1)(x - 4)$ (Before integrating, multiply brackets out)

19. $y = 3x - x^2$ (Factorize the function y to find where it cuts x axis)

20. $y = x(x - 2)^2$

21. Show that $y = 5$ cuts $y = x^2 - 2x + 2$ when $x = -1$ and $x = 3$. Sketch $y = x^2 - 2x + 2$ and $y = 5$ on the same graph. Find the area of the segment of $y = x^2 - 2x + 2$ cut off by the straight line $y = 5$.

22. Find the area of the segment of $y = x^2 - 6x + 9$ which is cut off by the straight line $y = 1$.

23. Sketch $y = x^5$. Show

$$\int_{-2}^{2} x^5 \, dx = 0$$

Use your sketch to explain why this is so.

For Problems 25–27 evaluate the area (Figure 8.35).

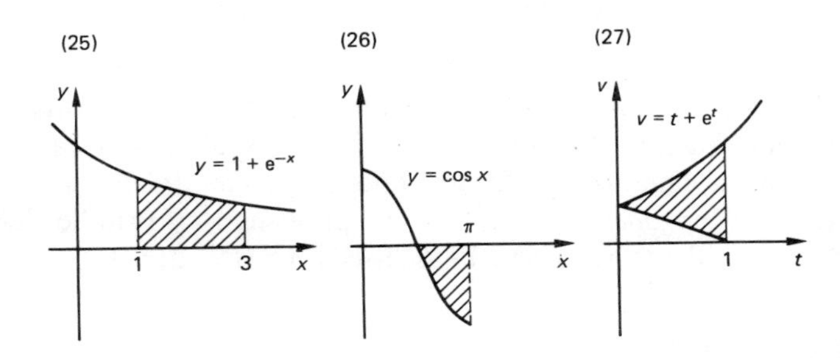

Figure 8.35

8.7 INTEGRATION BY SUBSTITUTION

Integration by substitution enables us to reduce a given integral to one with which we are familiar. The technique is very powerful, and covers a great range of problems. Unfortunately, it is not possible to give a general rule for choosing the required substitution, but this will come with experience gained through working examples. However general hints will be given where appropriate.

8.7.1 Indefinite Integrals by Substitution

Example 8.7.1

Find $\int e^{5x} \, dx$.

Solution

The substitution $u = 5x$ will reduce e^{5x} to e^u, so this seems a sensible choice. However we cannot find $\int e^u \, dx$ since this is asking to integrate a function of the variable u with respect to a different variable x.

Now $u = 5x$, so $du/dx = 5$ and hence $dx = (1/5)\, du$ (*).
So

$$\int e^{5x}\, dx = \int e^{u} \left(\frac{1}{5}\, du\right)$$

$$= \frac{1}{5} \int e^{u}\, du = \frac{1}{5} e^{u} + c$$

$$= \frac{1}{5} e^{5x} + c$$

In step (*) du/dx has been treated like a fraction. Although du/dx is not a fraction this step is valid, but we shall not explain why here. For an explanation see for example *Elementary Calculus – An Infinitesimal Approach* by H J Keisler, published by Prindle, Weber and Schmidt, Boston, Mass., 1976, ISBN 0 87150 215–1.

Example 8.7.2

Find $I = \int \cos(5x)\, dx$.

Solution

Let

$$u = 5x$$

so

$$\frac{du}{dx} = 5 \quad \text{and} \quad dx = \frac{1}{5}\, du$$

$$I = \int \cos u \left(\frac{1}{5}\, du\right) = \frac{1}{5} \int \cos u\, du$$

$$= \frac{1}{5} \sin u + c = \frac{1}{5} \sin(5x) + c$$

Example 8.7.3

Find $I = \int x(3x^2 - 1)^{1/2}\, dx$.

Solution

The most difficult part of the function to be integrated is $(3x^2 - 1)^{1/2}$.

This suggests the substitution $u = 3x^2 - 1$.
So

$$\frac{du}{dx} = 6x$$

Hence

$$du = 6x\,dx$$

or

$$x\,dx = \frac{1}{6}\,du$$

The integral can be rewritten to allow for substituting for the whole expression '$x\,dx$'.
So

$$I = \int x(3x^2 - 1)^{1/2}\,dx$$

$$= \int (3x^2 - 1)^{1/2}\,(x\,dx)$$

$$= \int u^{1/2}\left(\frac{1}{6}\,du\right) = \frac{1}{6}\int u^{1/2}\,du$$

$$= \frac{1}{6}\left(\frac{2}{3}u^{3/2}\right) + c$$

$$= \frac{1}{9}(3x^2 - 1)^{3/2} + c$$

Exercise 8.7.1

Find the following integrals:

1. $\int \sin 3x\,dx$
2. $\int \sqrt{x - 1}\,dx$
3. $\int \sqrt{2t - 1}\,dt$
4. $\int \sin\left(x + \frac{\pi}{3}\right)dx$

5. $\int \cos\left(2x - \frac{\pi}{3}\right) \mathrm{d}x$ 6. $\int x\mathrm{e}^{x^2} \mathrm{d}x$ (Let $u = x^2$)

7. $\int (x + 1)(x^2 + 2x - 1)^5 \mathrm{d}x$ (Let $u = x^2 + 2x - 1$)

8. $\int \frac{1}{(1 + t)^2} \mathrm{d}t$ 9. $\int \frac{1}{(3 - 2v)^3} \mathrm{d}v$
(Let $u = 3 - 2v$)

10. $\int (t + 2)(2t + 1)^{5/2} \mathrm{d}t$ 11. $\int \mathrm{e}^y(1 - \mathrm{e}^y)^{1/3} \mathrm{d}y$

12. $\int (1 + v)^{0.2} \mathrm{d}v$ 13. $\int (1 + 2v)^{0.2} \mathrm{d}v$

8.7.2 Integration of $1/x$ with respect to x

From Section 7.15 we know that

$$y = \ln x \Rightarrow \frac{\mathrm{d}y}{\mathrm{d}x} = \frac{1}{x}$$

From this result (and the fundamental theorem of calculus), it follows that

$$\boxed{\int \frac{1}{x} \mathrm{d}x = \ln x + c}$$

Example 8.7.4

Find $\int \frac{1}{2x + 1} \mathrm{d}x$.

Solution

Let $u = 2x + 1$ so $\mathrm{d}u/\mathrm{d}x = 2$ or $\mathrm{d}x = \frac{1}{2}\mathrm{d}u$

$$\int \frac{1}{2x + 1} \mathrm{d}x = \int \frac{1}{u} \tfrac{1}{2}\mathrm{d}u = \tfrac{1}{2} \ln u + c$$

$$= \tfrac{1}{2} \ln (2x + 1) + c$$
$$= \ln \sqrt{2x + 1} + c \qquad (*)$$
$$= \ln \sqrt{2x + 1} + \ln A$$
$$= \ln (A\sqrt{2x + 1})$$

Note: We could have stopped at (*), but often with logarithmic answers it is convenient to write the arbitrary constant c as $\ln A$ where A is an arbitrary positive constant. Then a single logarithmic answer may be obtained.

Example 8.7.5

Find $\displaystyle\int \frac{1}{1 - t}\, \mathrm{d}t$.

Solution

Let $u = 1 - t$, so $\mathrm{d}u/\mathrm{d}t = -1$ or $\mathrm{d}t = -\mathrm{d}u$.

$$\int \frac{1}{1 - t}\, \mathrm{d}t = \int \frac{1}{u} (-\mathrm{d}u) = -\int \frac{1}{u}\, \mathrm{d}u = -\ln u + c$$

$$= -\ln (1 - t) + \ln A = \ln \left(\frac{1}{1 - t}\right) + \ln A$$

$$= \ln \left(\frac{A}{1 - t}\right)$$

Example 8.7.6

Find $I = \displaystyle\int \frac{t}{1 + t^2}\, \mathrm{d}t$ by using the substitution $u = 1 + t^2$.

Solution

$$u = 1 + t^2 \Rightarrow \frac{\mathrm{d}u}{\mathrm{d}t} = 2t$$

so

$$t\mathrm{d}t = \tfrac{1}{2}\mathrm{d}u$$

Again substitute for the whole expression '$t\mathrm{d}t$'

$$I = \int \frac{t}{1+t^2}\,dt = \int \frac{1}{1+t^2}\,(t\,dt)$$

$$= \int \frac{1}{u}\,(\tfrac{1}{2}du) = \tfrac{1}{2}\ln u + c = \tfrac{1}{2}\ln(1+t^2) + c$$

Exercise 8.7.2

1. $\int \frac{1}{4x}\,dx$
2. $\int \frac{2}{x}\,dx$
3. $\int \frac{9}{x}\,dx$
4. $\int \frac{7}{1-2x}\,dx$
5. $\int \frac{1}{2t+2}\,dt$
6. $\int \frac{1}{3t-2}\,dt$
7. $\int \frac{1}{ax}\,dx$
8. $\int \frac{a}{x}\,dx$
9. $\int \frac{1}{ax+b}\,dx$
10. $\int \frac{a}{bx+c}\,dx$
11. $\int \frac{t^2}{1+t^3}\,dt$ (Let $u = 1 + t^3$)
12. $\int \frac{t^{1/2}}{4+t^{3/2}}$ (Let $u = 4 + t^{3/2}$)

8.7.3 Definite Integrals by Substitution

With indefinite integrals, it was necessary to substitute back to the original variable. This is not necessary with definite integrals but the limits must be changed when the initial substitution is made.

Example 8.7.7

Evaluate $I = \int_0^1 \frac{1}{3+4x}\,dx$

Solution

Step 1
Choose a suitable substitution. Let $u = 3 + 4x$, so $du/dx = 4$ and

$$dx = \tfrac{1}{4}du$$

Step 2
Compute new limits.

x	0	1
$u = 3 + 4x$	3	7

Step 3
Make the substitution, and integrate.

$$I = \int_{x=0}^{x=1} \frac{1}{3 + 4x}\,dx = \int_{u=3}^{u=7} \frac{1}{u}\,(\tfrac{1}{4}du)$$

It is very important to get the limits to match, that is $x = 0$ gives $u = 3$. The variables have been written on the limits to emphasize that when the substitution has been made. Normally the variables are omitted with the convention that they apply to the variable after the 'd', i.e. x or u.

$$\text{Therefore} \quad I = \frac{1}{4}\int_{3}^{7} \frac{1}{u}\,du = \left[\frac{1}{4}\ln u\right]_3^7 = \frac{1}{4}\ln\frac{7}{3}$$

Example 8.7.8

Find $I = \displaystyle\int_0^{\pi/2\omega} \cos(\omega t)\,dt$.

Solution

Step 1
Choose a suitable substitution.
Let $u = \omega t$, so $du/dt = \omega$ and $dt = (1/\omega)\,du$.

Step 2
Compute new limits.

t	0	$\dfrac{\pi}{2\omega}$
$u = \omega t$	0	$\dfrac{\pi}{2}$

Step 3
Make the substitution, and integrate.

$$I = \int_{t=0}^{t=\pi/2\omega} \cos(\omega t)\,dt = \int_{u=0}^{u=\pi/2} \cos u\left(\frac{1}{\omega}\,du\right)$$

$$= \frac{1}{\omega}\left[\sin u\right]_0^{\pi/2} = \frac{1}{\omega}(1 - 0) = \frac{1}{\omega}$$

Example 8.7.9

Evaluate $I = \displaystyle\int_1^2 xe^{-x^2}\,dx$.

Solution

Step 1
Choose a suitable substitution.
Let $u = x^2$, so $du/dx = 2x$ and $xdx = \frac{1}{2}du$.

Step 2
Compute new limits.

x	1	2
$u = x^2$	1	4

Step 3
Make the substitution and integrate

$$I = \int_1^2 xe^{-x^2}\,dx = \int_1^2 e^{-x^2}\,(xdx)$$

$$= \int_1^4 e^{-u}\,(\tfrac{1}{2}du) = \frac{1}{2}\int_1^4 e^{-u}\,du$$

$$= -\frac{1}{2}\left[e^{-u}\right]_1^4 = -\frac{1}{2}(e^{-4} - e^{-1}) = \frac{1}{2}(e^{-1} - e^{-4})$$

Example 8.7.10

Evaluate $I = \displaystyle\int_0^{\pi/3} \cos^2\theta\,\sin\theta\,d\theta$

Solution

Step 1
Choose a suitable substitution. Let $u = \cos\theta$. Then $du/d\theta = -\sin\theta$.
So $\sin\theta\,d\theta = -\,du$.

Step 2
Compute new limits.

θ	0	$\pi/3$
$u = \cos\theta$	1	0.5

Step 3
Make the substitution and integrate

$$I = \int_0^{\pi/3} \cos^2\theta\,(\sin\theta \mathrm{d}\theta)$$

$$= \int_1^{0.5} u^2\,(-\mathrm{d}u) \qquad \text{(note the order of the limits)}$$

$$= -\frac{1}{3}[u^3]_1^{0.5}$$

$$= -\frac{1}{3}(0.5^3 - 1^3) \simeq 0.292 \quad \text{(three decimal places)}$$

Exercise 8.7.3

Evaluate the following integrals:

1. $\displaystyle\int_0^{\pi/6} \sin 3x\,\mathrm{d}x$
2. $\displaystyle\int_1^3 \frac{1}{5x+3}\,\mathrm{d}x$
3. $\displaystyle\int_1^3 \frac{1}{5t+3}\,\mathrm{d}t$
4. $\displaystyle\int_1^2 \sqrt{2x-1}\,\mathrm{d}x$
5. $\displaystyle\int_1^2 \frac{t^2}{1+t^3}\,\mathrm{d}t$
6. $\displaystyle\int_0^3 t\sqrt{9-t^2}\,\mathrm{d}t$
7. $\displaystyle\int_0^{\pi/2} \sin^3\theta\cos\theta\,\mathrm{d}\theta$ [Let $u = \sin\theta$]
8. $\displaystyle\int_0^1 (x-1)\sqrt{x+1}\,\mathrm{d}x$
9. $\displaystyle\int_0^1 \frac{\mathrm{e}^{2t}}{(1+\mathrm{e}^{2t})^2}\,\mathrm{d}t$
10. $\displaystyle\int_0^{1/3} \cos\pi\theta\,\mathrm{d}\theta$

8.8 MORE ADVANCED SUBSTITUTIONS

8.8.1 Integrals of the form $\int \frac{f'(x)}{f(x)}\,dx$

This class of integrals leads to logarithmic answers. We have already covered some examples of these in the previous section.

They have the property that the integrand is a fraction where the numerator (top) is the derivative of the denominator (bottom).

If

$$I = \int \frac{f'(x)}{f(x)}\,dx, \text{ then let } u = f(x)$$

so

$$\frac{du}{dx} = f'(x) \quad \text{or} \quad du = f'(x)\,dx$$

$$I = \int \left[\frac{1}{f(x)}\right][f'(x)dx] = \int \frac{1}{u}\,du$$

$$= \ln u + c = \ln [f(x)] + c$$

$$\boxed{\int \frac{f'(x)}{f(x)}\,dx = \ln [f(x)] + c}$$

In practice it is best actually to carry out the substitution rather than use the formula direct. The method will also work when the derivative of the denominator (bottom) = a constant × numerator.

You should get used to recognizing this type of integral. For example

$$I = \int \frac{x^2}{1 + x^3}\,dx \quad \text{is of this form since}$$

$$\frac{d}{dx}(\text{denominator}) = \frac{d}{dx}(1 + x^3) = 3x^2 = \text{constant} \times (\text{numerator})$$

but

$$I = \int \frac{x^4}{1 + x^4}\,dx \quad \text{is } \textit{not} \text{ of this form since}$$

$$\frac{d}{dx}\text{(denominator)} = \frac{d}{dx}(1 + x^4) = 4x^3 \neq \text{constant} \times \text{(numerator)}$$

So this approach could not be used.

Example 8.8.1

Find $I = \displaystyle\int \frac{e^x - e^{-x}}{e^x + e^{-x}}\,dx$.

Solution

$$\frac{d}{dx}\text{(denominator)} = \frac{d}{dx}(e^x + e^{-x}) = e^x - e^{-x} = \text{numerator}$$

Let $u = e^x + e^{-x}$, then $du = (e^x - e^{-x})\,dx$
So

$$I = \int \frac{1}{e^x + e^{-x}}(e^x - e^{-x})\,dx$$

$$= \int \frac{1}{u}\,du = \ln u + c = \ln(e^x + e^{-x}) + c$$

Example 8.8.2

Find $I = \displaystyle\int \tan\theta\,d\theta$.

Solution

The integrand can be written as a fraction, so

$$I = \int \tan\theta\,d\theta = \int \frac{\sin\theta}{\cos\theta}\,d\theta$$

$$\frac{d}{d\theta}\text{(denominator)} = \frac{d}{d\theta}(\cos\theta) = -\sin\theta = -1 \times \text{numerator}$$

So let $u = \cos\theta$, then $du/d\theta = -\sin\theta$, and $\sin\theta\,d\theta = -du$.
Therefore

$$I = \int \frac{1}{u}(-du) = -\ln u + c$$

$$= -\ln(\cos\theta) + c = \ln(\cos\theta)^{-1} + c \quad \text{(by properties of logs)}$$

$$= \ln(\sec\theta) + c \quad \left[\text{since } (\cos\theta)^{-1} = \frac{1}{\cos\theta} = \sec\theta\right]$$

Exercise 8.8.1

1. $\displaystyle\int \cot\theta \, d\theta$

2. $\displaystyle\int \frac{x^{1/2}}{2 + x^{3/2}} \, dx$

3. $\displaystyle\int \tan 2\theta \, d\theta$

4. $\displaystyle\int \frac{\sin\theta}{1 + \cos\theta} \, d\theta$

5. $\displaystyle\int_0^1 \frac{t + 2}{t^2 + 4t + 1} \, dt$

6. $\displaystyle\int_0^{\pi/4} \frac{\sec^2\theta}{4 + 3\tan\theta} \, d\theta$

7. $\displaystyle\int \frac{e^{2x}}{e^{2x} + e^x} \, dx$ (divide top and bottom by e^x)

8. $\displaystyle\int \frac{x}{x + 2x^2} \, dx$

8.8.2 Trigonometric Substitutions Leading to Inverse Functions

Example 8.8.3

Find $I = \displaystyle\int \frac{1}{1 + t^2} \, dt$.

Solution

Let $t = \tan\theta$, then $dt/d\theta = \sec^2\theta$, so $dt = \sec^2\theta \, d\theta$.
So

$$I = \int \frac{1}{1 + \tan^2\theta} \sec^2\theta \, d\theta$$

$$= \int \frac{1}{\sec^2\theta} \sec^2\theta \, d\theta \quad \text{since } 1 + \tan^2\theta \equiv \sec^2\theta$$

$$= \int 1 \, d\theta = \theta + c = \arctan t + c$$

Example 8.8.4

Evaluate $I = \int_0^{0.5} \frac{1}{\sqrt{1 - 4x^2}}\,dx$.

Solution

In order to motivate the substituion, recall that $\cos^2 \theta + \sin^2 \theta \equiv 1$ so $1 - \sin^2 \theta \equiv \cos^2 \theta$ or $\sqrt{1 - \sin^2 \theta} \equiv \cos \theta$.

This suggests that

$$1 - 4x^2 = 1 - \sin^2 \theta$$

so

$$4x^2 = \sin^2 \theta$$

Let

$$x = \tfrac{1}{2} \sin \theta$$

Hence

$$dx = \tfrac{1}{2} \cos \theta \, d\theta$$

Now

$$x = \tfrac{1}{2} \sin \theta \Rightarrow \theta = \arcsin(2x))$$

x	0	0.5
$\theta = \arcsin(2x)$	$\arcsin 0 = 0$	$\arcsin 1 = \pi/2$

$$I = \int_0^{0.5} \frac{1}{\sqrt{1 - 4x^2}}\,dx = \int_0^{\pi/2} \frac{1}{\sqrt{1 - \sin^2 \theta}}\,\tfrac{1}{2} \cos \theta \, d\theta$$

$$= \frac{1}{2} \int_0^{\pi/2} \frac{\cos \theta}{\cos \theta}\,d\theta = \frac{1}{2} \int_0^{\pi/2} d\theta = \frac{1}{2} [\theta]_0^{\pi/2} = \frac{\pi}{4}$$

Exercise 8.8.2

1. $\int \frac{1}{1 + 9x^2}\,dx$
2. $\int \frac{1}{9 + x^2}\,dx$
3. $\int \frac{1}{4 + 9x^2}\,dx$
4. $\int \frac{1}{a^2 + b^2x^2}\,dx$

5. $\displaystyle\int \frac{1}{\sqrt{1-x^2}}\,dx$

6. $\displaystyle\int \frac{1}{\sqrt{25-x^2}}\,dx$

7. $\displaystyle\int \frac{1}{\sqrt{a^2-b^2x^2}}\,dx$

8. $\displaystyle\int_0^{\sqrt{3}/2} \frac{1}{1+4x^2}\,dx$

9. $\displaystyle\int_0^{0.125} \frac{1}{\sqrt{1-16t^2}}\,dt$

10. $\displaystyle\int_{1/\sqrt{2}}^{1} \frac{1}{\sqrt{2-t^2}}\,dt$

8.8.3 The Substitution $t = \tan \theta/2$ ('Half angle substitution')

This is a very useful substitution. Recall that if

$$t = \tan\frac{\theta}{2}$$

then

$$\frac{dt}{d\theta} = \frac{1}{2}\sec^2\frac{\theta}{2}$$

But

$$\sec^2\frac{\theta}{2} = 1 + \tan^2\frac{\theta}{2} = 1 + t^2$$

so

$$\frac{dt}{d\theta} = \tfrac{1}{2}(1 + t^2)$$

or

$$d\theta = \frac{2}{1+t^2}\,dt$$

Also if $t = \tan\dfrac{\theta}{2}$, then

$$\tan\theta = \tan\left[2\left(\frac{\theta}{2}\right)\right] = \frac{2\tan\dfrac{\theta}{2}}{1-\tan^2\dfrac{\theta}{2}} = \frac{2t}{1-t^2}$$

Sin θ, cos θ, etc can then be obtained from the triangle in Figure 8.35. The length of the hypotenuse h is given by $h^2 = (2t)^2 + (1 - t^2)^2 = (1 + t^2)^2$.

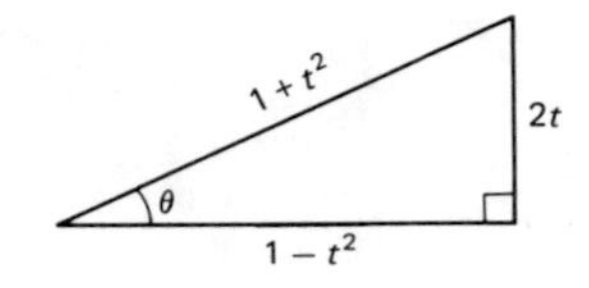

Figure 8.35

Example 8.8.5

Find $I = \int \sec\theta \, d\theta$.

Solution

Let $t = \tan\frac{\theta}{2}$, so $d\theta = \frac{2}{1 + t^2}\,dt$. Then from Figure 8.35,

$$\sec\theta = \frac{1 + t^2}{1 - t^2}$$

So

$$I = \int \frac{1 + t^2}{1 - t^2}\frac{2}{1 + t^2}\,dt = \int \frac{2}{1 - t^2}\,dt$$

$$= \int \frac{1}{1 + t} + \frac{1}{1 - t}\,dt \quad \text{by partial fractions}$$

$$= \ln(1 + t) - \ln(1 - t) + c$$

$$= \ln\left(\frac{1 + t}{1 - t}\right) + c = \ln\left[\frac{(1 + t)(1 + t)}{(1 - t)(1 + t)}\right] + c$$

$$= \ln\left(\frac{1 + t^2}{1 - t^2} + \frac{2t}{1 - t^2}\right) + c$$

Therefore $I = \ln(\sec\theta + \tan\theta) + c$

Example 8.8.6

Find $I = \int \frac{1}{3 + \cos 4\theta}\,d\theta$.

Solution

This integral has an integrand which is a function of 4θ rather than θ. So the substitution to try is

$$t = \tan 2\theta$$

since 2θ is half of 4θ

Hence $dt/d\theta = 2\sec^2 2\theta = 2(1 + \tan^2 2\theta) = 2(1 + t^2)$
So,

$$d\theta = \frac{dt}{2(1 + t^2)}$$

Also

$$\tan 4\theta = \frac{2\tan 2\theta}{1 - \tan^2 2\theta} = \frac{2t}{1 - t^2}$$

so

$$\cos 4\theta = \frac{1 - t^2}{1 + t^2}$$

$$3 + 4\cos 4\theta = \frac{3(1 + t^2) + 1 - t^2}{1 + t^2} = \frac{4 + 2t^2}{1 + t^2}$$

$$I = \int \frac{1 + t^2}{2(2 + t^2)} \frac{dt}{2(1 + t^2)} = \frac{1}{4}\int \frac{1}{2 + t^2}\, dt$$

$$= \frac{1}{4}\left[\frac{1}{\sqrt{2}} \arctan\left(\frac{t}{\sqrt{2}}\right)\right] + c$$

$$= \frac{\sqrt{2}}{8} \arctan\left(\frac{t}{\sqrt{2}}\right) + c = \frac{\sqrt{2}}{8} \arctan\left(\frac{1}{\sqrt{2}} \tan 2\theta\right) + c$$

Exercise 8.8.3

Use an appropriate 'half angle' tangent substitution on the following integrals.

1. $\int \operatorname{cosec} \theta \, d\theta$

2. $\int \frac{1}{1 + \sin \theta}\, d\theta$

8.8.4 A Useful Pair of Trigonometric Integrals

In order to evaluate $\int \cos^2 x \, dx$ and $\int \sin^2 x \, dx$ it is necessary to use the trigonometric identity

$$\cos 2x \equiv 2\cos^2 x - 1 \equiv 1 - 2\sin^2 x$$

which gives

$$\cos^2 x \equiv \tfrac{1}{2}(1 + \cos 2x) \quad \text{and} \quad \sin^2 x \equiv \tfrac{1}{2}(1 - \cos 2x)$$

Example 8.8.7

Evaluate $\int \cos^2 x \, dx$.

Solution

$$\int \cos^2 x \, dx = \tfrac{1}{2} \int (1 + \cos 2x) \, dx$$
$$= \tfrac{1}{2}(x + \tfrac{1}{2}\sin 2x) + C$$
$$= \tfrac{1}{2}x + \tfrac{1}{4}\sin 2x + C$$

Example 8.8.8

Evaluate $\int_0^{\pi/6} \sin^2(2x) \, dx$.

Solution

$\sin^2(2x) \equiv \tfrac{1}{2}[1 - \cos(4x)]$, so

$$\int_0^{\pi/6} \sin^2(2x) \, dx = \frac{1}{2}\int_0^{\pi/6} [1 - \cos(4x)] \, dx$$

$$= \frac{1}{2}\left[x - \frac{1}{4}\sin(4x)\right]_0^{\pi/6} = \frac{\pi}{12} - \frac{\sqrt{3}}{16}$$

Exercise 8.8.4

Find the following integrals:

1. $\int \sin^2 x \, dx$
2. $\int_0^{\pi/4} \cos^2(3x) \, dx$
3. $\int \cos^2(\tfrac{1}{2}x) \, dx$
4. $\int_{0.125}^{0.25} \sin^2(2\pi t) \, dt$

8.9 INTEGRATION BY PARTS

So far we have dealt with integrating products by making substitutions but this will not always work. Recall the formula for differentiating a product is (see Section 7.11)

$$\frac{\mathrm{d}}{\mathrm{d}x}(uv) = u\frac{\mathrm{d}v}{\mathrm{d}x} + v\frac{\mathrm{d}u}{\mathrm{d}x}$$

Integrate both sides with respect to x

$$\int \left[\frac{\mathrm{d}}{\mathrm{d}x}(uv)\right] \mathrm{d}x = \int u\frac{\mathrm{d}v}{\mathrm{d}x}\,\mathrm{d}x + \int v\frac{\mathrm{d}u}{\mathrm{d}x}\,\mathrm{d}x$$

so

$$uv = \int u\frac{\mathrm{d}v}{\mathrm{d}x}\,\mathrm{d}x + \int v\frac{\mathrm{d}u}{\mathrm{d}x}\,\mathrm{d}x$$

Rearranging gives

$$\boxed{\int u\frac{\mathrm{d}v}{\mathrm{d}x}\,\mathrm{d}x = uv - \int v\frac{\mathrm{d}u}{\mathrm{d}x}\,\mathrm{d}x}$$

This is the formula for integrating by parts. It exchanges the integral on the left hand side for a product of two functions and another integral. The idea is to ensure that this integral is easier than the one on the left hand side.

Example 8.9.1

Evaluate $I = \int x \sin x\, \mathrm{d}x$.

Solution

The integrand is a product of two functions: x and $\sin x$.

$$\int u\frac{\mathrm{d}v}{\mathrm{d}x}\,\mathrm{d}x = uv - \int v\frac{\mathrm{d}u}{\mathrm{d}x}\,\mathrm{d}x$$

(integrate: $\frac{\mathrm{d}v}{\mathrm{d}x}$; differentiate: u)

Note that u is differentiated to give $\mathrm{d}u/\mathrm{d}x$ while $\mathrm{d}v/\mathrm{d}x$ is integrated to give v.

It is easy to differentiate and integrate both x and $\sin x$, so the choice is not obvious at first. However as x differentiates to 1, choose this for u.

Let $u = x$ and $dv/dx = \sin x$

so

$$\frac{du}{dx} = 1 \quad \text{and} \quad v = -\cos x$$

$$\int x \sin x \, dx = -x \cos x - \int (-\cos x)/dx$$

$$= -x \cos x + \int \cos x \, dx$$

$$= -x \cos x + \sin x + c$$

Example 8.9.2

Evaluate $I = \int x^2 e^{-3x} \, dx$.

Solution

The choice here is $u = x^2$ since after differentiating twice this will give a constant.

So let $u = x^2$ and $dv/dx = e^{-3x}$

Hence

$$\frac{du}{dx} = 2x \quad \text{and} \quad v = -\frac{1}{3} e^{-3x}$$

$$\int u \frac{dv}{dx} dx = uv - \int v \frac{du}{dx} dx$$

$$\int x^2 e^{-3x} \, dx = -\frac{1}{3} x^2 e^{-3x} + \frac{2}{3} \int xe^{-3x} \, dx \qquad (*)$$

The last integral must be done by parts.

Let

$$I_2 = \int xe^{-3x} \, dx$$

Take

$$u = x \quad \text{and} \quad dv/dx = e^{-3x}$$

so

$$\frac{du}{dx} = 1 \quad \text{and} \quad v = -\frac{1}{3} e^{-3x}$$

$$I_2 = \int xe^{-3x}\,dx = -\frac{1}{3}xe^{-3x} + \frac{1}{3}\int e^{-3x}\,dx$$

$$= -\frac{1}{3}xe^{-3x} - \frac{1}{9}e^{-3x} + c$$

From (*)

$$I = -\frac{1}{3}x^2e^{-3x} + \frac{2}{3}I_2$$

$$= -\frac{1}{3}x^2e^{-3x} - \frac{2}{9}xe^{-3x} - \frac{2}{27}e^{-3x} + A$$

$$= -\frac{1}{27}(9x^2 + 6x + 2)e^{-3x} + A$$

(where A is an arbitrary constant).

Exercise 8.9.1

Evaluate the following integrals:

1. $\int x \cos x\,dx$
2. $\int xe^x\,dx$
3. $\int x \cos 2x\,dx$
4. $\int xe^{2x}\,dx$
5. $\int x\sqrt{x + 1}\,dx$
6. $\int x^2 \sin 3x\,dx$
7. $\int x^2e^{-x}\,dx$
8. $\int x \ln x\,dx$

8.9.1 Definite Integration by Parts

Make sure that the limits are written in at every stage — especially in the uv term.

$$\int_a^b u\frac{dv}{dx}\,dx = [uv]_a^b - \int_a^b v\frac{du}{dx}\,dx$$

Example 8.9.3

Find $\int_0^{\pi/6} x \sin 2x \, dx$.

Solution

Let $u = x$ and $dv/dx = \sin 2x$, so $du/dx = 1$ and $v = \frac{1}{2} \cos 2x$.

$$\int_0^{\pi/6} x \sin 2x \, dx = \left[-\frac{x}{2} \cos 2x \right]_0^{\pi/6} + \frac{1}{2} \int_0^{\pi/6} \cos 2x \, dx$$

$$= \left[-\frac{x}{2} \cos 2x \right]_0^{\pi/6} + \left[\frac{1}{4} \sin 2x \right]_0^{\pi/6}$$

$$= \left(-\frac{\pi}{12} \cos \frac{\pi}{3} - 0 \right) + \frac{1}{4} \left(\sin \frac{\pi}{3} - 0 \right)$$

$$= -\frac{\pi}{24} + \frac{\sqrt{3}}{8} = \frac{3\sqrt{3} - \pi}{24} \quad (\simeq 0.0856)$$

Example 8.9.4

Find $\int_0^1 x^2 e^{-4x} \, dx$.

Solution

Since x^2 is present two stages are required to differentiate this to a constant.

Let

$$I_1 = \int_0^1 x^2 e^{-4x} \, dx$$

and $u = x^2$ and $dv/dx = e^{-4x}$

$$\frac{du}{dx} = 2x \quad \text{and} \quad v = -\frac{1}{4} e^{-4x}$$

$$I_1 = \int_0^1 x^2 e^{-4x} \, dx = \left[-\frac{x^2}{4} e^{-4x} \right]_0^1 + \frac{1}{2} \int_0^1 x e^{-4x} \, dx$$

$$= -\frac{1}{4} e^{-4} + \frac{1}{2} I_2 \qquad (*)$$

where

$$I_2 = \int_0^1 x\mathrm{e}^{-4x}\,\mathrm{d}x$$

Let $u = x$ and $\mathrm{d}v/\mathrm{d}x = \mathrm{e}^{-4x}$ so $\mathrm{d}u/\mathrm{d}x = 1$ and $v = -\frac{1}{4}\mathrm{e}^{-4x}$

$$I_2 = \int_0^1 x\mathrm{e}^{-4x}\,\mathrm{d}x = \left[-\frac{x}{4}\mathrm{e}^{-4x}\right]_0^1 + \frac{1}{4}\int_0^1 \mathrm{e}^{-4x}\,\mathrm{d}x$$

$$= \left[-\frac{x}{4}\mathrm{e}^{-4x}\right]_0^1 - \left[\frac{1}{16}\mathrm{e}^{-4x}\right]_0^1$$

$$= -\frac{1}{4}\mathrm{e}^{-4} - \left[\frac{1}{16}\mathrm{e}^{-4} - \frac{1}{16}\right]$$

So from (*)

$$I_1 = -\frac{1}{4}\mathrm{e}^{-4} - \frac{1}{8} - \mathrm{e}^{-4} - \frac{1}{32}\mathrm{e}^{-4} + \frac{1}{32}$$

$$\int_0^1 x^2\mathrm{e}^{-4x} = \frac{1}{32}(1 - 13\mathrm{e}^{-4}) \simeq 0.0238$$

Exercise 8.9.2

Find the following integrals:

1. $\displaystyle\int_2^3 x\sqrt{x-1}\,\mathrm{d}x$

2. $\displaystyle\int_0^1 3x\mathrm{e}^{0.5x}\,\mathrm{d}x$

3. $\displaystyle\int_3^4 x\sqrt{2x-1}\,\mathrm{d}x$

4. $\displaystyle\int_0^{\pi/3} x\cos 3x\,\mathrm{d}x$

5. $\displaystyle\int_0^{\pi} x^2\sin(\tfrac{1}{2}x)\,\mathrm{d}x$

6. $\displaystyle\int_1^3 8x^2\mathrm{e}^{x-1}\,\mathrm{d}x$

7. $\displaystyle\int_1^2 \ln x\,\mathrm{d}x$ (Note $\ln x = 1 \times \ln x$, so take $u = \ln x$ and $\mathrm{d}v/\mathrm{d}x = 1$)

8.9.2 Further Techniques for Integration by Parts

In some cases after integrating by parts twice, the original integral is obtained. It is then possible to solve a simple equation and obtain the answer.

Example 8.9.5

Find $\int e^{-\frac{1}{2}x} \cos x \, dx$.

Solution

Let $I = \int e^{-\frac{1}{2}x} \cos x \, dx$

Let $u = \cos x$ and $dv/dx = e^{-\frac{1}{2}x}$
then $du/dx = -\sin x$ and $v = -2e^{-\frac{1}{2}x}$

$$\text{so } I = -2e^{-\frac{1}{2}x} \cos x - 2 \int e^{-\frac{1}{2}x} \sin x \, dx \qquad (*)$$

Let $I_2 = \int e^{-\frac{1}{2}x} \sin x \, dx$

and $u = \sin x$ and $dv/dx = e^{-\frac{1}{2}x}$
so $du/dx = \cos x$ and $v = -2e^{-\frac{1}{2}x}$

$$I_2 = -2e^{-\frac{1}{2}x} \sin x + 2\int e^{-\frac{1}{2}x} \cos x \, dx$$

$$I_2 = -2e^{-\frac{1}{2}x} \sin x + 2I$$

Combining this equation with (*) above gives

$$I = -2e^{-\frac{1}{2}x} \cos x - 2[-2e^{-\frac{1}{2}x} \sin x + 2I]$$

$$= -2e^{-\frac{1}{2}x} \cos x + 4e^{-\frac{1}{2}x} \sin x - 4I$$

$$\therefore 5I = 2e^{-\frac{1}{2}x} (2 \sin x - \cos x)$$

$$I = \frac{2}{5} e^{-\frac{1}{2}x} (2 \sin x - \cos x) + c$$

Example 8.9.6

Find $\int_0^{\pi/3} e^{3x} \sin 2x \, dx$.

Solution

This is a definite integral so the limits must be put in at each stage.

Let

$$I = \int_0^{\pi/3} e^{3x} \sin 2x \, dx$$

and $u = \sin 2x$ and $dv/dx = e^{3x}$
So

$$\frac{du}{dx} = 2 \cos 2x \quad \text{and} \quad v = \frac{1}{3} e^{3x}$$

$$I = \left[\frac{1}{3} e^{3x} \sin 2x\right]_0^{\pi/3} - \frac{2}{3}\int_0^{\pi/3} e^{3x} \cos 2x \, dx$$

$$= \frac{1}{3} e^{\pi} \sin\left(\frac{2\pi}{3}\right) - 0 - \frac{2}{3}\int_0^{\pi/3} e^{3x} \cos 2x \, dx$$

$$I = \frac{\sqrt{3}}{6} e^{\pi} - \frac{2}{3}\int_0^{\pi/3} e^{3x} \cos 2x \, dx \qquad (*)$$

Let

$$I_2 = \int_0^{\pi/3} e^{3x} \cos 2x \, dx$$

and $u = \cos 2x$ and $dv/dx = e^{3x}$
So

$$\frac{du}{dx} = -2 \sin 2x \quad \text{and} \quad v = \frac{1}{3} e^{3x}$$

$$I_2 = \left[\frac{1}{3} e^{3x} \cos 2x\right]_0^{\pi/3} + \frac{2}{3}\int_0^{\pi/3} e^{3x} \sin 2x \, dx$$

$$= \frac{1}{3} e^{\pi} \cos\left(\frac{2\pi}{3}\right) - \frac{1}{3} e^{0} \cos 0 + \frac{2}{3} I$$

$$= -\frac{1}{6} e^{\pi} - \frac{1}{3} + \frac{2}{3} I$$

Combining this result with (*) gives

$$I = \frac{\sqrt{3}}{6} e^{\pi} - \frac{2}{3}\left(-\frac{1}{6} e^{\pi} - \frac{1}{3} + \frac{2}{3} I\right)$$

$$= \frac{\sqrt{3}}{6} e^{\pi} + \frac{1}{9} e^{\pi} + \frac{2}{9} - \frac{4}{9} I$$

So

$$\frac{13}{9} I = \frac{e^{\pi}}{18} (3\sqrt{3} + 2) + \frac{2}{9}$$

Hence

$$I = \frac{e^{\pi}}{26} (3\sqrt{3} + 2) + \frac{2}{13}$$

Exercise 8.9.3

Find the following integrals:

1. $\int e^x \sin x \, dx$
2. $\int e^{-x} \cos x \, dx$
3. $\int e^{2x} \cos 4x \, dx$
4. $\int_0^{\pi} e^x \cos(\frac{1}{2}x) \, dx$
5. $\int_0^{\pi/6} e^{-x} \sin 2x \, dx$
6. $\int_{\pi/4}^{\pi/2} e^{2x} \sin \left(x + \frac{\pi}{4}\right) dx$

8.10 USE OF PARTIAL FRACTIONS IN INTEGRATION

The techniques for finding partial fractions were covered in Section 6.8.

Many integrals involving fractional integrands may be integrated by expressing the integral as the sum or difference of two or more partial fractions.

The partial fractions must themselves be integrated. Recall that there are three forms for partial fraction decomposition. Each form leads to a different type of integral.

Form 1

$$\int \frac{a}{b + cx} \, dx \Rightarrow \text{logarithmic answer (Section 8.7)}$$

or more generally

$$\int \frac{af'(x)}{f(x)}\,\mathrm{d}x \Rightarrow \text{ logarithmic answer (Section 8.8)}$$

Form 2

$$\int \frac{a}{(b + cx)^n}\,\mathrm{d}x \text{ where } n = 2, 3, \ldots$$

$\Rightarrow$ 'power' answer (Section 8.7)

Form 3

$$\int \frac{a}{b + cx^2}\,\mathrm{d}x$$

where $b + cx^2$ is an irreducible quadratic
$\Rightarrow$ 'arctan' answer (Section 8.8)

Example 8.10.1

Evaluate $I = \displaystyle\int \frac{3x + 9}{(x + 2)(x + 5)}\,\mathrm{d}x$.

Solution

Step 1
Split the integrand into partial fractions

$$\frac{3x + 9}{(x + 2)(x + 5)} \equiv \frac{1}{x + 2} + \frac{2}{x + 5}$$

Step 2
Integrate each partial fraction. Note in this case both fractions are Form 1 ('logarithmic').

$$\begin{aligned} I &= \int \frac{1}{x + 2}\,\mathrm{d}x + 2\int \frac{1}{x + 5}\,\mathrm{d}x \\ &= \ln(x + 2) + 2\ln(x + 5) + c \\ &= \ln(x + 2) + \ln(x + 5)^2 + \ln A \\ &= \ln[A(x + 2)(x + 5)^2] \end{aligned}$$

In the final answer it is usually best to combine the logarithms. If the answer is purely logarithmic the arbitrary constant c can be written as $\ln A$ where A is a (positive) arbitrary constant.

Example 8.10.2

Evaluate $I = \int \frac{4x - 7}{(x - 2)^2}\, \mathrm{d}x$.

Solution

Step 1
Split the integrand into partial fractions

$$\frac{4x - 7}{(x - 2)^2} = \frac{1}{(x - 2)^2} + \frac{4}{x - 2}$$

Step 2
Integrate each partial fraction. Note: here we have Form 2 ('power') and Form 1 ('logarithmic').

$$I = \int \frac{1}{(x - 2)^2}\, \mathrm{d}x + 4 \int \frac{1}{x - 2}\, \mathrm{d}x$$

$$= -\frac{1}{x - 2} + 4\ln(x - 2) + c$$

Example 8.10.3

Evaluate $I = \int \frac{2x^2 - x + 3}{x(1 + x^2)}\, \mathrm{d}x$.

Solution

Step 1
Split the integrand into partial fractions

$$\frac{2x^2 - x + 3}{x(1 + x^2)} = \frac{3}{x} - \frac{1 + x}{1 + x^2}$$

Step 2
Integrate each partial fraction. The first fraction is form 1. The second fraction has an irreducible denominator but as it stands it is *not* Form 3 ('arctan') as the numerator involves x as well as constants.

However

$$\frac{1 + x}{1 + x^2} = \frac{1}{1 + x^2} + \frac{x}{1 + x^2}$$

which gives a Form 3 fraction and a Form 1 ('logarithmic') fraction (since $d/dx\,(1 + x^2) = 2x$).

So

$$I = \int \frac{3}{x}\,dx - \int \frac{1}{1 + x^2}\,dx - \int \frac{x}{1 + x^2}\,dx$$

$$= 3 \ln x - \arctan x - \tfrac{1}{2}\ln(1 + x^2) + c$$

$$= \ln x^3 - \arctan x - \ln\sqrt{1 + x^2} + c$$

$$= \ln\left(\frac{x^3}{\sqrt{1 + x^2}}\right) - \arctan x + c$$

Exercise 8.10.1

Evaluate the following integrals:

1. $\displaystyle\int \frac{x - 5}{(x + 1)(x - 1)}\,dx$

2. $\displaystyle\int \frac{11x + 5}{(2x - 1)(x + 3)}\,dx$

3. $\displaystyle\int \frac{2x + 1}{(2x - 1)(2x + 3)}\,dx$

4. $\displaystyle\int \frac{4x^2 + 15x - 15}{x(x - 1)(x + 3)}\,dx$

5. $\displaystyle\int \frac{2x^2 - 9x - 1}{(3x + 1)(x^2 + 1)}\,dx$

6. $\displaystyle\int \frac{3x^2 + 2x - 1}{(x + 3)(x^2 + 1)}\,dx$

7. $\displaystyle\int \frac{x(2x + 7)}{(3x - 2)(x + 1)^2}\,dx$

8. $\displaystyle\int \frac{2x^2 + 6x + 17}{(x^2 + 4)(2x + 3)}\,dx$

9. $\displaystyle\int_1^2 \frac{5x + 17}{(x + 3)(x + 4)}\,dx$

10. $\displaystyle\int_0^{0.1} \frac{5}{(x+2)(1-2x)}\,\mathrm{d}x$

11. $\displaystyle\int_0^{3} \frac{4x^2+9x+3}{(x+2)(x+1)^2}\,\mathrm{d}x$

8.11 VOLUMES OF SOLIDS OF REVOLUTION

If a plane surface is rotated through 2π radians (or 360°) about an axis in its plane it generates a solid of revolution. For example, if the triangle OAB of Figure 8.36 is rotated through 2π radians about OA it generates a solid cone.

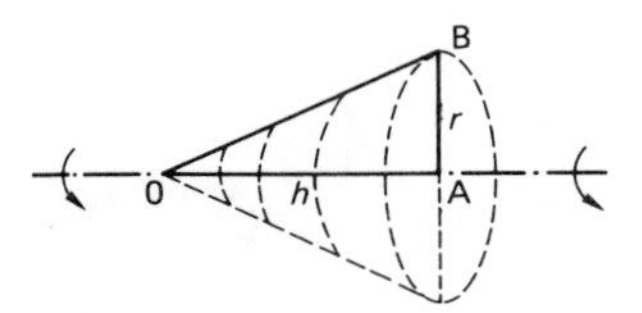

Figure 8.36

Let the cone have height h = OA, and base radius r = AB. The volume of this cone can be found by integration. Take OA as the x axis. A good approximation to the cone can be obtained by stacking discs on their sides, as shown in Figure 8.37.

Each of these discs is formed by 'spinning' a rectangle about the x axis. A typical rectangle is shown in Figure 8.37. It has height (radius) y and width Δx. When it is rotated through 2π radians (360°) about the x axis it generates a thin disc of volume

$$\Delta V = \pi y^2 \Delta x$$

Let V be the volume of the cone. Then

$$V \simeq \sum_{x=0}^{x=h} \Delta V = \sum_{x=0}^{x=h} \pi y^2 \Delta x$$

As the width of the discs decreases, we get a better approximation to the cone. So in the limit as $\Delta x \to 0$ we have

$$V = \lim_{\Delta x \to 0} \left(\sum_{x=0}^{x=h} \pi y^2 \Delta x \right)$$

Now this limit of a sum is an integral, see Section 8.3.
So

$$V = \int_0^h \pi y^2 \,\mathrm{d}x$$

To evaluate this integral we need y in terms of x. Since B has coordinates (h, r) the gradient of OB is r/h, and so the equation of OB is

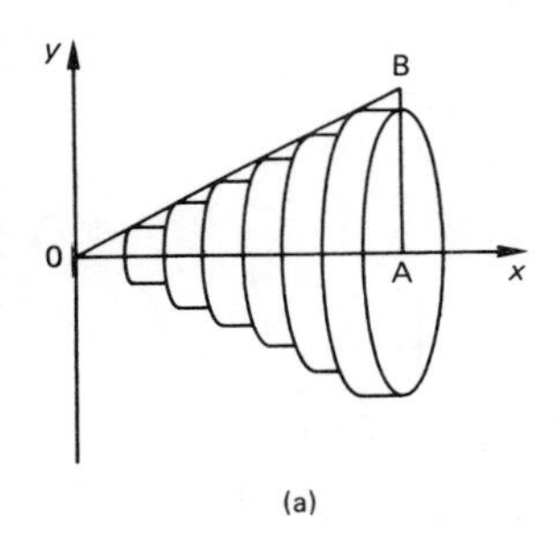

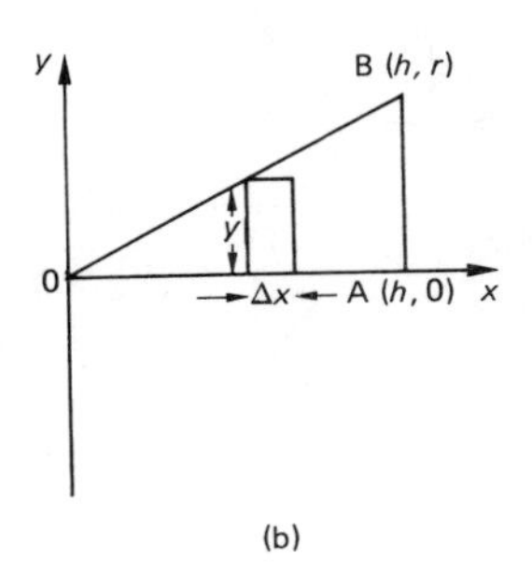

Figure 8.37 (a) Cone approximated by discs, (b) a 'typical' rectangle.

$$y = \frac{r}{h}x$$

$$V = \int_0^h \pi \left(\frac{r}{h}x\right)^2 dx$$

$$= \frac{\pi r^2}{h^2}\int_0^h x^2\,dx \quad \text{(since } \pi, r \text{ and } h \text{ are constants)}$$

$$= \frac{\pi r^2}{h^2}\left[\frac{x^3}{3}\right]_0^h$$

So

$$V = \frac{1}{3}\pi r^2 h$$

Example 8.11.1

Calculate the volume of a sphere of radius r.

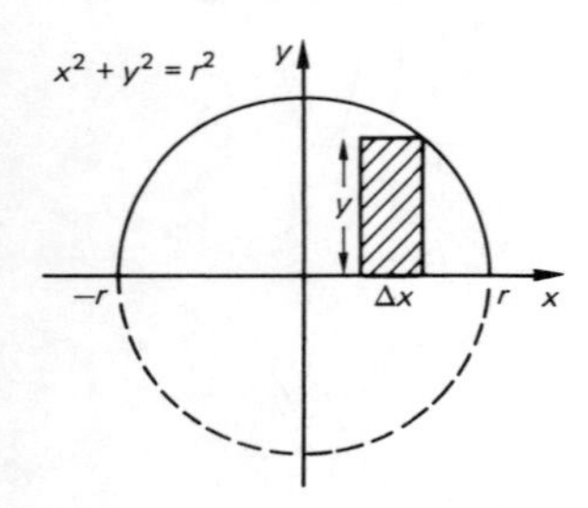

Figure 8.38

Solution

A sphere may be generated by 'spinning' a semicircle about its diameter. Set up axes as shown in Figure 8.38.

A typical rectangle is shown in the Figure 8.38. It has height y and width Δx. When it is rotated through 2π radians, about the x axis, it generates a disc of volume

$$\Delta V = \pi y^2 \Delta x$$

so the volume of the sphere

$$V \simeq \sum_{x=-r}^{x=r} \pi y^2 \Delta x$$

Take the limit as $\Delta x \to 0$

$$V = \lim_{\Delta x \to 0} \sum_{x=-r}^{x=r} \pi y^2 \Delta x = \int_{-r}^{r} \pi y^2\,dx$$

But $x^2 + y^2 = r^2$ is the equation of the semicircle

so

$$V = \pi \int_{-r}^{r} (r^2 - x^2)\,dx = \pi \left[r^2 x - \frac{x^3}{3} \right]_{-r}^{r}$$

$$= \pi \left[\left(r^3 - \frac{r^3}{3} \right) - \left(-r^3 + \frac{r^3}{3} \right) \right] = \frac{4}{3} \pi r^3$$

Example 8.11.2

Some lights have parabolic reflectors. A paraboloid is the solid generated when a parabola is rotated about its axis of symmetry. Find the volume of the paraboloid generated by rotating the parabola $y = x^2$ about its axis of symmetry. Cut the solid off at the plane where $y = 3$.

Solution

Refer to Figure 8.39. The y axis forms the axis of symmetry of the parabola. The paraboloid is formed by spinning the parabola about the y axis. So this time the discs are stacked 'vertically'. A typical disc is formed by rotating a rectangle of length x and height Δy through 2π radians about the y axis. So volume of a typical disc is

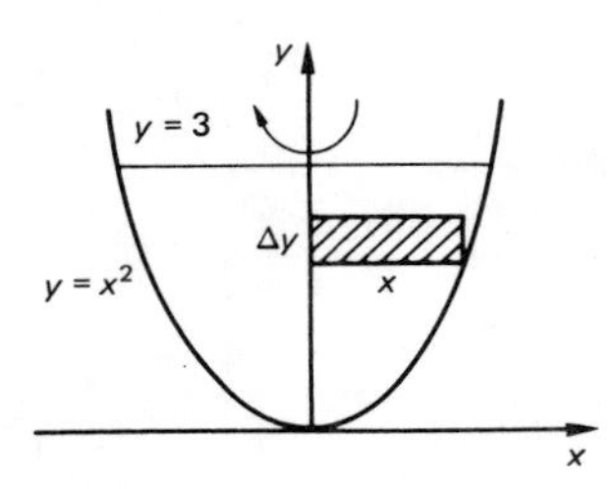

Figure 8.39

$$\Delta V = \pi x^2 \Delta y$$

The volume, V, of the paraboloid is given by

$$V \simeq \sum_{y=0}^{y=3} \pi x^2 \Delta y$$

So

$$V = \lim_{\Delta y \to 0} \sum_{y=0}^{y=3} \pi x^2 \Delta y$$

$$= \int_0^3 \pi x^2 dy$$

$$= \pi \int_0^3 y\,dy = \pi \left[\frac{y^2}{2} \right]_0^3 = \frac{9\pi}{2}$$

Exercise 8.11.1

For Problems 1–5 find the volume when the area bounded by the given lines is rotated through 2π radians about the x axis.

1. $y = x^3$, from $x = 0$ to $x = 1$

2. $y = \sin x$, from $x = 0$ to $x = \pi$
3. $y = e^{2x}$, from $x = 1$ to $x = 2$
4. $y = 1 + x^2$, from $x = 1$ to $x = 3$
5. $y = 1/x$, from $x = 1$ to $x = 2$

For Problems 6–8 find the volume when the area bounded by the given lines is rotated through 2π radians about the y axis.

6. $y = 1/x$, from $y = 1$ to $y = 2$
7. $y = \sqrt{x}$, from $y = 0$ to $y = 1$
8. $y = \ln x$, from $y = 1$ to $y = 1.5$
9. $y = 1/x$, the x axis and the lines $x = 1$ to $x = 2$ (work out the limits for y)
10. $y = x^3$, the x axis and the lines $x = 0$ and $x = 2$

8.12 CENTROIDS

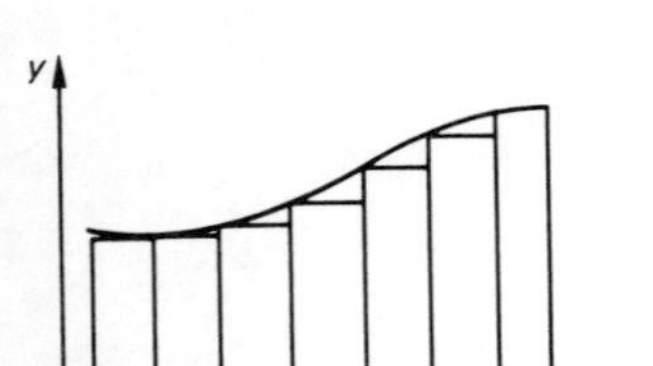

Figure 8.40

A flat, thin sheet of metal will balance in a horizontal position if it is supported directly below its centroid. (In practice balance would be difficult to achieve!)

So the centroid of a flat rectangular sheet of metal is at the meeting point of its diagonals. The centroid of a flat circular sheet of metal is at its geometrical centre.

The centroids of other plane figures could be obtained by experiment, but they may also be computed by integration.

In effect we approximate the plane figure by many thin rectangles, whose centroids are known, (Figure 8.40).

To understand the ideas behind finding centroids you must be familiar with the idea of moments. Consider the problem of getting a seesaw to balance horizontally. The seesaw is pivotted at P, a 55 kg mother sits at M and her 22 kg child sits at C (Figure 8.41).

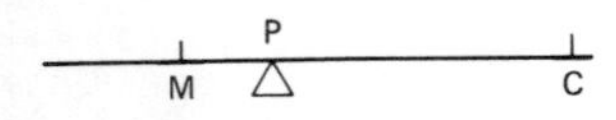

Figure 8.41

For balance

$$55 \times \text{MP} = 22 \times \text{CP} \qquad (8.12.1)$$

or

$$\frac{\text{MP}}{\text{CP}} = \frac{22}{55} = \frac{2}{5}$$

So if the child is 1 m from the pivot P, her mother should be 2/5 m ≡ 400 mm from P to achieve perfect balance.
Equation (8.12.1) uses moments.

The moment of a body about an axis is equal to the product of the mass of the body and the perpendicular distance of its centroid from the axis

The moment of a body measures its turning effect about the axis. The moment may be anticlockwise (usually taken as positive) or clockwise (usually taken as negative).
So in Figure 8.41, the moment of M about P is $+(55 \times \text{MP})$ and the moment of C about P is $-(22 \times \text{CP})$. For balance on the seesaw the sum of these moments must be zero, so

$$55 \times \text{MP} - 22 \times \text{CP} = 0$$

and this rearranges to give Equation (8.12.1).
Moments may now be applied to computing the position of the centroid of a body.

Example 8.12.1

Find the centroid of the plane body bounded by the curve $y = x^3$, the x axis and the lines $x = 0$ and $x = 2$.

Solution

The body is shown in Figure 8.42(a). Let its centroid be at $G(\bar{x}, \bar{y})$. The body may be approximated by rectangles (Figure 8.42(b)). A typical rectangle is shown in Figure 8.42(c).

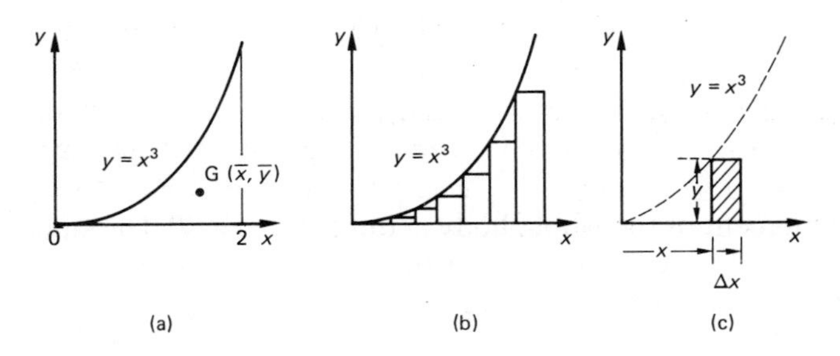

Figure 8.42

Step 1
Calculate the mass of the body. Let the density of the body be ρ per unit area. Then mass $m = \rho A$ where A is the area of the body. Now

$$A = \int_0^2 x^3 \, dx = \left[\frac{x^4}{4}\right]_0^2 = 4$$

So

$$\text{Mass } m = \rho A = 4\rho$$

Step 2
Find the position of $\bar{x}$ by taking moments about the y axis.

Step 2.1. Moment of the whole body about the y axis is

$$My = m\bar{x} = 4\rho\bar{x}$$

Step 2.2. Moment of a typical rectangle about the y axis is

ΔMy = mass of rectangle × distance of its centroid from the y axis.

Now referring to Figure 8.42(c).

Mass of rectangle = density × area
$= \rho y \Delta x$

The centroid of a rectangle is at its geometric centre. This is at a distance of $x + \Delta x/2$ from the y axis. Therefore the moment of the rectangle about the y axis is

$$\Delta My = \rho y \Delta x (x + \Delta x/2) \simeq \rho x y \Delta x$$

if we ignore the term in $(\Delta x)^2$ which will be small.

Step 2.3. Sum of the moments of the rectangles about the y axis is

$$\sum_{x=0}^{x=2} \Delta My = \sum_{x=0}^{x=2} \rho x y \Delta x$$

Step 2.4. The moment of the whole body about the y axis $\simeq$ The sum of the moments of the rectangles about the y axis. Now as $\Delta x \to 0$, the approximation gets better, until in the limit

Moment of whole body about y axis $= \lim\limits_{\Delta x \to 0}$ (sum of moments of rectangles about y axis)

$$4\rho\bar{x} = \lim_{\Delta x \to 0} \left(\sum_{x=0}^{x=2} \rho x y \Delta x \right)$$

So

$$4\rho\bar{x} = \int_0^2 \rho xy \, dx = \rho \int_0^2 xy \, dx \quad \text{since } \rho \text{ is constant}$$

and

$$\bar{x} = \frac{1}{4}\int_0^2 xy \, dx = \frac{1}{4}\int_0^2 x\,x^3 \, dx$$

$$= \frac{1}{4}\int_0^2 x^4 \, dx = \frac{1}{20}\left[x^5\right]_0^2 = \frac{8}{5}$$

Step 3
Find the position of $\bar{y}$ by taking moments about the x axis.

Step 3.1. Moment of whole body about the x axis is

$$M_x = m\bar{y} = 4\rho\bar{y}$$

Step 3.2. Moment of a typical rectangle about the x axis is

$$\Delta M_x = \text{mass of rectangle} \times \text{distance of its centroid from the } x \text{ axis}$$

$$= \rho y \Delta x \times \frac{y}{2} = \frac{1}{2}\rho y^2 \Delta x$$

Step 3.3. Sum of moments of the rectangles about the x axis is

$$\sum_{x=0}^{x=2} \tfrac{1}{2}\rho y^2 \Delta x$$

Step 3.4. Moment of whole body about x axis

$$= \lim_{\Delta x \to 0} (\text{sum of moments of the rectangles about the } x \text{ axis})$$

$$4\rho\bar{y} = \lim_{\Delta x \to 0}\left(\sum_{x=0}^{x=2} \tfrac{1}{2}\rho y^2 \Delta x\right)$$

$$\bar{y} = \frac{1}{8}\int_0^2 y^2 dx = \frac{1}{8}\int_0^2 x^6 \, dx$$

$$= \frac{1}{56}\left[x^7\right]_0^2 = \frac{16}{7}$$

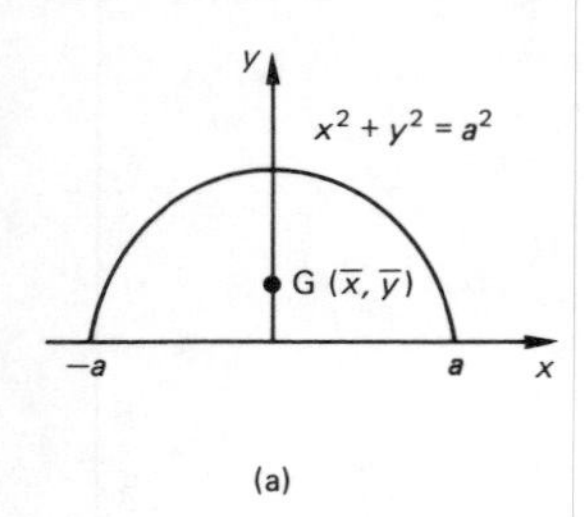

(a)

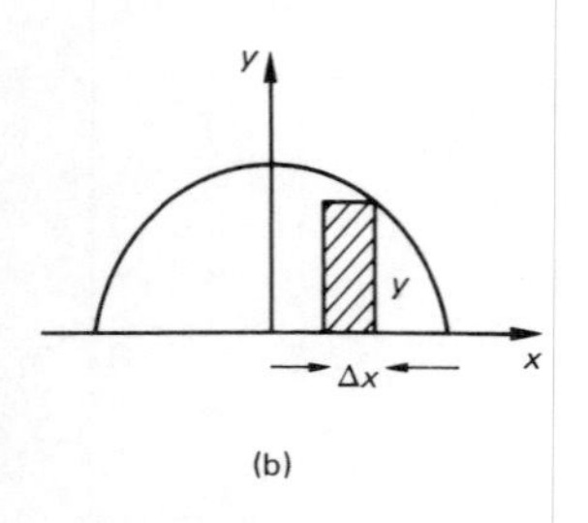

(b)

Figure 8.43

Step 4

Summary: the centroid G has coordinates $\left(\frac{8}{5}, \frac{16}{7}\right)$.

Example 8.12.2

Find the position of the centroid of a flat semicircular disc of radius a.

Solution

Where a figure has an axis of symmetry, the centroid must lie on it. Choose one of the coordinate axes to coincide with the axis of symmetry (Figure 8.43(a)). Let centroid be $G(\bar{x}, \bar{y})$.

Step 1
Calculate the mass of the semicircle. Let density be ρ units per unit area.
So, mass $m = \frac{1}{2}\pi\rho a^2$

Step 2
Find the position of $\bar{x}$. By symmetry $\bar{x} = 0$ (on y axis).

Step 3
Find the position of $\bar{y}$, by taking moments about the x axis.

Step 3.1. Moment of the whole body about the x axis is

$$M_x = (\tfrac{1}{2}\pi\rho a^2)\bar{y}$$

Step 3.2. Moment of a typical rectangle about the x axis is

$$\Delta M_x = (\rho y \Delta x)(\tfrac{1}{2}y) = \tfrac{1}{2}\rho y^2 \Delta x$$

(see Figure 8.43(b)).

Step 3.3. Sum of the moments of the rectangles about the x axis is

$$\sum_{x=-a}^{x=a} \tfrac{1}{2}\rho y^2 \Delta x$$

Step 3.4. Moment of the whole body about the x axis

$$= \lim_{\Delta x \to 0} \text{(sum of moments of the rectangles about the } x \text{ axis)}$$

So,

$$\tfrac{1}{2}\pi\rho a^2 \bar{y} = \lim_{\Delta x \to 0} \left(\sum_{x=-a}^{x=a} \tfrac{1}{2}\rho y^2 \Delta x \right)$$

and

$$\tfrac{1}{2}\pi\rho a^2\bar{y} = \tfrac{1}{2}\rho \int_{-a}^{a} y^2 \,dx$$

But on the semicircle $y^2 = a^2 - x^2$, so

$$\pi a^2\bar{y} = \int_{-a}^{a} (a^2 - x^2)\,dx = \left[a^2x - \frac{1}{3}x^3\right]_{-a}^{a}$$

$$= \frac{4}{3}a^3$$

Hence

$$\bar{y} = \frac{4a}{3\pi}$$

Step 4
The centroid G of a semicircular disc of radius a lies on the axis of symmetry at a distance of $4a/3\pi$ from the straight edge.

Exercise 8.12.1

Find the centroids of the following flat figures:

1. The area bounded by $y = x^2$, the x axis and the lines $x = 0$ and $x = 1$.
2. The area bounded by $y = x^3$, the y axis and the lines $y = 0$ and $y = 1$.
3. The area bounded by $y = \sin x$ and the x axis between $x = 0$ and $x = \pi$.
4. The area bounded by $y = \cos 2x$ and the x axis between $x = 0$ and $x = \pi/4$.
5. The area bounded by $y = e^{-0.5x}$, the x axis, the y axis and the line $x = 1$.

8.13 MISCELLANEOUS EXAMPLES

As you will appreciate integration is a vast topic which involves many different techniques. The following exercise is intended as practice at applying these techniques.

Exercise 8.13.1

1. Find the total area enclosed by the curve $y = \sin x$ and the x axis between $x = 0$ and $x = 2\pi$.

2. Find the total area enclosed by the curve $y = x(x - 1)(x - 2)$ and the x-axis between $x = 0$ and $x = 4$.

3. Use the trapezoidal rule with four intervals to find an approximation to the integral

$$\int_0^1 \frac{4}{(1 + x^2)}\,dx$$

For Problems 4–39, evaluate the integral.

4. $\displaystyle\int \frac{x}{x^2 + 1}\,dx$

5. $\displaystyle\int_0^1 \frac{x}{\sqrt{1 - x^2}}\,dx$

6. $\displaystyle\int te^{3t}\,dt$

7. $\displaystyle\int_0^1 te^{4t^2}\,dt$

8. $\displaystyle\int \frac{2}{(2 - x)(2 + x)}\,dx$

9. $\displaystyle\int \frac{x - 1}{\sqrt{1 - 9x^2}}\,dx$

10. $\displaystyle\int \sin x \cos x\,dx$

11. $\displaystyle\int_1^2 \frac{1}{t^2(1 + t)}\,dt$

12. $\displaystyle\int_0^a \sqrt{a^2 - u^2}\,du$

13. $\displaystyle\int \frac{3x^2}{x^3 - 3}\,dx$

14. $\displaystyle\int \frac{5}{x^2 + x - 6}\,dx$

15. $\displaystyle\int_0^1 \frac{3x^2 + 2x - 1}{x^3 + x^2 - x + 4}\,dx$

16. $\displaystyle\int \cos \pi x\,dx$

17. $\displaystyle\int_0^2 e^{(4t-1)}\,dt$

18. $\displaystyle\int_1^4 \frac{dx}{\sqrt{5 - x}}$

19. $\displaystyle\int x^2 \sin x\,dx$

20. $\displaystyle\int x \sin^2 x\,dx$

21. $\displaystyle\int \frac{1}{4 + t^2}\,dt$

22. $\displaystyle\int \frac{t}{4 + t^2}\,dt$

23. $\displaystyle\int \frac{1}{4 - t^2}\,dt$

24. $\displaystyle\int \frac{t}{4 - t^2}\,dt$

25. $\displaystyle\int \frac{u + 1}{u}\,du$

26. $\displaystyle\int \frac{u}{u + 1}\,du$

27. $\displaystyle\int \frac{t^2 + 2}{t}\,dt$

28. $\displaystyle\int \frac{t^2 + 2}{t^2}\,dt$

29. $\displaystyle\int \frac{t^2}{t^2 + 2}\,dt$

30. $\displaystyle\int \frac{t}{t^2 + 2}\,dt$

31. $\displaystyle\int (1 - x^2)\,dx$

32. $\displaystyle\int \sqrt{1 - x^2}\,dx$ (Let $x = \sin u$)

33. $\displaystyle\int \sqrt{1 - x}\,dx$

34. $\displaystyle\int \frac{1}{5 - 6t}\,dt$

35. $\displaystyle\int \frac{1}{(5 - 6t)^2}\,dt$

36. $\displaystyle\int \frac{1}{(5 - 6t)^{3/2}}\,dt$

37. $\displaystyle\int u(u^2 + 1)\,du$

38. $\displaystyle\int u^2(u^2 + 1)\,du$

39. $\displaystyle\int u^2\sqrt{u^2 + 1}\,du$

OBJECTIVE TEST 8

1. If the velocity v of a particle at time t is $v = 3t^2$, find its displacement s at time t.

2. Find $\displaystyle\int_0^{\pi} \sin(\tfrac{1}{4}x)\,dx$

3. Find $\displaystyle\int_0^1 xe^x\,dx$

4. Find $\displaystyle\int_0^1 xe^{x^2}\,dx$

5. Find $\displaystyle\int \frac{t}{64 + t^2}\,\mathrm{d}t$

6. Find $\displaystyle\int \frac{1}{64 + t^2}\,\mathrm{d}t$

7. Find $\displaystyle\int \frac{1}{64 - t^2}\,\mathrm{d}t$

8. Find $\displaystyle\int \cos^2(\pi x)\,\mathrm{d}x$

9 DIFFERENTIAL EQUATIONS

OBJECTIVES

When you have finished this chapter you should be able to

1. identify a first order differential equation
2. sketch the direction field for a first order differential equation
3. given initial conditions sketch a particular solution curve on the direction field plot
4. obtain the general solution to a first order differential equation by separating the variables
5. use initial conditions to obtain a particular solution from the general solution
6. set up differential equations
7. use simple numerical methods to obtain approximate solutions to differential equations

9.1 INTRODUCTION

In many engineering applications, mathematical models for the physical process lead to differential equations.

For example, if a body of mass M hanging from a stiff spring (Figure 9.1) is pulled downward from its equilibrium position, it will oscillate. Its distance x from the equilibrium position will be given by

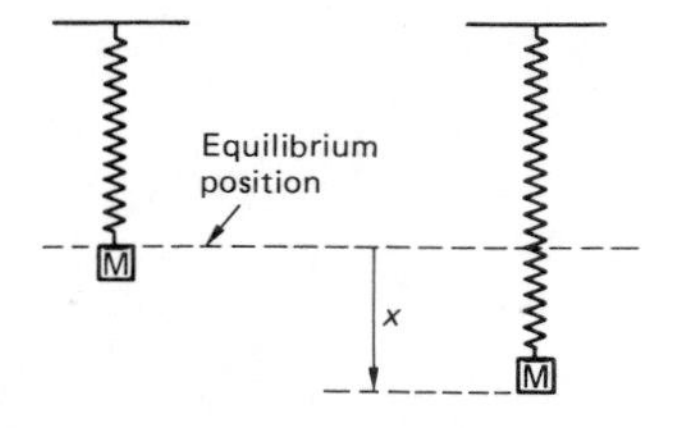

Figure 9.1

$$M\frac{d^2x}{dt^2} + A\frac{dx}{dt} + Bx = 0 \qquad (9.1.1)$$

where A and B are constants whose values depend on the damping effect and stiffness of the spring, and t is the time.

Equation (9.1.1) is a differential equation. In order to solve it, x must be found as a function of time. For example we might have

$$x = e^{-kt}(P\cos\omega t + Q\sin\omega t)$$

where k and ω are constants depending on M, A and B; P and Q are constants depending on the point of release.

Other examples of differential equations include:

$$\frac{dT}{dt} = -k(T - T_0) \tag{9.1.2}$$

which describes the temperature T of a liquid at time t as it cools to room temperature T_0 and

$$A\frac{d^2y}{dx^2} + By = \frac{1}{2}w(x^2 - lx) \tag{9.1.3}$$

which describes the vertical displacement y of a beam of length l, which carries a uniform load of weight w per unit length (Figure 9.2).

Figure 9.2

The differential equations (9.1.1) and (9.1.3) are called second order differential equations, since the highest derivatives which occur are the second derivatives,

$$\frac{d^2x}{dt^2} \quad \text{and} \quad \frac{d^2y}{dx^2}$$

Equation (9.1.2) is called a first order differential equation, since the highest derivative occurring is the first derivative, dT/dt.

In this chapter only first order differential equations are considered.

9.2 GENERAL AND PARTICULAR SOLUTIONS

To solve a differential equation such as

$$\frac{dy}{dt} = 1 - 3e^{-t} \tag{9.2.1}$$

we need to find a function $y = f(t)$ which satisfies (9.2.1). In fact, (9.2.1) has many solutions, for example

$$y = t + 3e^{-t} \tag{9.2.2}$$

$$\text{or} \quad y = t + 3e^{-t} - 1 \tag{9.2.3}$$

$$\text{or} \quad y = t + 3e^{-t} + 0.75 \tag{9.2.4}$$

Differentiating any of (9.2.2), (9.2.3) or (9.2.4) gives

$$\frac{dy}{dt} = 1 - 3e^{-t}$$

which is the differential equation (9.2.1). So (9.2.2), (9.2.3) and (9.2.4) are all solutions of (9.2.1). In fact there are an infinite number of solutions to (9.2.1) given by

$$y = t + 3e^{-t} + A \tag{9.2.5}$$

where A is a constant. (9.2.5) is called the *general solution* of the differential equation (9.2.1), and it defines an infinite family of functions which satisfy (9.2.1). Some members of this family of solutions are shown in Figure 9.3.

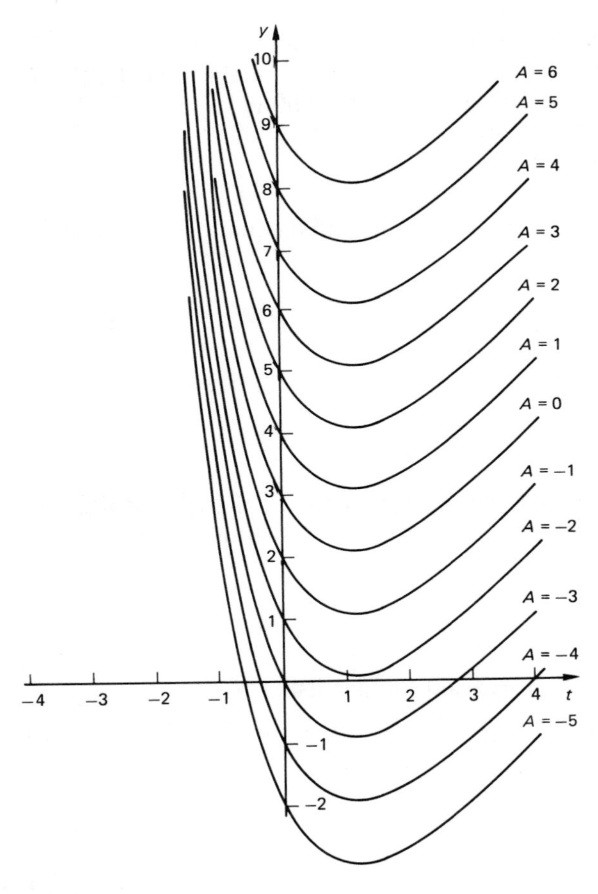

Figure 9.3 Solution curves for the differential equation $dy/dt = 1 - 3e^{-t}$

By assigning a particular value to the constant A in Equation (9.2.5), a *particular solution* to the differential equation is obtained. For example: if $A = 0$, then the particular solution is $y = t + 3e^{-t}$; if $A = -1$, then the particular solution is $y = t + 3e^{-t} - 1$.

In the case of equation (9.2.1), the general solution (9.2.5) is easily obtained by integration. Unfortunately it is not usually this easy to solve differential equations, and the techniques of solving differential equations form a large branch of mathematics. Two methods for solving first order differential equations are dealt with in Section 9.4.

9.2.1 Obtaining a particular solution from the general solution

To obtain a particular solution from the general solution (9.2.5), it is necessary to assign a value to the constant of integration A. This is done by providing initial or boundary conditions.

Example 9.2.1

Find the particular solution to the differential equation (9.2.1) which passes through the origin.

Solution

The general solution is

$$y = t + 3e^{-t} + A$$

The initial condition is that $y = 0$ when $t = 0$ (this ensures the solution curve passes through the origin). Substituting these values into the general solution gives

$$0 = 0 + 3e^{0} + A$$

Therefore $A = -3e^{0} = -3$ and the particular solution is

$$y = t + 3e^{-t} - 3$$

Summary
The *general solution* to a differential equation consists of a family of equations which satisfy the differential equation. For a first order differential equation, the general solution involves one arbitrary constant.
Given initial or boundary conditions, it is possible to find a value for the arbitrary constant in the general solution and so obtain a *particular solution*.

Example 9.2.2

Confirm that

$$y = Ae^{-\frac{1}{2}x^2} - 1 \tag{9.2.6}$$

is the general solution to

$$\frac{dy}{dx} = -x(1 + y) \tag{9.2.7}$$

and find the particular solution to the differential equation (9.2.6) given that $y = 1$ when $x = 0$.

Solution

$$y = Ae^{-\frac{1}{2}x^2} - 1$$

so

$$\frac{dy}{dx} = -xAe^{-\frac{1}{2}x^2} \tag{9.2.8}$$

Now the right hand side of the differential equation (9.2.7) is

$$-x(1 + y) = -x(1 + Ae^{-\frac{1}{2}x^2} - 1) = -xAe^{-\frac{1}{2}x^2} \tag{9.2.9}$$

So from (9.2.8) and (9.2.9), Equation (9.2.6) is the general solution to the differential equation (9.2.7).

Given the initial conditions $y = 1$ when $x = 0$, substituting into (9.2.6) gives

$$1 = Ae^0 - 1$$

Therefore $Ae^0 = 2$, so $A = 2$. The particular solution is

$$y = 2e^{-\frac{1}{2}x^2} - 1$$

Exercise 9.2.1

In each problem confirm by differentiation that (b) is the general solution of the differential equation (a). Obtain the particular solution defined by the initial conditions (c).

1. (a) $\dfrac{dy}{dx} = \dfrac{y(x + 2)}{x}$ (b) $y = Ax^2e^x$

(c) $y = 2\text{e}$ when $x = 1$

2. (a) $\dfrac{\text{d}y}{\text{d}x} = 2\text{e}^{-y}$ (b) $y = \ln(2x + A)$

(c) $y = \ln 3$ when $x = 0$

3. (a) $\dfrac{\text{d}y}{\text{d}x} = 2\text{e}^{-x}$ (b) $y = A - 2\text{e}^{-x}$

(c) $y = 3$ when $x = 0$

4. (a) $\dfrac{\text{d}y}{\text{d}t} = \dfrac{1}{1 + t^2}$ (b) $y = \arctan t + A$

(c) $y = \pi/2$ when $t = 0$

5. (a) $\dfrac{\text{d}x}{\text{d}t} = \dfrac{x}{t - 1}$ (b) $x = A(t - 1)$

(c) $x = 4$ when $t = 0$

9.3 SKETCHING DIRECTION FIELDS

A first order differential equation such as

$$\frac{\text{d}y}{\text{d}x} = x^2y$$

or

$$\frac{\text{d}y}{\text{d}x} = y + \text{e}^x$$

can be regarded as an expression which defines a gradient at every point in the plane. For example, if

$$\frac{\text{d}y}{\text{d}x} = \frac{x + 1}{y^2 + 1}$$

then when $x = 0$ and $y = 0$

$$\frac{\text{d}y}{\text{d}x} = \frac{1}{1} = 1$$

and when $x = 1$ and $y = 2$

$$\frac{dy}{dx} = \frac{1 + 1}{4 + 1} = \frac{2}{5}$$

The gradient may be represented at any point by a small line having the calculated gradient, as shown in the following examples.

Example 9.3.1

A circuit consists of a 20 ohm resistor in series with an inductance of 10 mH (Figure 9.4). When the switch S is closed, a current i will start to flow in the circuit. The rate of increase of this current is given by

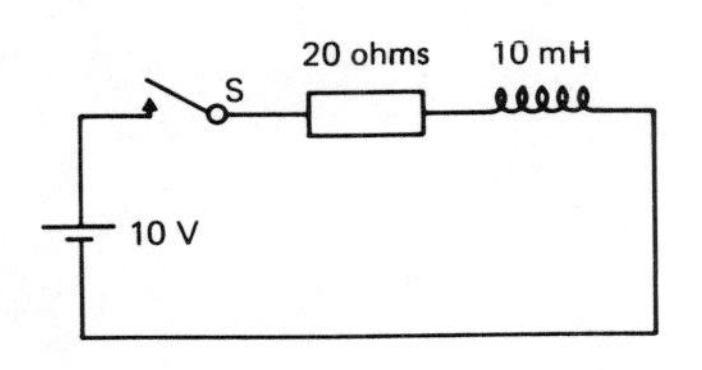

Figure 9.4

$$\frac{di}{dt} = 1 - 2i$$

where i = current in amperes and t = time in milliseconds.

A sketch graph of the solution to this equation may be obtained as follows: A graph which shows the current i (amperes) plotted against time t (milliseconds) is required. The differential equation

$$\frac{di}{dt} = 1 - 2i$$

makes it possible to calculate the gradient at any point (t, i). For example, if $t = 0$ ms and $i = 0.5$ A, $di/dt = 1 - 2 \times 0.5 = 0$.

Since di/dt is a function of i alone, it is easy to compute values for the gradient at a selection of points.

We will asume that the switch is closed at time $t = 0$ ms. The current flowing will be positive, so calculate the gradient for $i = 0$ A to $i = 1.0$ A in steps of 0.1 A. The results are summarized in Table 9.1 and plotted in Figure 9.5(a), which gives the direction field for the equation.

Table 9.1 Values of $di/dt = 1 - 2i$

i	0.0	0.1	0.2	0.3	0.4	0.5	0.6	0.7	0.8	0.9	1.0
$\frac{di}{dt}$	1.0	0.8	0.6	0.4	0.2	0.0	−0.2	−0.4	−0.6	−0.8	−1.0

Starting from $t = 0$, $i = 0$, a curve can be sketched which follows the direction field (Figure 9.5(b)). It can never cross the line $i = 0.5$,

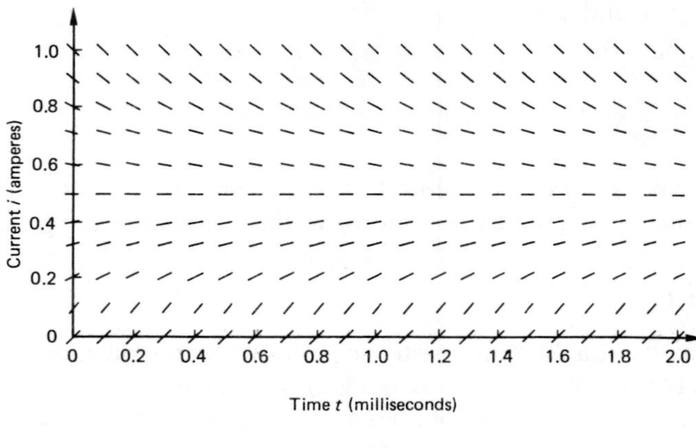

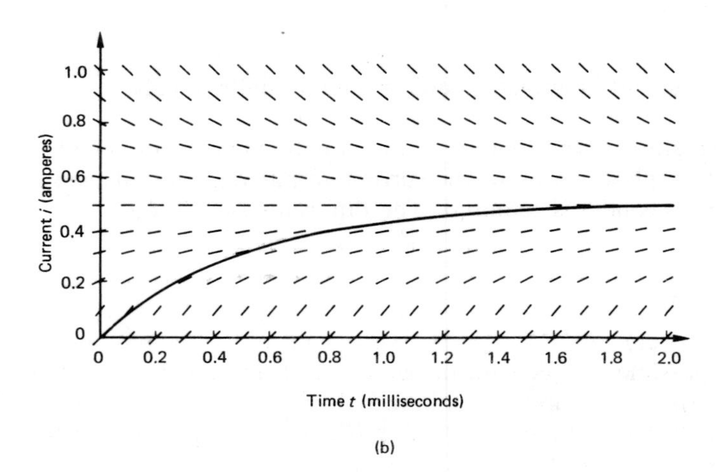

Figure 9.5(a) Direction field for $di/dt = 1 - 2i$ for $0 \leqslant t \leqslant 2$, $0 \leqslant i \leqslant 1$

since on this line $di/dt = 0$. So as time t increases, the current i approaches a steady state value of $i = 0.5$ A.

This curve is the particular solution starting from the initial condition $t = 0$, $i = 0$.

Example 9.3.2

Sketch the direction field defined by the equation

$$\frac{dy}{dx} = \frac{x}{1 - y}$$

Take values as follows: $-2 \leqslant x \leqslant 2$ and $-1 \leqslant y \leqslant 3$. Sketch the solution curves which start from (0, 2), (0, 2.5), (0, 3).

Solution

In this example, when $y = 1$ the gradient is of the form

$$\frac{dy}{dx} = \frac{x}{1-1} = \frac{x}{0}$$

So if $x = 0$, $dy/dx = 0/0$ which is undefined. But if $x \neq 0$, for example $x = 1$, $dy/dx = 1/0$ which gives an infinite gradient. So if $y = 1$ and $x \neq 0$, we will write $dy/dx = \infty$.

The gradients at points where $y \neq 1$ can be calculated with no trouble. A summary of the gradients for the required range of values is given in Table 9.2, and the direction field is shown in Figure 9.6.

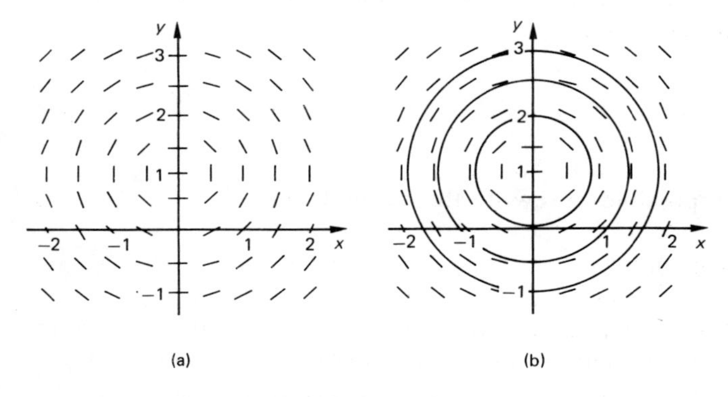

Figure 9.6(a) Direction field for $dy/dx = x/(1-y)$
(b) Solution curves for $dy/dx = x/(1-y)$

Table 9.2 Values for $\dfrac{dy}{dx} = \dfrac{x}{1-y}$

y \ x	−2.0	−1.0	0.0	1.0	2.0
3.0	1.00	0.50	0	−0.50	−1.00
2.0	2.00	1.00	0	−1.00	−2.00
1.0	∞	∞	U	∞	∞
0.0	−2.00	−1.00	0	1.00	2.00
−1.0	−1.00	−0.50	0	0.50	1.00

Exercise 9.3.1

Sketch the direction fields, and family of solution curves for the following differential equations:

1. $\dfrac{dy}{dx} = 2y$
2. $\dfrac{dy}{dx} = 3x$
3. $\dfrac{dy}{dx} = 2 + xy$
4. $\dfrac{dy}{dx} = 9 - y^2$

9.4 ANALYTICAL METHODS FOR SOLVING FIRST ORDER DIFFERENTIAL EQUATIONS

9.4.1 Solution by Separation of Variables

Let $f(x, y)$ be a function of x and y, then a first order differential equation can be written

$$\frac{dy}{dx} = f(x, y)$$

It is often possible to rewrite this in the form

$$g(y)dy = h(x)dx$$

where $g(y)$ is a function of y alone and $h(x)$ is a function of x alone. The equation can then be solved by integrating each side, giving

$$\int g(y)dy = \int h(x)dx$$

Solving first order differential equations by separation of variables
1. Rewrite the differential equation as $g(y)dy = h(x)dx$ if possible.
2. If step (1) is possible, solve the equation by integrating both sides.

The following examples illustrate this method.

Example 9.4.1

Obtain the general solution to the equation

$$\frac{dy}{dx} = \frac{x}{1 - y}$$

whose solutions were sketched in Figure 9.6.

Solution

$$\frac{dy}{dx} = \frac{x}{1 - y}$$

Separate variables

$$(1 - y)dy = xdx$$

Integrate each side

$$\int (1 - y)dy = \int xdx$$

$$y - \frac{y^2}{2} = \frac{x^2}{2} + A \qquad \text{where } A \text{ is an arbitrary constant}$$

So

$$x^2 + y^2 - 2y = B \qquad \text{where } B \text{ is an arbitrary constant}$$

$$x^2 + (y - 1)^2 = B + 1 \qquad \text{after completing the square for } y^2 - 2y$$

$$x^2 + (y - 1)^2 = C \qquad \text{where } C \text{ is an arbitrary constant.}$$

Provided $C > 0$, this represents a family of circles centre (0, 1), radius $\sqrt{C}$, so the solutions sketched in Figure 9.6 were correct.

If $C = 0$, the solution would represent the point (0, 1), but this is not a solution to the equation since the gradient is not defined at (0, 1).

If $C < 0$ the solutions are not real, and so are not solutions to the differential equation.

Example 9.4.2

Obtain an analytical solution to the differential equation of Example 9.3.1, that is solve

$$\frac{di}{dt} = 1 - 2i$$

given that $i = 0$ A when $t = 0$ ms.

Solution

$$\frac{di}{dt} = 1 - 2i$$

Separating variables,

$$\left(\frac{1}{1-2i}\right) \mathrm{d}i = \mathrm{d}t$$

Initially $i = 0$ and $t = 0$, and the current i at time t is required. So integrate both sides of the last equation using the appropriate initial and final limits for i and t.

$$\int_0^i \frac{1}{1-2i}\,\mathrm{d}i = \int_0^t \mathrm{d}t$$

$$-\frac{1}{2}\,[\ln(1-2i)]_0^i = [t]_0^t$$

$$\ln(1-2i) - \ln(1-0) = -2(t-0)$$

So

$$\ln(1-2i) = -2t$$
$$1 - 2i = \mathrm{e}^{-2t}$$
$$i = 0.5(1 - \mathrm{e}^{-2t})$$

So the sketch of the solution of the equation (Figure 9.5) was correct. For example, when $t = 0$, $i = 0.5(1 - \mathrm{e}^0) = 0.5(1 - 1) = 0$, and as $t \to \infty$, $\mathrm{e}^{-2t} \to 0$, so $i \to 0.5(1 - 0) = 0.5$ which means the current i has a steady state value of 0.5 A.

Exercise 9.4.1

Find the general solutions for Problems 1–10.

1. $\dfrac{\mathrm{d}y}{\mathrm{d}x} = \dfrac{3y}{x}$
2. $\dfrac{\mathrm{d}y}{\mathrm{d}x} = \dfrac{x}{y}$
3. $\dfrac{\mathrm{d}y}{\mathrm{d}x} = -\sin x$
4. $\dfrac{\mathrm{d}y}{\mathrm{d}t} = ky$

   where k is a constant
5. $2\dfrac{\mathrm{d}y}{\mathrm{d}t} = y$
6. $\dfrac{\mathrm{d}y}{\mathrm{d}t} = 2y(t+2)$
7. $\dfrac{\mathrm{d}x}{\mathrm{d}t} - t \operatorname{cosec} x = 0$
8. $\dfrac{\mathrm{d}x}{\mathrm{d}t} - t^2\mathrm{e}^x = 0$

9. $\dfrac{dx}{dt} - x^2e^t = 0$ 10. $\dfrac{dy}{dx} - \cos 2x \sin^2 y = 0$

Find the particular solutions for Problems 11–20.

11. $\dfrac{dy}{dx} = 8x$ given that $y = 1$ when $x = 0$

12. $\dfrac{dy}{dx} = -2y$ given that $y = 3$ when $x = 0$

13. $\dfrac{dy}{dt} = -\cos 2\omega t$ given that $y = 1$ when $t = 0$

14. $\dfrac{dy}{dt} = te^{3y}$ given $y = 0$ when $t = 0$

15. $\dfrac{dy}{dt} = 2t(1 + y^2)$ given $y = 1$ when $t = 0$

16. $(1 - 4t^2)^{1/2}x\dfrac{dx}{dt} - 1 = 0$ given $x = 0$ when $t = 0$

17. $e^{2x+t}\dfrac{dx}{dt} - t = 0$ given $x = \ln 2$ when $t = -1$

18. $\dfrac{dy}{dx} = \dfrac{\sin \omega x}{\cos 2\omega y}$ given that $y = 0$ when $x = 0$

19. $x(1 - t^2)\dfrac{dx}{dt} = (3 + t)(1 + x^2)$ given that $x = \sqrt{3}$ when $t = 0$

20. $y\dfrac{dy}{dx} = e^x \sec y$ given $y = 0$ when $x = 0$

9.5 SETTING UP DIFFERENTIAL EQUATIONS

Often it is necessary to set up and solve a differential equation from a description of a physical situation.

Phrases which involve a rate of change of a variable will translate to derivatives with respect to time. For example if t is time, V is volume and S is surface area, the rate of change of volume is $\mathrm{d}V/\mathrm{d}t$ and the rate of change of surface area is $\mathrm{d}S/\mathrm{d}t$.

Saying that 'the rate of change of volume V is proportional to the radius r', would translate as

$$\frac{\mathrm{d}V}{\mathrm{d}t} \propto r \text{ or } \frac{\mathrm{d}V}{\mathrm{d}t} = kr \text{ where } k = \text{constant}$$

'The rate of change of S is inversely proportional to time' translates as

$$\frac{\mathrm{d}S}{\mathrm{d}t} \propto \frac{1}{t} \quad \text{or} \quad \frac{\mathrm{d}S}{\mathrm{d}t} = \frac{A}{t} \quad \text{where } A = \text{constant}$$

In setting up problems, carefully define all variables and their units.

The next two examples illustrate the type of problem involved.

Example 9.5.1

The initial temperature of a cup of tea is 53 °C, and after 5 minutes its temperature is 45 °C. From Newton's law of cooling, it is known that the rate of cooling of a body is proportional to the temperature difference between the body and its surroundings. Use this to predict the temperature of the tea after a further 5 minutes, given that the room temperature was a constant 21 °C.

Solution

Let t = time in minutes, and T = temperature difference between tea and the room at time t. T is measured in degrees Celsius.

From Newton's law of cooling

$$\frac{\mathrm{d}T}{\mathrm{d}t} \propto T \text{ or } \frac{\mathrm{d}T}{\mathrm{d}t} = aT \text{ where } a = \text{constant}$$

and $\mathrm{d}T/\mathrm{d}t$ is measured in °C per minute.

So

$$\frac{\mathrm{d}T}{T} = a\ \mathrm{d}t$$

$$\ln T = at + b \text{ where } b = \text{constant.}$$

To find the values of the constants a and b, use the initial condition and the condition after 5 minutes.

$t = 0$: $\ln(53 - 21) = b \qquad \therefore b = \ln 32$

$t = 5$: $\ln(45 - 21) = 5t + b \therefore t = \frac{1}{5}(\ln 24 - \ln 32) = \frac{1}{5} \ln \frac{3}{4}$

$$\ln T = \left(\frac{1}{5} \ln \frac{3}{4}\right) t + \ln 32$$

$$T = 32e^{(1/5 \ln 3/4)t}$$

So when $t = 10$ minutes, the temperature difference is

$T = 32e^{2 \ln 3/4} = 18$ °C.

Therefore the temperature of the tea after 10 minutes is

21 °C + 18 °C = 39 °C

Example 9.5.2

It is known that in a vacuum a free-falling body will fall with a constant acceleration. However, the assumption of a vacuum is not sufficiently realistic for many applications.

One model suggests that a free falling body experiences a force equal to its weight and an air resistance which is proportional to its velocity. Taking this constant of proportionality equal to 0.29, find the velocity and displacement of a free falling body of mass 10 kg at time t, and show the velocity approaches a limiting value. (Take $g = 9.81$ ms^{-2}).

Solution

Let t = time (s), s = displacement (m), v = velocity (ms^{-1}), m = mass (kg), g = acceleration due to gravity (ms^{-2}).

Then by Newton's Law

$$m \frac{dv}{dt} = mg - kv$$

where $k = 0.29$

$$\int \frac{1}{mg - kv} dv = \frac{1}{m} \int dt$$

$$-\frac{1}{k}\ln(mg - kv) = \frac{t}{m} + A$$

When $t = 0$, $v = 0$, so

$$A = -\frac{1}{k}\ln(mg)$$

Hence

$$\frac{t}{m} = \frac{1}{k}[\ln(mg) - \ln(mg - kv)]$$

$$\frac{kt}{m} = \ln\left(\frac{mg}{mg - kv}\right)$$

$$\frac{mg - kv}{mg} = e^{-kt/m}$$

$$v = \frac{mg}{k}(1 - e^{-kt/m})$$

So for $m = 10$, $g = 9.81$, $k = 0.29$

$$v \simeq 338.3\,(1 - e^{-0.029t})$$

As $t \to \infty$, $v \to 338.3\ \text{ms}^{-1}$, so the velocity approaches a limiting value.

$$V = \frac{dS}{dt}, \text{ so } \frac{dS}{dt} = 338.3\,(1 - e^{-0.029t})$$

$$s = \int 338.3\,(1 - e^{-0.029t})\,dt$$

$$= 338.3t + 11664.7e^{-0.029t} + B$$

But $s = 0$ when $t = 0$, so

$$0 = 11664.7 + B$$

$$s = 338.3t + 11664.7\,(e^{-0.29t} - 1)$$

Example 9.5.3

A rain water soakaway consists of a cylindrical hole of depth 2 m and diameter 2 m. It is filled with gravel, which occupies 75% of its volume; the remaining 25% can hold rain water. The space available for water is uniformly distributed within the soakaway.

Water in the soakaway is absorbed into the soil at a rate which is equal to one tenth of the surface area of the soil around the soakaway.

A roof of area 200 m^2 drains into the soakaway. Initially the soakaway is empty. It then starts to rain at a constant rate of 8 mm per hour. How long will it take for the soakaway to overflow?

Solution

Let t = time (h), h = depth of water in soakway (m), V = volume of water in soakaway (m^3), S = surface area of soil surrounding the soakaway (m^2).

Rain falling at 8 mmh^{-1} onto a 200 m^2 roof gives a rate of increase of volume of 1.6 m^3h^{-1}.

Now, $S = \pi 1^2 + 2\pi 1h = \pi + 2\pi h$

So water soaks away at a rate of $0.1(\pi + 2\pi h)$ m^3h^{-1}.

Hence, $\dfrac{dV}{dt} = 1.6 - 0.1\pi(1 + 2h)$

Volume of water V = $0.25\pi h$, so $dV/dh = 0.25\pi$.
By the chain rule

$$\frac{dh}{dt} = \frac{dh}{dV}\frac{dV}{dt}$$

So

$$\frac{dh}{dt} = \frac{1.6 - 0.1\pi - 0.2\pi h}{0.25\pi}$$

$$\simeq 1.637 - 0.8h$$

$$\int \frac{dh}{1.637 - 0.8h} = \int dt$$

$$-1.25 \ln(1.637 - 0.8h) = t + C$$

Initially $t = 0$ and $h = 0$ so $C = -1.25 \ln(1.637)$.

$$t = -1.25 \ln\left(\frac{1.637 - 0.8h}{1.637}\right) = -1.25 \ln(1 - 0.489h)$$

For overflow $h = 2$, so $t = -1.25 \ln(0.023) \simeq 4.737$ hours. So the soakaway overflows after 4 hours 44 minutes.

Exercise 9.5.1

1. A water storage tank has a rectangular cross-section. Its base is 1 m by 1.25 m and its depth is 1 m. Water flows into the tank at a constant rate of 0.005 m^3s^{-1}. It flows out at a rate proportional to the depth of water in the tank. The constant of proportionality is 0.001 m^2s^{-1}.

   The tank is initially half full. What is the depth of water after 1 minute? How long does it take for the tank to overflow?
2. When a radioactive substance decays, its rate of decay is proportional to the remaining mass. Given that the initial mass is 1 kg and that after 50 seconds the mass is 0.53 kg, find an expression for the mass remaining at time t.
3. The difference between the temperature of a liquid and its surroundings is T °C.

   The liquid is being heated by a source which makes T increase at a rate of αt, where α is a constant and t = time in seconds. At the same time the liquid is losing heat through its container which causes T to decrease at a rate βT where β is a constant.

   Initially the liquid is at the temperature of its surroundings. Show that

$$T = \frac{\alpha}{\beta^2}[\beta t - 1 + e^{-\beta t}].$$

9.6 NUMERICAL METHODS FOR FINDING SOLUTIONS TO DIFFERENTIAL EQUATIONS

Although there are many more analytical techniques for solving differential equations than the one which we have considered, it is often difficult or even impossible to solve a differential equation by analytical methods. When this situation arises, it is necessary to use numerical methods.

Numerical methods will only yield approximations to particular solutions, but nevertheless they are very useful in practical situations where numerical solutions are desired.

In this section two simple numerical methods will be considered. Access to a programmable calculator or a computer would be very useful in this section.

9.6.1 Euler's Step by Step Method

The 'geometry' of the method

Suppose we have a direction field defined by the differential equation

$$\frac{\mathrm{d}y}{\mathrm{d}x} = f(x, y)$$

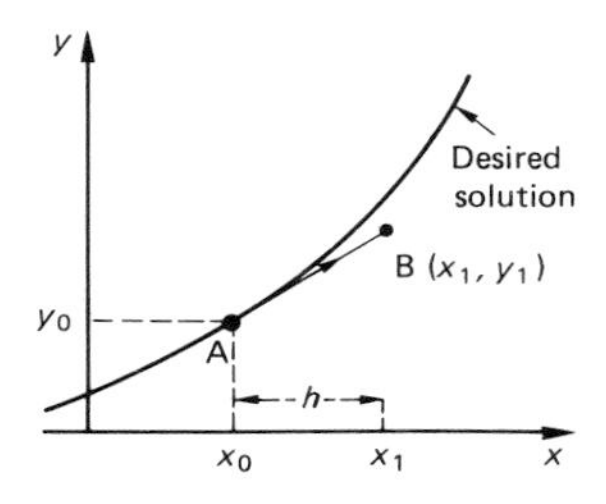

Figure 9.7

Then starting from an initial condition (boundary condition) at the point (x_0, y_0) we can estimate the value of y at a point where $x_1 = x_0 + h$, as follows:

We know the direction at A so we may proceed to a new point B along a straight line in that direction (Figure 9.7).

Having reached B, we know the direction at B, so we may proceed to C along that direction (Figure 9.8).

The method may be repeated indefinitely.

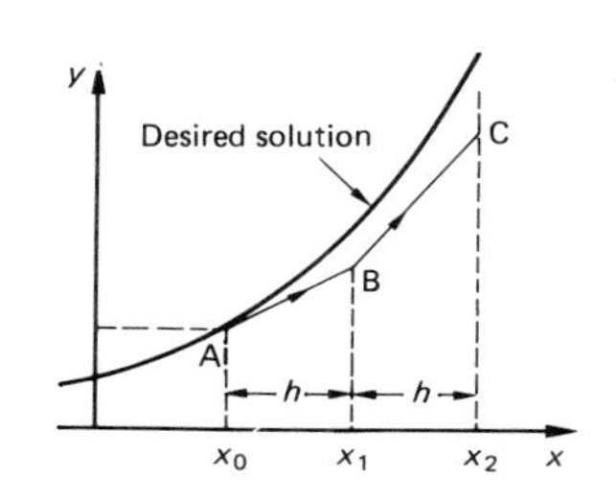

Figure 9.8

General formula

Consider the step from stage n to stage $n + 1$. Refer to Figure 9.9.

In Δ PQR,

$$\tan \theta = \frac{\text{QR}}{\text{PR}} = \frac{\text{QR}}{h}$$

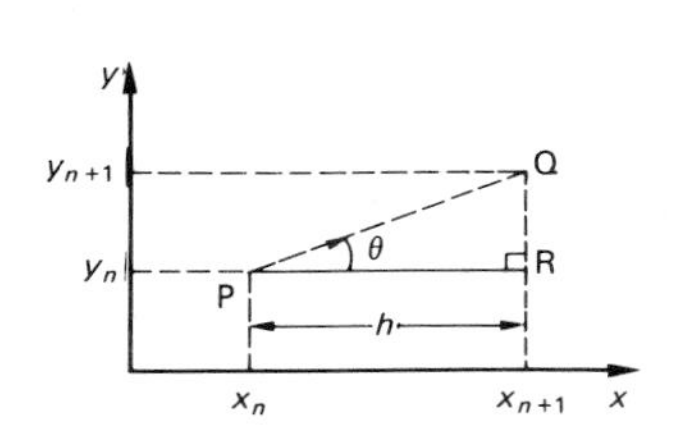

Figure 9.9

So

$$\text{QR} = h \tan \theta$$

But $\tan \theta = \mathrm{d}y/\mathrm{d}x = f(x_n, y_n)$ at P (from the differential equation)

$$\text{QR} = hf(x_n, y_n)$$

$$y_{n+1} = y_n + hf(x_n, y_n) \qquad (9.6.1)$$

Example 9.6.1

Refer to the differential equation of Example 9.3.2.

$$\frac{dy}{dx} = \frac{x}{1 - y}$$

Starting from the initial condition $x = 0$, $y = 0$ estimate y when $x = 1.0$. Take a step length of (a) 0.1 and (b) 0.01.

Solution

(a) Step length of 0.1
In the notation of Equation (9.6.1), $h = 0.1$ and $f(x) = x/1 - y$.

$$\therefore y_{n+1} = y_n + 0.1 \left(\frac{x_n}{1 - y_n}\right)$$

$n = 10$ steps are required to reach the final value of $x = 1$. The results are summarized in Table 9.3. The correct value is $y = 1$. So there is a very large error of 39.9%.

Table 9.3

n	0	1	2	3	4	5	6	7	8	9	10
x_n	0.000	0.100	0.200	0.300	0.400	0.500	0.600	0.700	0.800	0.900	1.000
y_n	0.000	0.000	0.010	0.030	0.061	0.104	0.160	0.231	0.322	0.440	0.601

(b) Step length of 0.01
With $h = 0.01$, the formula for estimating y is

$$y_{n+1} = y_n + 0.01 \left(\frac{x_n}{1 - y_n}\right)$$

$n = 100$ steps are required to reach $x = 1.0$. Every tenth result is shown in the Table 9.4. Even now with 100 steps, the estimate of 0.843 compares poorly with the exact value of 1.000 (an error of 15.7%).

Table 9.4

n	0	10	20	30	40	50	60	70	80	90	100
x_n	0.000	0.100	0.200	0.300	0.400	0.500	0.600	0.700	0.800	0.900	1.000
y_n	0.000	0.005	0.019	0.044	0.081	0.131	0.196	0.280	0.391	0.548	0.843

The reason for this poor performance can be easily explained geometrically. Euler's method always uses tangent approximations.

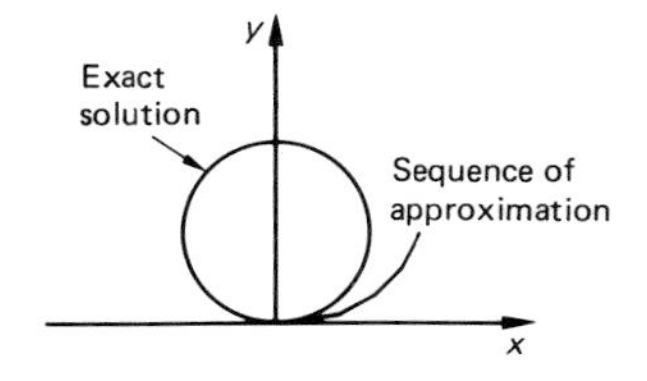

Figure 9.10

The correct solution is a circle (see Example 9.4.1). Thus the approximations are bound to spiral out, even if a small step length is used (Figure 9.10).

Example 9.6.2

Refer to the differential equation of Example 9.4.2, $\mathrm{d}i/\mathrm{d}t = 1 - 2i$ with initial condition $i = 10$ when $t = 0$. Use Euler's step by step method to estimate i when $t = 0$. Take a step length of $t = 0.1$ ms.

Solution

Adapting the notation of Equation (9.6.1), with $h = 0.1$

$$i_{n+1} = i_n + 0.1(1 - 2i_n)$$

so

$$i_{n+1} = 0.8i_n + 0.1$$

Start with $i_0 = 0$. When $t = 1$, $n = 10$. The results are summarized in Table 9.5. The exact answer (see Example 9.4.2) is $i = 0.432$ A at $t = 1$ ms. So the numerical method gives an error of 3.2%.

Table 9.5

n	0	1	2	3	4	5	6	7	8	9	10
t	0.000	0.100	0.200	0.300	0.400	0.500	0.600	0.700	0.800	0.900	1.000
i_n	0.000	0.100	0.180	0.244	0.295	0.336	0.369	0.395	0.416	0.433	0.446

9.6.2 The improved Euler method

Some of the errors in using the Euler method in Examples 9.6.1 and 9.6.2 were unacceptably large. However it is possible to improve the method in a fairly simple way.

SUBSTEP 1: Compute y_1^* *the auxiliary value of y*

$$\frac{\mathrm{d}y}{\mathrm{d}x} = f(x, y)$$

So gradient at A is $f(x_0, y_0)$. Hence auxiliary value $y_1^* = y_0 + hf(x_0, y_0)$ (Figure 9.11).

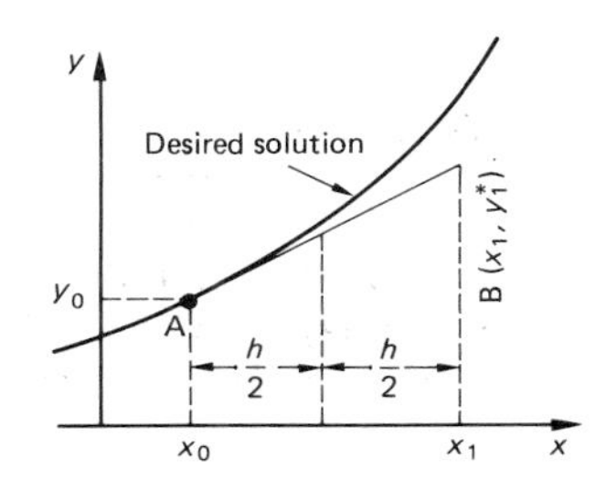

Figure 9.11

SUBSTEP 2: Compute y the new estimate for y
At B (x_1, y_1^*), the value of the gradient is computed from $f(x_1, y_1^*)$.

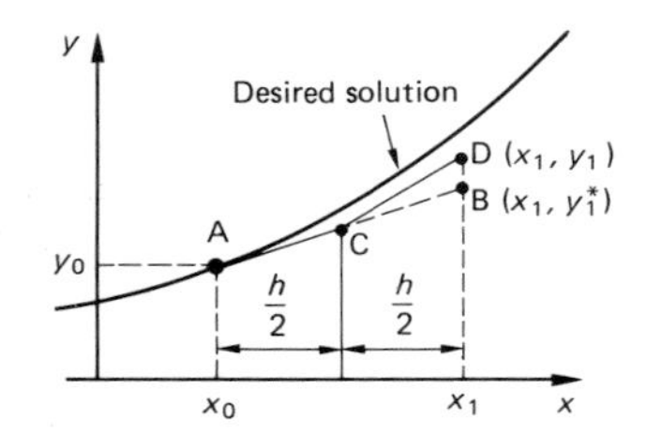

Figure 9.12

The step of size h is divided into two halves (Figure 9.12). Over the first half interval from A to C, the gradient is $f(x_0, y_0)$ and over the second half interval from C to D the gradient is $f(x_1, y_1^*)$.

So

$$y_1 = y_0 + \frac{h}{2} f(x_0, y_0) + \frac{h}{2} f(x_1, y_1^*)$$

or

$$y_1 = y_0 + \frac{h}{2} [f(x_0, y_0) + f(x_1, y_1^*)]$$

The general process from step n to step $n + 1$ is summarized below.

> *Improved Euler method*
> To obtain an approximation to the solution of the differential equation
>
> $$\frac{dy}{dx} = f(x, y)$$
>
> starting from (x_0, y_0), with step length h.
> 1. Calculate auxiliary value
>
> $$y_{n+1}^* = y_n + hf(x_n, y_n)$$
>
> 2. Calculate the $(n + 1)$th approximation by
>
> $$y_{n+1} = y_n + \frac{h}{2} [f(x_n, y_n) + f(x_{n+1}, y_{n+1}^*)]$$

Example 9.6.3

Apply the improved Euler method with a step length of $h = 0.1$ to the differential equation of Example 9.4.2.

$$\frac{di}{dt} = 1 - 2i$$

with initial condition $i = 0$ when $t = 0$.

Solution

This equation is of the form

$$\frac{\mathrm{d}i}{\mathrm{d}t} = f(t, i) = 1 - 2i$$

The auxiliary value is given by

$$i^*_{n+1} = i_n + 0.1(1 - 2i_n)$$

so

$$i^*_{n+1} = 0.80i_n + 0.1$$

Now

$$f(t_n, i_n) = 1 - 2i_n$$

$$\begin{aligned} f(t_{n+1}, i^*_{n+1}) &= 1 - 2(0.80i_n + 0.1) \\ &= 0.80 - 1.60i_n \end{aligned}$$

$$f(t_n, i_n) + f(t_{n+1}, i^*_{n+1}) = 1.80 - 3.60i_n$$

$$i_{n+1} = i_n + \frac{0.1}{2}(1.80 - 3.60i_n)$$

$$i_{n+1} = 0.820i_n + 0.090$$

The approximations are shown in Table 9.6. The true value is $i = 0.432$ A, so this is in error by only 0.2%. This is an improvement on the unmodified Euler method of Example 9.6.2.

Table 9.6

n	0	1	2	3	4	5	6	7	8	9	10
t_n	0	0.1	0.2	0.3	0.4	0.5	0.6	0.7	0.8	0.9	1.0
i_n	0.000	0.090	0.164	0.224	0.274	0.315	0.348	0.375	0.398	0.416	0.431

Example 9.6.4

Apply the improved Euler method with a step length of $h = 0.05$ to the differential equation

$$\frac{\mathrm{d}y}{\mathrm{d}x} = y + \mathrm{e}^{-x}\sin x$$

Given that $y = 1$ when $x = 0$, estimate y when $x = 1$.

Solution

$f(x, y) = y + e^{-x} \sin x$, and $h = 0.05$

So

(1) $y^*_{n+1} = y_n + 0.05f(x_n, y_n)$

and

(2) $y_{n+1} = y_n + 0.025[f(x_n, y_n) + f(x_{n+1}, y^*_{n+1})]$

Hence given $x_0 = 0$, $y_0 = 1$

$$
\begin{aligned}
(1)\ y^*_1 &= y_0 + 0.05f(x_0, y_0) \\
&= 1 + 0.05(1 + e^{-0} \sin 0) \\
&= 1.050 \\
x_1 &= 0.050
\end{aligned}
$$

(2) Remember to work in radians!

$$
\begin{aligned}
y_1 &= y_0 + 0.025(f^*(x_0, y_0) + f(x_1, y^*_1)) \\
&= 1 + 0.025[f(0, 1) + f(0.050, 1.050)] \\
&= 1.052
\end{aligned}
$$

Table 9.7 summarizes the results; only every fourth result is printed. The estimate for y when x equals 1.000 is 3.096.

Table 9.7

n	0	4	8	12	16	20
x_n	0.000	0.200	0.400	0.600	0.800	1.000
y_n	1.000	1.240	1.562	1.971	2.478	3.096

Exercise 9.6.1

You are recommended to use a computer or a programmable calculator for this exercise.

1. Use Euler's method with a step length of 0.05 on

$$\frac{dy}{dx} = y + e^{-x} \sin x$$

given $y = 1$ when $x = 0$, to estimate y when $x = 1$.

2. Use Euler's method with a step length of 0.05 on

$$\frac{dy}{dx} = 1 + x \sin y$$

given $y = 0.5$ when $x = 0$, to estimate y when $x = 0.5$. (Remember to use radians.)

3. Repeat Problem 2, but use the improved Euler method with step lengths of 0.1 and 0.05.

OBJECTIVE TEST 9

1. If $dy/dx = \cos 2x$, find y.
2. If $dv/dt = 3vt$, find v.
3. If $i\, di/dt = te^{t^2}$, find i.
4. If $dy/dx = -y/(x + 1)^2$, find y given that $y = e$ when $x = 0$.
5. If $dy/dx = x^2 + e^y$ and $y = 1$ when $x = 0$, use Euler's method with one step to estimate the value of y when $x = 0.05$.

10 NUMERICAL ANALYSIS

OBJECTIVES

After studying this chapter you should be able to

1. estimate the error interval associated with arithmetic calculations and in the evaluation of functions
2. solve equations using iterative methods
3. understand the importance of convergence of a numerical method
4. state and use the Newton–Raphson method
5. state and use Simpson's rule for numerical integration

10.1 INTRODUCTION

Much of this book has been concerned with the algebra and calculus required to solve problems in science and in engineering. Mostly we have concentrated on analytical (or exact) methods. However, real problems in science and engineering often lead to mathematical problems which cannot be solved exactly and we have to look for numerical (or approximate) methods of approach. For example, a simple looking integral such as

$$\int_{-1}^{1} e^{x^2}\, dx$$

does not have an analytical solution in a simple form. We could solve such an integral using the Trapezoidal rule introduced in Chapter 8.

Numerical methods of solving problems are now common and computer software packages and hand calculator routines are available for doing the 'number crunching'. It is important however to know what these routines are doing and some of the problems associated with numerical methods. We have adopted the practice when writing this book to include some numerical methods where they naturally fit. For example, in Chapter 2 we have introduced the Formula Iteration Method for solving equations, in Chapter 8 we have introduced the Trapezoidal rule for integration and in Chapter 9 Euler's method was used for solving first order differential equations. In this chapter we continue the development of numerical methods of solving equations of integration and introduce the important idea of *convergence*.

We begin this chapter by looking at the propagation of errors in calculations. In all numerical work, the numbers are either usually rounded off which introduces an approximation to the exact value or if the data is taken from an experiment then the numbers are subject to experimental errors. When using these approximate numbers in calculations the errors may combine and grow. It is important to keep track of the errors.

10.2 ERRORS AND ERROR PROPAGATION

10.2.1 Types of error

So far in this text we have said little about the occurrence of errors or approximations in the numbers that we are dealing with. In many practical problems, it is unlikely that the measurements we are using are exact. When taking readings from apparatus in experiments we can only read as accuractely as our eyes will permit, for example, can you measure a length from a ruler to within 0.5 mm? Another source of approximation occurs as soon as a calculation is started on a calculator. For example, if an expression involves 1/3 then a calculator storing eight digits can only store 0.3333333 and this is only an approximation to 1/3. In long calculations these approximations could build up making the final answer quite incorrect.

In this section we introduce the arithmetic of errors and their propagation.

Errors may occur in several different ways.

(a) In expressing a number as a decimal with a convenient number of decimal places, a *rounding error* is often introduced. For example, the number 0.33 is an approximation to 1/3, the rounding error is $(0.33 - 1/3)$.
(b) Errors may occur due to the mathematical method being used. For example, Euler's method is a numerical (approximate) scheme for solving a first order differential equation. Errors are introduced at each step because the solution curve is being replaced by a set of straight lines.
(c) There may be errors occurring in the data itself, perhaps due to experimental measurement.
(d) Human error is obvious and common. These can be reduced by checking and cross-checking the calculations.

We shall study errors of types (a) and (b) and their propagation. Those of type (c) require a statistical analysis.

10.2.2 Absolute and relative errors

Suppose that in a calculation the exact value of a quantity is denoted by the symbol x, and the approximate value is X. Then the *absolute error* in the quantity is the absolute value of the difference between the approximate and exact value of the quantity, i.e.

$$\boxed{e_{\text{abs}} = |X - x|} \qquad (= \text{approx} - \text{exact})$$

Note that often it is the magnitude (or size) of the error that we are interested in.

The absolute error is not necessarily a good indication of the error in a measurement or calculation. For example, two measurements involving time may be one minute and one hour, each with an absolute error of one second. The absolute error is one second in each case. However, a one second error is more significant in 60 (1 minute) than in 3600 (1 hour).

A more useful indicator of the error is called the *relative error* which is defined by the ratio of the absolute error to the absolute value of the quantity.

i.e. $$\boxed{e_{\text{rel}} = \frac{e_{\text{abs}}}{|x|}}$$

For the above example, the relative error in one minute is 1/60 and the relative error in 1 hour is 1/3600. We can now deduce mathematically that the former is more significant than the latter. In practice, of course, we do not know x but only the approximation X. Then it is possible to show that

$$e_{\text{rel}} \leqslant \frac{e_{\text{abs}}}{|X| - e_{\text{abs}}}$$

Relative errors are often quoted as percentages. The percentage error in x is $100e_{\text{rel}}$. For example, the percentage error of 1 second in 1 minute is $100/60 = 1\frac{2}{3}\%$.

When giving a measurement in an experiment or the answer to a calculation it is good practice to include some indication of the accuracy involved. For example, suppose that in using a ruler to measure a length of wire the reading is 75 mm and you are confident that you can read to within 0.5 mm. Then the length of the wire would be in the range of 74.5 mm to 75.5 mm. The absolute error in the length is then

$$-0.5 \leqslant e \leqslant 0.5$$

and this interval is called the *error interval*. In general a length l with this error interval would be in the range

$$L - 0.5 \leqslant l \leqslant L + 0.5$$

so that the length of wire is written with its accuracy limits as

$$75 - 0.5 \leqslant \text{length} \leqslant 75 + 0.5$$

The 75 shows the measured value and the 0.5 shows the absolute error.

Exercise 10.2.1

1. The scales in a laboratory are said to be accurate if they estimate the mass of an object to within 10 grams of the object's exact mass.
   (a) What is the absolute error in the mass of any object?
   (b) What are the relative and percentage errors in each of the following masses: 100 grams, 1 kilogram, 1 tonne?
   (c) In using the scales, I read the mass of an object as 137 grams. In what range of values does the correct mass lie?

2. Suppose the number $x = 3.14$ is the result of rounding a number to two decimal places. What is the range of values for the correct number x?

10.2.3 Propagation of errors in calculations

When two numbers are multiplied together, the magnitude of the rounding error in the answer may be larger than the rounding errors in the two original numbers. For example, consider the product 3.1×1.2 where each number is correct to one decimal place. The product 3.1×1.2 is 3.72. However 3.1 could lie anywhere in the range 3.05 and 3.15 and 1.2 could lie between 1.15 and 1.25. So the product lies in the range

$$3.05 \times 1.15 \text{ and } 3.15 \times 1.25$$

i.e. $$3.5075 \text{ and } 3.9375$$

If we round these answers then all we can say is that 3.1×1.2 lies between 3.5 and 3.9, so that the error interval is now approximately

$-0.2 < e < 0.2$. The analysis of the growth of errors in calculations and numerical methods is an important area of a branch of mathematics called *numerical analysis*.

Errors will propagate through all the basic operations of addition, subtraction, multiplication and division and by the use of functions.

There are general formulas for calculating errors that propagate in arithmetic operations. However with fast cheap calculators it is usually easier to calculate the error bounds numerically instead of algebraically, as the following example illustrates.

Example 10.2.1

Estimate the maximum error in the calculation

$$\frac{(4.2 \times 3.9) - 2.7}{(5.1 + 9.5)}$$

if each number is correct to one decimal place.

Solution

The largest value of the expression occurs when the numerator is a maximum and the denominator is a minimum. This is

$$\frac{(4.25 \times 3.95) - 2.65}{(5.05 + 9.45)} = 0.975$$

Similarly the smallest value of the expression is given by

$$\frac{(4.15 \times 3.85) - 2.75}{(5.15 + 9.55)} = 0.900$$

Evaluating the expression as it stands gives 0.937 so that the error interval is

$$0.937 - 0.037 \leqslant \frac{(4.2 \times 3.9) - 2.7}{(5.1 + 9.5)} \leqslant 0.937 + 0.037$$

This method of approach is much easier than using the general error formulas. As a result of the calculation in this example, at best we can claim that the answer is 1.0 correct to one significant figure.

Exercise 10.2.2

If the numbers in the following calculations are rounded off, find the maximum possible error in the answers and hence give the answer to the appropriate accuracy.

(a) $3.172 - 4.675 + 7.002$
(b) 8.312×2.875
(c) $4.62 \div 1.91$
(d) $5.371 \times 9.023 \div 7.12$
(e) $(16.3 + 7.01) \times 14.26$
(f) $\dfrac{(0.463 + 4.312) \times (7.031 - 3.412)}{(16.47 - 9.26)}$

10.2.4 Errors in evaluating functions

Suppose that $Y = f(X)$ is the approximate value of $y = f(x)$ where f is some function. Then if ϵ is the absolute error in x, the absolute error in y is

$$\begin{aligned} Y - y &= f(X) - f(x) \\ &= f(X) - f(X - \epsilon) \\ &= \epsilon \left[\frac{f(X) - f(X - \epsilon)}{\epsilon} \right] \\ &\simeq \epsilon f'(X) \end{aligned}$$

using the definition of the derivative in Chapter 7. It should come as no surprise that the error should involve the derivative of f since differentiation is about small changes and the propagation of errors involves small changes in the given data.

Example 10.2.2

Find the absolute error in evaluating sin(0.31) if 0.31 is rounded to two decimal places.

Solution

The absolute error in 0.31 is 0.005. Hence the absolute error in sin(0.31) is given by

$$\epsilon = 0.005 \cos(0.31) = 0.0048$$

since the derivative of $\sin(x)$ is $\cos(x)$.

Exercise 10.2.3

If the numbers in the following are rounded off, estimate the error in evaluating the following and hence give the answer to the appropriate accuracy.

(a) $e^{1.32}$ (b) $\cos(1.5)$
(c) $\ln(2.634)$ (d) $e^{(2.1)^2}$
(e) $3.42e^{1.32}$

10.3 SOLVING THE EQUATION $f(x) = 0$

10.3.1 Iterative methods

In Chapter 2 we introduced the method of formula iteration to find the roots of a polynomial and in subsequent chapters we have used the same method to solve more equations of the form $f(x) = 0$. We have seen that for some iterative formulas the calculations converged and for others they diverged. The difference is associated with the propagation of small changes. In this section, we identify a means of describing mathematically at the outset of the problem how good an iteration formula is by defining a *rate of convergence*.

In general an iterative method for finding the roots of $f(x) = 0$ is a formula written in the form

$$X_{n+1} = g(X_n)$$

The new value, X_{n+1}, is calculated directly from the old value X_n using an appropriate function $g(x)$. Clearly $g(x)$ must in some way be associated with $f(x)$.

The convergence of numerical methods in problem solving is an important area in numerical analysis. It is crucial to know at the outset of a problem whether a numerical method, such as one to solve an algebraic or differential equation, converges or diverges. We look at differential equations in the next section.

Consider the iterative formula $X_{n+1} = g(X_n)$ for solving $f(x) = 0$. Clearly at a solution $x = g(x)$. Suppose that an iteration X_n is close to the exact solution x. Then the next iteration is

$$X_{n+1} = g(X_n) = g(\epsilon + x)$$

where $X_n - x = \epsilon$.

The difference between X_{n+1} and the exact solution x is

$$\begin{aligned} X_{n+1} - x &= g(\epsilon + x) - g(x) \\ &\simeq \epsilon g'(x) \end{aligned}$$

Now if the iterative method is to converge it is important that $| X_{n+1} - x | < | \epsilon |$, i.e. the new value of X_{n+1} must be closer to the exact value x than the old value X_n.

A necessary and sufficient condition for convergence of such an iterative method is that $|g'(x)| < 1$ at the root x. Clearly in practice we do not know the value of x, but it is sufficient that $|g'(X_n)| < 1$ for values of X_n in the neighbourhood of the exact solution.

Consider the two iterative methods for solving the equation $x^3 + 2x - 4 = 0$.

(a) $X_{n+1} = (4 - X_n^3)/2$, for which $g(x) = (4 - x^3)/2$ and $g'(x) = -3/2\,x^2$.
For all values of x between 1 and 2 (where the solution occurs) $|g'(x)| \geqslant 1.5$. So the method diverges.

(b) $X_{n+1} = (4 - 2X_n)^{1/3}$, for which $g(x) = (4 - 2x)^{1/3}$ and $g'(x) = -\frac{2}{3}(4 - 2x)^{-2/3}$.
For all values of x between $x = 1$ and $x = 2.9$ [or more exactly $x = 2 - (1/2\sqrt{27})$] $|g'(x)| \leqslant 1$ so that the iterative method converges to the solution $x = 1.1795$ (to four decimal places).

Exercise 10.3.1

1. The following iterative methods have been suggested in solving the equation $x^2 - 4x + 2 = 0$ near the roots $x = 3.4$ and $x = 0.6$

   (a) $X_{n+1} = \dfrac{X_n^2 + 2}{4}$ (b) $X_{n+1} = \sqrt{2 - 4X_n}$

   (c) $X_{n+1} = \dfrac{2}{4 - X_n}$ (d) $X_{n+1} = \dfrac{4X_n - 2}{X_n}$

   Which method would you not be able to use to find each solution? Which do you think is the quickest method for each of the solutions?

2. Show that the equation $x = e^{-x}$ has a solution between $x = 0$ and $x = 1$. Which of the following iterative methods would you use to find the solution?

   (a) $X_{n+1} = e^{-X_n}$ (b) $X_{n+1} = -\ln(X_n)$

   Hence find the solution correct to three decimal places.

3. By finding a rearrangement, choose an appropriate iterative method to solve each of the following equations:
   (a) $x + 2\sin(x) = 1$
   (b) $x^2 - 1 = 2e^{-x}$
   (c) $x^3 - 2x^2 + 3x - 1 = 0$

10.3.2 The Newton–Raphson iterative formula

We now introduce an iterative method which is fairly simple to use and which converges fairly rapidly for many functions. It is called the *Newton–Raphson iterative formula*.

Consider the problem of solving the equation $f(x) = 0$. Suppose that $x = X_0$ is an initial guess at the solution. Figure 10.1 shows a graph of $y = f(x)$.

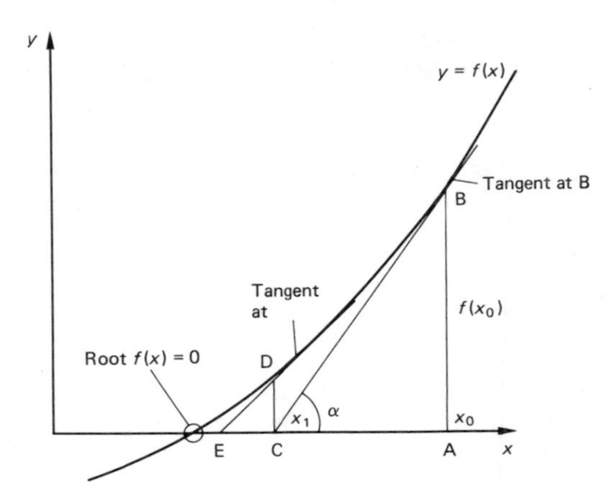

Figure 10.1

Now we devise an iterative method in the following way:

Step 1: start at A ($x = X_0$) and travel up to meet the curve of $y = f(x)$ at point B with coordinates X_0, $f(X_0)$.

Step 2: travel down the tangent at B to $y = f(x)$ to meet the x-axis at point C when $x = X_1$.

Now repeat Steps 1 and 2. Figure 10.1 suggests that this process will eventually arrive at the root (although of course it may take many 'ups and downs' to actually get there!). So let us derive an algebraic formula for this geometric description of the iterative process.

Consider the triangle ABC. The height of the triangle AB is $f(X_0)$ and the length of the base is $(X_0 - X_1)$. Now

$$\tan \alpha = \frac{\text{AB}}{\text{AC}} = \frac{f(X_0)}{X_0 - X_1}$$

But from the geometric interpretation of the derivative we know that $\tan \alpha = f'(X_0)$. Hence

$$f'(X_0) = \frac{f(X_0)}{X_0 - X_1}$$

which yields the formula

$$X_1 = X_0 - \frac{f(X_0)}{f'(X_0)} .$$

Similarly triangle CDE generates X_2 from X_1 using the formula

$$X_2 = X_1 - \frac{f(X_1)}{f'(X_1)}$$

In general if $x = X_n$ is the nth iterate, then the next one will be given by

$$\boxed{X_{n+1} = X_n - \frac{f(X_n)}{f'(X_n)}} \qquad (10.3.1)$$

This is called the *Newton–Raphson iterative* formula for solving $f(x) = 0$.

It converges if $| g'(x) | < 1$. Differentiating $g(x)$ we have

$$g'(x) = 1 - \left\{\frac{[f'(x)^2 - [f(x)f''(x)]}{[f'(x)^2]}\right\}$$

$$= \frac{f(x)f''(x)}{[f'(x)]^2} .$$

The condition for convergence is

$$\left|\frac{f(X_n)f''(X_n)}{[f'(X_n)]^2}\right| < 1.$$

Now at the solution itself $f(x) = 0$ so that $g'(x) = 0$. Hence at first sight equation (10.3.1) looks like a very good iterative method for solving $f(x) = 0$. But there are problems. A major difficulty occurs if $f'(x) = 0$ at or near the solution of $f(x) = 0$. For such functions the iterative formula will diverge.

Convergence is fairly rapid provided that $f'(X_n)$ is large but becomes slower as $f'(X_n)$ approaches zero. In latter cases it is necessary to compute $f(X_n)$ and $f'(X_n)$ for use in (10.3.1) very accurately. If $f'(x) = 0$ at the solution of $f(x) = 0$ then the Newton–Raphson method *cannot* be used, and another approach, such as a rearrangement of $f(x) = 0$, must be sought.

Example 10.3

Use the Newton–Raphson method to find the solution

$$x^3 + 2x - 4 = 0$$

correct to four decimal places.

Solution

In this example $f(x) = x^3 + 2x - 4$ so that $f'(x) = 3x^2 + 2$. Equation (10.3.1) then gives the iterative formula

$$X_{n+1} = X_n - \frac{X_n^3 + 2X_n - 4}{3X_n^2 + 2}$$

The sequence of results generated by this formula using the initial value $X_0 = 1$ is shown in the table below

X_n	$f(X_n)$	$f'(X_n)$	X_{n+1}
1	−1	5	1.2
1.2	0.0128	6.32	1.179 7
1.179 7	0.001 18	6.18	1.179 5
1.179 5	−0.000 056	6.17	1.179 5

We stop the iteration process when the 'new value' X_{n+1} and the 'old value' X_n are the same to the required accuracy, in this case four decimal places. The solution of $x^3 + 2x - 4 = 0$ is 1.1795 correct to four decimal places.

Exercise 10.3.2

Use the Newton–Raphson method to solve the following equations:

(a) $x^2 - 4x + 2 = 0$ for the root near 0.6
(b) $x = e^{-x}$
(c) $x + 2\sin(x) = 1$
(d) $x^2 - 1 = 2e^{-x}$
(e) $x^3 - 2x^2 + 3x - 1 = 0$

10.4 NUMERICAL INTEGRATION

The definite integral of many functions cannot be found analytically, for instance if the function is a set of values from an experiment then the function to be integrated is of unknown form. In such problems we devise numerical methods of solutions; the process of forming a definite integral is often called *integration by quadrature*.

In Chapter 8 we introduced one simple numerical method for integration called the *trapezoidal rule*. It is based on approximating the function by a set of straight lines (Figure 10.2).

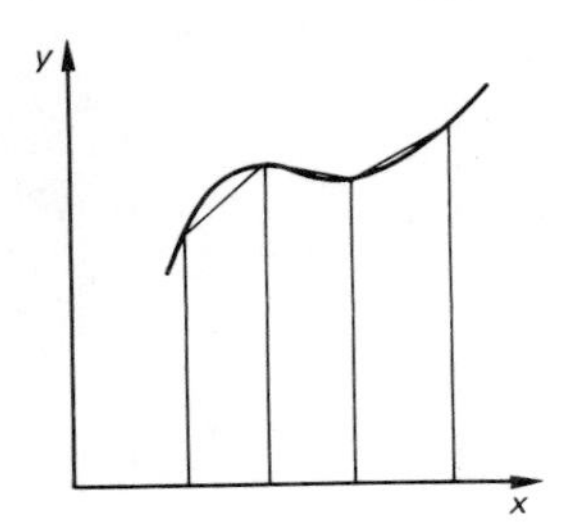

Figure 10.2

The formula for the trapezoidal rule is

$$\int_a^b f(x)\,dx \simeq \frac{h}{2}(f_0 + 2f_1 + \ldots + 2f_{n-1} + f_n)$$

where the interval $[a, b]$ is split into n intervals each of step length h, and f_i is the value of f at the ith step $x_i = a + ih$.

The trapezoidal rule is just one of many methods which involve replacing the integrand $f(x)$ by a set of known polynomial functions. The straight lines in Figure 10.2 are linear polynomials. Essentially the method involves dividing the range of the limits $[a, b]$ into subintervals usually of the same width. Within each subinterval the function is approximated by a polynomial function which can then be integrated. Table 10.1 summarizes two common methods.

Table 10.1

Degree of polynomial	No. of coordinate points to find polynomial coefficients	Name of method
1	2 (any number of subintervals)	trapezoidal
2	3 (even number of subintervals)	Simpson's rule

We shall illustrate the idea by looking at Simpson's rule. Consider the definite integral

$$I = \int_a^b f(x)\,dx$$

Divide the range of integration into an *even* number of subintervals and define a set of quadratic polynomials such that

$$p(x) = f(x) \text{ at the points } x_{i-1},\ x_i \text{ and } x_{i+1}$$

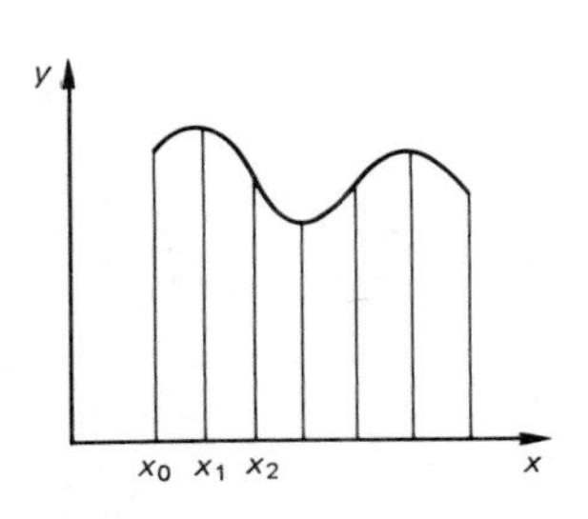

Figure 10.3

Figure 10.3 shows how a function $f(x)$ is approximated by three quadratic polynomials taken over six subintervals.

Consider the first quadratic through the points x_0, x_1 and x_2. If the polynomial $p(x) = ax^2 + bx + c$, then

$$\left.\begin{aligned} ax_0^2 + bx_0 + c &= f(x_0) \\ ax_1^2 + bx_1 + c &= f(x_1) \\ ax_2^2 + bx_2 + c &= f(x_2) \end{aligned}\right\} \qquad (10.4.1)$$

Now if we integrate $p(x)$ over the subinterval $[x_0, x_2]$ and simplify the result we obtain

$$\int_{x_0}^{x_2} f(x)\,dx \simeq \int_{x_0}^{x_2} p(x)\,dx$$

$$= \frac{a}{3}(x_2 - x_0)^3 + \frac{b}{2}(x_2 - x_0)^2 + c(x_2 - x_0) \qquad (10.4.2)$$

Solving Equations (10.4.1) for a, b and c and substituting into Equation (10.4.2) gives the formula

$$\int_{x_0}^{x_2} f(x)\,dx \simeq \frac{h}{3}[f(x_0) + 4f(x_1) + f(x_2)]$$

(The algebra involved is rather tedious so the result is stated and not derived.)

Adding each contribution for each trio of coordinate points gives *Simpson's rule*.

$$\int_a^b f(x)\,dx \simeq \frac{h}{3}[f(x_0) + 4f(x_1) + 2f(x_2) + 4f(x_3) + \ldots + 2f(x_{n-2}) + 4f(x_{n-1}) + f(x_n)]$$

$$= \frac{h}{3}\{f(x_0) + f(x_n) + 4[f(x_1) + f(x_3) + \ldots + f(x_{n-1})] + 2[f(x_2) + f(x_4) + \ldots + f(x_{n-2})]\}.$$

This formula is probably best remembered as

$$I \simeq \frac{h}{3}(\text{first} + \text{last} + 4 \times \text{odd ordinates} + 2 \times \text{even ordinates})$$

Example 10.4

Use Simpson's rule with step length 0.5 to evaluate the integral

$$\int_0^2 e^x\,dx.$$

Solution

With $h = 0.5$ we have four subintervals, these and values of the function $f = e^x$ at the end points are shown in Figure 10.4.

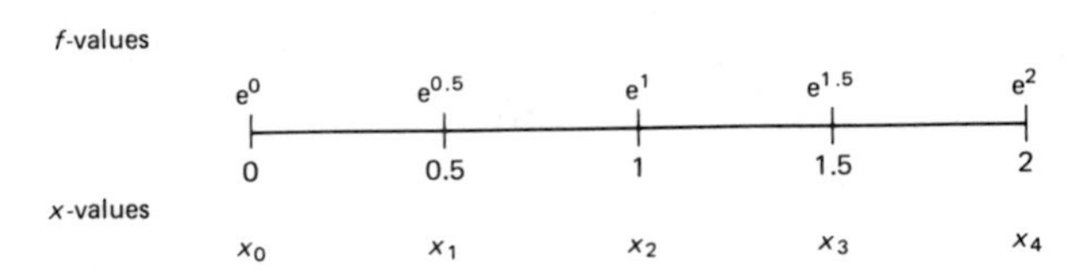

Figure 10.4

Using Simpson's rule with $h = 0.5$ gives

$$\int_0^2 e^x \, dx \simeq \frac{0.5}{3}[e^0 + e^2 + 4(e^{0.5} + e^{1.5}) + 2(e^1)]$$

$$= 6.3912 \text{ (rounding to four decimal places)}$$

Note that the exact value of the integral is $e^2 - e^0 = 6.3891$ (to four decimal places). Simpson's rule gives the answer correct to one decimal place in this case. To obtain an improved approximation using a numerical integration method we would continually reduce the step length h until two solutions were the same to the required degree of accuracy. For the integral in this example, Table 10.2 shows the approximate value as the step length is continually halved. For $h = 0.125$ and $h = 0.0625$, the approximations are equal to four decimal places so we deduce that to four decimal places

$$\int_0^2 e^x \, dx = 6.3891$$

Table 10.2

h	$\int_0^2 e^x \, dx$ using Simpson's rule
0.5	6.3912
0.25	6.3892
0.125	6.3891
0.0625	6.3891

Although somewhat tedious to carry out numerical integrations by hand they can be incorporated into computer software giving efficient ways of evaluating integrals numerically. Several modern calculators have Simpson's rule available at the press of a key.

Exercise 10.4.1

1. Use Simpson's rule to approximate the value of the following integrals:

(a) $\int_{-1}^{1} e^{x^2}\,dx$ step length 0.25

(b) $\int_{0}^{1} \frac{x^2}{1 + x^4}\,dx$ step length 0.1

(c) $\int_{1}^{2} \frac{1}{x}\,dx$ with 10 subintervals

(d) $\int_{0}^{1} \frac{4}{1 + x^2}\,dx$ with 4 subintervals

(e) $\int_{9}^{10} f(x)\,dx$ where $f(x)$ are experimental values given in Table 10.3.

Table 10.3

x	9.0	9.25	9.5	9.75	10.0
$f(x)$	0.1111	0.1081	0.1053	0.1026	0.1000

2. During the launch of a rocket the speed was noted every second for the first ten seconds of its flight and the Table 10.4 obtained. Use Simpson's rule to estimate the distance travelled by the rocket during the first 10 seconds.

Table 10.4

Time (s)	0	1	2	3	4	5	6	7	8	9	10
Speed (km/h)	0	32	80	128	176	224	272	320	368	400	448

3. A body of mass 2 kg is acted upon by a variable force $F(x)$ newtons. After travelling distance D its speed is given by the integral

$$V^2 = \int_{0}^{D} F(x)\,dx$$

If the force F is given in Table 10.5 use Simpson's rule to estimate the speed at the end of 40 metres.

Table 10.5

x(m)	0	5	10	15	20	25	30	35	40
F(N)	90	62	45	34	26	19	15	10	8

4. In an experiment it is necessary to estimate accurately the mean temperature $\overline{\theta}$ of a body over a period of time T. The quantity $\overline{\theta}$ is defined by the formula

$$\overline{\theta} = \frac{1}{T}\int_0^T \theta \, dt$$

where θ is the temperature of the body at time t. Over a 5-minute period θ was obtained every 30 s (Table 10.6). Obtain the value of $\overline{\theta}$ over this 5-minute period using Simpson's rule. How do these results compare with the simple average of the 11 temperature values?

Table 10.6

Time (min)	0.0	0.5	1.0	1.5	2.0	2.5	3.0	3.5	4.0	4.5	5.0
θ (°C)	79	71	64	58	52	47	42	38	34	31	28

OBJECTIVE TEST 10

1. The numbers in the following calculations have been rounded to two decimal places. Find the maximum possible error in the calculation and hence given an answer to the appropriate accuracy.

$$\frac{(14.32 + 0.69) \times (13.91 - 4.68)}{(6.21 - 3.98)}$$

2. The following iterative methods have been suggested for solving the equation $x^2 - 5.3x + 1.7 = 0$ near the roots $x = 0.3$ and $x = 5$.

(a) $X_{n+1} = \sqrt{5.3X_n - 1.7}$

(b) $X_{n+1} = \dfrac{X_n^2 + 1.7}{5.3}$

(c) $X_{n+1} = \dfrac{5.3X_n - 1.7}{X_n}$

(d) $X_{n+1} = \dfrac{1.7}{5.3 - X_n}$

Which method would you not be able to use to find each root? Which do you think is the quickest method for each of the roots? Hence find each root correct to four decimal places.

3. Use the Newton–Raphson method to solve the equation in Problem 2 above.

4. Use Simpson's rule to approximate the value of the following integrals:

(a) $\displaystyle\int_{1.5}^{3} \frac{x}{1 - x^3}\,dx$ step length 0.25

(b) $\displaystyle\int_{1}^{3} f(x)\,dx$ where $f(x)$ are experimental values given in Table 10.7

Table 10.7

x	1.0	1.5	2.0	2.5	3.0
$f(x)$	0.317	0.382	0.416	0.452	0.513

11 VECTORS

OBJECTIVES

When you have completed your study of this chapter you should be able to

1. recognize vector and scalar quantities
2. find the sum of two vectors geometrically
3. write two dimensional vectors in component form and add vectors algebraically
4. apply vector methods to simple problems involving forces, displacements and velocities
5. write the equation of a straight line in vector form
6. find the velocity and acceleration vectors from the position vector and vice versa using calculus

11.1 WHAT IS A VECTOR?

In many physical problems there occur certain quantities in which a direction is specified. For example, in specifying the position of Plymouth relative to Exeter, we would say that Plymouth is 45 miles west of Exeter. In this statement we have specified a distance of '45 miles' and a direction 'west'. Not all quantities have a direction associated with them. For example, if the temperature of a rod is −5 °C then the quantity 'temperature' only has a size or a value not a direction.

Exercise 11.1.1

Which of the following italicized quantities have a direction associated with them?

(a) The *time* for an object to travel 6 metres was 5 seconds
(b) The *force* of gravity on an object of mass 2 kg is 19.6 newtons towards the centre of the earth
(c) The *velocity* of a car is 131 mph due north on the M1
(d) The *area* of a sheet of A4 paper is 615 square cm
(e) The friction *force* of a sliding block is 2 newtons and opposes the direction of motion of the block

(f) The *pressure* in my car tyres is 32 psi
(g) The *displacement* of Portsmouth relative to Oxford is 107 kilometres on a bearing of 167°
(h) The *mass* of an apple is 0.1 kg
(i) My *bank balance* is £257.01
(j) The *speed* of Concorde is Mach 2
(k) The *velocity* of a car on the A38 at Ivybridge was 70 mph due East

---

In each part of this exercise, the quantity has a size or *magnitude*, for example, in (a) the magnitude of the temperature is −5 °C and in (b) the *force* of gravity has magnitude 19.6 newtons. The major difference between the quantities in the exercise are that some have a direction (e.g. force, velocity, displacement) and the others do not (e.g. temperature, area, pressure, mass, bank balance, speed, time).

Any quantity having both a magnitude and a direction is called a *vector quantity* or just a *vector*. Any quantity having a magnitude and no direction specified is called a *scalar quantity* or just a *scalar*.

To represent vectors geometrically on a diagram we use arrows. The length of the arrow is chosen according to some scale to represent the magnitude of the vector and the direction of the arrow is referred to a reference direction. For example, the velocity of 13 km/h in a direction 060° could be represented by the arrow shown in Figure 11.1. The reference direction here is due north.

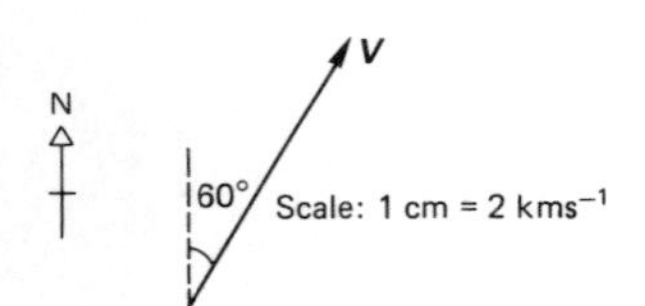

Figure 11.1

To represent vectors algebraically we use an underlined letter or, in textbooks, a vector is often printed using bold type face. For example, the velocity in Figure 11.1 is represented as $\boldsymbol{v}$.

It is very important to distinguish between vectors and scalars. They are different objects. The symbol v for the velocity in Figure 11.1 represents the magnitude of velocity i.e. v is speed whereas $\boldsymbol{v}$ is velocity. A notation often used for magnitude is $|\boldsymbol{v}|$, so that you will often see velocity (vector) $\boldsymbol{v}$ and speed (scalar) $|\boldsymbol{v}|$ or v.

Example 11.1

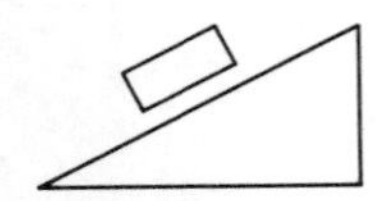

Figure 11.2

A block of mass 2 kg slides down a slope as shown in Figure 11.2. Two forces that act on the block are the force of gravity of magnitude 20 newtons and the force of friction of magnitude 8 newtons. Draw a diagram to a suitable scale representing these two forces by arrows.

Solution

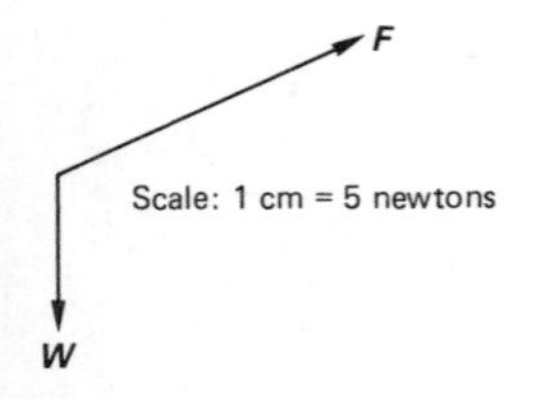

Figure 11.3

Before we can draw the diagram, we need to identify the direction of the two forces. We know that the force of gravity acts towards the centre of the earth and we usually denote this direction as 'vertically downwards'. The friction force acts so as to oppose the motion of the block, so it acts up the slope. Both forces are shown in Figure 11.3.

Exercise 11.1.2

1. The displacement of Birmingham from Derby is 57 km on a bearing 210°. The displacement of Leicester from Derby is 32 km on a bearing 135°. Draw a diagram to a suitable scale representing these two displacements by arrows.
2. A stone of mass 3 kg is dropped down a well. At some instant, the air resistance acting on the stone is 16 newtons and the force of gravity is 30 newtons. Draw a diagram to a suitable scale showing the two forces acting on the stone.
3. Represent on a diagram to a suitable scale the vectors in Exercise 11.1.1.

In practice, we often do not need to draw arrows very accurately to a scale, we just use a sketch diagram showing the relative directions of the vectors. For example, for all the forces on a sliding block (Figure 11.2), we would use Figure 11.4 in which the arrows are not drawn to scale.

In this chapter we develop an algebra of vectors allowing us to add and hence apply vectors in science and engineering applications.

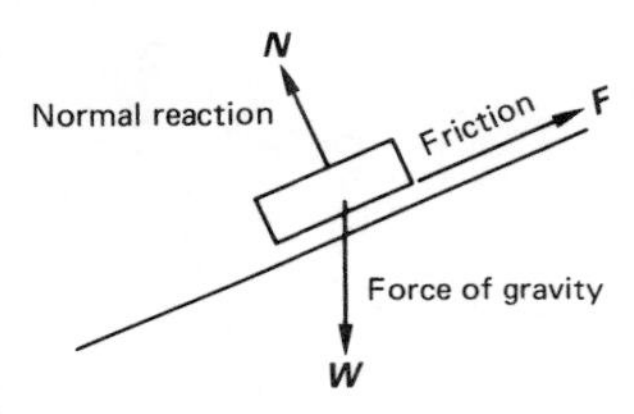

Figure 11.4

11.2 ALGEBRA OF VECTORS

To add two values of a scalar quantity is straightforward, we use the usual rules for adding numbers. For example, if the air temperature increases from −5 °C by 7 °C then we would write $-5 + 7 = 2$, giving a new temperature of 2 °C.

Now we ask the question: how do we add two vectors representing the same quantity? In other words if $\boldsymbol{a}$ and $\boldsymbol{b}$ are two displacements what do we mean by $\boldsymbol{a} + \boldsymbol{b} = \boldsymbol{c}$?

First of all we look at the meaning of the equality of two vectors. A vector is a (physical) quantity having both magnitude and a definite direction in space. So two vectors are equal if

(a) they represent the same quantity
(b) they have the same magnitude, and
(c) they point in the same direction

The first requirement is fairly obvious in that, for example, a force vector cannot equal a velocity vector! The second and third requirements mean that the arrows representing the two vectors must have the same length, must be parallel and have the same sense.

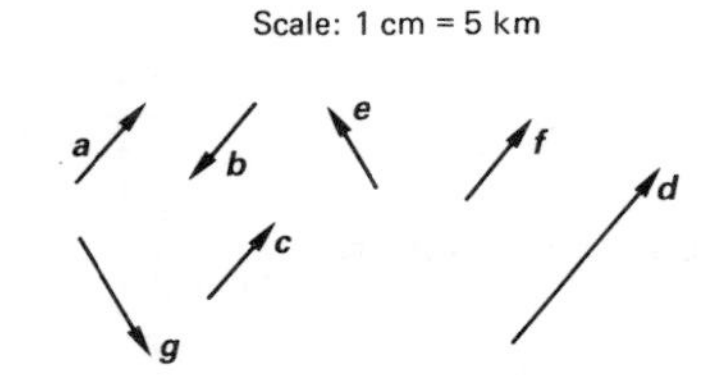

Figure 11.5

Example 11.2.1

Which of the displacement vectors shown in Figure 11.5 are equal?

Solution

Vectors $\boldsymbol{a}$, $\boldsymbol{c}$ and $\boldsymbol{f}$ are equal and algebraically we can write $\boldsymbol{a} = \boldsymbol{c}$, $\boldsymbol{c} = \boldsymbol{f}$, $\boldsymbol{a} = \boldsymbol{f}$.

Note that $\boldsymbol{a}$ is not equal to $\boldsymbol{b}$ because they have the opposite sense, so we write $\boldsymbol{b} = -\boldsymbol{a}$. Note also that vector $\boldsymbol{d}$ has the same direction as $\boldsymbol{a}$ but is 2.5 times as long (i.e. $|\boldsymbol{d}| = 2.5|\boldsymbol{a}|$). We write this algebraically as $\boldsymbol{d} = 2.5\boldsymbol{a}$.

Three points come out of this example:

1. For two vectors to be equal they do *not* have to start at the same point. Provided they have the same magnitude and direction they are equal; where we draw them is irrelevant. For example, a velocity of 50 mph due north on the M1 equals (in the vector sense) a velocity of 50 mph due north on the M92. (Although two vectors are equal if they have the same magnitude and direction irrespective of where we draw them, it is important to realize that their effects might be different. For example, a velocity of 50 mph due North on the M1 in England will lead to a different journey compared to the same velocity vector on the M92 in Scotland!)
2. A *negative* vector, $-\boldsymbol{a}$, has the same magnitude but opposite direction to $\boldsymbol{a}$.
3. A positive scalar times a vector changes the magnitude of the vector, so that if $\boldsymbol{b} = \alpha\boldsymbol{a}$ $(\alpha > 0)$ then $\boldsymbol{b}$ and $\boldsymbol{a}$ have the same direction but $|\boldsymbol{b}|$ is $\alpha|\boldsymbol{a}|$, i.e. the magnitude of $\boldsymbol{b}$ is α times the magnitude of $\boldsymbol{a}$.

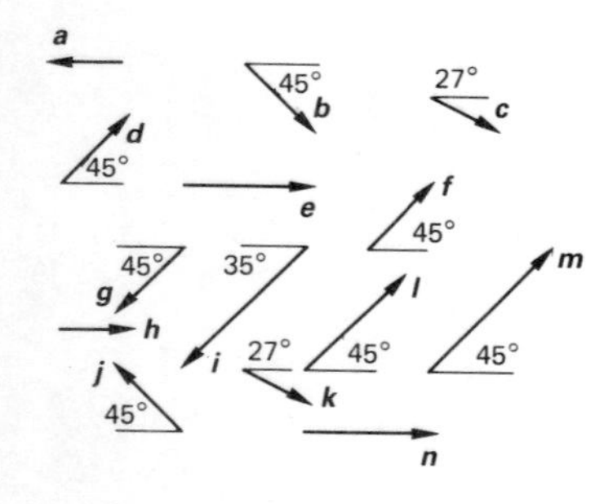

Figure 11.6

Exercise 11.2.1

1. Which of the force vectors shown in Figure 11.6 are equal?
2. For the vectors in Figure 11.6 write down algebraic formulas relating

   (a) $\boldsymbol{a}$ and $\boldsymbol{e}$ (b) $\boldsymbol{a}$ and $\boldsymbol{h}$ (c) $\boldsymbol{a}$ and $\boldsymbol{n}$
   (d) $\boldsymbol{d}$ and $\boldsymbol{m}$ (e) $\boldsymbol{d}$ and $\boldsymbol{g}$ (f) $\boldsymbol{d}$ and $\boldsymbol{i}$
   (g) $\boldsymbol{d}$ and $\boldsymbol{l}$ (h) $\boldsymbol{j}$ and $\boldsymbol{b}$

3. Which of the following vectors are equal?

   (a) a force of magnitude 5 newtons vertically upwards
   (b) a velocity of magnitude 5 ms^{-1} due north
   (c) a displacement of magnitude 5 metres due north
   (d) the force of gravity of 5 newtons
   (e) the wind velocity 5 ms^{-1} due north
   (f) air resistance of 5 newtons on a ball falling vertically

4. For each of the force vectors in Figure 11.6 write down their magnitude and direction. (Scale: 4 mm = 1 N)

To add two vectors, $\boldsymbol{a}$ and $\boldsymbol{b}$, we find the single vector, $\boldsymbol{c}$, that would have the same effect as $\boldsymbol{a}$ followed by $\boldsymbol{b}$. We then write this algebraically as

$$\boldsymbol{a} + \boldsymbol{b} = \boldsymbol{c}$$

The vector $\boldsymbol{c}$ is often called the *resultant* of $\boldsymbol{a}$ and $\boldsymbol{b}$. The mathematical definition of vector addition comes from the way physical vectors such as displacement, force and velocity add. Consider two simple displacement vectors. Suppose that you walk 4 km due east and then 3 km due north, then you finish up 5 km on a bearing 053° (to the nearest degree). This is shown in Figure 11.7, in which $\boldsymbol{a}$ is the displacement 4 km due east and $\boldsymbol{b}$ is the displacement 3 km due north.

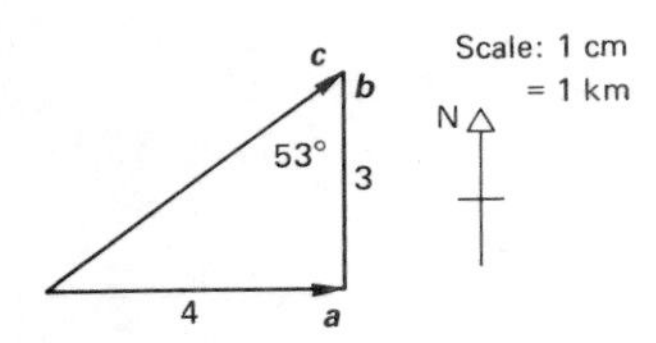

Figure 11.7

The effect of $\boldsymbol{a}$ followed by $\boldsymbol{b}$, i.e. the resultant of $\boldsymbol{a}$ and $\boldsymbol{b}$, is the vector $\boldsymbol{c}$ which forms the third side of the triangle. In general then we define the resultant (or sum) of two vectors $\boldsymbol{a}$ and $\boldsymbol{b}$ in the following way.

The triangle rule
To add two vectors $\boldsymbol{a}$ and $\boldsymbol{b}$ we place the tail of one arrow at the head of the other and then the vector that forms the third side of the triangle is the resultant $\boldsymbol{a} + \boldsymbol{b}$

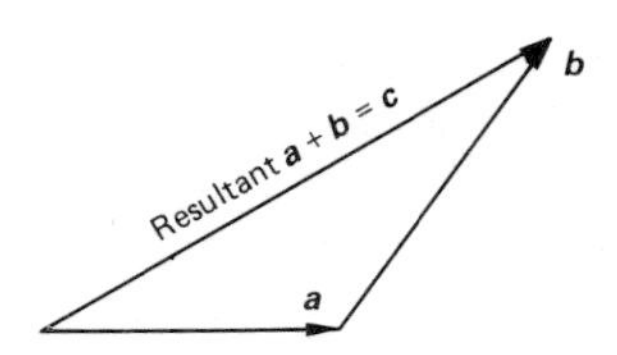

Figure 11.8 The triangle rule of vector addition

An alternative and equivalent definition of the resultant of two forces is the parallelogram rule. Suppose that an object experiences two forces each of magnitude 15 newtons and perpendicular to each other (Figure 11.9). Then we would expect the effect (or resultant force) to act along the line OC formed by drawing the parallelogram (a square in this special case) OACB.

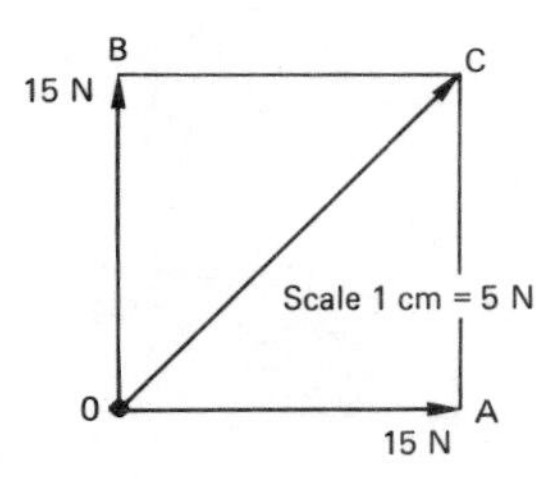

Figure 11.9

The parallelogram rule
If we place the tails of the arrows together and complete a parallelogram then the diagonal of the parallelogram represents the resultant $\boldsymbol{a} + \boldsymbol{b}$

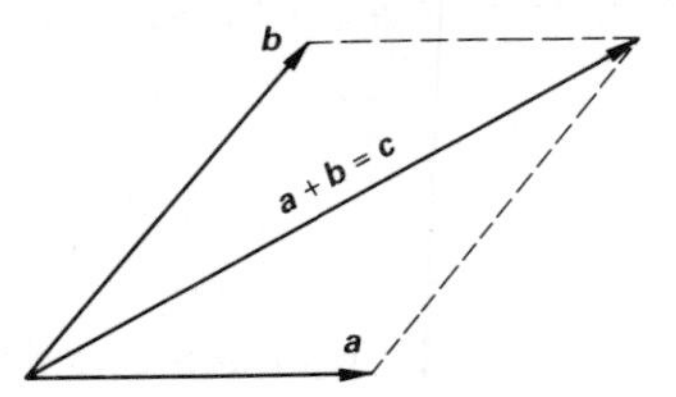

Figure 11.10 The parallelogram rule of vector addition

The parallelogram rule and triangle rule lead to exactly the same vector $\boldsymbol{c}$.

Example 11.2.2

A ship heads on a bearing of 100° at a speed of 18 knots while the current is 4 knots from a bearing of 220°. From a diagram, find the speed and actual direction of the ship.

Solution

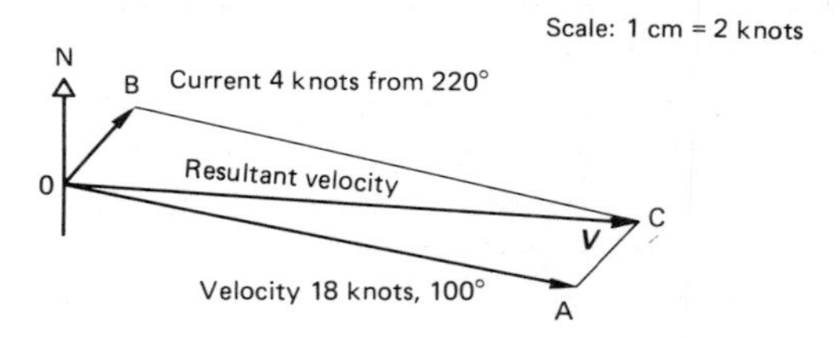

Figure 11.11

The resultant velocity of the ship has magnitude 20.5 knots on a bearing of 090°. (In this problem we have used the parallelogram rule to find the resultant because it is easier to draw two bearings from the single point O. We could equally have used the triangle rule by drawing in AC instead of OB.)

Example 11.2.3

$\boldsymbol{a}$ is the vector of magnitude 4 units due east and $\boldsymbol{b}$ is the vector of magnitude 3 units due south. Draw a diagram showing the vectors $\boldsymbol{a} - \boldsymbol{b}$ and $-2\boldsymbol{a} - 3\boldsymbol{b}$.

Solution

Using the normal rules of algebra

$$\boldsymbol{a} - \boldsymbol{b} = \boldsymbol{a} + (-\boldsymbol{b})$$

So we need to add $\boldsymbol{a}$ and $(-\boldsymbol{b})$: and

$$-2\boldsymbol{a} + 3\boldsymbol{b} = (-2\boldsymbol{a}) + 3\boldsymbol{b}$$

Hence we need to add $-2\boldsymbol{a}$ and $3\boldsymbol{b}$. These vectors are shown in Figure 11.12.

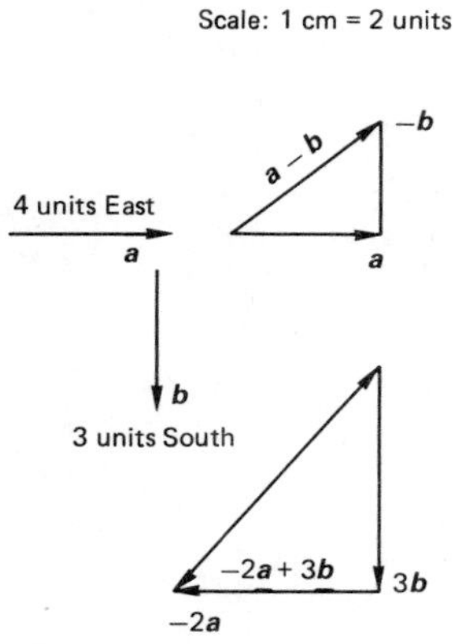

Figure 11.12

Exercise 11.2.2

1. For the vectors in Figure 11.6 draw diagrams to show the following resultant vectors:

   (a) $\boldsymbol{a} + \boldsymbol{d}$ (b) $\boldsymbol{b} + \boldsymbol{e}$ (c) $\boldsymbol{g} + \boldsymbol{i}$ (d) $\boldsymbol{m} + \boldsymbol{n}$
   (e) $\boldsymbol{c} + \boldsymbol{e}$ (f) $\boldsymbol{m} - \boldsymbol{l}$ (g) $\boldsymbol{f} - \boldsymbol{c}$ (h) $\boldsymbol{e} - \boldsymbol{b}$

2. The position vectors of two points A and B are shown in Figure 11.13. Draw a diagram showing the points C and D given by $2\boldsymbol{a} + 3\boldsymbol{b}$ and $\boldsymbol{a} - 2\boldsymbol{b}$ respectively.
3. Two forces of magnitude 5 newtons and 12 newtons act on an object. The direction of the forces are at right angles to each other. Find the magnitude of the resultant force and its direction relative to the larger force.
4. Forces of magnitude 7 newtons and 10 newtons act on an object, and the angle between their directions is 50°. Find the magnitude of the resultant force and its direction relative to the smaller force. Show on a diagram the magnitude and direction of the force needed so that the object remains at rest.
5. An aircraft travels on a bearing of 60° for 200 miles and then on a bearing of 210° for 160 miles. How far is it from its starting point and on what bearing?

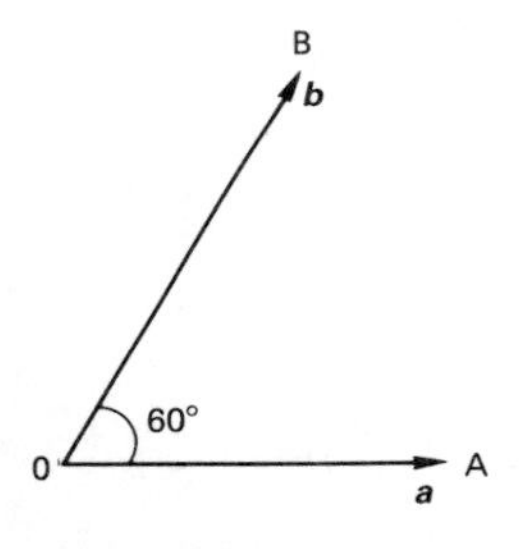

Figure 11.13

With the meaning of equality and addition of vectors, we have the beginning of an algebra of vectors. There are two important general properties which come out of all of our examples.

Property 1: Vector addition is commutative

$$\boldsymbol{a} + \boldsymbol{b} = \boldsymbol{b} + \boldsymbol{a}$$

Property 2: Vector addition is associative

$$\boldsymbol{a} + (\boldsymbol{b} + \boldsymbol{c}) = (\boldsymbol{a} + \boldsymbol{b}) + \boldsymbol{c}$$

You are familiar with these rules for ordinary numbers but it is always important to check them for a quantity such as a vector.

Exercise 11.2.3

Given the three vectors $\boldsymbol{a}$, $\boldsymbol{b}$ and $\boldsymbol{c}$ in Table 11.1, by drawing an accurate scale drawing, show that the commutative and associative laws hold.

Table 11.1

Vector	Magnitude	Bearing
$\boldsymbol{a}$	5	030°
$\boldsymbol{b}$	7	090°
$\boldsymbol{c}$	3	150°

11.3 VECTORS AND COMPONENTS

11.3.1 Components of a vector

You probably solved the exercises in Section 11.2 by drawing accurate scale drawings. This is not only tedious but somewhat inaccurate depending on your drawing skill. It is more convenient to solve problems using vector algebra instead of vector geometry.

Given a vector (in two dimensions) we can split the vector into two vectors at right angles to each other whose sum is the original vector. The two perpendicular directions are normally the horizontal and vertical directions coinciding with the x and y-axes. As an example consider the vector $\boldsymbol{a}$ of magnitude 15 in a direction of 40° with the x-axis as shown in Figure 11.14.

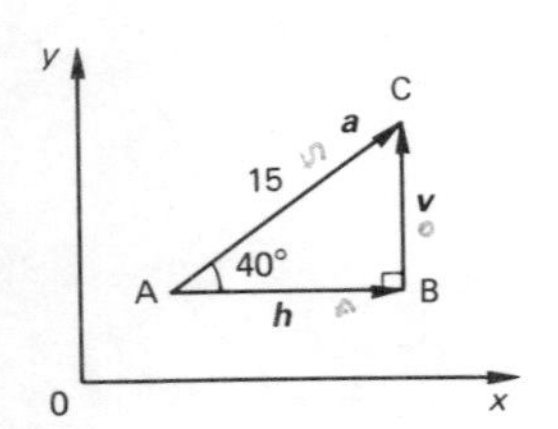

Figure 11.14

The vector $\boldsymbol{a}$ acts along AC. Triangle ABC is a right-angled triangle with AB parallel to Ox and BC parallel to Oy. (Given a line AC we can always construct such a triangle.) Let the vector of length AB and direction from A to B be $\boldsymbol{h}$ and let the vector of length BC and direction from B to C be $\boldsymbol{v}$, then using the definition of vector addition we have

$$\boldsymbol{a} = \boldsymbol{h} + \boldsymbol{v}$$

$\boldsymbol{h}$ and $\boldsymbol{v}$ are called the (Cartesian) *component vectors* of $\boldsymbol{a}$.

Using the cosine definition for triangle ABC

$$AB = |\boldsymbol{h}| = 15 \cos 40^\circ$$

$$BC = |\boldsymbol{v}| = 15 \cos 50^\circ$$

These are called the (Cartesian) *components* of the vector $\boldsymbol{a}$. A notation adopted to denote Cartesian components of a vector is the column

$$\boldsymbol{a} = \begin{pmatrix} 15 \cos 40^\circ \\ 15 \cos 50^\circ \end{pmatrix}$$

This is called the *Cartesian component* form of $\boldsymbol{a}$.

Exercise 11.3.1

1. A ship is travelling at 20 knots on a bearing 060°. Write down the Cartesian component form of the velocity vector of the ship.
2. An aircraft is flying at 250 mph on a bearing 020°. Find the Cartesian component form of its velocity vector.
3. A force, $\boldsymbol{F}$, of magnitude 100 newtons acts vertically upwards on an object. A second force, $\boldsymbol{G}$, of magnitude 75 newtons acts horizontally (to the right) on the object. Write down $\boldsymbol{F}$ and $\boldsymbol{G}$ in Cartesian component form.

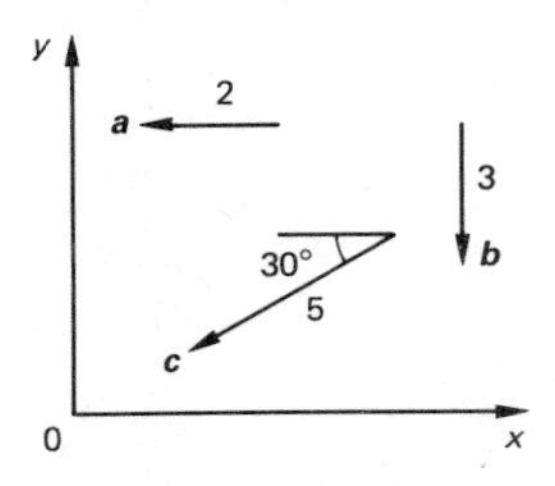

Figure 11.15

Example 11.3.1

Find the component form for the three vectors shown in Figure 11.15.

Solution

What makes these vectors different from the vectors in Exercise 11.3.1 is that $\boldsymbol{a}$, $\boldsymbol{b}$ and $\boldsymbol{c}$ do not point in the directions of the axes. We have to be careful with the signs.

Consider vector $\boldsymbol{a}$. Clearly we cannot write it as $\begin{bmatrix} 2 \\ 0 \end{bmatrix}$ because this vector points in the x-direction. However $\boldsymbol{a}$ and the vector $\begin{bmatrix} 2 \\ 0 \end{bmatrix}$ have the same magnitude, but are opposite in direction. So from Section 11.2 we have

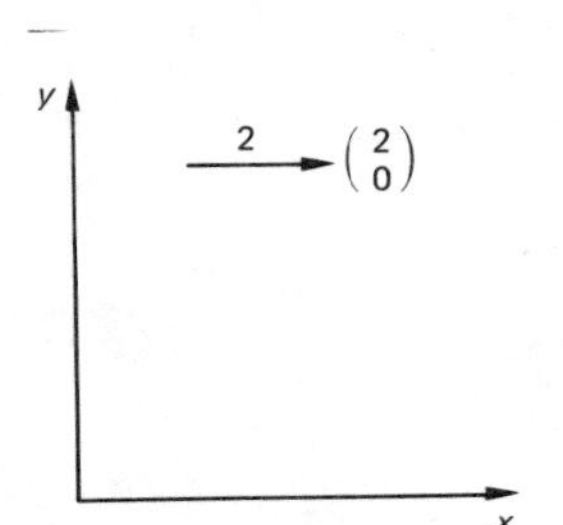

Figure 11.16

$$\boldsymbol{a} = -\begin{bmatrix} 2 \\ 0 \end{bmatrix} = \begin{bmatrix} -2 \\ 0 \end{bmatrix}$$

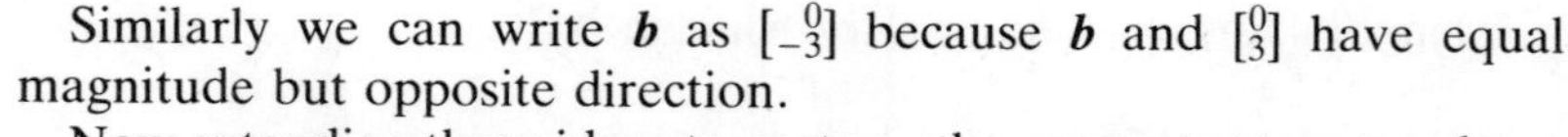

Similarly we can write $\boldsymbol{b}$ as $\left[\begin{smallmatrix}0\\-3\end{smallmatrix}\right]$ because $\boldsymbol{b}$ and $\left[\begin{smallmatrix}0\\3\end{smallmatrix}\right]$ have equal magnitude but opposite direction.

Now extending these ideas to vector $\boldsymbol{c}$ the component vectors have magnitude 5 cos 30° and 5 cos 60° but point in the opposite directions to the x and y-axes. Hence we write

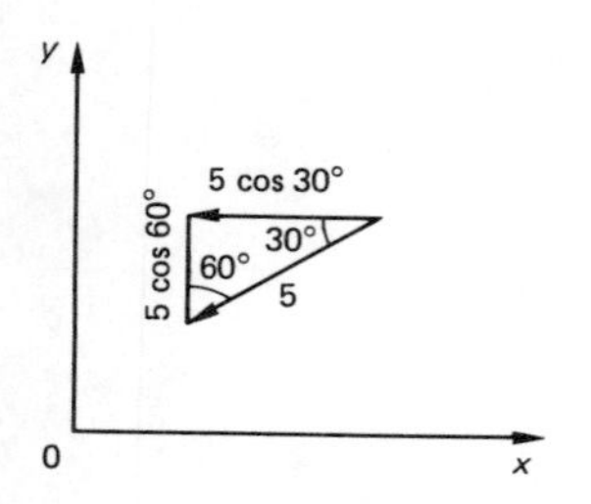

Figure 11.17

$$\boldsymbol{c} = -\begin{bmatrix} 5\cos 30° \\ 5\cos 60° \end{bmatrix} = \begin{bmatrix} -5\cos 30° \\ -5\cos 60° \end{bmatrix}$$

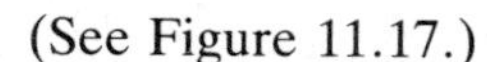

(See Figure 11.17.)

In general then, if the arrowhead of the component vector is in the opposite direction to the appropriate axis the component has a negative sign attached to it.

Exercise 11.3.2

1. A ship is travelling at 30 knots on a bearing 210°. Write down the Cartesian component form of the velocity vector of the ship.
2. The force of gravity on a block at rest on an inclined plane of angle 20° to the horizontal has magnitude 30 newtons. Find the components of this force in directions parallel and perpendicular to the plane.
3. Write down the vectors in Figure 11.6 in Cartesian component form. (Scale: 4 mm = 1 N.)

To summarize,

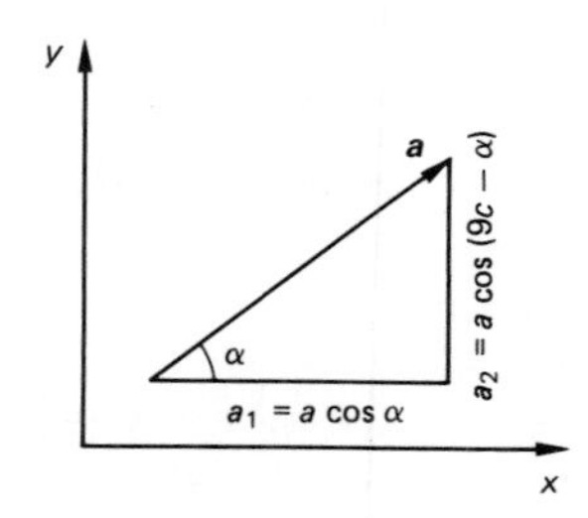

Figure 11.18 The Cartesian component form of a vector **a** is given algebraically by

$$\boldsymbol{a} = \begin{bmatrix} a\cos(\alpha) \\ a\cos(90-\alpha) \end{bmatrix} = \begin{bmatrix} a_1 \\ a_2 \end{bmatrix}$$

where α is the angle between the vector and the x direction as shown.

Note: Some texts will use

$$\begin{bmatrix} a\cos\alpha \\ a\sin\alpha \end{bmatrix}$$

and this is equivalent to the notation above because $\sin\alpha = \cos(90 - \alpha)$. We have adopted the use of cosine for two reasons:

1. We recommend that you get into the habit of looking for the angle between the vector and the x or y-axes and then use cosine.
2. In three dimensions you must always use the cosine of the angle between the vector and the x, y and z axes. It is a good idea to get into good habits at this early stage.

Finally we note from Figure 11.18 that if we are given a vector in Cartesian component form

$$\boldsymbol{a} = \begin{bmatrix} a_1 \\ a_2 \end{bmatrix}$$

then its magnitude is $|\boldsymbol{a}| = \sqrt{a_1^2 + a_2^2}$ and its direction α is

$$\cos^{-1}\left(\frac{a_1}{\sqrt{a_1^2 + a_2^2}}\right).$$

Exercise 11.3.3

Find the magnitude and direction of the following vectors given in Cartesian component form:

$$\boldsymbol{a} = \begin{bmatrix} 3 \\ 4 \end{bmatrix}, \quad \boldsymbol{b} = \begin{bmatrix} 5 \\ 12 \end{bmatrix}, \quad \boldsymbol{c} = \begin{bmatrix} -3 \\ 4 \end{bmatrix}, \quad \boldsymbol{d} = \begin{bmatrix} -1 \\ -2 \end{bmatrix},$$

$$\boldsymbol{e} = \begin{bmatrix} -1 \\ -1 \end{bmatrix}, \quad \boldsymbol{f} = \begin{bmatrix} 1 \\ -3 \end{bmatrix}$$

[Hint: it often helps to draw a sketch diagram showing the vector.]

11.3.2 Adding vectors using components

We have seen that given a vector, we can 'resolve' it into two components; adding two vectors simply involves adding the components.

Example 11.3.2

A ship travels 100 km on a bearing of 045° followed by 180 km on a bearing of 120°. Find the distance and bearing of the ship from its original position.

Solution

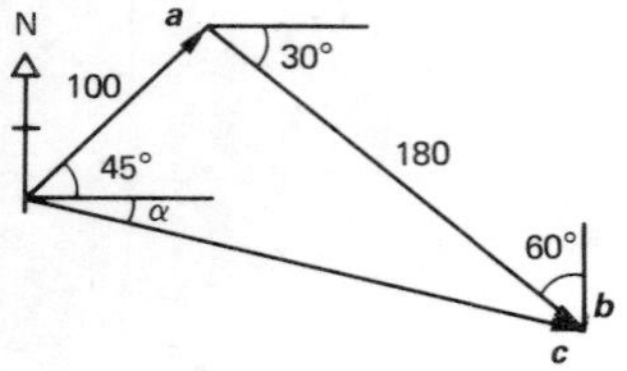

Figure 11.19

Even though we now add vectors algebraically in component form, a sketch diagram is always useful in problem solving.

The component form of the first displacement $\boldsymbol{a}$ in Figure 11.19 is

$$\boldsymbol{a} = \begin{bmatrix} 100 \cos 45° \\ 100 \cos 45° \end{bmatrix}$$

The component form of the second displacement $\boldsymbol{b}$ is

$$\boldsymbol{b} = \begin{bmatrix} 180 \cos 30° \\ -180 \cos 60° \end{bmatrix}$$

The resultant of $\boldsymbol{a}$ and $\boldsymbol{b}$ is then

$$\boldsymbol{c} = \boldsymbol{a} + \boldsymbol{b} = \begin{bmatrix} 100 \cos 45° \\ 100 \cos 45° \end{bmatrix} + \begin{bmatrix} 180 \cos 30° \\ -180 \cos 60° \end{bmatrix}$$

$$= \begin{bmatrix} 100 \cos 45° + 180 \cos 30° \\ 100 \cos 45° - 180 \cos 60° \end{bmatrix}$$

$$= \begin{bmatrix} 226.6 \\ -19.3 \end{bmatrix}$$

The vector $\boldsymbol{c}$ is the displacement of the ship after the two journeys.

The magnitude of $\boldsymbol{c} = \sqrt{226.6^2 + 19.3^2} = 227.4$ km and its direction is the bearing $090° + \alpha$ where $\cos \alpha = 226.6/227.4$, i.e. the bearing 094.8°.

Exercise 11.3.4

1. Solve problems 3, 4, 5 of Exercise 11.2.3 using the component form.
2. For the vectors in Exercise 11.3.3 find the magnitude and direction of

   (a) $\boldsymbol{a} + \boldsymbol{b}$ (b) $\boldsymbol{b} + \boldsymbol{c}$ (c) $\boldsymbol{a} + \boldsymbol{c}$ (d) $\boldsymbol{c} + \boldsymbol{f}$
   (e) $\boldsymbol{d} - \boldsymbol{e}$ (f) $\boldsymbol{c} - \boldsymbol{a}$ (g) $\boldsymbol{b} + \boldsymbol{e}$ (l) $\boldsymbol{f} - \boldsymbol{e}$
3. Figure 11.20 shows various forces acting on four objects. For each object write down in Cartesian compo-

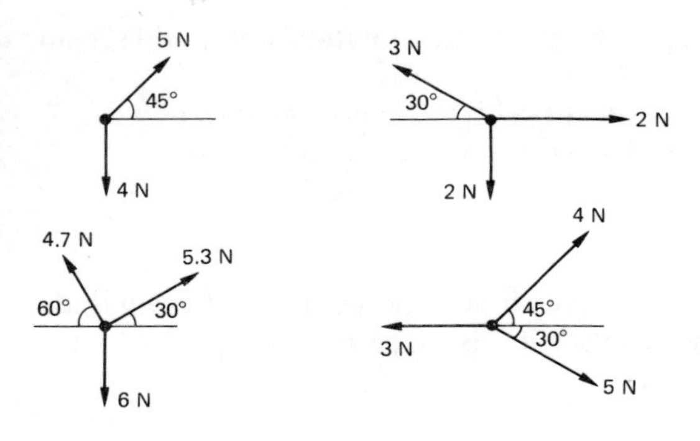

Figure 11.20

nent form the resultant force acting on it. Find the magnitude and direction of each of these resultant forces.

4. Three vectors $\boldsymbol{a}$, $\boldsymbol{b}$ and $\boldsymbol{c}$ are given in component form by

$$\begin{bmatrix} 1 \\ 3 \end{bmatrix}, \begin{bmatrix} 2 \\ -5 \end{bmatrix}, \begin{bmatrix} -2 \\ 4 \end{bmatrix}$$

respectively.

(a) Write down in component form the vectors $\boldsymbol{a} + \boldsymbol{b}$, $\boldsymbol{b} + \boldsymbol{c}$, $\boldsymbol{a} - \boldsymbol{b}$, $3\boldsymbol{a} + 2\boldsymbol{b}$ and $-2\boldsymbol{a} - 3\boldsymbol{b} + 4\boldsymbol{c}$.

(b) Calculate the magnitude and direction of each vector in part (a).

11.3.3 Unit vectors and components

An alternative and widely-used notation for component vectors uses the idea of two special vectors each with magnitude 1 and directions along the x and y-axis. A vector with magnitude 1 is called a *unit vector*. The two special unit vectors along the x and y-axes are usually denoted by $\boldsymbol{i}$ and $\boldsymbol{j}$ respectively (Figure 11.21).

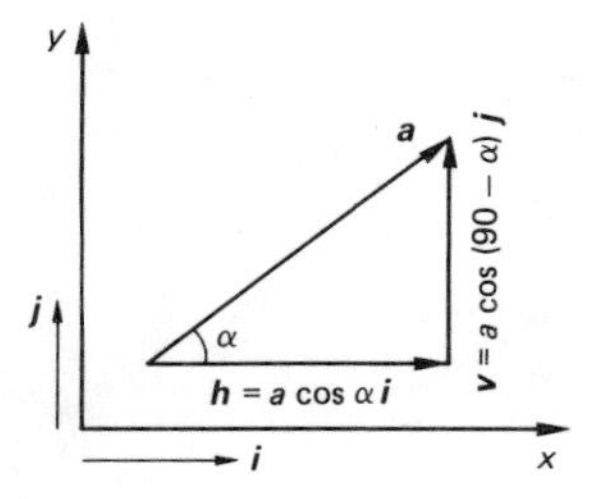

Figure 11.21

The component vectors $\boldsymbol{h}$ and $\boldsymbol{v}$ can be written in terms of $\boldsymbol{i}$ and $\boldsymbol{j}$ quite easily as

$$\boldsymbol{h} = (a \cos \alpha)\boldsymbol{i} \text{ and } \boldsymbol{v} = a[\cos(90 - \alpha)]\boldsymbol{j}$$

Hence in terms of $\boldsymbol{i}$ and $\boldsymbol{j}$ we have the *Cartesian component* form of $\boldsymbol{a}$ as

$$\boxed{\boldsymbol{a} = a \cos \alpha\boldsymbol{i} + a \cos(90 - \alpha)\boldsymbol{j}}$$

If we compare this idea with the column notation we should see an easy relationship

$$a = \begin{bmatrix} a_1 \\ a_2 \end{bmatrix} = a_1 i + a_2 j$$

You should become familiar with both notations. The magnitude of a is $\sqrt{a_1^2 + a_2^2}$ and the direction is given by $\alpha = \cos^{-1}(a_1/a)$.

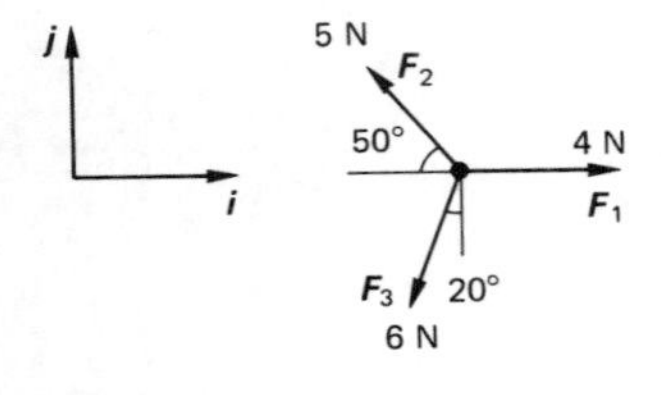

Figure 11.22

Example 11.3.4

Three forces act on an object with magnitudes in the directions shown in Figure 11.22. Write each force in terms of i and j and hence find the resultant force on the object.

Solution

Taking each force in turn we have

$$F_1 = 4i$$

$$F_2 = -5 \cos 50° \, i + 5 \cos 40° \, j$$

$$F_3 = -6 \cos 70° \, i - 6 \cos 20° \, j$$

The resultant force is then

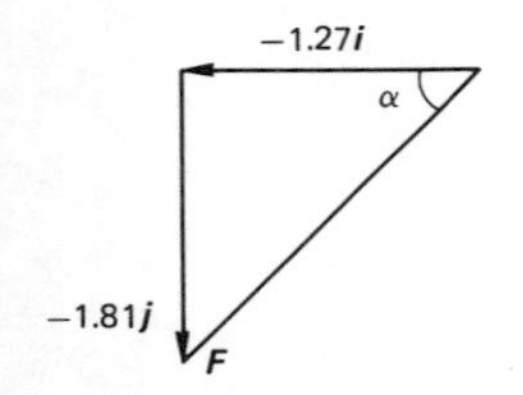

Figure 11.23

$$F = F_1 + F_2 + F_3$$

$$F = (4 - 5 \cos 50° - 6 \cos 70°)i + (5 \cos 40° - 6 \cos 20°)j$$

$$= -1.27i + -1.81j$$

The magnitude of F is $\sqrt{1.27^2 + 1.81^2} = 2.21$ and its direction is $\alpha = 55°$ as shown in Figure 11.23.

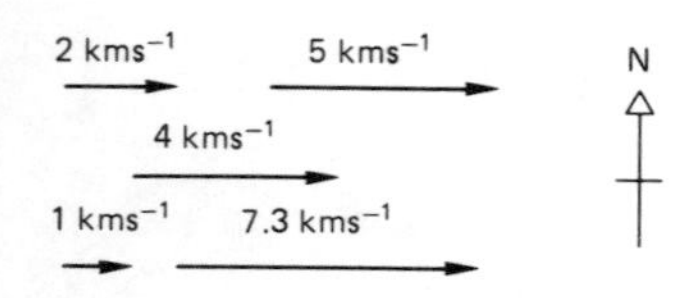

Figure 11.24

Exercise 11.3.5

1. Write down the velocity vectors shown in Figure 11.24 in terms of a unit vector i pointing east.
2. Write down the Cartesian component form of the four vectors shown below in terms of i and j.

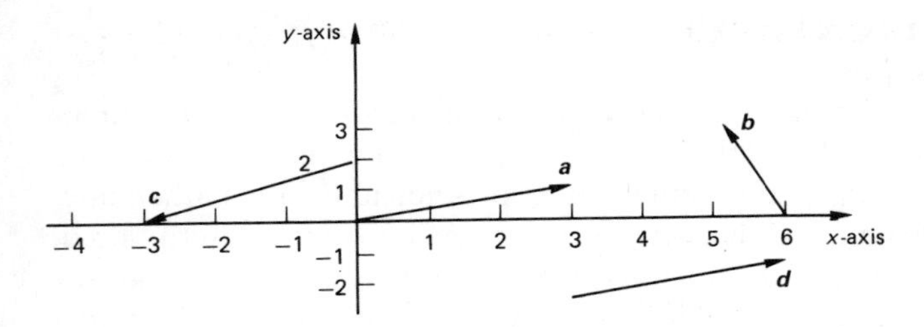

Figure 11.25

3. Calculate the magnitude of the following vectors.

$$\boldsymbol{a} = 5\boldsymbol{i},\ \boldsymbol{b} = 2\boldsymbol{i} + 3\boldsymbol{j} \text{ and } \boldsymbol{c} = 3\boldsymbol{i} + 4\boldsymbol{j}$$

Show each of these vectors on a diagram by arrows

(a) starting at the origin,
(b) starting at the point $(-1, 2)$.

4. Three vectors $\boldsymbol{a}$, $\boldsymbol{b}$ and $\boldsymbol{c}$ are given in component form by $\boldsymbol{i} + 3\boldsymbol{j}$, $2\boldsymbol{i} - 5\boldsymbol{j}$ and $-2\boldsymbol{i} + 4\boldsymbol{j}$ respectively.

   (a) Write down in component form the vectors $\boldsymbol{a} + \boldsymbol{b}$, $\boldsymbol{b} + \boldsymbol{c}$, $\boldsymbol{a} - \boldsymbol{b}$, $3\boldsymbol{a} + 2\boldsymbol{b}$ and $-2\boldsymbol{a} - 3\boldsymbol{b} + 4\boldsymbol{c}$.
   (b) Calculate the magnitude and direction of each of the vectors in part (a).

5. Figure 11.26 shows various forces acting on objects. For each object write down the resultant force acting on it in terms of unit vectors $\boldsymbol{i}$ (horizontally) and $\boldsymbol{j}$ (vertically).

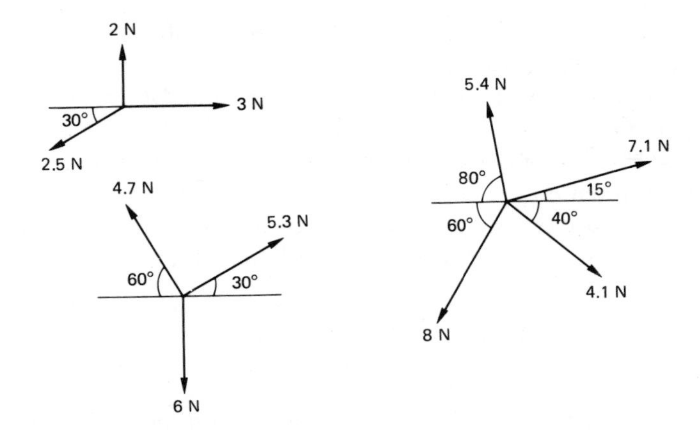

Figure 11.26

11.4 APPLICATIONS

In this chapter we have seen applications of vectors to solve problems involving physical quantities such as displacements, velocities and forces. There are many applications in other areas of mathematics and in this section we illustrate the use of vectors in the description of motion of objects. In engineering and science, vectors are an important tool. For example, you may learn how to express circular motion in terms of vectors and thus to analyse the motion of many different mechanisms, such as cranks and pistons etc. In electrical engineering, vector notation and vector manipulation help in alternating current problems.

11.4.1 The position vector

An important vector with wide applications in geometry and mechanics is the *position vector*. Thus far, in previous sections, we have not specified particular positions of vectors on the page; in fact we have said that any two arrows of the same length and direction are identical no matter where they are drawn. Such vectors are called *free vectors* in that there is no specified point of application. Often it is useful to be able to define the position of a point in space P in terms of the fixed origin O. In doing so geometrically we would draw an arrow from the origin O to the point P, the displacement vector from O to P is called *the position vector of P relative to O*. We often denote the position vector by the symbol $\boldsymbol{r}$, or by $\overrightarrow{OP}$.

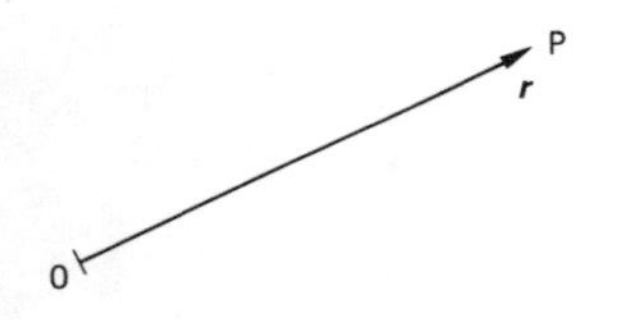

Figure 11.27

Once the origin O has been chosen the position vector of a point is unique, so that two identical position vectors represent the same point. The position vector is an example of a *bound vector*, it has a specified point of application, i.e. the origin.

If we have various points A, B, C, say, then we often denote the respective position vectors $\overrightarrow{OA}$, $\overrightarrow{OB}$ and $\overrightarrow{OC}$ by small letters $\boldsymbol{a}$, $\boldsymbol{b}$ and $\boldsymbol{c}$.

The use of position vectors does not necessarily require the use of coordinate axes and components. However often it is convenient to introduce coordinate axes centred on the origin O. For example, the point in the x–y plane with coordinates (2, −3) has position vector

$$\boldsymbol{a} = 2\boldsymbol{i} - 3\boldsymbol{j} \text{ or } \boldsymbol{a} = \begin{bmatrix} 2 \\ -3 \end{bmatrix}$$

depending on which Cartesian representation we are using. Note that the Cartesian components of the position vector of a point are the same as the Cartesian coordinates of the point.

Exercise 11.4.1

1. Write down in terms of $\boldsymbol{i}$ and $\boldsymbol{j}$ the position vectors of the four points whose Cartesian coordinates are:

   A(0, 1) B(1, 0) C(−2, 3) D(−1, −2)

   Draw a diagram showing the four vectors.

2. Which of the vectors shown in Figure 11.28 are equal to the vector $3\boldsymbol{i} + 4\boldsymbol{j}$? Which vector is the position vector of the point (3, 4)?

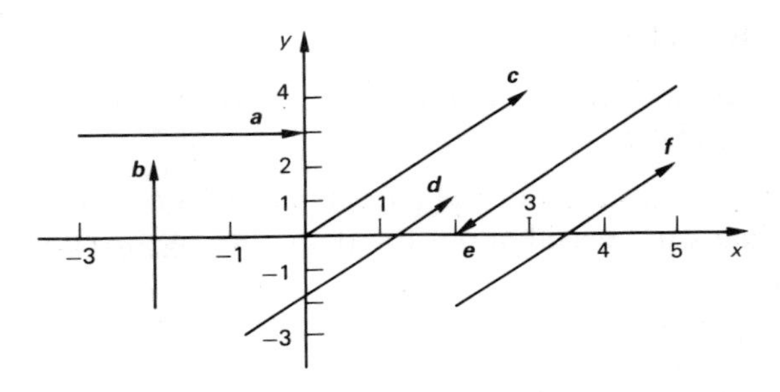

Figure 11.28

3. A, B and C are the points (1, 2), (5, 1) and (7, 8) respectively. Write down in terms of $\boldsymbol{i}$ and $\boldsymbol{j}$ the position vectors of these three points. Hence using the triangle rule of vector addition deduce the form of the displacement vectors $\overrightarrow{AB}$, $\overrightarrow{OA}$ and $\overrightarrow{CA}$.

4. Given the points A(4, 2) and B(5, 4) write down the position vectors of A and B and hence the displacement vector $\overrightarrow{AB}$. Find the magnitude and direction of this displacement vector.

5. For the three points A(4, −2), B(0, −10) and C(−6, −22), show that the displacement vectors $\overrightarrow{AB}$ and $\overrightarrow{BC}$ are parallel and hence deduce that the three points all lie on a straight line. (Such points are said to be *collinear*.)

6. Consider the displacement vector $2\boldsymbol{i} + 3\boldsymbol{j}$. On a diagram show this vector three times, once as a position vector and then twice as any other displacement.

11.4.3 Motion of objects

Newtonian mechanics is a subject which is concerned with the familiar everyday world of objects larger than atoms, which move at speeds less than a few million metres per second. To study the motion of such objects we need to define the position, velocity and acceleration of the object at any instant of time. In one dimension, we represent the position of the object at any time t by one variable $x(t)$ and the velocity and acceleration of the object can be written in terms of x by $v = \mathrm{d}x/\mathrm{d}t$ and $a = \mathrm{d}^2x/\mathrm{d}t^2$. In two and three dimensions, the use of vectors is a powerful tool providing a very general formulation that applies in all coordinate systems.

Example 11.4.1

The position vector of a particle is given by $\boldsymbol{r} = \boldsymbol{b} + \boldsymbol{u}t$ where $\boldsymbol{b} = \boldsymbol{i} + \boldsymbol{j}$ and $\boldsymbol{u} = 3\boldsymbol{i} + 4\boldsymbol{j}$.

(a) Plot the position of the particle at times $t = 0$, 1, 2, 3 and 4.
(b) What can be said about the path and the speed of the particle?

Solution

(a) See Figure 11.29.

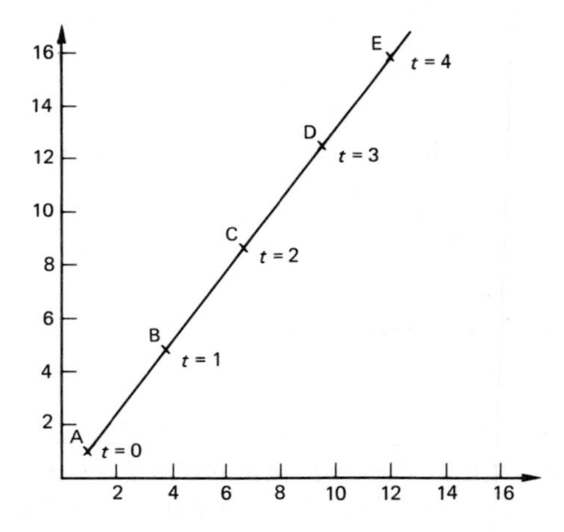

Figure 11.29

(b) We can see from Figure 11.29 that the path of the particle is a straight line.

The speed of the particle can be deduced from the distance travelled in each time interval of 1 second. We have

$$\frac{\text{distance AB}}{\text{1 second}} = \frac{5}{1} = 5 \text{ ms}^{-1}$$

$$\frac{\text{distance BC}}{\text{1 second}} = \frac{5}{1} = 5 \text{ ms}^{-1}$$

and so on. The speed of the particle is constant at 5 ms^{-1} which is the magnitude of the vector $\boldsymbol{u}$.

This example gives a clue that we can find the velocity of an object in terms of its position vector $\boldsymbol{r}$. In one dimension the velocity $v(t)$ is just the rate of change of $x(t)$. If we generalize this to two (and three) dimensions we will have

$$\text{velocity vector} \quad \boldsymbol{v} = \frac{d\boldsymbol{r}}{dt} = \frac{dx}{dt}\boldsymbol{i} + \frac{dy}{dt}\boldsymbol{j}$$

$$\text{acceleration vector } \boldsymbol{a} = \frac{d\boldsymbol{v}}{dt} = \frac{d^2\boldsymbol{r}}{dt^2} = \frac{d^2x}{dt^2}\boldsymbol{i} + \frac{d^2y}{dt^2}\boldsymbol{j}$$

To find the velocity and acceleration vectors from $\boldsymbol{r}$ requires differentiation; however if we start with the acceleration vector (which is often the case) then integration gives the velocity and position vectors.

If we apply these formulas to the position vector in Example 11.4.1 we have

$$\boldsymbol{r} = \boldsymbol{b} + \boldsymbol{u}t$$

$$\boldsymbol{v} = \frac{d\boldsymbol{r}}{dt} = \boldsymbol{0} + \boldsymbol{u} \text{ since } \boldsymbol{b} \text{ and } \boldsymbol{u} \text{ are constant vectors}$$

$$\boldsymbol{a} = \frac{d\boldsymbol{v}}{dt} = \boldsymbol{0} \qquad \text{since } \boldsymbol{u} \text{ is a constant vector}$$

So the particle travels with constant velocity $\boldsymbol{u} = 3\boldsymbol{i} + 4\boldsymbol{j}$, i.e. a velocity with magnitude $|\boldsymbol{u}| = \sqrt{3^2 + 4^2} = 5$. Note the *magnitude of velocity* is called *speed*; velocity is a vector quantity with magnitude and direction whereas speed is a scalar quantity.

Exercise 11.4.2

1. In this exercise, the position vector of a particle varies with time. In each case, find the velocity and acceleration vectors at time $t = 0$ and $t = 1$.

(a) $\boldsymbol{r}(t) = (t^3 + 3t)\boldsymbol{i} + 4t^2\boldsymbol{j}$
(b) $\boldsymbol{r}(t) = 10t\boldsymbol{i} + (2 + 10t - 4.9t^2)\boldsymbol{j}$
(c) $\boldsymbol{r}(t) = u\boldsymbol{i} + (h + wt - gt^2/2)\boldsymbol{j}$

where u, h, w and g are constants. In each case what is the speed of the particle at time $t = 0$?

2. A particle's acceleration is given by $\boldsymbol{a} = 2t\boldsymbol{i} - 3\boldsymbol{j}$, find the expressions for the velocity and position vectors if the particle starts at the origin with velocity $2\boldsymbol{j}$.
3. A particle moves in a circular path with position vector $\boldsymbol{r} = d\cos(wt)\boldsymbol{i} + d\sin(wt)\boldsymbol{j}$. Find the velocity and acceleration vectors $\boldsymbol{v}$ and $\boldsymbol{a}$. Show that $\boldsymbol{v}$ is perpendicular to $\boldsymbol{r}$ and that $\boldsymbol{a}$ is parallel to $\boldsymbol{r}$.

OBJECTIVE TEST 11

1. The following is a list of some physical quantities. Decide which are scalar and which are vector quantities:

   area, temperature, velocity, energy, force, displacement, time, acceleration.

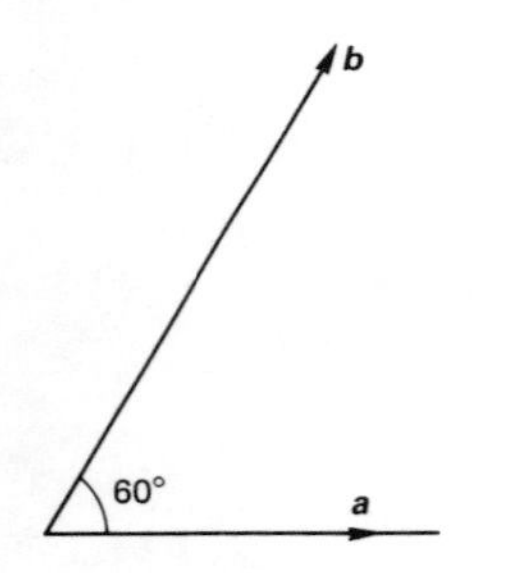

Figure 11.30

2. The vectors $\boldsymbol{a}$ and $\boldsymbol{b}$ are represented by the arrows shown in Figure 11.30. The magnitude of $\boldsymbol{a}$ and $\boldsymbol{b}$ are 3 and 5 units respectively. Draw a diagram to scale to show the vectors $\boldsymbol{a} + \boldsymbol{b}$, $\boldsymbol{a} - \boldsymbol{b}$, $\boldsymbol{a} + 2\boldsymbol{b}$ and $2\boldsymbol{a} + \frac{1}{2}\boldsymbol{b}$.
3. Write the vectors $\boldsymbol{a}$ and $\boldsymbol{b}$ in Exercise 2 in Cartesian component form using the column and unit vector notations. Find the component form of $\boldsymbol{a} + \boldsymbol{b}$, $\boldsymbol{a} - \boldsymbol{b}$ and $2\boldsymbol{a} - 3\boldsymbol{b}$. Calculate the magnitude and direction of these three vectors.

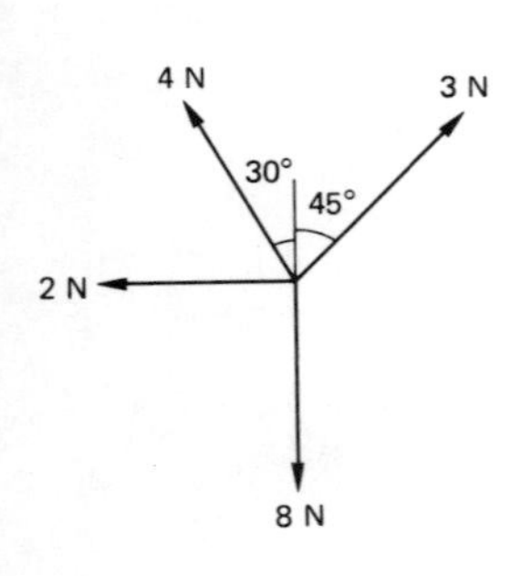

Figure 11.31

4. An object suspended by two strings experiences four forces, as shown in Figure 11.31. Write each vector in Cartesian component form and find the resultant force on the object. Hence find the magnitude and direction of this resultant force.
5. A ship heads on the bearing 220° and has a speed of 25 knots in still water, sails in a sea with a current speed 10 knots from due south. Find the actual speed and bearing of the ship.
6. A walker on Dartmoor walks from base camp for 16 km on a bearing 060° and then 12 km on a bearing 190°. How far and on what bearing is the walker from base camp?

7. The velocity vector of a particle is given by

$$\boldsymbol{v}(t) = (3t^2 + 1)\boldsymbol{i} - (4t + 2)\boldsymbol{j}$$

Find the acceleration and position vectors if the particle starts at the origin.

12 MATRICES

OBJECTIVES

When you have finished this chapter you should be able to

1. add and subtract matrices
2. multiply matrices
3. state that in general $AB \neq BA$
4. write down the 2×2 and 3×3 identity matrices
5. rewrite simultaneous equations in matrix form
6. calculate the determinant of a 2×2 and 3×3 matrix
7. calculate the inverse of a 2×2 and 3×3 matrix when the inverse exists
8. solve simultaneous equations by inverse matrix methods
9. solve simultaneous equations by elimination
10. understand why simultaneous equations do not always have a unique solution
11. find when simultaneous equations have no solutions
12. find when simultaneous equations have infinitely many solutions, and find such solutions

12.1 INTRODUCTION

In mathematics it is often necessary to solve simultaneous equations. For example, the electric currents I_1 and I_2 flowing in two branches of a circuit may be given by

$$\left.\begin{aligned} 5I_1 - 2I_2 &= 6 \\ -2I_1 + 6I_2 &= 0 \end{aligned}\right\} \tag{12.1.1}$$

The values that we obtain for the currents I_1 and I_2 depend on the values of the coefficients, i.e. they depend on

$$\begin{bmatrix} 5 & -2 \\ -2 & 6 \end{bmatrix} \text{ and } \begin{bmatrix} 6 \\ 0 \end{bmatrix}$$

If we change either of these tables of numbers, the solutions for I_1 and I_2 will change. The table

$$\begin{bmatrix} 5 & -2 \\ -2 & 6 \end{bmatrix}$$

is called a 2 × 2 matrix and the table

$$\begin{bmatrix} 6 \\ 0 \end{bmatrix}$$

is called a 2 × 1 matrix, or column vector.

The simultaneous equations

$$\left.\begin{aligned} 5x - 2y &= 6 \\ -2x + 6y &= 0 \end{aligned}\right\} \qquad (12.1.2)$$

have the same matrix and right hand side column vector as system (12.1.1). So the solutions for x and y are the same as the solutions for I_1 and I_2.

However the simultaneous equations

$$\left.\begin{aligned} 4x - 2y &= 6 \\ -2x + 6y &= 0 \end{aligned}\right\} \qquad (12.1.3)$$

have matrix

$$\begin{bmatrix} 4 & -2 \\ -2 & 6 \end{bmatrix}$$

and column vector

$$\begin{bmatrix} 6 \\ 0 \end{bmatrix}$$

The matrix for equation (12.1.3) is different from that of equations (12.1.1) or (12.1.2), and the solutions will be different as well.

12.2 ADDITION AND SUBTRACTION OF MATRICES

A matrix is a rectangular array or table of numbers, for example

a 2 × 2 matrix $\begin{bmatrix} 1 & -1 \\ 2 & 1 \end{bmatrix}$ ← 2 rows

↑ 2 columns

a 2 × 3 matrix $\begin{bmatrix} 4 & -2 & -2 \\ 8 & -1 & 0 \end{bmatrix}$ 2 rows

3 columns

A matrix with m rows and n columns is referred to as an m by n matrix. The notation for the entries of an m by n matrix is

$$A = \begin{bmatrix} a_{11} & a_{12} & \cdots & a_{1n} \\ a_{21} & a_{22} & \cdots & a_{2n} \\ \cdot & \cdot & & \cdot \\ \cdot & \cdot & & \cdot \\ \cdot & \cdot & & \cdot \\ a_{m1} & a_{m2} & \cdots & a_{mn} \end{bmatrix} \begin{matrix} \leftarrow \text{row } 1 \\ \leftarrow \text{row } 2 \\ \cdot \\ \cdot \\ \cdot \\ \leftarrow \text{row } m \end{matrix}$$

↓ col 1, ↓ col 2, ↓ col n

The entries a_{11}, a_{12}, . . . are read 'a one one', 'a one two', . . . The notation is slightly different from that used in computer science where a_{11} is written $A(1, 1)$.

A matrix A and a matrix B are equal if an only if (1) A and B both have the same number of rows (say m rows) and the same number of columns (say n columns) and (2) all their corresponding entries are equal, that is $a_{ij} = b_{ij}$ for $i = 1$ to m and $j = 1$ to n.

An $m \times 1$ matrix

$$\begin{bmatrix} a_{11} \\ a_{21} \\ \cdot \\ \cdot \\ \cdot \\ a_{m1} \end{bmatrix}$$

is a column vector with m rows and a $1 \times n$ matrix

$$a_{11} \quad a_{12} \quad \ldots \quad a_{1n}$$

is a row vector with n columns.

12.2.1 Addition of matrices

Provided the matrices have the same number of rows and the same number of columns the following rule applies:

To add matrices, add the corresponding entries

Example 12.2.1

$$\begin{bmatrix} 4 & -1 & 7 \\ 0 & 1 & 2 \end{bmatrix} + \begin{bmatrix} 1 & 2 & 0 \\ 5 & -1 & 2 \end{bmatrix} = \begin{bmatrix} 5 & 1 & 7 \\ 5 & 0 & 4 \end{bmatrix}$$

$$\begin{bmatrix} a_{11} & a_{12} \\ a_{21} & a_{22} \end{bmatrix} + \begin{bmatrix} b_{11} & b_{12} \\ b_{21} & b_{22} \end{bmatrix} = \begin{bmatrix} (a_{11} + b_{11}) & (a_{12} + b_{12}) \\ (a_{21} + b_{21}) & (a_{22} + b_{22}) \end{bmatrix}$$

12.2.2 Subtraction of matrices

Provided the matrices have the same number of rows and the same number of columns the following rule applies:

To subtract matrices, subtract the corresponding entries

Example 12.2.2

$$\begin{bmatrix} 4 & -1 & 7 \\ 0 & 1 & 2 \end{bmatrix} - \begin{bmatrix} 1 & 2 & 0 \\ 5 & -1 & 2 \end{bmatrix} = \begin{bmatrix} 3 & -3 & 7 \\ -5 & 2 & 0 \end{bmatrix}$$

$$\begin{bmatrix} a_{11} & a_{12} \\ a_{21} & a_{22} \end{bmatrix} - \begin{bmatrix} b_{11} & b_{12} \\ b_{21} & b_{22} \end{bmatrix} = \begin{bmatrix} (a_{11} - b_{11}) & (a_{12} - b_{12}) \\ (a_{21} - b_{21}) & (a_{22} - b_{22}) \end{bmatrix}$$

Exercise 12.2.1

Let

$$A = \begin{bmatrix} 4 & -1 \\ 5 & -2 \end{bmatrix}, B = \begin{bmatrix} -1 & 0 \\ 3 & 2 \end{bmatrix}, C = \begin{bmatrix} 0 & 1 \\ 3 & -1 \end{bmatrix}.$$

Find

(1) $A + B$, (2) $B + C$, (3) $A + B + C$, (4) $A - B$, (5) $B - A$, (6) $C - A$

12.3 PRODUCTS OF MATRICES

12.3.1 Product of a scalar and a matrix

> *Rule*
> To multiply a matrix A by a scalar c, multiply each entry in A by c

Example 12.3.1

$$2\begin{bmatrix} 0 & 3 & 2 \\ 1 & 2 & 1 \end{bmatrix} = \begin{bmatrix} 0 & 6 & 4 \\ 2 & 4 & 2 \end{bmatrix}$$

In practice this rule is often used ‘in reverse’ to factorize a common factor from every entry of a matrix.

Example 12.3.2

$$\begin{bmatrix} 12 & 36 \\ 24 & -18 \end{bmatrix} = 6\begin{bmatrix} 2 & 6 \\ 4 & -3 \end{bmatrix}$$

12.3.2 Multiplication of matrices

Let A and B be matrices. In order to calculate the product AB the matrix A must have the same number of columns as the matrix B has rows. If this holds, then the matrix AB is calculated by taking the products of every row of A with every column of B in a special way.

Example 12.3.3

Let

$$A = \begin{bmatrix} 2 & 1 \\ 3 & -2 \end{bmatrix} \text{ and } B = \begin{bmatrix} 0 & 4 & 5 \\ -1 & 3 & 7 \end{bmatrix}.$$

Since A has two columns and B has two rows, it is possible to form the product AB. Let $C = AB$. Then for example the entry c_{12} in the matrix C is the product of row 1 of A with column 2 of B:

$$\begin{bmatrix} \boxed{2 \quad 1} \\ 3 \quad -2 \end{bmatrix}\begin{bmatrix} 0 & \boxed{\begin{matrix} 4 \\ 3 \end{matrix}} & 5 \\ -1 & & 7 \end{bmatrix}$$

So $c_{12} = a_{11}b_{12} + a_{12}b_{22} = 2 \times 4 + 1 \times 3 = 11$. This is called the *scalar product* of a row and column. Therefore

$$C = AB = \begin{bmatrix} 2 & 1 \\ 3 & -2 \end{bmatrix} \begin{bmatrix} 0 & 4 & 5 \\ -1 & 3 & 7 \end{bmatrix} = \begin{bmatrix} -1 & 11 & 17 \\ 2 & 6 & 1 \end{bmatrix}$$

Rule

To multiply an $(m \times n)$ matrix A by an $(n \times p)$ matrix B

(1) The number of columns of A must equal the number of rows of B

(2) The entry c_{ij} in the product matrix $C = AB$ is:
c_{ij} = scalar product of row i in matrix A by column j in matrix B
$c_{ij} = a_{i1}b_{1j} + a_{i2}b_{2j} + \ldots + a_{in}b_{nj}$

Example 12.3.4

In general, $AB \neq BA$. That is multiplication of matrices is not commutative. For example,

$$\begin{bmatrix} 1 & 0 \\ 1 & 3 \end{bmatrix} \begin{bmatrix} 4 & 1 \\ 0 & -1 \end{bmatrix} = \begin{bmatrix} 4 & 1 \\ 4 & -2 \end{bmatrix}$$

$$\begin{bmatrix} 4 & -1 \\ 0 & -1 \end{bmatrix} \begin{bmatrix} 1 & 0 \\ 1 & 3 \end{bmatrix} = \begin{bmatrix} 5 & 3 \\ -1 & -3 \end{bmatrix}$$

12.3.3 The identity matrix I

The 2×2 identity matrix is

$$I = \begin{bmatrix} 1 & 0 \\ 0 & 1 \end{bmatrix}$$

It has the property that $IM = MI = M$ for any 2×2 matrix M.

Example 12.3.5

$$\begin{bmatrix} 1 & 0 \\ 0 & 1 \end{bmatrix} \begin{bmatrix} a & b \\ c & d \end{bmatrix} = \begin{bmatrix} a & b \\ c & d \end{bmatrix}$$

The 3×3 identity matrix is

$$I = \begin{bmatrix} 1 & 0 & 0 \\ 0 & 1 & 0 \\ 0 & 0 & 1 \end{bmatrix}$$

It has been the property that $IM = MI = M$ for any 3×3 matrix M.

In general the $n \times n$ identity matrix I has all entries on the main (top left to bottom right) diagonal equal to 1, and all other entries 0.

The identity matrix I plays the same role in the algebra of matrices as the number 1 plays in the algebra of numbers, since $1x = x1 = x$ for all numbers x, and $IM = MI = M$ for all $n \times n$ matrices M.

Exercise 12.3.1

Find the following matrices:

1. $3\begin{bmatrix} 1 \\ 2 \\ -1 \end{bmatrix}$ 2. $\frac{1}{10}\begin{bmatrix} 4 & -1 \\ -3 & -7 \end{bmatrix}$ 3. $c\begin{bmatrix} p & q \\ r & s \end{bmatrix}$

Factorize a common factor from the following matrices:

4. $\begin{bmatrix} -6 & 3 \\ 9 & -12 \end{bmatrix}$ 5. $\begin{bmatrix} \frac{1}{2} & \frac{1}{8} & -\frac{1}{4} \end{bmatrix}$

6. $\begin{bmatrix} 3x & x^2 \\ ax & -x^3 \\ 10x & 2x \end{bmatrix}$

Find the products of the following matrices:

7. $\begin{bmatrix} 9 & -1 \\ 8 & 4 \end{bmatrix}\begin{bmatrix} -2 \\ 1 \end{bmatrix}$

8. $\begin{bmatrix} 2 & 0 & 3 \\ -1 & 1 & 2 \\ 1 & 0 & 1 \end{bmatrix}\begin{bmatrix} 4 \\ -1 \\ 3 \end{bmatrix}$

9. $\begin{bmatrix} 3 & 8 \\ -1 & 2 \end{bmatrix}\begin{bmatrix} x \\ y \end{bmatrix}$

10. $\begin{bmatrix} a & b \end{bmatrix}\begin{bmatrix} 1 & 2 & 3 \\ 2 & 3 & 4 \end{bmatrix}$

11. $\begin{bmatrix} -0.3 & 0.2 \\ 0.4 & 0.5 \end{bmatrix} \begin{bmatrix} 2 & -1 \\ 3 & -4 \end{bmatrix}$

12. $10 \begin{bmatrix} -0.3 & 0.2 \\ 0.4 & 0.5 \end{bmatrix} \begin{bmatrix} 2 & -1 \\ 3 & -4 \end{bmatrix}$

13. $\begin{bmatrix} 1 & 0 & 0 \\ 0 & 1 & 0 \\ 0 & 0 & 1 \end{bmatrix} \begin{bmatrix} x \\ y \\ z \end{bmatrix}$

14. $\begin{bmatrix} -8 & 1 & 0 \\ 2 & 3 & 1 \\ 4 & 5 & -2 \end{bmatrix} \begin{bmatrix} 1 & 0 & 0 \\ 0 & 1 & 0 \\ 1 & 0 & 0 \end{bmatrix}$

15. If $A = \begin{bmatrix} a_{11} & a_{12} & a_{13} \\ a_{21} & a_{22} & a_{23} \\ a_{31} & a_{32} & a_{33} \end{bmatrix}$ and I is the 3×3 identity matrix, prove that $AI = IA = A$.

16. If $\begin{bmatrix} a & -1 \\ 3 & b \end{bmatrix} \begin{bmatrix} 1 & 2 \\ 0 & 3 \end{bmatrix} = \begin{bmatrix} 6 & 9 \\ 3 & 1 \end{bmatrix}$, find a and b.

17. If $\begin{bmatrix} 2 & 3 \\ m & 2 \end{bmatrix} \begin{bmatrix} x \\ 1 \end{bmatrix} = \begin{bmatrix} 13 \\ 12 \end{bmatrix}$, find m and x.

12.4 MATRICES AND SIMULTANEOUS EQUATIONS

Consider the problem of finding x and y in the matrix equation

$$\begin{bmatrix} 2 & -1 \\ 3 & 1 \end{bmatrix} \begin{bmatrix} x \\ y \end{bmatrix} = \begin{bmatrix} -5 \\ 0 \end{bmatrix}$$

Multiplying out the left hand side gives

$$\begin{bmatrix} 2x - y \\ 3x + y \end{bmatrix} = \begin{bmatrix} -5 \\ 0 \end{bmatrix} \qquad (12.4.1)$$

Using the rule for equality of vectors (Section 11.2) leads to two simultaneous equations in two unknowns,

$$\left.\begin{aligned} 2x - y &= -5 \\ 3x + y &= 0 \end{aligned}\right\} \qquad (12.4.2)$$

So the matrix system (12.4.1) is equivalent to the system of simultaneous equations (12.4.2). These equations may be solved to give $x = -1$ and $y = 3$.

Given a system of simultaneous equations it is also possible to convert them into matrix form.

Example 12.4.1

$$\begin{aligned} 2x + 5y &= 19 \\ 3x + y &= 9 \end{aligned}$$

is equivalent to the matrix equation

$$\begin{bmatrix} 2 & 5 \\ 3 & 1 \end{bmatrix} \begin{bmatrix} x \\ y \end{bmatrix} = \begin{bmatrix} 19 \\ 9 \end{bmatrix}$$

Note that the left hand side of the second equation is $3x + y$ which means $3x + 1y$. In the matrix this becomes 3 1.

Example 12.4.2

Convert

$$\begin{aligned} x - y + 2z &= 1 \\ 2x \qquad + z &= 1 \\ -x + 3y + 4z &= -3 \end{aligned}$$

to matrix form.

Solution

The second equation has the y term missing. It must be rewritten as

$$2x + 0y + 1z = 1.$$

The matrix equation is

$$\begin{bmatrix} 1 & -1 & 2 \\ 2 & 0 & 1 \\ -1 & 3 & 4 \end{bmatrix} \begin{bmatrix} x \\ y \\ z \end{bmatrix} = \begin{bmatrix} 1 \\ 1 \\ -3 \end{bmatrix}$$

Exercise 12.4.1

Write the following as matrix equations (but do not solve).

1. $3u + v = 6$
   $-u + 2v = 5$

2. $2p + q + r = 0$
   $q - r = -2$

2. $2x + y = 1$
   $y - x = 4$

12.5 DETERMINANTS

12.5.1 2 × 2 Determinants and Cramer's Rule

Consider the problem of solving two simultaneous equations in the two unknowns x and y.

$$ax + by = p \tag{12.5.1}$$

$$cx + dy = q \tag{12.5.2}$$

Multiply (12.5.1) by d

$$adx + dby = dp$$

Multiply (12.5.2) by b

$$bcx + bdy = bq$$

Subtract,

$$(ad - bc)x = dp - bq$$

So provided that $ad - bc \neq 0$

$$x = \frac{dp - bq}{ad - bc}$$

Similarly

$$y = \frac{aq - cp}{ad - bc}$$

The expression $ad - bc$ is called the *determinant* of the matrix

$$M = \begin{bmatrix} a & b \\ c & d \end{bmatrix}$$

which is the matrix associated with the two simultaneous equations (12.5.1) and (12.5.2). We write

$$\det M = \begin{vmatrix} a & b \\ c & d \end{vmatrix} = ad - bc$$

using straight bars instead of brackets to distinguish it from the matrix. The determinant of a matrix is a number (scalar).

The solutions to the simultaneous equations (12.5.1) and (12.5.2) may be expressed entirely in terms of determinants using Cramer's rule. Note that if we let

$$dp - bq = \begin{vmatrix} p & b \\ q & d \end{vmatrix} = D_x$$

and

$$aq - cp = \begin{vmatrix} a & p \\ c & q \end{vmatrix} = D_y$$

then D_x is the determinant obtained by replacing column 1 of matrix M by $\left[\begin{smallmatrix} p \\ q \end{smallmatrix}\right]$ and D_y is the determinant obtained by replacing column 2 of matrix M by $\left[\begin{smallmatrix} p \\ q \end{smallmatrix}\right]$.

Cramer's rule

The solution to the matrix equation

$$M\,\boldsymbol{x} = \boldsymbol{p}, \text{ or}$$

$$\begin{bmatrix} a & b \\ c & d \end{bmatrix} \begin{bmatrix} x \\ y \end{bmatrix} = \begin{bmatrix} p \\ q \end{bmatrix}$$

is given by

$$x = \frac{\begin{vmatrix} p & b \\ q & d \end{vmatrix}}{\det M} \text{ and } y = \frac{\begin{vmatrix} a & p \\ c & q \end{vmatrix}}{\det M}$$

provided that $\det M = ad - bc \neq 0$

Example 12.5.1

Use Cramer's rule to find the solutions to

$$\begin{aligned} 2x + y &= 1 \\ 4x - 3y &= 12 \end{aligned}$$

Solution

In matrix form the simultaneous equations are

$$\begin{bmatrix} 2 & 1 \\ 4 & -3 \end{bmatrix} \begin{bmatrix} x \\ y \end{bmatrix} = \begin{bmatrix} 1 \\ 12 \end{bmatrix}$$

or $M\boldsymbol{x} = \boldsymbol{p}$, where

$$M = \begin{bmatrix} 2 & 1 \\ 4 & -3 \end{bmatrix}, \boldsymbol{x} = \begin{bmatrix} \mathrm{x} \\ y \end{bmatrix} \text{ and } \boldsymbol{p} = \begin{bmatrix} 1 \\ 12 \end{bmatrix}$$

Therefore

$$\det M = \begin{vmatrix} 2 & 1 \\ 4 & -3 \end{vmatrix} = -6 - 4 = -10 \neq 0$$

so the system has a unique solution.

$$D_x = \begin{vmatrix} 1 & 1 \\ 12 & -3 \end{vmatrix} = -15 \text{ and } D_y = \begin{vmatrix} 2 & 1 \\ 4 & 12 \end{vmatrix} = 20$$

So

$$x = \frac{D_x}{\det M} = \frac{-15}{-10} = 1.5, \text{ and } y = \frac{D_y}{\det M} = \frac{20}{-10} = -2.0$$

Exercise 12.5.1

Evaluate the following determinants:

1. $\begin{vmatrix} 1 & 0 \\ 0 & 1 \end{vmatrix}$
2. $\begin{vmatrix} 2 & 3 \\ x & y \end{vmatrix}$
3. $\begin{vmatrix} \cos\theta & -\sin\theta \\ \sin\theta & \cos\theta \end{vmatrix}$
4. $\begin{vmatrix} 2 & 3 \\ 6 & 9 \end{vmatrix}$
5. $\begin{vmatrix} \sec\theta & \tan\theta \\ \tan\theta & \sec\theta \end{vmatrix}$

Write the following sets of simultaneous equations in matrix form. Find the determinant of the matrix. If this is non-zero solve the equations by Cramer's rule.

6. $x + y = 3$
   $2x + 5y = 6$

7. $-x + 3y = 7$
   $3x + y = 9$

8. $3x + y = 0$
   $2x + 2y = 5$

9. $4u - 5y = 1.8$
   $2u + 3y = 10.8$

10. $0.50s - 0.30t = 1.30$
    $1.25s - 0.75t = 0.25$

12.5.2 3 × 3 Determinants

3×3 determinants are most easily evaluated in terms of 2×2 determinants. Let M be the 3×3 matrix

$$M = \begin{bmatrix} m_{11} & m_{12} & m_{13} \\ m_{21} & m_{22} & m_{23} \\ m_{31} & m_{32} & m_{33} \end{bmatrix}$$

Then its determinant is written

$$\det M = \begin{vmatrix} m_{11} & m_{12} & m_{13} \\ m_{21} & m_{22} & m_{23} \\ m_{31} & m_{32} & m_{33} \end{vmatrix}$$

Associate the following sign pattern with the 3×3 determinant:

$$M = \begin{vmatrix} + & - & + \\ - & + & - \\ + & - & + \end{vmatrix}$$

This sign pattern is given by the rule: the sign associated with m_{ij} is $(-1)^{i+j}$. For example m_{12} is associated with $(-1)^{1+2} = (-1)^3 = -1$, or simply '$-$'.

12.5.3 Expanding det M along the first row

$$\det M = \begin{vmatrix} m_{11} & m_{12} & m_{13} \\ m_{21} & m_{22} & m_{23} \\ m_{31} & m_{32} & m_{33} \end{vmatrix}$$

$$= +m_{11}\begin{vmatrix} m_{22} & m_{23} \\ m_{32} & m_{33} \end{vmatrix} - m_{12}\begin{vmatrix} m_{21} & m_{23} \\ m_{31} & m_{33} \end{vmatrix}$$

$$+ m_{13}\begin{vmatrix} m_{21} & m_{22} \\ m_{31} & m_{32} \end{vmatrix}$$

There are two points to note about this expansion: (1) Each element from the first row of det M is multiplied by a 2×2 determinant, or minor. The minor of m_{11} is obtained by taking det M and deleting the row and column containing m_{11}. The minor for m_{11} is

$$\begin{vmatrix} \not{m}_{11} & \not{m}_{12} & \not{m}_{13} \\ \not{m}_{21} & m_{22} & m_{23} \\ \not{m}_{31} & m_{32} & m_{33} \end{vmatrix} = \begin{vmatrix} m_{22} & m_{23} \\ m_{32} & m_{33} \end{vmatrix}$$

The minors of m_{12} and m_{13} are obtained in a similar way. (2) Each element from the first row is associated with its sign from the sign pattern determinant (12.5.2).

Example 12.5.2

$$\begin{vmatrix} 2 & 1 & 2 \\ 1 & 0 & -1 \\ 4 & 3 & 7 \end{vmatrix} = +2\begin{vmatrix} 0 & -1 \\ 3 & 7 \end{vmatrix} -1\begin{vmatrix} 1 & -1 \\ 4 & 7 \end{vmatrix} +2\begin{vmatrix} 1 & 0 \\ 4 & 3 \end{vmatrix}$$

$$= 2(0 + 3) - 1(7 + 4) + 2(3 - 0) = 1$$

12.5.4 Expanding det M along other rows or columns

The determinant det M may be expanded along any other row or column using the sign pattern and appropriate 2×2 minors. For example, expanding along the second column,

$$\det M = \begin{vmatrix} m_{11} & m_{12} & m_{13} \\ m_{21} & m_{22} & m_{23} \\ m_{31} & m_{32} & m_{33} \end{vmatrix}$$

$$= -m_{12}\begin{vmatrix} m_{21} & m_{23} \\ m_{31} & m_{33} \end{vmatrix} + m_{22}\begin{vmatrix} m_{11} & m_{13} \\ m_{31} & m_{33} \end{vmatrix}$$

$$- m_{32}\begin{vmatrix} m_{11} & m_{13} \\ m_{21} & m_{23} \end{vmatrix}$$

Example 12.5.3

Find the value of the following determinant by expanding along the second column:

$$\begin{vmatrix} 2 & 1 & 2 \\ 1 & 0 & -1 \\ 4 & 3 & 7 \end{vmatrix}$$

Solution

$$\begin{vmatrix} 2 & 1 & 2 \\ 1 & 0 & -1 \\ 4 & 3 & 7 \end{vmatrix} = -1\begin{vmatrix} 1 & -1 \\ 4 & 7 \end{vmatrix} + 0\begin{vmatrix} 2 & 2 \\ 4 & 7 \end{vmatrix} - 3\begin{vmatrix} 2 & 2 \\ 1 & -1 \end{vmatrix}$$

$$= -1(7 + 4) + 0 - 3(-2 - 2) = 1$$

This is the same answer as before (Example 12.5.2), but less work was involved due to expanding along a column containing a zero.

Exercise 12.5.2

Evaluate the following determinants.

1. $\begin{vmatrix} -1 & 4 & 1 \\ 0 & 0 & 3 \\ 2 & 1 & 3 \end{vmatrix}$

2. $\begin{vmatrix} 1 & 0 & 0 \\ 0 & 1 & 0 \\ 0 & 0 & 1 \end{vmatrix}$

3. $\begin{vmatrix} 0.5 & 0.3 & 0 \\ 0.1 & -0.4 & 0.1 \\ 0.4 & 0.6 & 0 \end{vmatrix}$

4. $\begin{vmatrix} 3 & 0 & 2 \\ 0 & -1 & 1 \\ 5 & 1 & 1 \end{vmatrix}$

5. $\begin{vmatrix} 1 & 2 & 3 \\ 2 & 3 & 1 \\ 3 & 1 & 2 \end{vmatrix}$

6. $\begin{vmatrix} 1 & 0 & 0 \\ 0 & x & y \\ 0 & x^2 & y^2 \end{vmatrix}$

7. Find a so that $\begin{vmatrix} 3 & -4 & 5 \\ 1 & 0 & 11 \\ -1 & 2 & a \end{vmatrix} = 0$

12.5.5 Properties of 3 × 3 determinants

3 × 3 determinants have many properties which reduce the effort of computation. The most useful ones are summarized below. The rules below are stated for rows, but they are also valid for columns.

> *Rule (1)*: If a multiple of any row is added or subtracted from any other row the value of the determinant is unchanged

> The above rule also applies to columns

Example 12.5.4

$$\begin{vmatrix} -2 & 3 & -1 \\ 1 & -1 & 5 \\ 4 & 2 & 3 \end{vmatrix} = \begin{vmatrix} 0 & 1 & 9 \\ 1 & -1 & 5 \\ 4 & 2 & 3 \end{vmatrix} \quad \text{add } 2 \times \text{row } 2 \text{ to row } 1$$

$$= \begin{vmatrix} 0 & 1 & 9 \\ 1 & -1 & 5 \\ 0 & 6 & -17 \end{vmatrix} \quad \text{subtract } 4 \times \text{row } 2 \text{ from row } 3$$

$$= -1 \begin{vmatrix} 1 & 9 \\ 6 & -17 \end{vmatrix} \quad \text{expanding along column } 1$$

$$= 71$$

> *Rule (2)*: If one row of a determinant is multiplied by a constant c then the value of the determinant is multiplied by c.
> For example,
>
> $$\begin{vmatrix} m_{11} & m_{12} & m_{13} \\ cm_{21} & cm_{22} & cm_{23} \\ m_{31} & m_{32} & m_{33} \end{vmatrix} = c \begin{vmatrix} m_{11} & m_{12} & m_{13} \\ m_{21} & m_{22} & m_{23} \\ m_{31} & m_{32} & m_{33} \end{vmatrix}$$
>
> This rule is most useful for factorizing constants from rows.

> The above rule also applies to columns

Example 12.5.5

From Example 12.5.2

$$\begin{vmatrix} 2 & 1 & 2 \\ 1 & 0 & -1 \\ 4 & 3 & 7 \end{vmatrix} = 1$$

so

$$\begin{vmatrix} 2 & 1 & 2 \\ 3 & 0 & -3 \\ 4 & 3 & 7 \end{vmatrix} = 3 \begin{vmatrix} 2 & 1 & 2 \\ 1 & 0 & -1 \\ 4 & 3 & 7 \end{vmatrix} = 3 \times 1 = 3$$

Example 12.5.6

Rule (2) can be used repeatedly on rows or columns to remove common factors. For example,

$$\begin{vmatrix} 0.4 & -0.1 & 0.3 \\ -0.2 & 0.2 & -0.4 \\ 0.1 & 0.6 & -0.5 \end{vmatrix}$$

$$= \frac{1}{10} \begin{vmatrix} 4 & -1 & 3 \\ -0.2 & 0.2 & -0.4 \\ 0.1 & 0.6 & -0.5 \end{vmatrix} \quad \text{Rule (2) on row 1}$$

$$= \frac{1}{100} \begin{vmatrix} 4 & -1 & 3 \\ -2 & 2 & -4 \\ 0.1 & 0.6 & -0.5 \end{vmatrix} \quad \text{Rule (2) on row 2}$$

$$= \frac{1}{1000} \begin{vmatrix} 4 & -1 & 3 \\ -2 & 2 & -4 \\ 1 & 6 & -5 \end{vmatrix} \quad \text{Rule (2) on row 3}$$

$$= \frac{1}{1000} \begin{vmatrix} 4 & 3 & 3 \\ -2 & 0 & -4 \\ 1 & 7 & -5 \end{vmatrix} \quad \text{add col 1 to col 2}$$

$$= \frac{1}{1000} \begin{vmatrix} 4 & 3 & -5 \\ -2 & 0 & 0 \\ 1 & 7 & -7 \end{vmatrix} \quad \text{subtract } 2 \times \text{col 1 from col 3}$$

$$= \left(\frac{1}{1000}\right)(2)\begin{vmatrix} 3 & -5 \\ 7 & -7 \end{vmatrix} \qquad \text{expand along row 2}$$

$$= \frac{1}{500}(-21 + 35) = 0.028$$

Exercise 12.5.3

For Problems 1–3 evaluate the determinant.

1. $\begin{vmatrix} 4 & -11 & 0 \\ 2 & 1 & 0 \\ -1 & 3 & 0 \end{vmatrix}$ 2. $\begin{vmatrix} 2 & 4 & 3 \\ 2 & 1 & 5 \\ 2 & 4 & 3 \end{vmatrix}$ 3. $\begin{vmatrix} 4 & 1 & -2 \\ 4 & 6 & -1 \\ -4 & 3 & 2 \end{vmatrix}$

4. Prove that if any column of a 3×3 determinant det A consists entirely of zeros, then det $A = 0$.
5. Prove that if any two rows of a 3×3 determinant det A are the same, then det $A = 0$.
6. Prove that if one column of a 3×3 determinant det A is a multiple of another column, then det $A = 0$.

For Problems 7–10 evaluate the determinant.

7. $\begin{vmatrix} 30 & -40 & 20 \\ -6 & 9 & 9 \\ -2 & 8 & 6 \end{vmatrix}$ 8. $\begin{vmatrix} \frac{1}{2} & -\frac{1}{4} & \frac{3}{4} \\ \frac{2}{3} & 0 & -\frac{1}{3} \\ \frac{4}{5} & -\frac{1}{5} & -\frac{2}{5} \end{vmatrix}$

9. $\begin{vmatrix} a & b & c \\ a+b & b+c & c+a \\ b & c & a \end{vmatrix}$ 10. $\begin{vmatrix} 1 & 1 & 1 \\ x & y & z \\ x^2 & yz & xz \end{vmatrix}$

12.5.6 Cramer's rule for 3 × 3 systems

Consider the simultaneous equations

$$m_{11}x + m_{12}y + m_{13}z = c_1$$
$$m_{21}x + m_{22}y + m_{23}z = c_2$$
$$m_{31}x + m_{32}y + m_{33}z = c_3$$

or $\boldsymbol{Mx} = \boldsymbol{c}$ in matrix form where

$$M = \begin{bmatrix} m_{11} & m_{12} & m_{13} \\ m_{21} & m_{22} & m_{23} \\ m_{31} & m_{32} & m_{33} \end{bmatrix}, \quad \boldsymbol{x} = \begin{bmatrix} x \\ y \\ z \end{bmatrix} \quad \text{and } \boldsymbol{c} = \begin{bmatrix} c_1 \\ c_2 \\ c_3 \end{bmatrix}$$

Provided that $\det M \neq 0$, the solutions are given by

$$x = \frac{D_x}{\det M}, \; y = \frac{D_y}{\det M}, \; z = \frac{D_z}{\det M}$$

where

$$D_x = \begin{vmatrix} c_1 & m_{12} & m_{13} \\ c_2 & m_{22} & m_{23} \\ c_3 & m_{32} & m_{33} \end{vmatrix}$$

that is the determinant obtained from M by replacing the first (x) column by c.

Similarly

$$D_y = \begin{vmatrix} m_{11} & c_1 & m_{13} \\ m_{21} & c_2 & m_{23} \\ m_{31} & c_3 & m_{33} \end{vmatrix} \quad \text{and } D_z = \begin{vmatrix} m_{11} & m_{12} & c_1 \\ m_{21} & m_{22} & c_2 \\ m_{31} & m_{32} & c_3 \end{vmatrix}$$

Example 12.5.7

Solve

$$\begin{aligned} 2y + z &= -2 \\ 2x + 3y + 3z &= 5 \\ x + y - z &= 3 \end{aligned}$$

Solution

In matrix notation the simultaneous equations are $\boldsymbol{Mx} = \boldsymbol{c}$, where

$$M = \begin{bmatrix} 0 & 2 & 1 \\ 2 & 3 & 3 \\ 1 & 1 & -1 \end{bmatrix}, \; \boldsymbol{x} = \begin{bmatrix} x \\ y \\ z \end{bmatrix} \quad \text{and } \boldsymbol{c} = \begin{bmatrix} -2 \\ 5 \\ 3 \end{bmatrix}$$

So

$$\det M = \begin{vmatrix} 0 & 2 & 1 \\ 2 & 3 & 3 \\ 1 & 1 & -1 \end{vmatrix} = \begin{vmatrix} 0 & 2 & 1 \\ 0 & 1 & 5 \\ 1 & 1 & -1 \end{vmatrix} = 1 \begin{vmatrix} 2 & 1 \\ 1 & 5 \end{vmatrix} = 9$$

Since $\det M \neq 0$, the equations have a unique solution.

$$D_x = \begin{vmatrix} -2 & 2 & 1 \\ 5 & 3 & 3 \\ 3 & 1 & -1 \end{vmatrix} = \begin{vmatrix} -2 & 2 & 1 \\ 11 & -3 & 0 \\ 1 & 3 & 0 \end{vmatrix} = 1 \begin{vmatrix} 11 & -3 \\ 1 & 3 \end{vmatrix} = 36$$

$$D_y = \begin{vmatrix} 0 & -2 & 1 \\ 2 & 5 & 3 \\ 1 & 3 & -1 \end{vmatrix} = \begin{vmatrix} 0 & -2 & 1 \\ 2 & 5 & 3 \\ 1 & 1 & 0 \end{vmatrix} = -9$$

$$D_z = \begin{vmatrix} 0 & 2 & -2 \\ 2 & 3 & 3 \\ 1 & 1 & 3 \end{vmatrix} = \begin{vmatrix} 0 & 2 & -2 \\ 0 & 1 & -1 \\ 1 & 1 & 3 \end{vmatrix} = 0$$

So

$$x = \frac{D_x}{\det M} = \frac{36}{9} = 4, \ y = \frac{D_y}{\det M} = \frac{-9}{9} = -1,$$

and

$$z = \frac{D_z}{\det M} = 0$$

Exercise 12.5.4

Use Cramer's rule to solve

1. $$\begin{aligned} x + 5y - z &= 3.2 \\ 2x - y + 2z &= -0.1 \\ x - 2y - z &= -1.7 \end{aligned}$$

2. $$\begin{aligned} 2s + t + u &= 2 \\ -4s - 3t + u &= 3 \\ s \qquad + 2u &= 4.5 \end{aligned}$$

3. $$\begin{aligned} 2a \qquad + 3c &= -1 \\ a \qquad + c &= -1 \\ 3a + 2b + c &= -1 \end{aligned}$$

12.6 INVERSE MATRICES

12.6.1 The Inverse of a 2 × 2 matrix

The 2 × 2 identity matrix

$$I = \begin{bmatrix} 1 & 0 \\ 0 & 1 \end{bmatrix}$$

has the property that $IM = MI = M$ for any 2 × 2 matrix M. Recall that I has similar properties to 1 in the algebra of numbers.

Given any non-zero number x it is always possible to find a number x^{-1} such that $xx^{-1} = x^{-1}x = 1$. x^{-1} is called the inverse of x.

It is reasonable to ask if a similar situation exists for matrices. In fact the situation for matrices is more complicated; not every non-zero matrix has an inverse. However the following result holds.

Let M be a 2 × 2 matrix. Then provided $\det M \neq 0$, the matrix M has an inverse M^{-1} with the property that

$$MM^{-1} = M^{-1}M = \mathrm{I}$$

When an inverse exists it can be calculated using the following method.

Finding the inverse of a 2 × 2 matrix
Let $M =$

$$\begin{bmatrix} a & b \\ c & d \end{bmatrix}$$

be a 2 × 2 matrix
(1) Calculate $\det M$
(2) If $\det M = 0$, then M has no inverse
(3) If $\det M \neq 0$, then

$$M^{-1} = \frac{1}{\det M}\begin{bmatrix} d & -b \\ -c & a \end{bmatrix}$$

Proof

$$MM^{-1} = \begin{bmatrix} a & b \\ c & d \end{bmatrix} \frac{1}{\det M}\begin{bmatrix} d & -b \\ -c & a \end{bmatrix}$$

$$= \frac{1}{\det M} \begin{bmatrix} a & b \\ c & d \end{bmatrix} \begin{bmatrix} d & -b \\ -c & a \end{bmatrix}$$

$$= \frac{1}{\det M} \begin{bmatrix} ad - bc & 0 \\ 0 & -bc + ad \end{bmatrix}$$

$$= \frac{1}{\det M} \begin{bmatrix} \det M & 0 \\ 0 & \det M \end{bmatrix} = \begin{bmatrix} 1 & 0 \\ 0 & 1 \end{bmatrix}$$

Example 12.6.1

If

$$M = \begin{bmatrix} 2 & 3 \\ 2 & 4 \end{bmatrix},$$

then

$$\det M = \begin{vmatrix} 2 & 3 \\ 2 & 4 \end{vmatrix} = 2 \neq 0.$$

So

$$M^{-1} = \frac{1}{2} \begin{bmatrix} 4 & -3 \\ -2 & 2 \end{bmatrix}$$

12.6.2 The inverse of a 3 × 3 matrix

A 3 × 3 matrix has an inverse provided that its determinant does not equal zero. However the inverse of a 3 × 3 matrix is more complicated to compute than that for a 2 × 2 matrix. Before giving the formula for the inverse it is necessary to define some other matrices.

The transpose of a matrix

If

$$M = \begin{bmatrix} m_{11} & m_{12} & m_{13} \\ m_{21} & m_{22} & m_{23} \\ m_{31} & m_{32} & m_{33} \end{bmatrix}$$

then its *transpose* M^{T} is obtained from M by writing its rows as columns, or equivalently reflecting in the top left to bottom right diagonal. This gives $(m_{ij})^{\mathrm{T}} = m_{ji}$. So

$$M^{\mathrm{T}} = \begin{bmatrix} m_{11} & m_{21} & m_{31} \\ m_{12} & m_{22} & m_{32} \\ m_{13} & m_{23} & m_{33} \end{bmatrix}$$

For example

$$\begin{bmatrix} 1 & 2 & 8 \\ -1 & 3 & 5 \\ 7 & 0 & 4 \end{bmatrix}^{\mathrm{T}} = \begin{bmatrix} 1 & -1 & 7 \\ 2 & 3 & 0 \\ 8 & 5 & 4 \end{bmatrix}$$

12.6.4 Minors and cofactors

Each element of a matrix M has a number associated with it, called its *minor*. The minor of m_{ij} is written M_{ij}, and it is the 2×2 determinant obtained from $\det M$ by deleting row i and column j (see Section 12.5).

The cofactor of the entry m_{ij} in the matrix M is a number equal to $(-1)^{i+j}M_{ij}$.

Example 12.6.2

The matrix of cofactors of

$$\begin{bmatrix} 1 & 0 & 2 \\ 3 & -4 & 1 \\ -5 & -2 & 1 \end{bmatrix}$$

is obtained by replacing each element in the matrix by its cofactor. This gives

$$\begin{bmatrix} +\begin{vmatrix} -4 & 1 \\ -2 & 1 \end{vmatrix} & -\begin{vmatrix} 3 & 1 \\ -5 & 1 \end{vmatrix} & +\begin{vmatrix} 3 & -4 \\ -5 & -2 \end{vmatrix} \\ -\begin{vmatrix} 0 & 2 \\ -2 & 1 \end{vmatrix} & +\begin{vmatrix} 1 & 2 \\ -5 & 1 \end{vmatrix} & -\begin{vmatrix} 1 & 0 \\ -5 & -2 \end{vmatrix} \\ +\begin{vmatrix} 0 & 2 \\ -4 & 1 \end{vmatrix} & -\begin{vmatrix} 1 & 2 \\ 3 & 1 \end{vmatrix} & +\begin{vmatrix} 1 & 0 \\ 3 & -4 \end{vmatrix} \end{bmatrix}$$

$$= \begin{bmatrix} -2 & -8 & -26 \\ -4 & 11 & 2 \\ 8 & 5 & -4 \end{bmatrix}$$

12.6.5 The adjoint matrix

The *adjoint matrix* of M, Adj M, is the transposed matrix of cofactors of M.

Example 12.6.3

Find the adjoint of

$$\begin{bmatrix} 1 & 0 & 2 \\ 3 & -4 & 1 \\ -5 & -2 & 1 \end{bmatrix}$$

Solution

Using the last Example 12.5.2,

$$\text{Adj } M = \begin{bmatrix} -2 & -8 & -26 \\ -4 & 11 & 2 \\ 8 & 5 & -4 \end{bmatrix}^{\mathrm{T}} = \begin{bmatrix} -2 & -4 & 8 \\ -8 & 11 & 5 \\ -26 & 2 & -4 \end{bmatrix}$$

12.6.6 Finding the Inverse of a 3 × 3 matrix

Finding the inverse of a 3 × 3 matrix M
(1) Calculate det M.
(2) If det $M = 0$, then M has no inverse.
(3) If det $M \neq 0$, calculate Adj M, the transposed matrix of the cofactors of M.

(4) $M^{-1} = \dfrac{1}{\det M} \text{Adj } M$

Example 12.6.4

Find the inverse of

$$M = \begin{bmatrix} 1 & 2 & 2 \\ 2 & 10 & 5 \\ 1 & 3 & 3 \end{bmatrix}$$

Solution

$$\det M = \begin{vmatrix} 1 & 2 & 2 \\ 2 & 10 & 5 \\ 1 & 3 & 3 \end{vmatrix} = 5 \neq 0,$$

so M has an inverse.

$$\text{Adj } M = \begin{bmatrix} +\begin{vmatrix} 10 & 5 \\ 3 & 3 \end{vmatrix} & -\begin{vmatrix} 2 & 5 \\ 1 & 3 \end{vmatrix} & +\begin{vmatrix} 2 & 10 \\ 1 & 3 \end{vmatrix} \\ -\begin{vmatrix} 2 & 2 \\ 3 & 3 \end{vmatrix} & +\begin{vmatrix} 1 & 2 \\ 1 & 3 \end{vmatrix} & -\begin{vmatrix} 1 & 2 \\ 1 & 3 \end{vmatrix} \\ +\begin{vmatrix} 2 & 2 \\ 10 & 5 \end{vmatrix} & -\begin{vmatrix} 1 & 2 \\ 2 & 5 \end{vmatrix} & +\begin{vmatrix} 1 & 2 \\ 2 & 10 \end{vmatrix} \end{bmatrix}^{T}$$

$$= \begin{bmatrix} 15 & -1 & -4 \\ 0 & 1 & -1 \\ -10 & -1 & 6 \end{bmatrix}^{T} = \begin{bmatrix} 15 & 0 & -10 \\ -1 & 1 & -1 \\ -4 & -1 & 6 \end{bmatrix}$$

$$M^{-1} = \frac{1}{\det M} \text{Adj } M = \frac{1}{5} \begin{bmatrix} 15 & 0 & -10 \\ -1 & 1 & -1 \\ -4 & -1 & 6 \end{bmatrix}$$

Exercise 12.6.1

For each matrix, calculate its determinant. If this is non zero find the inverse of the matrix.

1. $\begin{bmatrix} -1 & 2 \\ 3 & 6 \end{bmatrix}$

2. $\begin{bmatrix} \sqrt{3} & 1 \\ -1 & \sqrt{3} \end{bmatrix}$

3. $\begin{bmatrix} 2 & -3 & 1 \\ 1 & 4 & 1 \\ 4 & 5 & 3 \end{bmatrix}$ 4. $\begin{bmatrix} 1 & 0 & -1 \\ 1 & 1 & 2 \\ 0 & 1 & 2 \end{bmatrix}$

5. $\begin{bmatrix} a & 0 & 0 \\ 0 & b & 0 \\ 0 & 0 & c \end{bmatrix}$ 6. $\begin{bmatrix} 1 & k & m \\ 0 & 1 & n \\ 0 & 0 & 1 \end{bmatrix}$

12.6.7 Solving simultaneous equations by using inverse matrices

The 2 × 2 system of simultaneous equations

$$m_{11}x + m_{12}y = c_1$$
$$m_{21}x + m_{22}y = c_2$$

may be written in matrix form as

$$M\boldsymbol{x} = \boldsymbol{c} \qquad (12.6.1)$$

where

$$M = \begin{bmatrix} m_{11} & m_{12} \\ m_{21} & m_{22} \end{bmatrix}, \ \boldsymbol{x} = \begin{bmatrix} x \\ y \end{bmatrix}, \text{ and } \boldsymbol{c} = \begin{bmatrix} c_1 \\ c_2 \end{bmatrix}$$

If M has an inverse M^{-1}, then multiplying both sides of (12.6.1) by M^{-1}

$$M^{-1}M\boldsymbol{x} = M^{-1}\boldsymbol{c}$$

so

$$I\boldsymbol{x} = M^{-1}\boldsymbol{c}$$

Therefore the solution is $\boldsymbol{x} = M^{-1}\boldsymbol{c}$.

Example 12.6.5

Solve

$$2x + y = 1$$
$$4x + 3y = 7$$

by matrix methods.

Solution

This system is equivalent to $\boldsymbol{Mx} = \boldsymbol{c}$, where

$$M = \begin{bmatrix} 2 & 1 \\ 4 & 3 \end{bmatrix}, \boldsymbol{x} = \begin{bmatrix} x \\ y \end{bmatrix}, \text{ and } \boldsymbol{c} = \begin{bmatrix} 1 \\ 7 \end{bmatrix}.$$

$$\det M = \begin{vmatrix} 2 & 1 \\ 4 & 3 \end{vmatrix} = 2, \text{ so } M^{-1} = \frac{1}{2}\begin{bmatrix} 3 & -1 \\ -4 & 2 \end{bmatrix}$$

So

$$\boldsymbol{x} = \begin{bmatrix} x \\ y \end{bmatrix} = \frac{1}{2}\begin{bmatrix} 3 & -1 \\ -4 & 2 \end{bmatrix}\begin{bmatrix} 1 \\ 7 \end{bmatrix} = \frac{1}{2}\begin{bmatrix} -4 \\ 10 \end{bmatrix} = \begin{bmatrix} -2 \\ 5 \end{bmatrix}$$

Hence $x = -2$ and $y = 5$.

These ideas are also applicable to 3×3 systems.

Example 12.6.6

Solve

$$\begin{aligned} x - 2y &= 1 \\ y + 3z &= 2 \\ x - z &= -2 \end{aligned}$$

Solution

In matrix form this is $\boldsymbol{Mx} = \boldsymbol{c}$, where

$$M = \begin{bmatrix} 1 & -2 & 0 \\ 0 & 1 & 3 \\ 1 & 0 & -1 \end{bmatrix}, \boldsymbol{x} = \begin{bmatrix} x \\ y \\ z \end{bmatrix}, \boldsymbol{c} = \begin{bmatrix} 1 \\ 2 \\ -2 \end{bmatrix}$$

$$\det M = -7, \text{ and Adj } M = \begin{bmatrix} -1 & 3 & -1 \\ -2 & -1 & -2 \\ -6 & -3 & 1 \end{bmatrix}^{T}$$

so

$$M^{-1} = \frac{1}{7}\begin{bmatrix} 1 & 2 & 6 \\ -3 & 1 & 3 \\ 1 & 2 & -1 \end{bmatrix}$$

$$\mathbf{x} = M^{-1}\mathbf{c} = \frac{1}{7}\begin{bmatrix} 1 & 2 & 6 \\ -3 & 1 & 3 \\ 1 & 2 & -1 \end{bmatrix}\begin{bmatrix} 1 \\ 2 \\ -2 \end{bmatrix} = \frac{1}{7}\begin{bmatrix} -7 \\ -7 \\ 7 \end{bmatrix}$$

Therefore $x = -1$, $y = -1$, $z = 1$.

Example 12.6.7

Solve

$$\begin{aligned} x - 2y \qquad &= 2 \\ y + 3z &= 9 \\ x \qquad - z &= -1 \end{aligned}$$

Solution

This system has the same matrix as Example 12.6.6, but the right hand side is different so

$$\mathbf{c} = \begin{bmatrix} 2 \\ 9 \\ -1 \end{bmatrix}$$

$$\mathbf{x} = M^{-1}\mathbf{c} = \frac{1}{7}\begin{bmatrix} 1 & 2 & 6 \\ -3 & 1 & 3 \\ 1 & 2 & -1 \end{bmatrix}\begin{bmatrix} 2 \\ 9 \\ -1 \end{bmatrix} = \frac{1}{7}\begin{bmatrix} 14 \\ 0 \\ 21 \end{bmatrix}$$

Therefore $x = 2$, $y = 0$, $z = 3$.

Exercise 12.6.2

Solve the following by matrix methods.

1. $x - 2y = -4$
   $5x + 2y = 16$
2. $s - 2t = -4$
   $5s + 2t = 16$
3. $x - 2y = 13$
   $5x + 2y = 17$
4. $-3x + 2y = 13$
   $3x + 4y = 17$
5. $6x + 2y - z = 5$
   $x + y - 2z = 4$
   $3x + 2y + z = 3$
6. $6p + 2q - r = 2.6$
   $p - q - 2r = -0.6$
   $3p + 2q + r = 2.1$

12.7 SOLVING SIMULTANEOUS EQUATIONS BY ELIMINATION

The system

$$2x + y = 1 \tag{1}$$

$$4x + 3y = 7 \tag{2}$$

can be solved by eliminating x from Equation (2) by subtracting twice Equation (1).

$$2x + y = 1 \qquad (1)' = (1)$$

$$y = 5 \qquad (2)' = (2) - 2(1)$$

Substituting $y = 5$ in (1) gives

$2x + 5 = 1$, so $x = -2$.

It is not really necessary to carry the variables x and y throughout this process of elimination. It is better to set the working out in a tabular format, together with a check sum for each row. The row check sum is the sum of all the coefficients in the row. This acts as a check on arithmetic, and helps to avoid errors in the elimination process. For example if

(new row 2) = (old row 2) $-$ 2 $\times$ (old row 1)

or

$(2)' = (2) - 2 \times (1)$

then the check sums should reflect this, so the check sum of $(2)'$ should equal check sum $((2) - 2 \times (1))$.

Equation	x	y	c	Row check sum
(1)	2	1	1	4
(2)	4	3	7	14
$(1)' = (1)$	2	1	1	4
$(2)' = (2) - 2(1)$	0	1	5	$6 = 14 - 2 \times 4$

From Equation $(2)'$, $y = 5$. Substituting in Equation (1), $2x + 5 = 1$, therefore $x = -2$.

Elimination is most useful for 3 × 3 systems. Given the system

$$m_{11}x + m_{12}y + m_{13}z = c_1 \quad (1)$$

$$m_{21}x + m_{22}y + m_{23}z = c_2 \quad (2)$$

$$m_{31}x + m_{32}y + m_{33}z = c_3 \quad (3)$$

the aim is to use elimination to reduce it to an upper triangular system of the form

$$n_{11}x + n_{12}y + n_{13}z = d_1 \quad (1)'$$

$$n_{22}y + n_{23}z = d_2 \quad (2)'$$

$$n_{33}z = d_3 \quad (3)'$$

The final coefficients $n_{11}, \ldots, n_{33}$ and d_1, d_2 and d_3 are found during the process of elimination. z can be found from Equation (3)′, then after substituting for z in Equation (2)′ y can be found. Finally substitution for y and z in Equation (1)′ gives the value of x.

In the process of elimination any of the following steps are allowed.

Valid steps in solving equations by elimination
(1) A multiple of any equation may be added to (or subtracted from) any other equation.
(2) Any two equations may be exchanged.

3 × 3 elimination is best carried out with a tabular layout.

Example 12.7.1

Solve

$$x - y + 2z = 8 \quad (1)$$

$$3x - y - z = 8 \quad (2)$$

$$2x + 3y + 5z = 13 \quad (3)$$

by elimination.

Solution

The equations are numbered using the following convention: 'b_1r_1' means 'block 1 row 1', 'b_2r_3' means 'block 2 row 3'.

Equation	x	y	z	c	Check sum
b_1r_1	1	−1	2	8	10
b_1r_2	3	−1	−1	8	9
b_1r_3	2	3	5	13	23
$b_2r_1 = b_1r_1$	1	−1	2	8	10
$b_2r_2 = b_1r_2 - 3b_1r_1$	0	2	−7	−16	$-21 = 9 - 3(10)$
$b_2r_3 = b_1r_3 - 2b_1r_1$	0	5	1	−3	$3 = 23 - 2(10)$
$b_3r_1 = b_2r_1$	1	−1	2	8	10
$b_3r_2 = b_2r_2$	0	2	−7	−16	−21
$b_3r_3 = b_2r_3 - \frac{5}{2}b_2r_2$	0	0	$\frac{37}{2}$	37	$\frac{111}{2} = 3 - \frac{5}{2}(-21)$

From Equation b_3r_3:

$$\frac{37}{2}z = 37, \text{ so } z = 2.$$

Substitute for z in b_3r_2: $2y = -16 + 7z = -2$, so $y = -1$.
Substitute for y and z in b_3r_1: $x = 8 + y - 2z = 3$.
So the solutions are $x = 3$, $y = -1$, $z = 2$.

12.7.1 Pivotal equations, pivots and multipliers

Pivotal equation

At each stage, one equation is used to eliminate a given variable from all the other equations in its block. This equation is called the *pivotal equation*. In Example 12.7.1, b_1r_1 is the pivotal equation for block 1. It is used to eliminate x from the other two equations to give the equations in block 2.

Pivot

In the pivotal equation, the coefficient of the variable to be eliminated from the other equations is called the *pivot*. In Example 12.7.1, b_1r_1 is the pivotal equation for block 1, and the coefficient of x in this equation is the pivot. So 1 is the pivot for block 1.

Multipliers

At each stage of the elimination process, multiplies of the pivotal equation are added to all other equations in the block. These are the *multipliers*. In Example 12.7.1, in block 1, -3 is the multiplier for b_1r_2 and -2 is the multiplier for b_1r_3.

Elimination and rounding errors

In Example 12.7.1, all the coefficients were integers and it was possible to work with exact arithmetic, so there were no rounding errors. In most cases this is not possible, and approximations for coefficients such as $2/3 \approx 0.667$ will lead to rounding errors, which must be minimized.

To minimize rounding errors the pivotal equation must be chosen by the rule:

> *Choice of pivotal equation to minimize rounding errors*
> Examine all the coefficients of x, y and z within a given block. The one which has the largest absolute value is selected as the pivot. The pivotal equation contains the pivot.

Note that the constants c_1, c_2 and c_3 are *not* considered in this process.

Example 12.7.2

Solve

$$x + 2y + 4z = 2.5 \qquad (1)$$

$$x - 4y - 2z = -3.0 \qquad (2)$$

$$-6x + 5y + 3z = 2.25 \qquad (3)$$

by elimination.

Solution

To save space the first column of the table will contain the pivotal row multipliers only.

At each stage the pivot is shown in bold type, and working is to four decimal places.

	x	y	z	c	Check sum
0.1667	1.0000	2.000	4.0000	2.5000	9.5000
0.1667	1.0000	−4.0000	−2.0000	−3.0000	−8.0000
pivot	**−6.0000**	5.0000	3.0000	2.2500	4.2500
pivot	0.0000	2.8335	**4.5001**	2.8751	10.2087 ≈ 10.2085
0.3333	0.0000	−3.1665	−1.4999	−2.6249	−7.2913 ≈ −7.2915
–	−6.0000	5.0000	3.0000	2.2500	–
	0.0000	2.8335	4.5001	2.8751	–
	0.0000	−2.2221	0.0000	−1.6666	−3.8887 = −3.8887
	−6.0000	5.0000	3.0000	2.2500	–

The multipliers are calculated by dividing the pivotal column coefficient by the pivot, and then multiplying by −1. For example the first multiplier is

$$\left(\frac{1.0000}{-6.0000}\right) \times (-1) \approx 0.1667,$$

and for the second equation in block two the multiplier is

$$\left(\frac{-1.4999}{4.5001}\right) \times (-1) \approx 0.3333.$$

To carry out the elimination:

1. Multiply the pivotal equation by the multiplier.
2. Add the result of stage (1) to the present equation.

The solution is found from the last block of equations, which may be arranged in upper triangular form as

$$-6.0000x + 3.0000z + 5.0000y = 2.2500 \quad (1)$$

$$4.5001z + 2.8335y = 2.8751 \quad (2)$$

$$-2.2221y = -1.6666 \quad (3)$$

From (3), $y = \dfrac{-1.6666}{-2.2221} \approx 0.7500$

Substituting in (2),

$$z = \frac{2.8751 - 2.8335 \times 0.7500}{4.5001} \approx 0.1667$$

Substituting for y in z in (1),

$$x = \frac{2.2500 - 5.0000 \times 0.7500 - 3.000 \times 0.1667}{-6.0000} \approx 0.3334$$

These solutions compare favourably with the exact solutions, $x = 1/3$, $y = 3/4$, $z = 1/6$.

Exercise 12.7.1

Solve by elimination.

1. $$\begin{aligned} 3x - 2y - z &= 4 \\ 2x + y + z &= 7 \\ -x + 2y + 4z &= 12 \end{aligned}$$

2. $$\begin{aligned} x + y + z &= 0 \\ 2x - y + z &= -1.1 \\ x + 3y + 2z &= 0.9 \end{aligned}$$

3. $$\begin{aligned} 0.7u + 1.5v + 2.0w &= 6.45 \\ 1.0u - 2.0v + 0.5w &= 1.25 \\ 1.2u + 0.3v + 0.6w &= 2.85 \end{aligned}$$

4. $$\begin{aligned} -2p + 3q + 4r &= -7 \\ p + q + 2r &= 6 \\ 2p + 3q + 3r &= 13 \end{aligned}$$

12.8 SYSTEMS OF SIMULTANEOUS EQUATIONS WITH NO UNIQUE SOLUTIONS

Until this section all the simultaneous equations considered have had unique solutions. This is not always so. Sometimes there are infinitely many solutions, and sometimes there are no solutions at all.

Example 12.8.1

$$x + 2y = 5 \qquad (1)$$

$$3x + 6y = 15 \qquad (2)$$

The second equation is three times the first, so they are not independent. After elimination,

$$x + 2y = 5 \qquad (1)'$$

$$0x + 0y = 0 \qquad (2)' = (2) - 3(1)$$

Equation (2)′ states that $0 = 0$, which is true, but it is not very informative. Essentially there is only one equation

$$x + 2y = 5 \qquad (1)$$

which has an infinite number of solutions.

For example if $x = 3$, $2y = 5 - x = 5 - 3 = 2$ and if $y = 1$, $x = 5 - 2y = 5 - 6 = -1$. If $x = \lambda$, where λ is any real number, Equation (1) gives

$$y = \frac{5 - \lambda}{2}$$

so

$$x = \lambda,\ y = \frac{5 - \lambda}{2},\ \lambda \in \mathbb{R} \qquad \text{(GS1)}$$

The expression '$\lambda \in \mathbb{R}$' means that λ is any real number. This is called the general solution (GS) to the equations. By choosing any real value for λ, an infinite number of solutions are obtained, for example if $\lambda = 10$, $x = 10$ and $y = -2.5$. Alternatively, by taking $y = \mu$, where μ is a real number, Equation (1) gives the general solution

$$x = 5 - 2\mu,\ y = \mu,\ \mu \in \mathbb{R} \qquad \text{(GS2)}$$

By choosing any value for μ an infinite number of solutions are obtained. Although the general solutions (GS1) and (GS2) appear to be different, they yield exactly the same infinite set of solutions. For example,

$$\left\{\begin{matrix} \text{if } \lambda = 3, \text{ then } x = 3,\ y = 1 \\ \text{if } \mu = 1, \text{ then } x = 3,\ y = 1 \end{matrix}\right\}$$

Note that the determinant of the system is

$$\begin{vmatrix} 1 & 2 \\ 3 & 6 \end{vmatrix} = 0$$

Simultaneous equations with an infinite number of solutions always have a zero determinant.

Example 12.8.2

$$x + 2y = 5 \qquad (1)$$

$$3x + 6y = 10 \qquad (2)$$

This is the same as Example 12.7.1 except for the right hand side of Equation (2). This change makes the equations inconsistent. After elimination,

$$x + 2y = 5 \qquad (1)'$$

$$0x + 0y = -5 \qquad (2)' = (2) - 3(1)$$

Equation (2)′ states that $0 = -5$, which is a contradiction. So the original equations contradict each other, and there are no solutions. Note that the determinant of the system is

$$\begin{vmatrix} 1 & 2 \\ 3 & 6 \end{vmatrix} = 0$$

It is a general property that simultaneous equations with no solutions have a zero determinant. A similar situation arises for 3×3 systems of simultaneous equations.

> *The types of solutions to simultaneous equations*
> If there are n equations in n unknowns, then the system may be written in matrix form as
>
> $$\boldsymbol{Mx} = \boldsymbol{c}$$
>
> The equations have a unique solution if and only if $\det M \neq 0$.
> In cases where $\det M = 0$, the equations either (1) have an infinite number of solutions, or (2) have no solutions at all.
> Which of these cases hold may be found by elimination.

Systems for which $\det M = 0$ are called *singular*. Systems for which $\det M \neq 0$ are called *non-singular*.

Example 12.8.3

By using elimination investigate the solutions of

$$x - 4y + 8z = 30$$

$$2x + y + z = 9$$

$$x + 2y - 2z = -4$$

Solution

Since exact arithmetic will be used in this example, it is not necessary to choose the pivot to minimize rounding errors.

	Multiplier	x	y	z	c	Check sum
b_1r_1	pivot	1	−4	8	30	35
b_1r_2	−2	2	1	1	9	13
b_1r_3	−1	1	2	−2	−4	−3
b_2r_1	–	1	−4	8	30	35
b_2r_2	pivot	0	9	−15	−51	$-57 = 13 - 2 \times 35$
b_2r_3	−2/3	0	6	−10	−34	$-38 = -3 - 35$
b_3r_1		1	−4	8	30	35
b_3r_2		0	9	−15	−51	−57
b_3r_3		0	0	0	0	$0 = -38 - 2/3(-57)$

Equation b_3r_3 in the third block says $0 = 0$, which is true, but provides no useful information. However the system is consistent. There are two independent equations in three unknowns

$$x - 4y + 8z = 30 \qquad b_3r_1$$

$$3y - 5z = -17 \qquad b_3r_2 \div 3$$

taking $z = \lambda$, where $\lambda \in \mathbb{R}$, and using last equation

$$y = \frac{5\lambda - 17}{3}$$

Substituting these values into the first equation gives

$$x = \frac{22 - 4\lambda}{3}$$

So there is an infinite number of solutions given by

$$x = \frac{22 - 4\lambda}{3}, \; y = \frac{5\lambda - 17}{3}, \; z = \lambda, \; \lambda \in \mathbb{R}$$

To find particular solutions, choose values for λ. For example if $\lambda = 4$, $x = 2$, $y = 1$, $z = 4$.

Example 12.8.4

Investigate the solutions of

$$\begin{aligned} x - 2y + 2z &= 3 \\ 3x - y + z &= 2 \\ 2x + y - z &= 5 \end{aligned}$$

Solution

Again exact arithmetic will be used.

	Multiplier	x	y	z	c	Check sum
b_1r_1	pivot	1	−2	2	3	4
b_1r_2	−3	3	−1	1	2	5
b_1r_3	−2	2	1	−1	5	7
b_2r_1	–	1	−2	2	3	4
b_2r_2	pivot	0	5	−5	−7	−7 = 5 − 3 × 4
b_2r_3	−1	0	5	−5	−1	−1 = 7 − 2 × 4
b_3r_1		1	−2	2	3	4
b_3r_2		0	5	−5	−7	−7
b_3r_3		0	0	0	6	6 = −1 + 7

Equation b_3r_3 states that $0x + 0y + 0z = 6$, that is $0 = 6$, which is false. So the system of equations is inconsistent, and has no solutions.

Exercise 12.8.1

Solve the following simultaneous equations where possible.

1. $2s + 6t = 1$, $s + 3t = 3$
2. $6u - v = 3$, $12u - 2v = 6$
3. $x - 8y + 5z = 6$, $x + 2y - z = 0$, $3x - 4y + 3z = 7$
4. $2x + y + z = 1$, $4x - y - 2z = 3$, $2x - 2y - 3z = 2$
5. $u + v + w = 5$, $u + 2v - 2w = 2$, $u + 6v - 14w = -10$

12.9 APPLICATIONS TO ELECTRICAL NETWORKS

To determine the voltages and currents in an electric network the following rules are required. Using these rules the systems of simultaneous equations which describe the network can be set up and solved.

12.9.1 Ohm's law

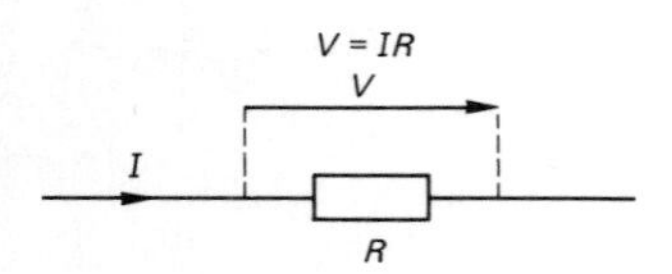

Figure 12.1

The voltage drop V (volts) across a resistor of resistance R (ohms) when a current of I (amperes) flows (Figure 12.1) is given by

$$V = IR$$

12.9.2 Kirchhoff's first law

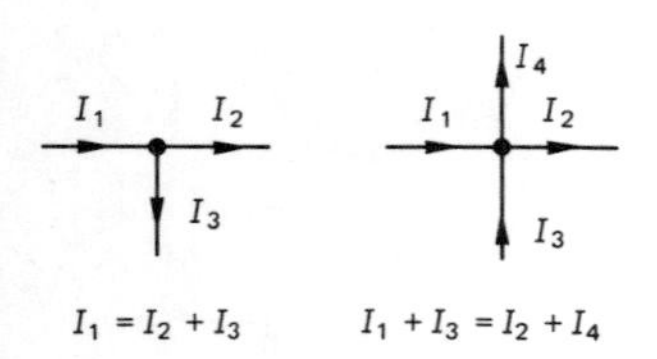

Figure 12.2

The sum of currents entering a junction = The sum of currents leaving the junction (see Figure 12.2)

12.9.3 Kirchhoff's second law

In any closed loop:
applied voltage = sum of voltage drops

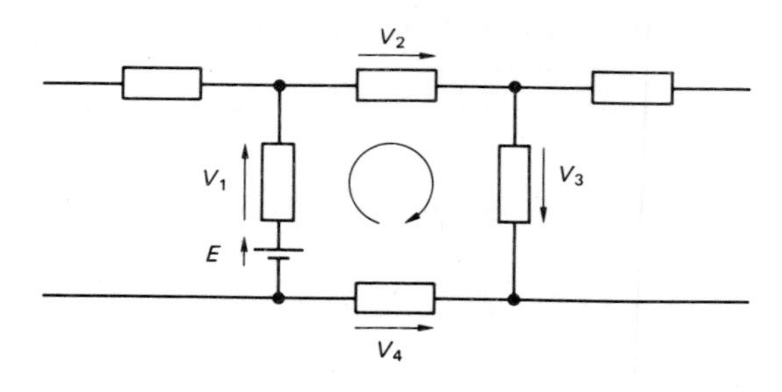

Figure 12.3

In Figure 12.3, going round the loop in a clockwise direction $E = V_1 + V_2 + V_3 - V_4$.

12.9.4 Mesh currents or loop currents

The analysis of networks can be greatly simplified by using mesh or loop currents instead of branch currents. Their use reduces the number of equations to be solved. The branch currents are easily computed from the loop currents (Figure 12.4). The loop currents are I_1, I_2 and I_3. The branch currents are shown in the branches.

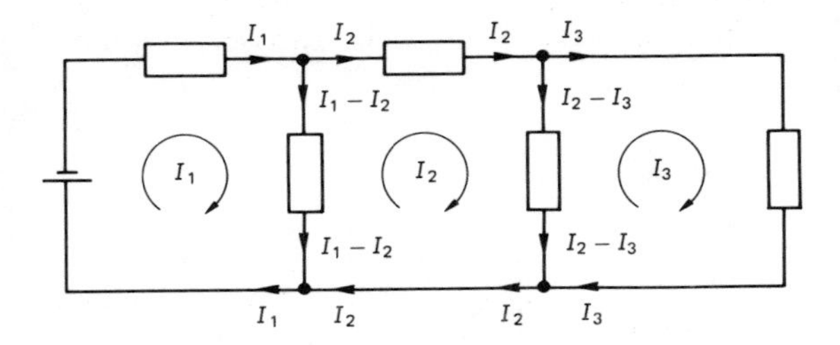

Figure 12.4

Example 12.9.1

Determine the currents flowing in each branch of the circuit of Figure 12.5. Also determine the voltage drop across each resistor.

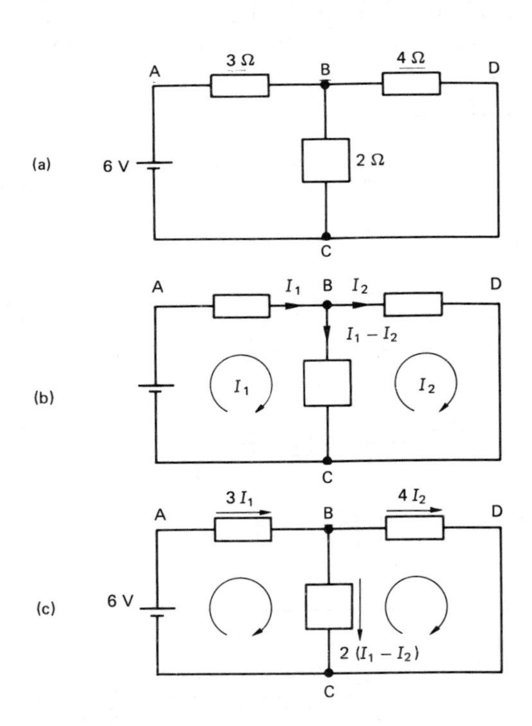

Figure 12.5 (a) The circuit, (b) loop and branch currents, (c) Voltage drops across resistors (using Ohm's Law).

Using Kirchhoff's second law in the third diagram of Figure 12.5

loop ABC $\quad 6 = 3I_1 + 2(I_1 - I_2)$

loop BDC $\quad -2(I_1 - I_2) + 4I_2 = 0$

So
$$5I_1 - 2I_2 = 6$$
$$-2I_1 + 6I_2 = 0$$

Solving these equations gives

$I_1 = 1.39$ A and $I_2 = 0.46$ A

	AB	BC	BD
Current (A)	1.39	0.92	0.46
Voltage (V)	4.15	1.85	1.85

Exercise 12.9.1

For the following circuits determine the currents flowing in each branch of the circuit, and the voltage drop across each resistor.

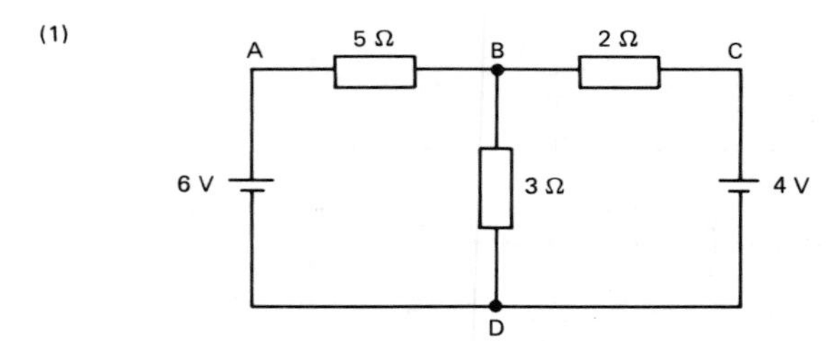

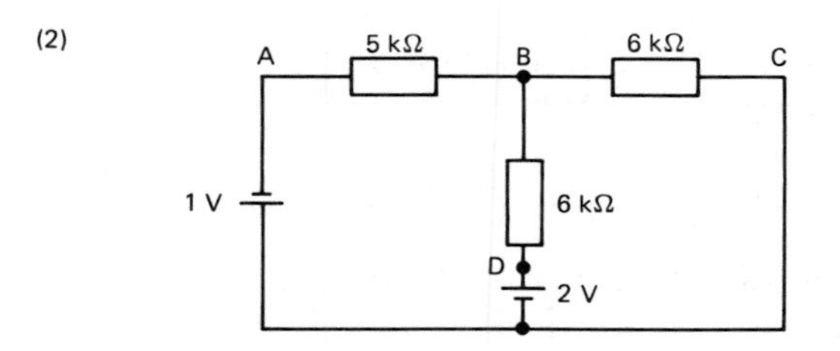

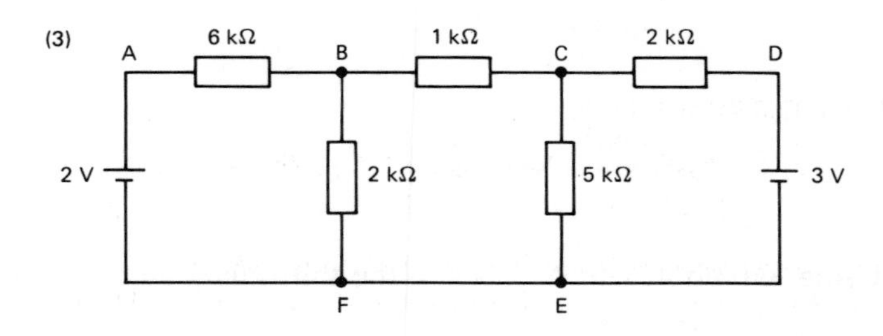

(4)

(5)

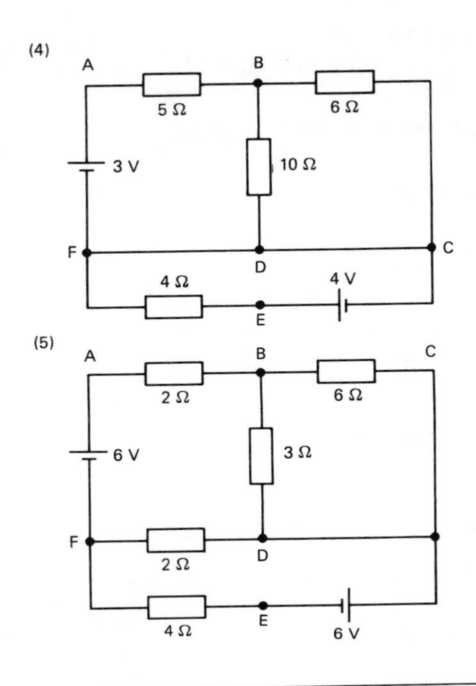

OBJECTIVE TEST 12

1. If $A = \begin{bmatrix} 0.1 & 0.4 \\ 0.3 & -0.5 \end{bmatrix}$ and $B = \begin{bmatrix} -0.7 & 0.8 \\ 0.5 & -0.1 \end{bmatrix}$

   find (a) $A + B$ (b) $A - 2B$ (c) AB (d) BA.
2. What is the 3×3 identity matrix I?
3. If $M = \begin{bmatrix} 0 & -1 & 7 \\ -4 & 6 & 8 \end{bmatrix}$ find MI.
4. Write the simultaneous equations

$$\begin{aligned} 8p - 2q &= 4 \\ -3p + q &= 9 \end{aligned}$$

   in matrix form.
5. Solve Problem 4 using Cramer's rule.
6. Find $\begin{vmatrix} 6 & -1 & 2 \\ 0 & 1 & 5 \\ 3 & 0 & 1 \end{vmatrix}$.

7. If $M = \begin{bmatrix} 0.2 & -0.3 \\ 0.6 & 0.1 \end{bmatrix}$ find det M and M^{-1}.

8. Use your result from Problem 7 to solve

$$\begin{aligned} 0.2x - 0.3y &= -3 \\ 0.6x + 0.1y &= 2 \end{aligned}$$

9. Use elimination to solve

$$\begin{aligned} 6x \qquad + 7z &= 10 \\ -x + 2y - z &= -1 \\ x - 4y \qquad &= -3 \end{aligned}$$

10. Show that the equations

$$\begin{aligned} 4u + 10v &= 6 \\ 6u + 15v &= 9 \end{aligned}$$

have infinitely many solutions. Find these solutions.

13 MATHEMATICAL MODELLING

OBJECTIVES

When you have finished studying this chapter you should be able to

1. distinguish between theoretical and empirical modelling
2. know the difference between a model and modelling
3. improve your problem solving skills

13.1 INTRODUCTION

Throughout this text we have used engineering and science applications to show the need for and the use of mathematics concerning problems in the physical world. The models in these applications are fairly standard and we have assumed that these models are familiar to you. Finding these models is a challenging activity. *Mathematical modelling* is a process by which we obtain a model, solve any mathematics associated with the model and use the model to describe or predict some feature concerning the real world. In this chapter we look briefly at this process of problem solving. We shall identify two classes of mathematical models, namely, *empirical models* and *theoretical models*.

To illustrate the ideas consider the problem of finding the current as a function of time for the electric circuit in Figure 13.1

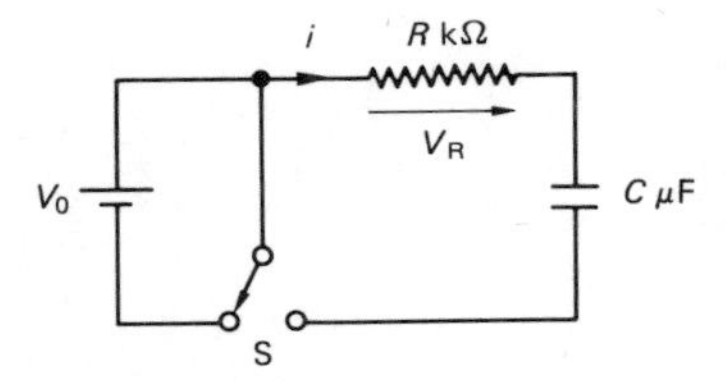

Figure 13.1

The circuit consists of a resistor of value R kΩ and a capacitor of value C microfarads. The capacitor is charged by the terminal voltage V_0 and the problem is to find the current in the circuit as a function of time t as the capacitor discharges through the resistor when the switch is thrown to position S.

One approach is to carry out an experiment to measure the current i at known time intervals. By plotting a graph, we can find a formula between i and t. Table 13.1 shows the results of such an experiment for $R = 100$ kΩ and $C = 10$ microfarads.

Table 13.1

Time, t (s)	10	20	30	40	50
Current, i (mA)	0.88	0.32	0.12	0.04	0.02

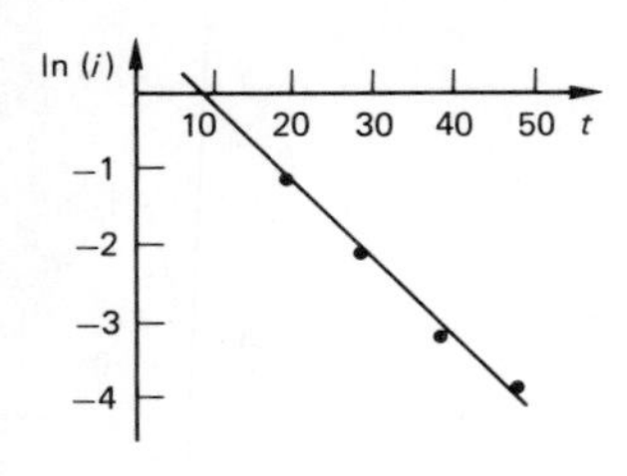

Figure 13.2

It is clear from the experimental data that the relationship between i and t is not linear. Using the techniques of Chapter 3, we look for an exponential model and draw a graph of $\ln(i)$ against t. This is shown in Figure 13.2.

The graph of $\ln(i)$ against t is a straight line so that we assume a relation of the form

$$i = Ae^{kt}$$

From the slope and the intercept we deduce that

$$i = 2.4e^{-0.1t}$$

This problem solving method is called empirical modelling and the formula $i = 2.4e^{-0.1t}$ is called an *empirical* (mathematical) model. We test this model by predicting future current values at some points in time and checking with more data. If the predictions are good (within acceptable experimental error) then we accept the model as a good one. One drawback of this problem solving method is that our model only holds for this particular electric circuit. If we could find a more general model then we would be able to apply it to other situations.

The second approach to solving the problem is called *theoretical* (mathematical) modelling and uses existing laws or models for electric circuits such as Kirchhoff's laws.

Suppose that the charge on the capacitor at some time t is Q and the current in the circuit is i.

define variables

Applying Kirchhoff's second law we obtain

formulate models

$$V_C + V_R = 0$$

where V_C and V_R are voltage drops across the capacitor and resistor.

Now $V_C = Q/C$ and $V_R = iR$ so that

$$\frac{Q}{C} + iR = 0 \text{ and } i = \frac{dQ}{dt}$$

The charge satisfies the first order differential equation

global model

$$\frac{dQ}{dt} + \frac{1}{RC}Q = 0$$

solve the mathematics

$$Q = Q_0 e^{-1/RC\,t}$$

and then $$i = -\frac{Q_0}{RC}e^{-1/RC\,t}$$

[The negative sign means that the discharging current is in the opposite direction to the charging current.]

If we substitute values for $R = 100\ \text{k}\Omega$, $C = 10\ \mu\text{F}$ and $Q_0 = V_0/C = 2.4$ then we have

$$i = 2.4\text{e}^{-0.1t}$$

validate the solution

To check that this is a good model we would collect data from an experiment and compare it with the theoretical model. In this case the model used is Kirchhoff's second law and we have produced a successful model for describing the current in the circuit.

In mathematical modelling the formulation of mathematical models is based on making *assumptions* and *simplifications*. For example, in the above example we are assuming that there is no loss of voltage in the circuit and that there is no build up of electric potential at any point in the circuit. Furthermore Kirchhoff's second law is a statement that energy is conserved in the circuit (an assumption). It is important that in problem solving you are aware of the assumptions that you are making and then you can state the limitations of your model.

The main stages of the mathematical modelling process are summarized in Figure 13.3 by three phases.

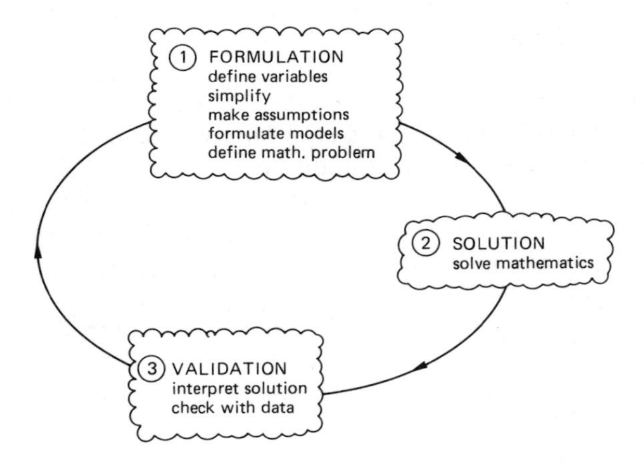

Figure 13.3

How these stages apply to the electric circuit problem are shown as marginal notes. Mostly in this text you have been learning the techniques for phase 2. Most people believe that this second phase is the hardest part of problem solving. Actually it's probably the easiest part. Phase 1, defining the problem and converting it into symbols is much harder.

In the next section we provide some fairly simple modelling problems. Have a go at them and discuss your approach with other students. Think about and list the assumptions you are making. This is a good habit to get into. In Section 13.3 we give our approach to two of these problems.

13.2 MODELLING PROBLEMS

Problem 1 Shed building

A man wishes to construct a garden shed and has decided that it should have a square base and also be box-shaped with a horizontal roof. Having assessed all his gardening equipment, he decides that he needs the shed to have a volume of 8 m^3. Given the high cost of wood (for the sides and top) and bearing in mind the work involved in painting the shed in future years, the man wishes to find what dimensions would minimize the area of wood to be used. (The base is just concrete.)

Establish a model and solve it.

Consider the model solution in terms of the practical problems involved in construction of a shed.

Problem 2 Traffic flow

When motorways are being repaired holdups often occur at busy times at the point where the traffic is reduced to a single lane. Bearing in mind aspects of safety and the desire to produce the maximum flow of traffic at peak periods, the traffic controller wishes to erect signs indicating the maximum speed, and the distance to be maintained between vehicles, whilst travelling in this single lane.

What recommendations would you make to the traffic controller on this matter?

Problem 3 Length of a toilet roll

Your group is provided with a roll of toilet paper, and each member of the group with one sheet of toilet paper. You are required to estimate the length of the toilet roll, without unwrapping it.

Problem 4 Car park layout

A rectangular car park which measures 100 m by 60 m has a single entrance/exit midway along one of the shortest sides. The layout of the parking spaces, each measuring 5 m by 3 m, needs to be designed to cater for a maximum number of cars (minimum aisle width 10 m).

Produce a model giving due consideration to the necessary assumptions and practical aspects of the problem.

Problem 5 How much tape is left?

Introduction

Most audio cassette recorders have a numerical tape counter which provides a numerical referencing to index a cassette tape for playback purposes. It is often convenient to be able to relate the number displayed on the tape counter with the playing time remaining.

Does the counter on the cassette recorder operate so that the elapsed time is directly proportional to the playing time? For example, for a C90 cassette (45 minutes playing time each side) the tape counter on my cassette deck goes from 0 to 600 for each side?

Does it take 15 minutes to reach 200?

When the counter reads 400 are there 15 minutes of tape playing time remaining?

Problem statement

For a particular cassette player equipped with a tape counter, formulate a model that describes the relationship between the counter reading and the amount of playing time that has elapsed.

Problem 6 Sleeping policemen

A common way of attempting to discourage motorists from driving at excessive speeds on housing estates, university campuses and other restricted road systems is the installation of road humps (often called 'sleeping policemen'). The road hump is assumed to be some form of surface irregularity which is sufficiently uncomfortable to car drivers and motor cyclists to compel them to slow to some small speed to pass over it. If such road humps are to be introduced in a road system, the highway engineer must decide on the spacing of road humps.

By developing a suitable mathematical model, advise the engineer about the effects of the spacing or road humps.

Problem 7 Sprinklers

Plan the layout of a system of sprinklers for a large grassed area such as a cricket oval. Occasionally Britain has a hot dry summer when water has to be rationed. Unfortunately grass has to be watered regularly at such times. Examine the performance of some appropriate sprinklers and then show how they should be laid out and used to ensure adequate watering for least cost and little inconvenience.

13.3 Solutions to Problems 1 and 2

Problem 1 Shed building

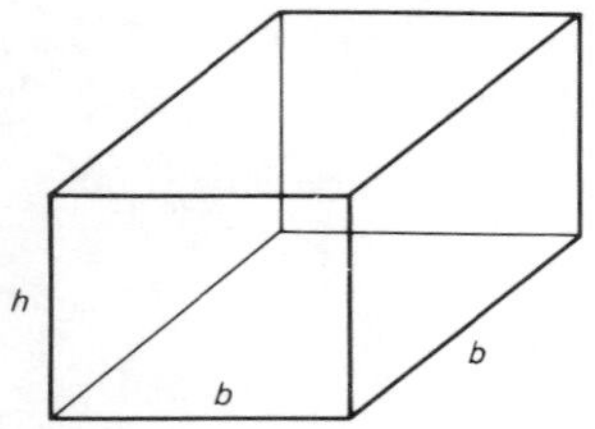

Formulation

The shape of the shed is decided for us. It is a rectangular box with a square base. Suppose that the length of the base is b and the height is h.

h
b
b

Figure 13.4

The volume of the box is b^2h and this is given as 8 m^3 so

$$b^2h = 8$$

The area of wood to be used for four sides and a flat top is

$$A = 4bh + b^2$$

Substituting for h we have

$$A = \frac{32}{b} + b^2$$

This is the mathematical model for the area of wood.

Mathematical problem

The mathematical problem is to minimize the function A.

Solution of the mathematics

Using the calculus techniques of Chapter 3 then A is a minimum when $dA/db = 0$.

$$A = \frac{32}{b} + b^2$$

$$\frac{dA}{db} = -\frac{32}{b^2} + 2b = 0$$

Solving for b we have

$$b = \sqrt[3]{16} = 2\sqrt[3]{2} = 2.52 \text{ (in metres)}$$

Solving for h,

$$h = \frac{8}{b^2} = 1.26 \text{ m}$$

Interpretation

The mathematics tells us that the height of the shed should ideally be 1.26 m with a square base of side length 2.52 m. This uses 19.05 m^2 of wood. Is this a realistic size of shed? The average height of a man is roughly 1.8 m so it could prove difficult to use effectively all the volume in the shed if its height is 1.26 m.

If we propose a height of 1.8 m then the square base has size 2.11 m and the area of wood used is 19.64 m^2. This is an extra 3% of wood. Perhaps this is a small price to pay for more comfort. On the other hand the area of the base is reduced from 6.35 m^2 to 4.45 m^2, a reduction of nearly 30%. It is the floor area of a shed that is most useful not "air space".

What conclusions would you come to?

Problem 2 Traffic flow

The solution to this problem given here is not intended as the 'correct' solution, other approaches are possible and we would encourage a side discussion of the various solutions that you may have found. The different approaches will inevitably depend on the assumptions made.

Formulation

In order to simplify the problem we make the following assumptions.

1. There is a steady flow of traffic along the single lane implying constant speed of vehicles at the same distance apart.
2. Assume that the vehicles have the same length.
3. Assume that the separation distance between cars is the recommended Highway Code stopping distance.

Let the speed and length of each vehicle be v (ms^{-1}) and L (metres) respectively and the separation distance be d (metres).

The time between each vehicle is $T = (d + L)/V$ seconds so that flow rate, i.e. the number of cars travelling along the single carriageway and passing a fixed point per second, is $1/T$.

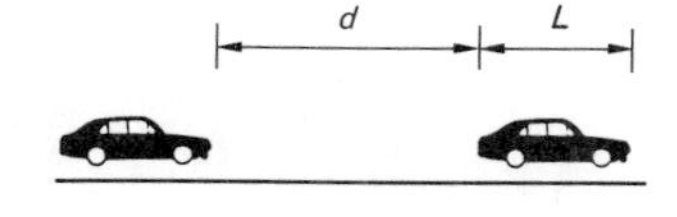

Figure 13.5

$$\text{Flow rate } F = \frac{V}{d + L} \text{ vehicles per second} \qquad (1)$$

This is the main mathematical model for the problem. However the separation distance d depends on V.

Data

The Highway Code gives the stopping distances as a table. This is summarized in Table 13.2.

Table 13.2

Speed, u (mph)	Thinking distance, T (m)	Braking distance, B (m)	Stopping distance, d (m)
20	6	6	12
30	9	14	23
40	12	24	36
50	15	38	53
60	18	55	73
70	21	75	96

To convert this table into a formula relating d and u we adopt an empirical approach, drawing graphs of thinking distance and braking distance against speed.

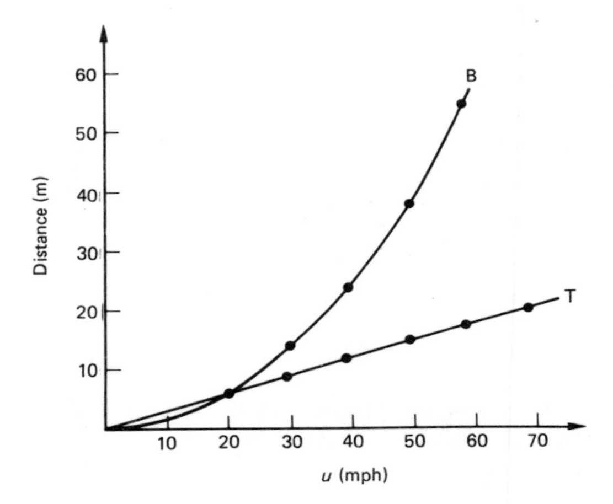

Figure 13.6

From these graphs we deduce the formula

$$d = 0.3u + 0.015u^2$$

In this formula, u is in mph and d is in metres. Converting to ms^{-1} we have

$$1 \text{ mph} = 1 \times \frac{8}{5} \times \frac{1000}{60^2} \text{ ms}^{-1}$$

$$= \frac{4}{9}\ \mathrm{ms}^{-1}$$

Replacing u by 9/4 V we have

$$d = 0.675V + 0.076V^2 \tag{2}$$

Substituting for d from Equation (2) into the flow rate formula in Equation (1) gives the model for flow rate as

$$F = \frac{V}{L + 0.675V + 0.076V^2}$$

Note that in formulating this model we have used the techniques of theoretical and empirical modelling.

Mathematical problem

The problem is to find the value of V which gives a maximum for F.

Solution of the mathematics

Using calculus to maximize F,

$$\frac{\mathrm{d}F}{\mathrm{d}V} = \frac{(L + 0.675V + 0.076V^2) - V(0.675 + 0.152V)}{(L + 0.675V + 0.076V^2)^2}$$

$$= \frac{L - 0.076V^2}{(L + 0.675V + 0.076V^2)^2}$$

$\mathrm{d}F/\mathrm{d}V = 0$ when $L - 0.076V^2 = 0$ which reduces to

$$V = \sqrt{\left(\frac{L}{0.076}\right)} = 3.63\sqrt{L}$$

[It can be checked that $\mathrm{d}^2F/\mathrm{d}V^2 < 0$ when $V = 3.63\sqrt{L}$ so that F is a maximum at this speed.]

Interpretation

Table 13.3 shows the results for various sized vehicles. This table suggests that for cars a speed of about 17 mph with a separation of around two car lengths will maximize the traffic flow.

Table 1.1

Vehicle length L(m)	Speed (ms^{-1})	Speed (mph)	Separation (m)	Flow rate (veh/hr)
Small car 4	7.26	16.3	8.9	2026
Medium car 5	8.12	18.3	10.5	1886
Long lorry 15	14.06	31.6	24.5	1281

Is it reasonable to expect car drivers to travel at 17 mph?

The answer to this question is: probably not! Observing traffic approaching a motorway contraflow where a two lane carriageway reduces to one, suggests that vehicles are driven as fast as possible (within the law?) and very close together. This often leads to long traffic jams. Clearly if drivers realized that a slower speed and a separation distance as recommended by the highway code would greatly increase the traffic flow rate then there would be less traffic jams.

So how can we improve the model so that more realistic results are possible? The assumption that is most questionable is number 3. Cars do not travel at the recommended stopping distances given in the Highway Code.

Possible amendments are suggested below:

1. The separation distance equals thinking distance, i.e. $d = 0.675\ V$. This leads to the flow rate increasing indefinitely with V, so that the maximum flow rate occurs at the speed limit of 70 mph. This is as unrealistic as the original model.
2. The separation distance equals the braking distance, i.e. $d = 0.076V^2$. This leads to the same result as in the original model, $V = 3.63\sqrt{L}$. So no improvement to the recommended speed but a more realistic separation distance.

Exercise

Assume a model for the separation distance as

$$d = 0.675V + K0.096V^2$$

where K is a parameter between 0 and 1. The separation distance is then the thinking distance plus a proportion of the braking distance.

APPENDIX 1: UNITS

The basic SI units needed in this text are based on the following fundamental quantities shown in Table A1.1

Table A1.1

Quantity	*SI unit*	*Other common units*
Length	metre, m	
Mass	kilogram, kg	gram $= \frac{1}{1000}$ kg
Time	second, s	
Temperature	kelvin, K	°C
Angle	radian, rad	degree
Electric current	ampere, A	

The units associated with other quantities are based on the fundamental quantities shown in Table A1.2.

Table A1.2

Quantity	*SI unit*	*Relationship to fundamental units*
Velocity	ms^{-1}	
Acceleration	ms^{-2}	
Force	newton, N	$N = kgms^{-2}$
Work, energy	joule, J	$J = Nm = kgm^2s^{-2}$
Pressure	pascal, Pa	$Pa = Nm^{-2} = kgm^{-1}s^{-2}$
Density	kgm^{-3}	
Electric charge	coulomb, C	$C = As$
Electric potential	volts, V	$V = J/C = kgm^2A^{-1}s^{-3}$

APPENDIX 2: MATHEMATICAL FORMULAS

PROPERTIES OF EXPONENTIALS AND LOGARITHMS

$$a^m a^n = a^{m+n}$$

$$\frac{a^m}{a^n} = a^{m-n}$$

$$(a^m)^n = a^{mn}$$

$$\log_a x = \frac{\log_b x}{\log_b a}$$

$$\log(xy) = \log x + \log y$$

$$\log\left(\frac{x}{y}\right) = \log x - \log y$$

$$\log(x^n) = n \log x$$

QUADRATIC EQUATIONS

If $ax^2 + bx + c = 0$ then

$$x = \frac{-b \pm \sqrt{b^2 - 4ac}}{2a}$$

THE BINOMIAL THEOREM

For n a non negative integer

$$(a + b)^n = a^n + {}^nC_1 a^{n-1} b + {}^nC_2 a^{n-2} b^2 + \ldots + b^n$$

where

$$^nC_r = \frac{n!}{(n-r)!r!}$$

For n a real number and $|x| < 1$

$$(1+x)^n = 1 + nx + \frac{n(n-1)}{2!}x^2 + \frac{n(n-1)(n-2)}{3!}x^3 + \ldots$$

ARITHMETIC AND GEOMETRIC PROGRESSIONS

For an arithmetic progression with first term a and common difference d

nth term $t_n = a + (n-1)d$, and

Sum of n terms $S_n = n/2[2a + (n-1)d]$

For a geometric progression with first term a and common ratio ρ

nth term $t_n = a\rho^{n-1}$, and

Sum of first n terms $S_n = \dfrac{a(1-\rho^n)}{1-\rho}$

If $|\rho| < 1$ then

Sum to infinity $S_\infty = \dfrac{a}{1-\rho}$

DIFFERENTIATION

General rules

The chain rule

$$\frac{dy}{dx} = \frac{dy}{du} \cdot \frac{du}{dx}$$

The product rule

$$y = uv \Rightarrow \frac{dy}{dx} = u\frac{dv}{dx} + v\frac{du}{dx}$$

The quotient rule

$$y = \frac{u}{v} \Rightarrow \frac{dy}{dx} = \frac{v\frac{du}{dx} - u\frac{dv}{dx}}{v^2}$$

Some useful derivatives

Note in the following table a and b are constants

y	$\frac{dy}{dx}$
a	0
x	1
x^n	nx^{n-1}
ax^n	anx^{n-1}
$\sin x$	$\cos x$
$\cos x$	$-\sin x$
$\tan x$	$\sec^2 x$
$\sec x$	$\sec x \tan x$
$\operatorname{cosec} x$	$-\operatorname{cosec} x \cot x$
$\cot x$	$-\operatorname{cosec}^2 x$
e^x	e^x
$\ln x$	$\frac{1}{x}$
$\sin(ax + b)$	$a \cos(ax + b)$
$\cos(ax + b)$	$-a \sin(ax + b)$
e^{ax+b}	ae^{ax+b}
$\ln(ax + b)$	$\frac{a}{ax + b}$

INTEGRATION

Some Useful Integrals

Note in the following table a and b are constants and c is an arbitrary constant.

$$\int a \, dx = ax + c$$

$$\int x \, dx = \frac{x^2}{2} + c$$

$$\int x^n \, dx = \frac{1}{n+1} x^{n+1} + c, \text{ provided } n \neq -1$$

$$\int ax^n \, dx = \frac{a}{n+1} x^{n+1} + c, \text{ provided } n \neq -1$$

$$\int \sin x \, dx = -\cos x + c$$

$$\int \cos x \, dx = \sin x + c$$

$$\int e^x \, dx = e^x + c$$

$$\int \frac{1}{x} \, dx = \ln x + c$$

$$\int \sin(ax + b) \, dx = -\frac{1}{a} \cos (ax + b) + c$$

$$\int \cos (ax + b) \, dx = \frac{1}{a} \sin (ax + b) + c$$

$$\int e^{ax+b} \, dx = \frac{1}{a} e^{ax+b} + c$$

$$\int \frac{1}{ax + b} \, dx = \frac{1}{a} \ln(ax + b) + c$$

$$\int \frac{1}{\sqrt{a^2 - b^2x^2}}\,\mathrm{d}x = \frac{1}{b}\sin^{-1}\left(\frac{bx}{a}\right) + c$$

$$\int \frac{1}{a^2 + b^2x^2}\,\mathrm{d}x = \frac{1}{ab}\tan^{-1}\left(\frac{bx}{a}\right) + c$$

Integration by parts

$$\int u\frac{\mathrm{d}v}{\mathrm{d}x}\,\mathrm{d}x = uv - \int v\frac{\mathrm{d}u}{\mathrm{d}x}\,\mathrm{d}x$$

General logarithmic form

$$\int \frac{f'(x)}{f(x)}\,\mathrm{d}x = \ln[f(x)] + c$$

Numerical or approximate integration

In the following the interval of integration is from $x = a$ to $x = b$.

This interval $[a, b]$ is subdivided into n equal subintervals/strips. Each strip has width $h = (b - a)/n$.

$$x_0 = a,\ x_1 = a + h,\ \ldots,\ x_n = b$$

$$y_0 = f(x_0),\ y_1 = f(x_1),\ \ldots$$

The trapezium rule

$$\int_a^b f(x)\,\mathrm{d}x \simeq \frac{h}{2}[y_0 + y_n + 2(y_1 + y_2 + \ldots + y_{n-1})]$$

Simpson's rule

$$\int_a^b f(x)\,\mathrm{d}x \simeq \frac{h}{3}[y_0 + y_n + 4(y_1 + y_3 + \ldots + y_{n-1}) + 2(y_2 + y_4 + \ldots + y_{n-2})]$$

NUMERICAL MATHEMATICS

The Newton–Raphson Iterative Method for approximating to a root of $f(x) = 0$.

Let the nth approximation be x_n and the $(n + 1)$th approximation be x_{n+1}, then:

$$x_{n+1} = x_n - \frac{f(x_n)}{f'(x_n)}$$

TRIGONOMETRIC IDENTITIES

Basic definitions

$$\operatorname{cosec} \theta \equiv \frac{1}{\sin \theta}$$

$$\sec \theta \equiv \frac{1}{\cos \theta}$$

$$\cot \theta \equiv \frac{1}{\tan \theta}$$

Pythagorean formulas

$$\cos^2 \theta + \sin^2 \equiv 1$$

$$1 + \tan^2 \theta \equiv \sec^2 \theta$$

$$\cot^2 \theta + 1 \equiv \operatorname{cosec}^2 \theta$$

Addition formulas

$$\sin (A \pm B) \equiv \sin A \cos B \pm \cos A \sin B$$

$$\cos (A \pm B) \equiv \cos A \cos B \mp \sin A \sin B$$

$$\tan (A \pm B) \equiv \frac{\tan A \pm \tan B}{1 \mp \tan A \tan B}$$

Double angle formulas

$$\sin 2\theta \equiv 2 \sin \theta \cos \theta$$

$$\cos 2\theta \equiv \begin{cases} \cos^2 \theta - \sin^2 \theta \\ 2\cos^2 \theta - 1 \\ 1 - 2 \sin^2 \theta \end{cases}$$

$$\tan 2\theta \equiv \frac{2 \tan \theta}{1 - \tan^2 \theta}$$

$$\sin^2 \theta \equiv \frac{1}{2} (1 - \cos 2\theta)$$

$$\cos^2 \theta \equiv \frac{1}{2} (1 + \cos 2\theta)$$

VECTORS

Two dimensional vectors

Magnitude

If $\boldsymbol{a} = a_1\boldsymbol{i} + a_2\boldsymbol{j}$ then

$$|\,\boldsymbol{a}\,| = a = \sqrt{a_1^2 + a_2^2}$$

Three dimensional vectors

Magnitude

If $\boldsymbol{a} = a_1\boldsymbol{i} + a_2\boldsymbol{j} + a_3\boldsymbol{k}$ then

$$|\,\boldsymbol{a}\,| = a = \sqrt{a_1^2 + a_2^2 + a_2^3}$$

MATRICES

2 × 2 matrices

Identity matrix

$$\boldsymbol{I} = \begin{bmatrix} 1 & 0 \\ 0 & 1 \end{bmatrix}$$

Determinant

If

$$\boldsymbol{A} = \begin{bmatrix} a & b \\ c & d \end{bmatrix}$$

then

$$\det \boldsymbol{A} = |\boldsymbol{A}| = \begin{vmatrix} a & b \\ c & d \end{vmatrix} = ad - bc$$

Inverse

If

$$\boldsymbol{A} = \begin{bmatrix} a & b \\ c & d \end{bmatrix} \quad \text{and } \det \boldsymbol{A} \neq 0 \text{ then}$$

$$\boldsymbol{A}^{-1} = \frac{1}{\det \boldsymbol{A}} \begin{bmatrix} d & -b \\ -c & a \end{bmatrix}$$

ANSWERS TO THE EXERCISES

CHAPTER 1

Exercise 1.2.1

1. (a) x: −1 0 1 2 3 4 5

   (b) x: −1 0 1 2 3 4 5

   (c) x: −2 −1 0 1 2 3

   (d) x: −3 −2 −1 0 1 2 3

   (e) x: −4 −3 −2 −1 0 1 2 3

   (f) −3 −2 −1 0 1 2

2. (a) $-3 \leqslant x \leqslant 2$
   (b) $-1 \leqslant t < 5$
   (c) $-0.2 < u < 0.3$

Exercise 1.2.2

1. (a) $x > -3$ and $x > 1$ so $-3 < x < 1$
   (b) $x < -3$ and $x > -6$ so $-6 < x < -3$
   (c) $x \geqslant -2$ and $x \geqslant -3$ so $x \geqslant -2$

Exercise 1.3.1

1. Independent variable — resistance;
   dependent variable — current

Exercise 1.3.2

1. (a) Daylight hours

Jan Dec

Months

(b)

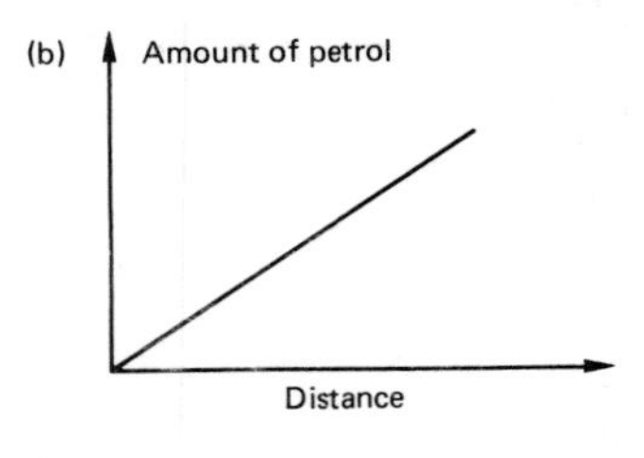

(c)

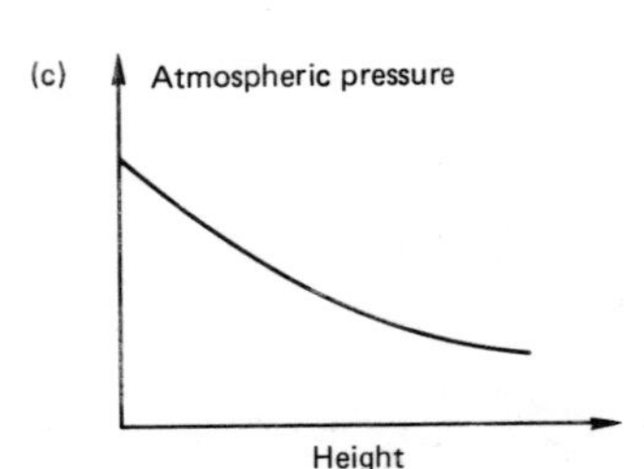

(d)

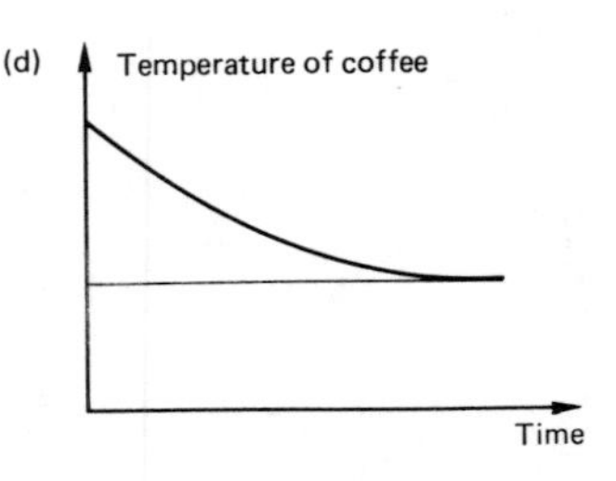

Exercise 1.3.3

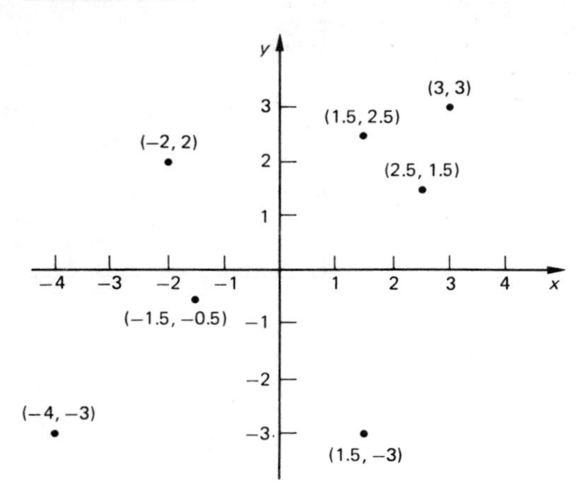

Exercise 1.3.4

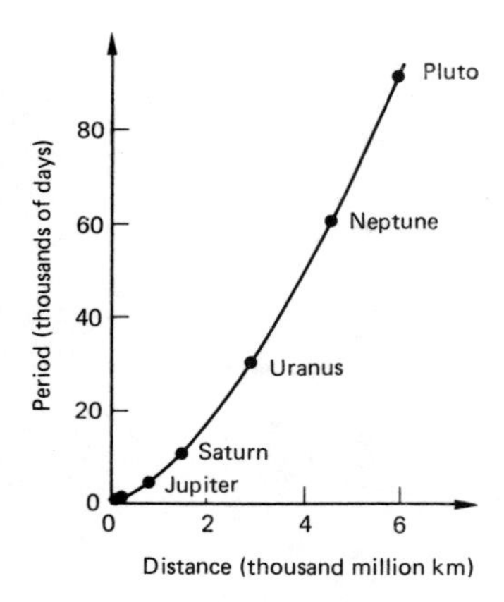

Exercise 1.4.1

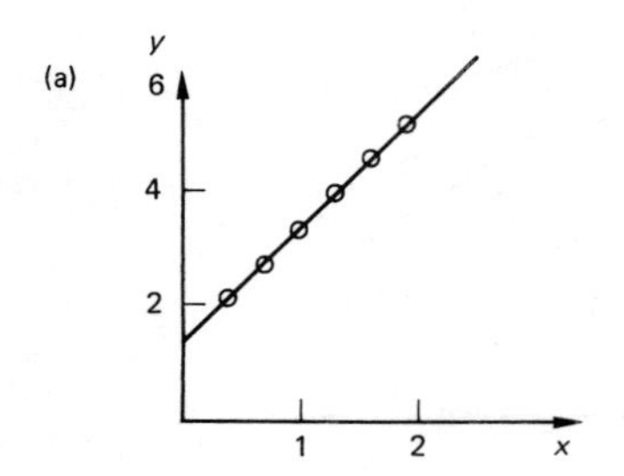

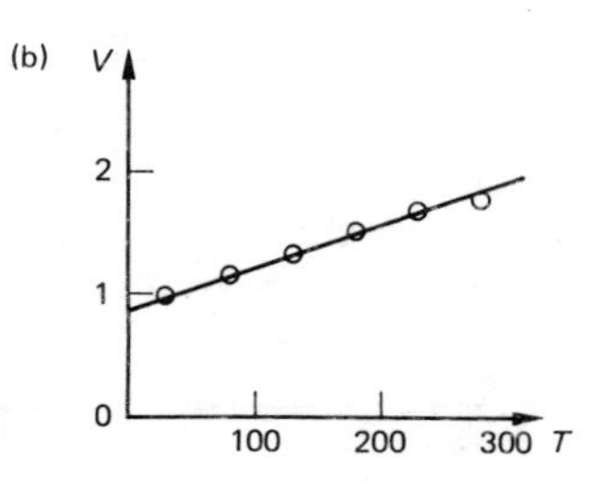

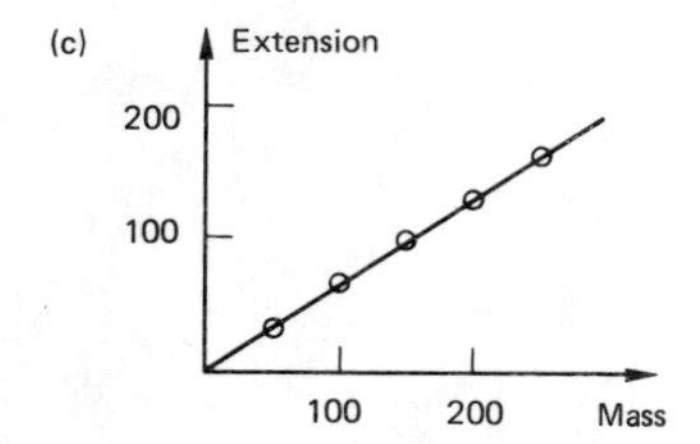

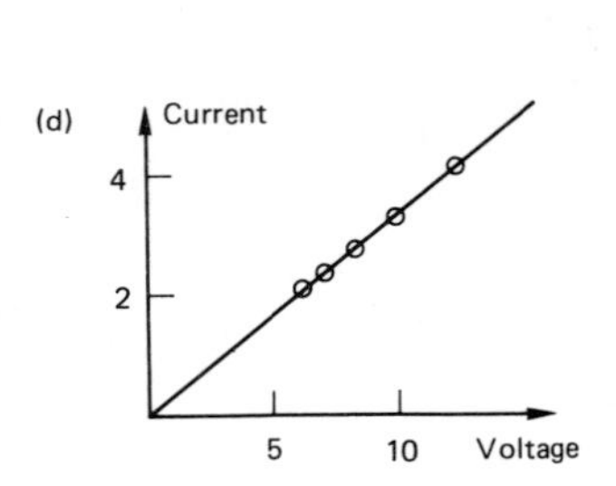

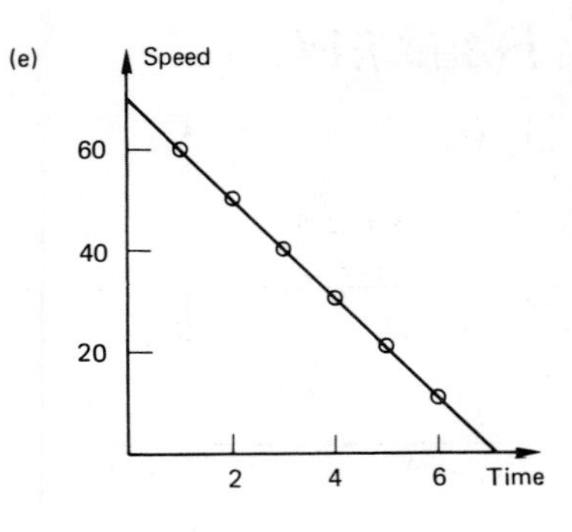

Exercise 1.4.2

1. (a) 3/3 = 1 (b) 1/3 = 0.3 (c) 6/6 = 1
(d) 1/4 = 0.25 (e) −1/2 = −0.5 (f) 1/5 = 0.2
(g) 1/2 = 0.5 (h) −2/5 = −0.4 (i) −1/4 = −0.25

Exercise 1.4.3

1.

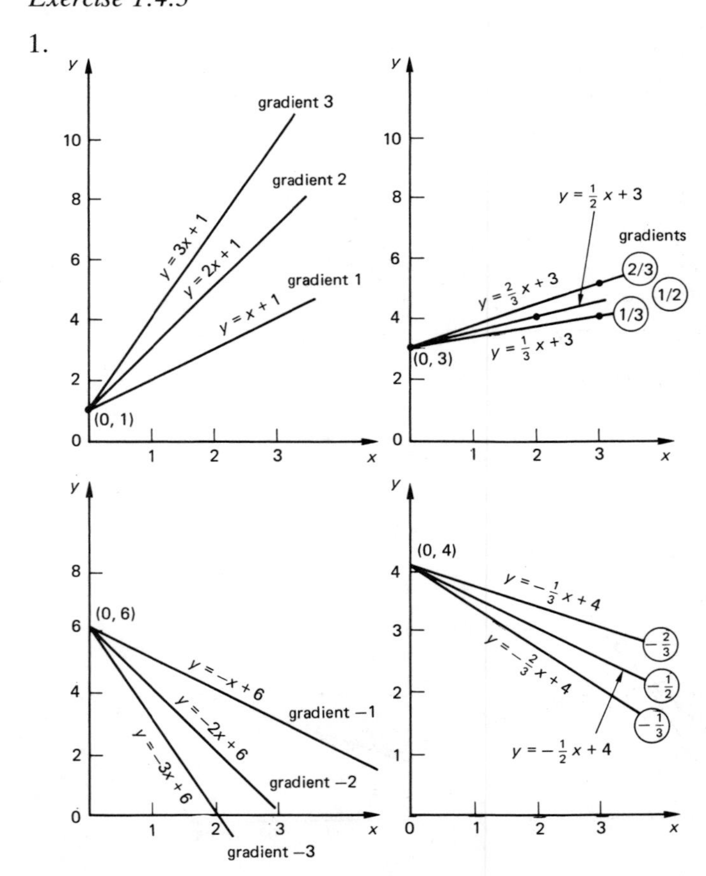

Exercise 1.4.4

1. (a) 1 (b) 3 (c) 6 (d) 4

Exercise 1.4.5

1.

Graph	*Gradient*	*Intercept*
(a)	2	1.5
(b)	0.0033	0.85
(c)	0.67	0.0
(d)	0.33	0.0
(e)	−9.8	70

Exercise 1.4.6

1. Graphs in Figure 1.17

(a) $y = x + 4$
(b) $y = 1/3\ x + 4$
(c) $y = x$
(d) $y = \frac{1}{4}x + 3$
(e) $y = -\frac{1}{2}x + 7$
(f) $y = 0.2x + 2$
(g) $y = 0.5x - 2$
(h) $y = -0.4x + 3$
(i) $y = -\frac{1}{4}x - 1$

Exercise 1.4.4

(a) $y = x + 1,\ y = 2x + 1,\ y = 3x + 1$
(b) $y = \frac{1}{2}x + 3,\ y = \frac{1}{3}x + 3,\ y = \frac{2}{3}x + 3$
(c) $y = -x + 6,\ y = -2x + 6,\ y = -3x + 6$
(d) $y = -\frac{1}{3}x + 4,\ y = -\frac{1}{2}x + 4,\ y = -\frac{2}{3}x + 4$

Exercise 1.4.7

1.

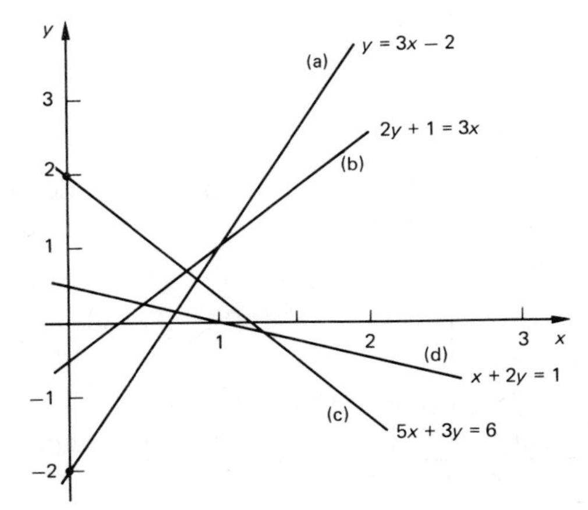

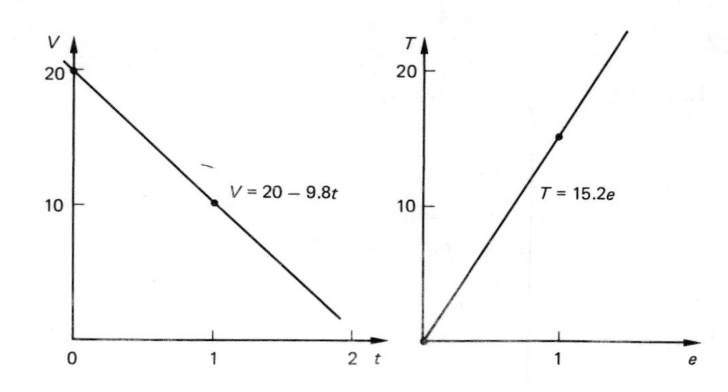

2. (a) 5/3 (b) 2 (c) 2/3
   (d) 5/3 (e) 12/4 = 3 (f) 6/3 = 2
3. (a) $y = 5/3\ x$ (b) $y = 2x$ (c) $y = 2/3\ x + 7/3$
   (d) $y = 5/3\ x - 1/3$ (e) $y = 3x - 41$ (f) $y = 2x - 1$

Exercise 1.5.1

1. (a)

Time	0	0.5	1.0	1.5	2.0	2.5	3.0
Amount	0	7.5	15	22.5	30	37.5	45

(b)

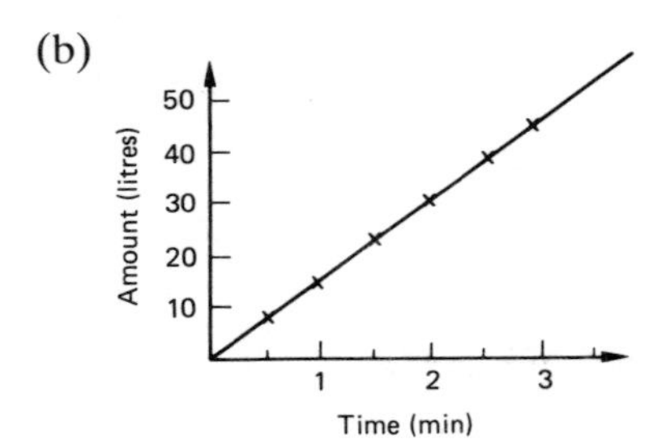

2. Rate of filling = 6 gallons per 100 sec = 0.06 gal/sec

Exercise 1.5.2

1. (a) Gradient = −5.8 gal/min
   (b) The tank is emptying
2. (a) Town driving is AB and EF. Motorway driving is BC and DE. Usually less fuel consumption on a motorway
   (b) CD is petrol refilling

Exercise 1.6.1

Straight line graphs through the origin for (1) and (5)
Note (3) and (4) are straight line graphs not through the origin

Exercise 1.6.2

Direct proportionality for (r, q) graph only

Exercise 1.6.3

1. Current = 17.3 amperes when voltage = 8 volts; voltage = 7.4 volts when current = 16 amperes
2. (a) and (c); $V = A \times D$ where A is the area of the base

Exercise 1.6.4

1. When volume is 250 cm^3, the mass is 200 grams. Mass = $k \times$ volume. $k = 0.8$ is called the density.
2. Yes, direct proportionality is a good model. $F = 0.41$ N

Exercise 1.7.1

1. (a) $W = 3$ when $L = 15$
   (b) $W = 1.125$ when $L = 40$
   (c) $L = 19.6$ when $W = 2.3$

2. If an inverse proportionality law is appropriate then the product of resistance and current is the same for each pair of data. It is a reasonable model in this case.

$$\text{current} = \frac{5}{\text{resistance}}$$

3. (a) $33\frac{1}{3}$% decrease
   (b) 100% increase
   (c) $16\frac{2}{3}$% decrease
   (d) 25% increase

Exercise 1.8.1

1. (a) $d = 176.4$ m
   (b) $d = 11.03$ m. When t is multiplied by k, d is multiplied by k^2.
2. $P = cu^3$. Multiply u by $2 \rightarrow$ multiply P by 8. Multiply u by $k \rightarrow$ multiply P by k^3. P is increased by 137%.

Exercise 1.8.2

1. $p = kT/V$
2. $P = c\sqrt{\dfrac{M}{k}}$

Exercise 1.9.1

1. (a) $R = 0.17T + 44.8$
   (b) (i) $R = 48.2$ ohms (ii) $T = -263.5$ °C

2. $l = 0.01T + 7.2$
   (a) (i) 7.45 cm (ii) 7.75 cm (iii) 7.95 cm
   (b) (i) at 0 °C, $l = 7.2$ cm
   (ii) at 200 °C, $l = 9.2$ cm
   (iii) when $l = 14.8$ cm, $T = 760$ °C

Exercise 1.9.2

1. In Exercise 1, extrapolation is being used. In Exercise 2(a), interpolation is being used. In Exercise 2(b), extrapolation is being used.

Objective Test 1

1. Gradient = 3.1, intercept = −4.2, $x = 3.1t - 4.2$

2.

	Gradient	*y-intercept*	*x-intercept*
(i)	2	5	−2.5
(ii)	3	−4	1.33
(iii)	−4	8	2
(iv)	−0.5	1.5	3

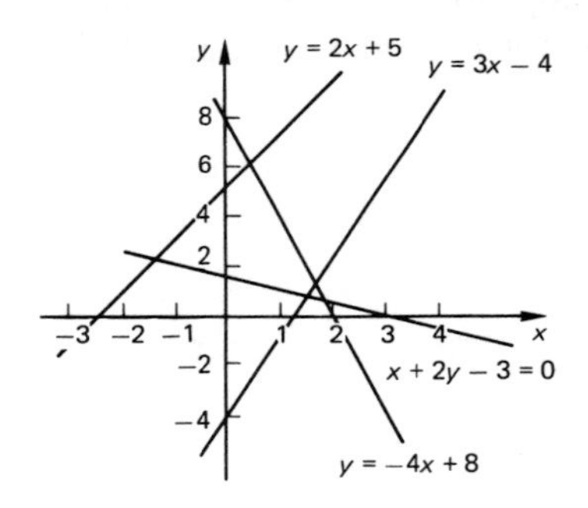

3. Q, R, A, D

4. (a) T vs x^2 (b) p vs $1/V$
   (c) W vs $1/l^2$ (d) T vs $\sqrt{L}$

5.

p	2.4	3.1	4.33
ρ	2.78×10^{-5}	3.59×10^{-5}	5.01×10^{-5}

   $P = 8.6 \times 10^4 \rho$

6.

C	5	1000	32
R	0. 2	0.001	0.03125

   $C = 1/R$

7. $V = c\sqrt{\dfrac{p}{\rho}}$

8. (a)

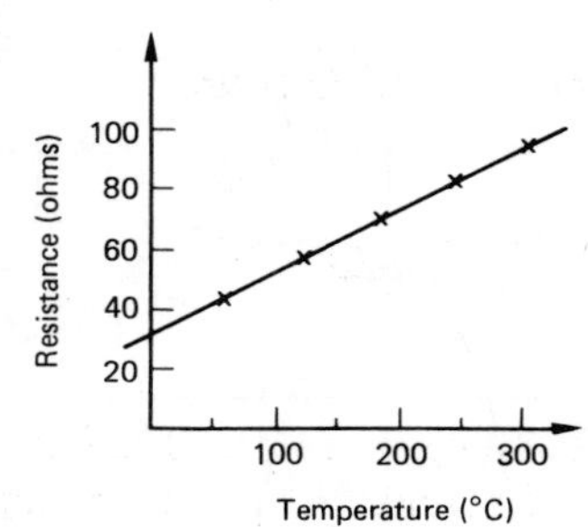

$R = 0.21T + 31.1$

(b) 0.21 ohm/°C

(c) (i) 52.1 ohm (ii) 115.1 ohm

(d) −148 °C

(e) c(i) because it uses interpolation.

CHAPTER 2

Exercise 2.1.1

1.

Quadratics	*Coefficients* $a =$	$b =$	$c =$
$2x^2 + 3x + 1$	2	3	1
$3x^2 - 5x - 3$	3	−5	−3
$7 + 4x + x^2$	1	4	7
$-2x^2 + 5$	−2	0	5
$-9x^2 + x$	−9	1	0
$-x^2 - 3x + 4$	−1	−3	4
$4 - 20t + 9.8t^2$	9.8	−20	4
$1 + v + 0.7v^2$	0.7	1	1
$u - 3u^2$	−3	1	0

Exercise 2.2.1

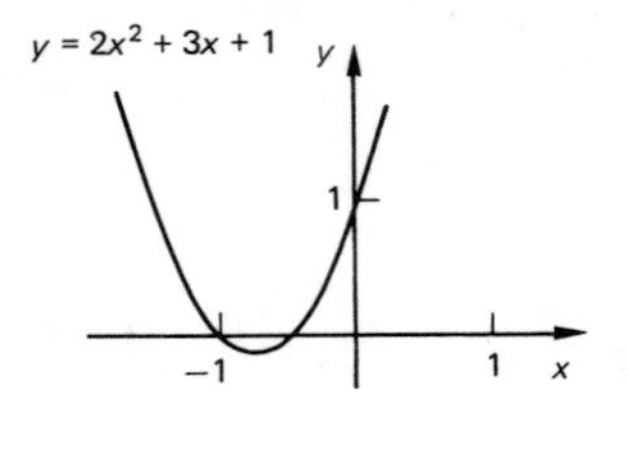

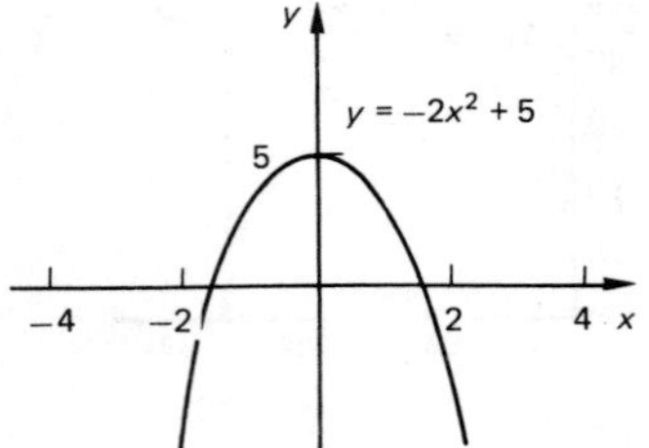

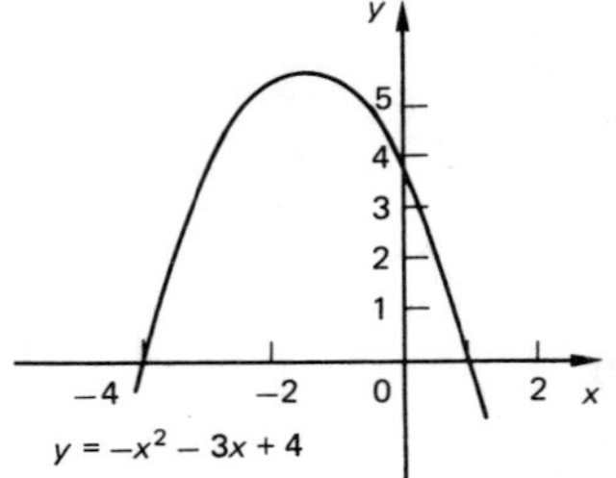

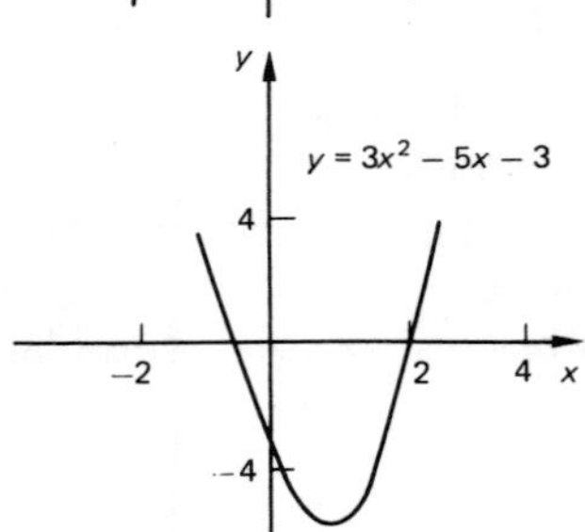

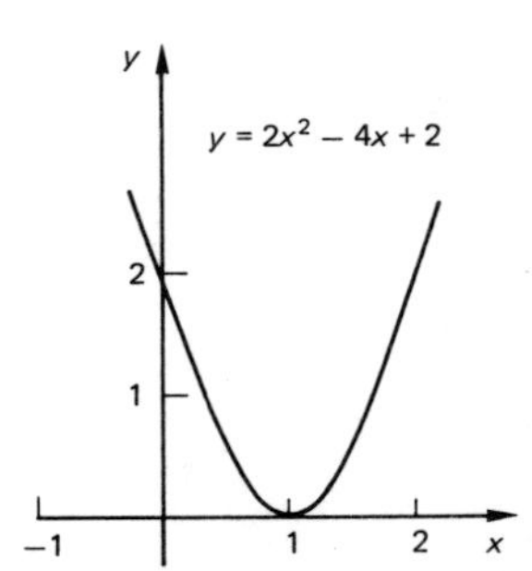

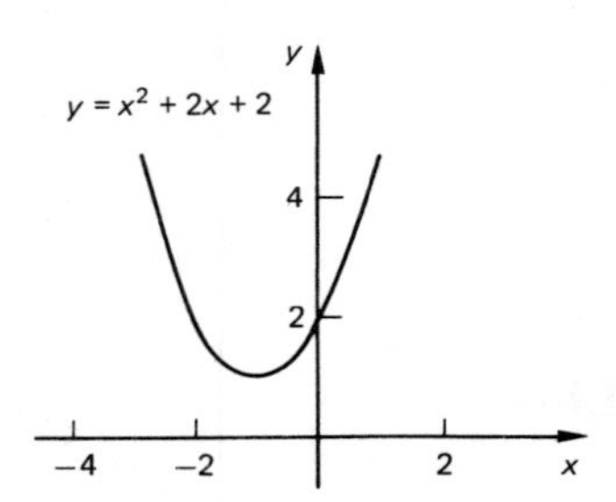

Exercise 2.2.2

1. (a) $-\frac{1}{2}, -1$ (b) $-1.6, 1.6$
   (c) $-4, 1$ (d) $2.1, -0.5$
   (e) -1 (f) no roots
2. (a) $0.85, -2.35$ (b) $-1.4, 1.4$
   (c) $-1, -2$ (d) $-2, 1$
3. (i) 2 m (ii) 2.04 s (iii) 4.18 s

Exercise 2.3.1

1. (a) $x^2 + 4x + 3$ (b) $x^2 - 3x + 2$
   (c) $x^2 + 2x + 1$ (d) $6x^2 + x - 2$
   (e) $4x^2 - 12x + 9$ (f) $2x^2 + 7x - 4$
   (g) $x^2 - 2x - 35$ (h) $x^2 - (u - v)x - uv$
   (i) $x^2 - u^2$ (j) $x^2 + 2x$
   (k) $6x^2 - 12x$

Exercise 2.3.2

1. (a) $x^2 - 2x + 1$ (b) $x^2 + 8x + 16$
   (c) $4x^2 - 12x + 9$ (d) $x^2 - 1$
   (e) $x^2 - 16$ (f) $4x^2 - 9$

Exercise 2.4.1

1. (a) $(x + 4)(x + 1)$ (b) $(x - 4)(x + 1)$
   (c) $(x + 4)(x - 3)$ (d) $(x + 5)(x + 2)$
   (e) $(2x + 5)(x + 1)$ (f) $(3x - 1)(3x + 1)$
   (g) $(x + 3)^2$ (h) $(x - 4)^2$
   (i) $(x - u)(x + u)$ (j) $(2x + u)^2$
   (k) $x(x + 4)$ (l) $x(3x - 10)$

Exercise 2.5.1

1. (a) $-1, -4$ (b) $-1, 4$
   (c) $-4, 3$ (d) $-2, -5$
   (e) $-2.5, -1$ (f) $\frac{1}{3}, -\frac{1}{3}$
   (g) $-3, -3$ (h) $4, 4$
   (i) $u, -u$ (j) $-u/2, -u/2$
   (k) $0, -4$ (l) $0, 10/3$

Exercise 2.6.1

1. (i) (a) $(x + 3)^2 - 9$ (b) $(x + 5)^2 - 25$
   (b) $(x + 5/2)^2 - 25/4$ (d) $(x + 7/2)^2 - 49/4$
2. (a) $-4, -2$ (b) $-1, -9$
   (c) $\dfrac{-5}{2} \pm \dfrac{\sqrt{21}}{2}$ (d) $\dfrac{-7}{2} \pm \dfrac{\sqrt{61}}{2}$

Exercise 2.6.2

1. (a) $2(x + 2)^2 - 8$ (b) $3\left(x + \dfrac{7}{6}\right)^2 - \dfrac{49}{12}$
   (c) $4\left(x + \dfrac{3}{2}\right)^2 - 12$ (d) $2\left(x - \dfrac{5}{4}\right)^2 - \dfrac{17}{8}$
2. (a) $x = -\dfrac{3}{2} \pm \sqrt{3}$ (b) $x = \dfrac{5}{4} \pm \dfrac{\sqrt{17}}{4}$

Exercise 2.6.3

1. (a) $-4, -2$ (b) $-1, -9$
   (c) $-\dfrac{5}{2} \pm \dfrac{\sqrt{21}}{2}$ (d) $-\dfrac{7}{2} \pm \dfrac{\sqrt{61}}{2}$
2. (a) $\dfrac{-1 \pm \sqrt{13}}{2}$ (b) $\dfrac{-1 \pm \sqrt{33}}{4}$

(c) $\frac{1}{3}$, 1 (d) -2

(e) $\pm\sqrt{5}$ (f) $\dfrac{-5 \pm \sqrt{37}}{2}$

Exercise 2.6.4

1. (a) 40 — two roots (b) -8 — no roots
   (c) 24 — two roots (d) -16 — no roots
   (e) 16 — two roots (f) 49 — two roots
2. $K = \pm\sqrt{12}$ gives one repeated root
   $K < -\sqrt{12}$ or $K > \sqrt{12}$ gives two roots
3. (i) 1.58 m or S = 8.42 m (ii) 2 m or 8 m
   zero bending moment
   (i) S = 0 m and 10 m (ii) S = 0 m and 10 m
4. T = 307.5 °C
5. (a) $w^2 - 3w - 4 = 0$ $w = 4$ (must be +ve)
   (b) $600w^2 + 3200w - 1 = 0$ $w = 0.00031$

Exercise 2.7.1

1.

Polynomial	*Degree*
$4x^5 + 3x^2 + 2x + 1$	5
$9x^7 + 8x^6 - 4x^3 + 2x$	7
$t^3 - 3t^2 + 4t + 1$	3
$u^4 + 3u^3 - 2u^2 + u + 7$	4
$0.1x^2 + 1.2x - 3.7$	2
$3.1x^{11} + 4x^7$	11

Exercise 2.7.2

1.

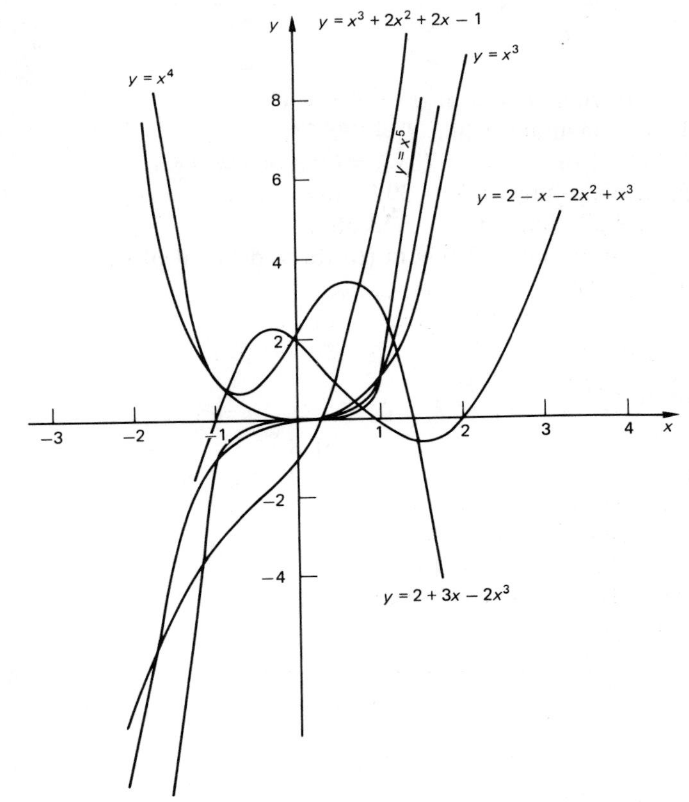

Exercise 2.7.3

1. −1, −0.5, 0.3, 2
2. −0.33, 0.33, 1.50

Exercise 2.8.1

1. 2
2. 3.732

Exercise 2.8.3

1. (a), (c) are no good near $x = 3$; (d) is quickest
2. (a) 0.38, 2.62
   (b) 1.21
   (c) 0.43

Exercise 2.9.1

1. 1. 2.18 s
   2. 243.7 k

3. 1612 K
4. $l_0 + \dfrac{2(M + m)g}{k}$
5. $V = 0$ when $T = 0$ or $T = 571.4$ °C
   V is a maximum when $T = 285.7$°C
6. (a) $i = 1$ or $i = 4$ (b) $i = 0.58$ or $i = 1.56$
7. $R = 5.73$ ohms or $R = 39.27$ ohms
   $r = 39.27$ ohms or $r = 5.73$ ohms
8. $x = 0.6$ m or $x = 1.639$ m (to three decimal places)
9. 0, 3, 6, 10

Objective Test 2

1. (a) 2 (b) 8
   (c) 1 (d) 3
2. (b), (d), (e)
3. (a) $(x - 4)(x + 4)$; $x = -4$ or $x = 4$
   (b) $(x - 2)(x - 1)$; $x = 1$ or $x = 2$
   (c) $(3x - 1)(x + 2)$; $x = -2$ or $x = 1/3$
4. $x = 1.4$ or $x = 3.6$

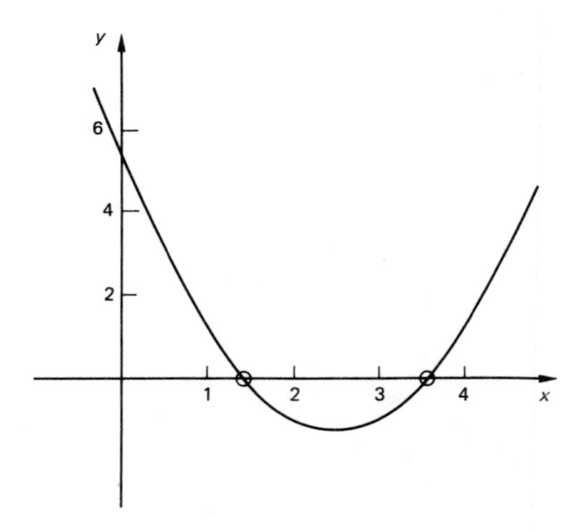

5. (a) $R = -0.573$ or $R = 0.873$ (to three decimal places)
   (b) $x = -2.679$ or $x = 1.679$ (to three decimal places)

6. (c), (a), (b)

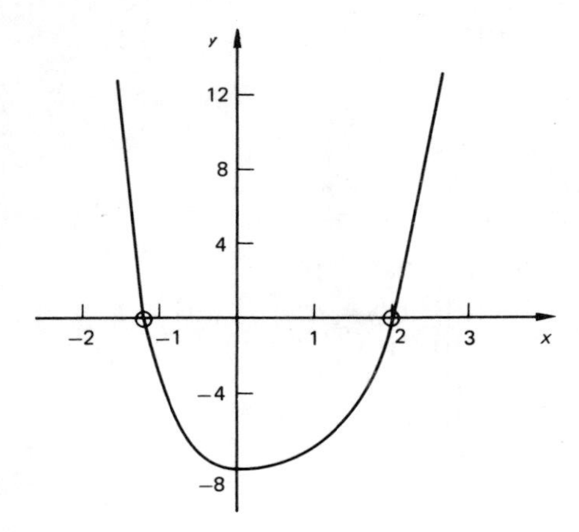

7. $x = -1.180$, $x = 2.000$

CHAPTER 3

Exercise 3.1.1

1. $r = \left(\dfrac{A}{4\pi}\right)^{1/2}$

Exercise 3.1.2

1. $r = \left(\dfrac{3v}{4\pi}\right)^{1/3}$

Exercise 3.1.3

1. $f = 0.316R^{-1/4}$

Exercise 3.2.1

1.

Number	*Base*	*Index*
7^3	7	3
21^{13}	21	13
a^4	a	4
2^n	2	n
a^m	a	m
5^7	5	7

Exercise 3.2.2

1. (a) 2^9 (b) 3^5 (c) a^6 (d) x^5

Exercise 3.2.3

1. (a) x^7 (b) x^{12} (c) $72x^5$ (d) a^2
   (e) $3k$ (f) x^6 (g) 4^6 (h) x^{15}
   (i) a^8b^{12}
2. (a) $x^5 + a^7$ (b) $6x^5 - 10x^{11}$
   (c) $-m^2 - 5m^3$ (d) $a^3b + a^2b^2 - ab^3$
3. (a) a^5 (b) $216x$
   (c) e^{n+m-p} (d) $4a^2bkt^5$
4. (a) 1 (b) 2
   (c) 1 (d) 24

Exercise 3.2.4

1. (a) 1/4 (b) 1/16 (c) 1/25
   (d) 1/27 (e) 1/4
2. (a) a^{-5} (b) x^{-3} (c) m^{-2}
   (d) p^2 (e) 6 (f) ab^{-1}
   (g) $e^{3x} + e^{-x}$ (h) 9 (i) $3\,x^{-1}\,y^2$
   (j) $a^3x^{-2}\,y^{-1}$ (k) $\frac{1}{9}c^2a^2b^{-2}$ (l) $m^{-3p}\,x^{3n}$

Exercise 3.2.5

1. (a) $x^{1/2}$ (b) $x^{3/2}$ (c) $x^{7/2}$
   (d) $x^{5/2}$ (e) $x^{1/2}$ (f) x^{-1}
1. (a) 2 (b) 8 (c) 3 (d) 2
   (e) 32 (f) 512 (g) 100 000 (h) $4x^3$
   (i) $\frac{1}{2}$ (j) 1 (k) 2 (l) 10 000

Exercise 3.2.7

1. (a) $k = 4$ (b) $k = 2$ (c) $r = 5/3$
   (d) $a = 2$ (e) $x = 0$ (f) $x = 3$
   (g) $a = 2/3$ (h) $x = 1/2$ (i) $a = 4$
   (j) $a = -5/2$ (k) $x = 3/2$ (l) $x = 0$

Exercise 3.3.1

Physical	*Variables*	*Rule*	*Domain*	*Range*
Spring-mass oscillator	T, m	$T = 1.13m^{1/2}$	$0 < m < \infty$	$0 < T < \infty$
Radiation heat loss rate	E, T	$E = K(T^4 - T_0^4)$	$T_0 \leqslant T < \infty$	$0 \leqslant E < \infty$
Newton's Law of gravitation	F, r	$F = \dfrac{GMm}{r^2}$	$0 < r < \infty$	$0 < F < \infty$

Exercise 3.3.2

1. (i) $F(1) = 1$ $F(3) = 79.601$ $F(0.61) = 0.997$
   (ii) $F(3.1) = 19.747$ $F(0) = 0$

Exercise 3.3.3

1. $g[f(x)] = 3x - 3$
   $f[g(t)] = 3(t - 3)$
   $f[f(x)] = 9x$
   $g[g(t)] = t - 6$
2. $f[g(x)] = (x - 2)^3$
   $g[h(x)] = \sqrt{x - 2}$
   $f\{g[h(x)]\} = (\sqrt{x} - 2)^3$
3. (a) Range of f is $0 < x < \infty$; range of g is $-\infty < x < \infty$
   (b) $f[g(x)] = \sqrt{(x - 4)}$ domain $[4, \infty)$ range $[0, \infty)$
   (c) $g[f(x)] = \sqrt{x} - 4$ domain $[0, \infty)$ range $[-4, \infty)$

Exercise 3.3.4

1. $f^{-1}(x) = \dfrac{(x - 4)}{3}$
2. $f^{-1}(x) = (x + 3)^{1/3} - 1$

Exercise 3.3.5

1. (a) $x = 2.5$ (b) $x = -15/7$
   (c) no asymptotes (d) $t = -2$, $t = 2$

Exercise 3.3.6

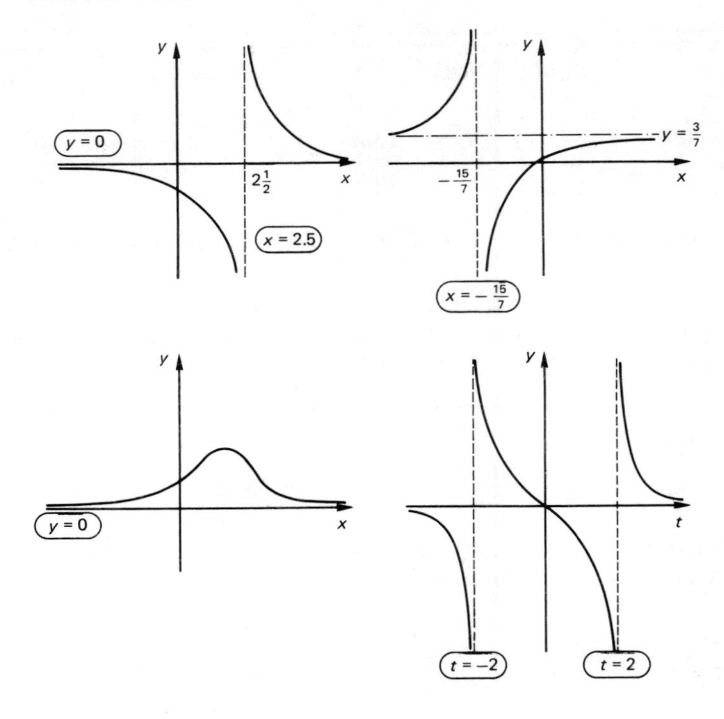

Exercise 3.4.1

1.

Function	*Rate of change*	
	$t = 1$	$t = 2$
$y = 3^t$	approx 3.3	approx 9.9
$y = 2.5^t$	approx 2.3	approx 5.7

Exercise 3.4.2

1. rate of change (i) when $t = 1$ is approx -0.3 and (ii) when $t = 3$ is approx -0.02

Exercise 3.4.3

1.

n	$(1 + 1/n)^n$
1	2
1.5	2.151657
2	2.25
3	2.370370
4	2.441406
5	2.488320
10	2.593743
100	2.704814
1000	2.716924
10000	2.718146

Exercise 3.4.4

1. (a) 1.0000 (b) 0.1353 (c) 20.0855
   (d) 5.4739 (e) 1.3634 (f) 0.6570
   (g) 2.7183 (h) 1.6487
3. It is not a reflection in any line!
4. 0.567

Exercise 3.5.1

1. (a) 4 (b) −2 (c) 1 (d) 3
   (e) 1 (f) −2 (g) 2 (h) 6
2. (a) $x = \log_3 7$ (b) $x = \log_6 4.1$ (c) $x = \log_{10} 12.2$
   (d) $x = \log_e 4.1$ (e) $x = \log_2 8.3$ (f) $x = \log_5 7.6$

Exercise 3.5.2

1. (a) 1.3979 (b) 1.0000 (c) 0.8633
   (d) 3.0000 (e) 3.2189 (f) 2.3026
   (g) 1.0000 (h) 1.9879 (i) 0.3222
   (j) 0.0000 (k) 0.0000 (l) 0.6931
3. (a) 0.4314 (b) 1.4183 (c) 0.6534
   (d) −1.0459 (e) 1.0492 (f) 1584.8932
   (g) 4.4817 (h) 37.9038 (i) 6.3246

Exercise 3.5.3

1. 60 dB
2. 0.100 Wm^{-2} to three decimal places

Exercise 3.5.4

1. $T = 500e^{-0.4t}$
2. $t = 0.0805$ seconds
3. $t = 22.65$ minutes
4. $p_2 = 348.51$ pascals
5. (a) 86071, 74082, 47237, 22313
   (b) $h = 4.62$ km
   (c) $h = 15.35$ km

Exercise 3.5.5

1. (a) $2 \log p + 3 \log v$
   (b) $\ln p - \ln v$
   (c) $\log_5 6 + \log_5 p + \log_5(a + b)$
   (d) $\ln 0.21 + \ln t - 1$
   (e) $\log s + \log t - 1/2 \log 3$
   (f) $\ln 0.1 + 2 \ln v$
2. (a) 1 (b) $\ln 3$ (c) $\log \left(\dfrac{5wv}{u}\right)$
   (d) $\ln (u^3/v^2)$ (e) $\log(xy^2/p^3)$
   (f) 0
3. (a) $x = 2.7381$ (b) $x = -0.5217$
   (c) $x = 0.5438$

Exercise 3.6.1

1. $k = 0.0133\ \text{s}^{-1}$
2. $T = 10.60$ h
3. (a) $x = 0.1652\ \text{min}^{-1}$ (b) 4.20 min
4. A, D, E
5. (a) 7 (b) $1/2^7 \simeq 0.0078$ (c) 0.9999 (almost all of it!)

Exercise 3.7.1

1. (a) $y = 3.4\, x^{2.7}$
   (b) $v = 1.7\, u^{-1.3}$
3. $\mu = 47.2\, T^{-2.06}$
4. $C = 0.12\, x^{-2}$

Exercise 3.7.2

3. $H = 16.7\, M^{0.92}$

Exercise 3.8.1

1. $v = 13.2\, e^{-0.1t}$
2. Graph of $\ln(k)$ against $1/T$. $k = 5.1 \times 10^{12} \times e^{-30300/T}$

Objective Test 3

1. (a) y^3/x^2 (b) $\dfrac{1}{ab^4}$ (c) $\dfrac{1}{2xy^2}$

2. Domain $[3, \infty]$, range $[0, \infty]$
3. $f^{-1}(x) = \frac{1}{3}\ln(x)$, domain $[1, 2.117]$, range $[0, 0.25]$

$g^{-1}(x) = \dfrac{(1-x)}{4}$, domain $[0, 1]$, range $[0, 0.25]$

$f[g(x)] = e^{3(1-4x)}$

4. (a) $x = 0.4720$ (b) $x = 0$
   (c) $x = 20.09$ (d) $x = 626.48$
   (e) $x = 7.5546$ (f) $x = 2.1133$
5. Half-life = 5599 yr, age = 3480 yr
6. $y = 3\,e^{-4x}$, $s = 4.2\,t^{1.7}$

CHAPTER 4

Exercise 4.2.1

1. (a) 140° (b) 290° (c) 50° (d) 233° (e) 300°
2. (a) −160° (b) −161° (c) −22° (d) −45° (e) −80°

Exercise 4.3.1

2. 0.5, −0.8660; −0.5, −0.8660; −0.5, 0.8660; −0.5, 0.8660
3. 0, −1; −1, 0; 0, 1

Exercise 4.5.1

1. (b) 57.3, 573.0, 5729.6; −5729.6, −573.0, −57.3

Exercise 4.6.1

4. −0.342 5. 0.458 6. −1.192
7. −0.7 8. −0.1 9. 0.39

Exercise 4.7.1

1. (a) ±98.2° (b) 7.2°, 172.8° (c) ±76.7°
   (d) −5.2°, −174.8° (e) −30°, −150° (f) ±90°
2. (a) 46°48′, 133°12′ (b) ±66°44′
   (c) −23°35′, −156°25′ (d) ±112°1′

Exercise 4.8.1

1. (a) $14.5° + 360°n$; $165.5° + 360°n$
   (b) $42° + 180°n$
   (c) $\pm131.8° + 360°n$
   (d) $\pm99.6° + 360°n$
   (e) $-41.8° + 360°n$; $221.8° + 360°n$
   (f) $45° + 180°n$

Exercise 4.9.1

3. 360°   4. 180°   5. 180°, 480°, 5

Exercise 4.9.2

1. 2, 4; 360°, 180°
2. 2, 1; 360°, 720°; 1, 0.5
3. 24°, 66°, 204°, 246°
4. 240°, 3/2
5. (a) 72° 5
   (b) 1440° 0.25
   (c) 600° 3/5
   (d) 360/k k
   (e) 360/k k

Exercise 4.9.3

1.

	Amplitude	*Max*	*Min*	*y = 0 at*
(a)	1	1 at −270°, 90°	−1 at −90°, 270°	0°, ±180°, ±360°
(b)	2	2 at −270°, 90°	−2 at −90°, 270°	0°, ±180°, ±360°
(c)	0.5	0.5 at −270°, 90°	−0.5 at −90°, 270°	0°, ±180°, ±360°
(d)	1	1 at 0°, ±360°	−1 at ±180°	±90°, ±270°
(e)	3	3 at 0°, ±360°	−3 at ±180°	±90°, ±270°
(f)	$\frac{1}{4}$	$\frac{1}{4}$ at 0°, ±360°	$-\frac{1}{4}$ at ±180°	±90°, ±270°
(g)	∞	*n/a*	*n/a*	0°, ±180°, ±360°

2. (a) $\frac{1}{3}$, 360°   (b) 5, 360°   (c) ∞, 180°
3. (a) 5, 360°   (b) 10, 180°   (c) 7, 480°   (d) 1, 90°
4. (a) 2, 3, 120°   (b) 4, 2, 180°   (c) 4.3, $\frac{3}{4}$, 480°
   (d) 0.73, 0.21, 1714°

Exercise 4.9.4

1. Shift 10° to the left. Period 360°, amplitude 1
2. Shift 40° to the right. Period 360°, amplitude 1. sin(θ + 50°)
3. 0°, 180°, 360°; 155°, 335°; 65°, 245°

4. $\tan\theta$: asymptotes $\theta = \pm 90°$; zero at $\theta = 0°, \pm 180°$
   $\tan(\theta + 20°)$: asymptotes $\theta = -110°, 70°$; zero at $\theta = -20°$, $160°$
5. (a) $240°, 2, 30°$; $y = 2\sin(\frac{3}{2}\theta + 30°)$
   (b) $120°, 1.7, 180°$; $y = 1.7\sin(3\theta + 180°)$

Exercise 4.10.1

1. (a) 73.4°, 106.6°, 253.4°, 286.6°
   (b) 53.7°, 126.3°, 233.7°, 306.3°
   (c) 35.3°, 144.7°, 215.3°, 324.7°
   (d) 65.9°, 114.1°, 245.9°, 294.1°
   (e) no solutions
   (f) 40.9°, 139.1°, 220.9°, 319.1°
2. (a) 45°, 135°, 225°, 315°
   (b) 60°, 120°, 240°, 300°
   (c) 0°, 180°, 360°
   (d) 60°, 120°, 240°, 300°
3. (a) $\pm 30° + 180°n$ (b) $\pm 48.2° + 180°n$
   (c) $\pm 53.1° + 180°n$ (d) $\pm 54.7° + 180°n$
   (e) $\pm 54.7° + 180°n$ (f) $\pm 21.5° + 180°n$
   (g) $\pm 50.8° + 180°n$ (h) $45° + 90°n$
   (i) $\pm 62.7° + 180°n$

Exercise 4.10.2

1. (a) $24.3° + 180°n$, $65.7° + 180°n$
   (b) $247.4° + 360°n$
   (c) $\pm 46.2° + 120°n$
2. (a) 29.2°, 240.8° (b) no solutions
   (c) 45°, 225°
3. (a) −104.9, 126.9 (b) −159.0°, 21.0°
   (c) −40.0°, 80.0° (d) −133.9°, 46.1°

Exercise 4.11.1

1. (a) 7.2 (b) 96.4° (c) 81.9°
   (d) 130.9° (e) −23.6° (f) 34.3°
2. (a) 84°11′, 264°11′ (b) 116°34′, 243°26′
   (c) 142°44′, 322°44′ (d) 35°16′, 114°44′
   (e) 211°5′, 328°55′ (f) 66°25′, 293°35′
3. (a) ±48.2° (b) −154.2°, −25.8°
   (c) −36.9°, 143.1° (d) ±131.8°
   (e) 16.6°, 163.4° (f) −135.0°, 45.0°
4. (a) 360°; asymptotes at $x = 90°$ and $270°$
   (b) 360°; asymptotes at $x = 0°, 180°$ and $360°$
   (c) 180°; $\tan x$ has asymptotes at $x = 90°, 270°$ and $\cot x$ at $x = 0°, 180°, 360°$

Exercise 4.12.1

1. (a) 420° (b) −30° (c) 225° (d) 540°
2. (a) $\frac{5}{12}\pi$ (b) $\frac{5}{3}\pi$ (c) $-\frac{\pi}{6}$ (d) $\frac{5}{2}\pi$
3.

	Period	*Amplitude*	*Zero at*
(a)	$\frac{2}{3}\pi$	2	$0, \frac{\pi}{3}, \frac{2\pi}{3}, \pi, \frac{4\pi}{3}, \frac{5\pi}{3}, 2\pi$
(b)	2π	0.5	$\frac{\pi}{2}, \frac{3\pi}{2}$
(c)	π	5	$0, \frac{\pi}{2}, \pi, \frac{3\pi}{2}, 2\pi$

4. (a) $3\pi/2$ (b) $\pi/3, 5\pi/3$ (c) $\pi/3, 2\pi/3$
   (d) $3\pi/4, 7\pi/4$ (e) $\pi/3, 4\pi/3$
5.

	Period	*Amplitude*
(a)	$\frac{2\pi}{3}$	1
(b)	π	10
(c)	$\frac{8\pi}{3}$	0.5
(d)	3π	$\sqrt{2}$
(e)	$\frac{2\pi}{k}$	1
(f)	$\frac{2\pi}{k}$	a
(g)	1	b
(h)	$\frac{2}{n}$	1

Exercise 4.13.1

1.

	Period (seconds)	*Frequency (hertz)*
(a)	π	$\frac{1}{\pi}$
(b)	2	0.5
(c)	$\frac{2\pi}{k}$	$\frac{k}{2\pi}$
(d)	$\frac{2}{k}$	$\frac{k}{2}$

2. 4 cm, 1 Hz
3. $a = 30$ mm, $\omega = 4\pi$ Hz
4. $i = 75 \sin(120\pi t)$ mA
7. (c) (e) are true
8. Amplitude = 5°, period = 3 seconds, phase angle = 0.75^c,

   $\theta = 5 \cos\left(\frac{2\pi}{3} t\right)$

   $\theta = 5°$ when $t = 3n$ seconds, $n = 0, 1, 2, \ldots$

   $\theta = 2.5 \cos\left(\frac{4\pi}{3} t\right)$

Objective Test 4

1. 140°
2. $-12.7° + 180°n$, $102.7° + 180°n$
3. 0.0717, 1.50, 2.17, 3.59, 4.26, 5.69
4. $50.2° + 180°n$
5. 1080°
6. $\frac{8\pi}{5}$, 2
7. Translate 30° to the right
8. $x = \pm \frac{n\pi}{2}$
9. $\frac{5}{2\pi}$ Hz

CHAPTER 5

Exercise 5.2.1

1. (a) $\dfrac{\cos\theta}{\sin\theta}$ (b) $\dfrac{1}{\sin\theta}$ (c) $\dfrac{1}{\cos^2\theta}$

   (d) $\sin\theta\cos\theta$ (e) $\cos\theta$ (f) $\dfrac{1}{\sin\theta\cos\theta}$

   (g) $\dfrac{\sin\theta}{\cos\theta}$ (h) $\sin^2\theta - \cos^2\theta$

Exercise 5.2.2

1. (a) 77/85 (b) 13/85 (c) −36/85
   (d) 84/85 (e) −77/36 (f) 13/84

   (A + B) is in quadrant II
   (A − B) is in quadrant I

2. (a) 0.8912 (b) 0.4712 (c) −0.4535
   (d) 0.8820 (e) −1.9652 (f) 0.5343

   (A + B) is in quadrant II
   (A − B) is in quadrant I

Exercise 5.2.4

2. (a) $5\sin(x + 53.1°)$ (b) $13\sin(x + 22.6°)$
   (c) $\sqrt{2}\sin(x - 45°)$ (d) $\sqrt{13}\sin(x + 146.3°)$
   (e) $\sqrt{106}\sin(x + 150.9°)$ (f) $\sqrt{666}\sin(x + 54.5°)$

Exercise 5.2.5

1. (a) 14.5°, 90°, 165.5°, 270° (b) 68.5°, 291.5°
   (c) 90°, 180°, 270° (d) 0°, 45°, 180°, 225°, 360°
   (e) 53.8°, 159.9°, 233.8°, 339.9° (f) 96.8°, 263.2°
   (g) 0°, 360°, 126.9°
2. (a) 103.3°, 330.5° (b) 157.4°
   (c) 0°, 30°, 120°, 150°, 240°, 270°, 360°
   (d) 39.0°, 162.8°, 219.0°, 342.8°

Exercise 5.3.1

1.

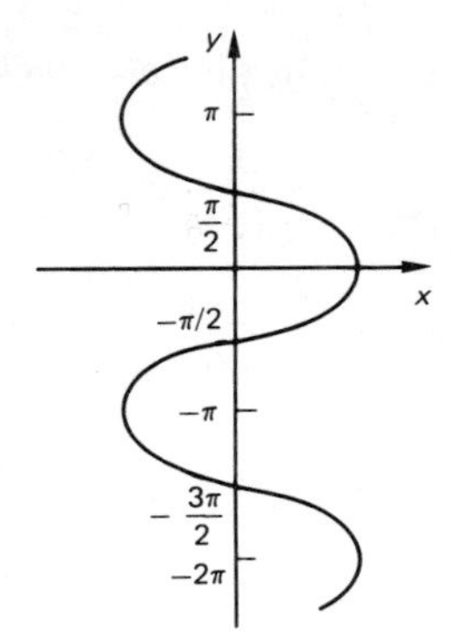

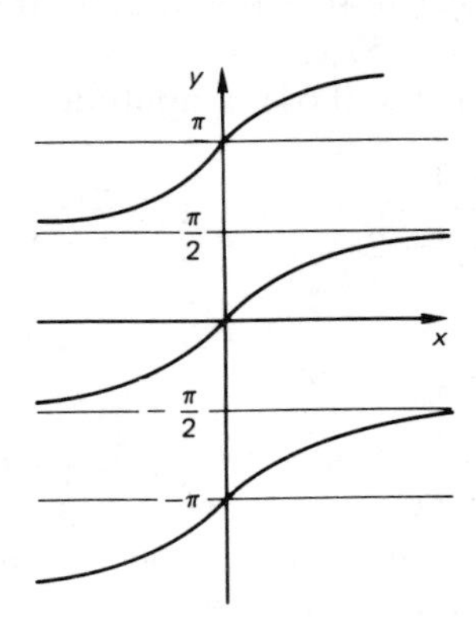

2. 23.6°, 156.4°, 383.6°, 516.4°, 743.6°, 876.4°

Exercise 5.3.2

1. (a) $f(x) = 30°$ or $\pi/6$ $\quad g(x) = 60°$ or $\pi|3$
   (b) $f(x) = -30°$ or $-\pi/6$ $\quad g(x = 120°$ or $2\pi/3$
   (c) $f(x) = 90°$ or $\pi/2$ $\quad g(x) = 0°$ or 0
   (d) $f(x) = 45°$ or $\pi/4$ $\quad g(x) = 45°$ or $\pi/4$
   (e) $f(x) = -38.3°$ or -0.67 $\quad g(x) = 128.3°$ or 2.24
   (f) $f(x) = -71.8°$ or -1.25 $\quad g(x) = 161.8°$ or 2.82
   (g) no function values
   (h) $f(x) = 11°$ or 0.19 $\quad g(x) = 79°$ or 1.38

2. (a) 1.7321 (b) 0.7141
   (c) −0.4364 (d) 0.8660
   (e) 0.2316 (f) no value

Exercise 5.4.1

1. $2a \sin(500t) \cos(5t)$, period $2\pi/5$, a is amplitude of original waves
2. $2a \sin 2ft \cos ft$
3. $2a \sin(\omega t + \varphi/2) \cos(\varphi/2)$
   Period is same as original wave. Amplitude is $2a \cos(\varphi/2)$ and phase $\varphi/2$.
   (a) $\varphi = 0$: wave is twice original amplitude.
   (b) $\varphi = \pi/2$: wave is amplitude $\sqrt{2}a$, and $\pi/4$ out of phase with original wave.
   (c) $\varphi = \pi$: two waves destroy each other.

Exercise 5.4.2

1. $x = A \sin \omega t$
5. $P = E^2 + I^2(R^2 + X^2) + 2EI\sqrt{R^2 + X^2}\sin(2t + \alpha)$ where $\alpha = \arctan(R/X)$

6. $x = 5.63 \sin(4t + 0.99)$, amplitude = 5.63 cm and period = 1.57 s. $t = 0.40$ s.
7. $50 \sin(5\pi t + 0.64)$, amplitude = 50, period = 0.4, phase = 0.64, $t = 0.059$.
8. 45°
9. $y = -\frac{1}{2}x + 2$

Exercise 5.5.1

1. (a) $C = 75°$, $b = 4.90$, $c = 6.69$
   (b) $C = 45°$, $b = 2.02$, $c = 3.39$
   (c) $A = 69.9°$, $b = 11.90$, $c = 6.36$
   (d) $B = 42.1°$, $C = 82.9°$, $c = 9.22$
   (e) $C = 22.0°$, $A = 117.0°$, $a = 0.29$
   (f) $C = 19.4°$, $B = 115.6°$, $b = 20.40$
   (g) $A = 48.2°$, $B = 15.8°$, $b = 1.24$
   (h) $B = 45.3°$, $C = 74.7°$, $c = 293$
2. $F_2 = 13.7$ newtons, angle = 42.6°
3. 104.6 metres
4. 036.9°
5. 18.7 metres.

Exercise 5.6.1

1. (a) $c = 6.29$, $B = 67.1°$, $A = 37.9°$
   (b) $c = 3.39$, $B = 107.7°$, $A = 29.2°$
   (c) $c = 8.58$, $B = 10.2°$, $A = 42.8°$
   (d) $b = 337$, $A = 30.1°$, $C = 39.9°$
   (e) $C = 116°$, $A = 22.7°$, $B = 41.3°$
   (f) $C = 21.3°$, $A = 4.8°$, $B = 153.9°$
   (g) $a = 17.7$, $B = 40.4°$, $C = 66.6°$
   (h) $b = 13.33$, $A = 32°$, $B = 43°$
2. 11.7 newtons, angle = 24.3°
3. Speed = 1119.5 knots, bearing = 267.1°
4. Distance = 62.6 miles, bearing = 078.2°

Objective Test 5

1. $\theta = \pi/3, 2.3$ or $\theta = 60°, 131.8°$
2. $\cos x = \dfrac{b \cos \alpha + a \sin \beta}{\cos(\alpha - \beta)}$, $\sin x = \dfrac{a \cos \beta - b \sin \alpha}{\cos(\alpha - \beta)}$

3. $x + y = 6 \sin 20t \cos 5t$

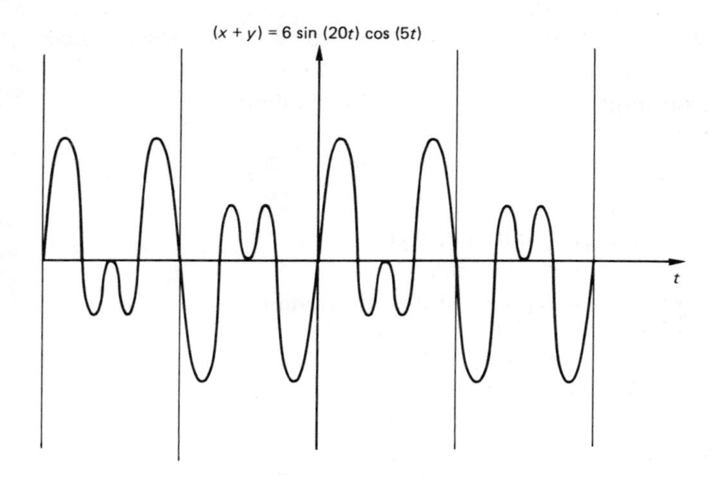

5. $5 \sin(2x + 36.9°)$; $x = 1.5°, 51.7°$
6. 29.6°, 87.4°, 6.3; 3.8, 45.2°, 94.8°

CHAPTER 6

Exercise 6.1.1

1. 1/10, 1/100, 1/1000
2. 1, 1/10, 1/100
3. 2.1, 2.01, 2.001
4. 4, 3.1, 3.01
5. 3/4, 3/16, 3/64
6. $-1, 1, -1$
7. $1, -1/10, 1/100$
8. 4, 2.9, 3.01
9. 3, 5, 7
10. $2\frac{1}{2}, 2, 1\frac{1}{2}$
11. $\left(\frac{1}{4}\right)^n$
12. $\left(\frac{1}{4}\right)^{n-1}$
13. $\left(\frac{1}{3}\right)^{n-1}$
14. $2(-1)^n$
15. $3(-1)^{n-1}$
16. $\left(\frac{1}{10}\right)^n$ or $(0.1)^n$
17. $5 + \left(\frac{1}{10}\right)^n$
18. $1 - \left(\frac{1}{10}\right)^n$
19. £6650, £4655, £3258.50; £9500$(0.7)^n$

Exercise 6.1.2

1. $\left(\frac{1}{4}\right)^{n-1}$, 0
2. $3 + \left(\frac{1}{2}\right)^{n}$, 3
3. $(-1)^{n-1}$, no limit
4. $5n$, no limit
5. $2\frac{1}{2} - \frac{n}{2}$, no limit
6. $10 + \left(\frac{1}{10}\right)^{n-1}$, 10
7. $0.8^n C_{\text{init}}$ where C_{init} = initial charge
8. $\sqrt{2}l$, $\left(\frac{1}{\sqrt{2}}\right)^{n-3} l$, the squares shrink to a point

Exercise 6.1.3

1. $\sum_{n=1}^{7} \left(\frac{1}{10}\right)^{n-1}$
2. $\sum_{n=1}^{8} \left(\frac{1}{2}\right)^{n}$
3. $\sum_{n=1}^{5} (2n - 1)$
4. $\sum_{n=1}^{10} (-1)^{n-1} 2n$
5. $\sum_{n=1}^{31} x^{n-1}$

Exercise 6.1.4

1. 5050
2. 2870
3. 18 496
4. 15 150
5. 2185
6. 180 125

Exercise 6.2.1

1. 29
2. 25.25
3. −7
4. 15, −2
5. 10
6. 101
7. 51

Exercise 6.2.2

1. 5050
2. 318.5
3. 7.5
4. 860, 1890, 1030
5. 825

Exercise 6.3.1

1. 1.2, 13.437 (three decimal places
2. −0.8, 13.422 (three decimal places)
3. 15th term
4. 7 seconds
5. £8192
6. 21.5%

Exercise 6.3.2

1. 453.320 (three decimal places)
2. 5.601 (three decimal places)
3. 32.675 (three decimal places)
4. 1/4
5. 26/21, 4

Exercise 6.4.1

1. $\frac{5}{9}$
2. $\frac{37}{99}$
3. $\frac{417}{999}$
4. $5\frac{18}{99}$
5. $5\frac{17}{90}$
6. $3\frac{442}{495}$

Exercise 6.4.2

1. 75
2. 5
3. Does not exist
4. $-3\frac{1}{3}$
5. Does not exist
6. 8
7. $3\frac{1}{3}$ m, $\frac{10}{17}$ m from A
8. 20 units, (8, 4)
9. 1.316 m (three decimal place), $\frac{25}{19}(1 - 0.62^n)$, $n = 11$
10. 60π cm

Exercise 6.5.1

1. $q = x^2 - x - 1$; $r = 2$
2. $q = 2x^2 + x - 3$; $r = 0$
3. $q = 6x^2 - 2x + 5$; $r = 8$
4. $q = 3x - 4$; $r = 20x - 21$
5. $q = -t^2 - t - 1$; $r = 1$
6. $q = \frac{1}{2}v - \frac{3}{4}$; $r = 2\frac{1}{4}$

Exercise 6.6.1

1. 5
2. 8
3. 0
4. -1
5. -48
6. Yes
7. Yes
8. No
9. Yes
10. No
11. $(x - 2)(x + 3)(x - 4)$
12. $(t + 1)(t + 2)(t + 6)$

Exercise 6.6.2

1. $2\frac{10}{27}$
2. 2
3. $-1\frac{5}{8}$
4. No
5. Yes
6. Yes
7. $(x - 2)(2x + 1)^2$
8. $(x + 1)(x - 3)(2x - 1)$
9. $(t + 1)(t - 2)(3t - 2)(t + 4)$
10. $(y - 1)(y - 2)(2y + 1)(3y - 2)$

Exercise 6.7.1

1. $x^2 + 9$
2. $(x + 3)(x - 3)$
3. $(x - 1)(x^2 + x + 1)$
4. $(x + 1)(x^2 - x + 1)$
5. $(x - 3)(x^2 + 2)$
6. $(t + 1)(t - 1)^2(t + 2)$
7. $(y + 3)(2y^2 + y + 3)$
8. $(3z - 1)(z^2 + 2z + 3)$

Exercise 6.8.1

1. Improper
2. Improper
3. Improper
4. Proper
5. Proper
6. Improper
7. $4 + \dfrac{2}{x - 1}$
8. $3 - \dfrac{(3t - 3)}{t^2 + 3t + 1}$
9. $s + 2 + \dfrac{3}{s - 1}$
10. $2y - 1 + \dfrac{1 - y}{1 + 2y - y^2}$

Exercise 6.8.2

1. $\dfrac{1}{x + 1} + \dfrac{1}{x - 1}$
2. $\dfrac{2}{1 - 2t} + \dfrac{3}{t + 1}$
3. $\dfrac{4}{t - 1} - \dfrac{3}{t + 2}$
4. $\dfrac{2}{x + 1} - \dfrac{4}{x - 1} + \dfrac{1}{x - 2}$
5. $\dfrac{1}{2(x - 1)} - \dfrac{1}{2(x + 1)}$

Exercise 6.8.3

1. $\dfrac{1}{t - 3} + \dfrac{1}{t + 2} - \dfrac{1}{(t + 2)^2}$
2. $\dfrac{1}{2(2x + 1)^2} + \dfrac{1}{2(2x + 3)}$
3. $\dfrac{2}{x} + \dfrac{5}{x^2} + \dfrac{1}{x - 1}$
4. $\dfrac{1}{t^3} + \dfrac{2}{t} + \dfrac{1}{t + 1}$
5. $\dfrac{3}{x^2} - \dfrac{1}{x} + \dfrac{1}{(x - 1)^2} - \dfrac{1}{x - 1}$

Exercise 6.8.4

1. $\dfrac{x}{x^2 + 2} + \dfrac{4}{x + 1}$
2. $\dfrac{3}{t} - \dfrac{1 + 2t}{t^2 + 3}$
3. $\dfrac{2}{s + 1} + \dfrac{1}{s + 2} + \dfrac{1 - s}{s^2 + 1}$
4. $\dfrac{1}{v} + \dfrac{2v - 3}{v^2 + 1}$
5. $\dfrac{1}{t} - \dfrac{4}{t^2} + \dfrac{3}{t^2 + 1}$

Exercise 6.8.5

1. $2v - 1 + \frac{3}{v} - \frac{1}{v+1}$
2. $2 + \frac{1}{t+1} - \frac{3}{(t+1)^2}$
3. $x + \frac{1-x}{x^2+2} + \frac{3}{x}$
4. $5 + \frac{2}{x-2} - \frac{1}{x-1}$

Exercise 6.8.6

1. $\frac{1}{6(x-3)} - \frac{1}{6(x+3)}$
2. $\frac{1}{2(v+1)} - \frac{1}{2(v+3)}$
3. $\frac{3}{x-2} + \frac{1}{x+2}$
4. $\frac{2}{v+3} - \frac{1}{2v+1}$
5. $\frac{1}{2x} - \frac{1}{3(x+1)} + \frac{1}{6(x+2)}$
6. $1 - \frac{2}{t} + \frac{3}{t-1}$
7. $\frac{2}{t} - \frac{t}{t^2+1}$
8. $\frac{1-2x}{x^2+1} + \frac{1}{(x+1)^2} - \frac{2}{x+1}$

Exercise 6.9.1

1. $x^6 - 6x^5y + 15x^4y^2 - 20x^3y^3 + 15x^2y^4 - 6xy^5 + y^6$
2. $a^7 + 7a^6b + 21a^5b^2 + 35a^4b^3 + 35a^3b^4 + 21a^2b^5 + 7ab^6 + b^7$
3. $243p^5 + 405p^4q + 270p^3q^2 + 90p^2q^3 + 15pq^4 + q^5$
4. $81x^4 - 216x^3y + 216x^2y^2 - 96xy^3 + 16y^4$
5. $1 - 5x^2 + 10x^4 - 10x^6 + 5x^8 - x^{10}$
6. $81v^4 - 54v^3 + 27/2\ v^2 - 3/2\ v + 1/16$
7. $1 - 3/x + 3/x^2 - 1/x^3$
8. $t^6 + 6t^4 + 15t^2 + 20 + 15/t^2 + 6/t^4 + 1/t^6$

Exercise 6.9.2

1. 24
2. 1
3. 3 628 800
4. 840
5. 30 240
6. 9
7. 11.10
8. 11.5
9. n
10. $n(n-1)(n-2)$

Exercise 6.9.3

1. 15
2. 56
3. 252
4. 12
5. 3
6. 21

Exercise 6.9.4

1. $u^3 + 6u^2v + 12uv^2 + 8v^3$
2. $x^4 - 2x^3 + \frac{3}{2}x^2 - \frac{1}{2}x + \frac{1}{16}$

3. $32t^{10} + 80t^7 + 80t^4 + 40t + \dfrac{10}{t^2} + \dfrac{1}{t^5}$

4. $11520s^8t^2$
5. $924x^6y^6$
6. $-48384v^5$

7. $\dfrac{21}{32}p^5$
8. $\dfrac{15}{4}$
9. 490

Exercise 6.10.1

1. $1 + 4x + 12x^2 + 32x^3$; $|x| < \frac{1}{2}$
2. $2^{1/5}\left(1 + \dfrac{1}{10}y - \dfrac{1}{50}y^2 + \dfrac{3}{500}y^3\right)$; $|y| < 2$
3. $3^{2/5}\left(1 - \dfrac{4}{15}t - \dfrac{4}{75}t^2 - \dfrac{64}{3375}t^3\right)$; $|t| < 3/2$
4. $x^{1/2} - \dfrac{1}{2}x^{-1/2}y - \dfrac{1}{8}x^{-3/2}y^2 - \dfrac{1}{16}x^{5/2}y^3$; $|y| < |x|$
5. $10 + \dfrac{1}{20}x - \dfrac{1}{8000}x^2$; 10.149
6. 3.936
7. $\dfrac{1}{2} + \dfrac{x}{4} - \dfrac{x^2}{8} + \dfrac{x^3}{16}$

Objective Test 6

1. $\dfrac{3(-1)^n}{2^{n-1}}$
2. 1
3. 16, 5
4. 1.92 mm
5. $-3\dfrac{1}{32}$
6. $(x - 2)(x + 3)(2x - 1)$
7. $\dfrac{x}{x^2 + 2} + \dfrac{1}{2x + 3}$
8. $2x + \dfrac{1}{x} - \dfrac{3}{2x + 1}$
9. $16x^4 - 16x^3y + 6x^2y^2 - xy^3 + \dfrac{1}{16}y^4$
10. $2 - \dfrac{t}{12} - \dfrac{t^2}{288}$

CHAPTER 7

Exercise 7.1.1

The complete list of average velocities (ms^{-1}) is

0.7, 2.1, 3.8, 5.7, 8.1, 11.3, 15.1, 20.1.

Exercise 7.1.2

1. Yes. We shall see that the instantaneous rate of change is obtained from the average rate of change as time steps decrease in size.
2. Average decay rates are correct to three decimal places.

0.118	0.104	0.092	0.081	0.071	0.063
0.056	0.049	0.043	0.038	0.034	0.030

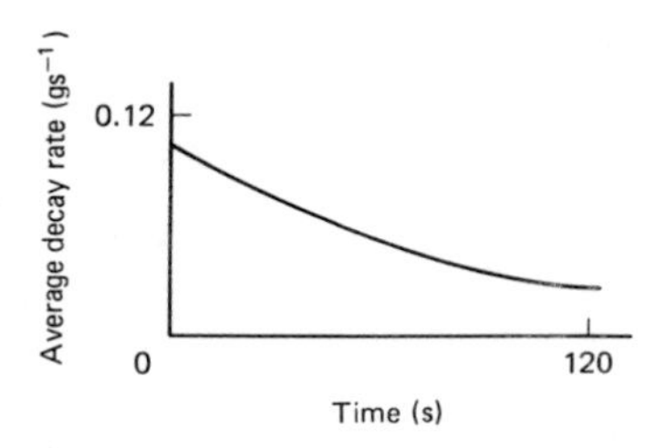

Exercise 7.3.1

1. Instantaneous velocity is 7 ms^{-1}.
2. Instantaneous rate of change of current is -0.4 As^{-1}.

Exercise 7.3.2

2. Gradient of $f(t) = t^2$ is 4; gradient of $f(t) = t^5$ is 80

Exercise 7.4.1

1. $-2t$ 2. $9t^2 - 8t$ 3. $x + 1$

Exercise 7.6.1

1. $5x^4$ 2. $-3x^{-4}$ 3. $-t^{-2}$ 4. $\frac{1}{2} t^{-1/2}$

5. $\dfrac{3}{2} t^{1/2}$ 6. $1.1v^{0.1}$ 7. $-1.1v^{-2.1}$ 8. $\dfrac{1}{3} u^{-2/3}$

9. $-\dfrac{1}{4} u^{-5/4}$ 10. $1x^0 = 1$

Exercise 7.6.2

1. $4x^3 - 3x^2 + 8x - 8$
2. $-\frac{1}{2}x^{-1/2}$
3. $-5x^{-2} + 6x^{-3} - 16x^{-5}$
4. $\frac{1}{2}t^{-1/2} + 3$
5. $24t^{-4}$
6. $-1.2v^{-2.2}$
7. $\frac{4}{3}u^{-2/3} + \frac{7}{2}u^{-1/2}$
8. $-\frac{2}{3}t^{-1/3} + t^{-3/2}$
9. $6t - \frac{1}{4}$
10. $2\pi t$

Exercise 7.7.1

1. $\frac{3}{2}x^{1/2}$
2. $-\frac{1}{2}x^{-3/2}$
3. $1 - \frac{1}{2}x^{-1/2}$
4. $3 - \frac{7}{x^2}$
5. $2.1x^{1.1} - 1.1x^{0.1}$

Exercise 7.7.2

1. $6t$
2. -2
3. 0
4. $\frac{2}{t^3}$
5. $6t^{-4}$
6. $-\frac{2}{9}x^{-5/3}$
7. 2
8. $x + 1$
9. $\frac{4}{9}x^{-7/3}$
10. 0

Exercise 7.8.1

1. $\frac{1}{6}$
2. $-\frac{1}{9}$
4. $\left(\frac{1}{3}, \frac{65}{27}\right), \left(\frac{3}{5}, \frac{59}{25}\right)$
6. $12x - 12$; $x = 1$ m
7. 1.832 sm^{-1}
8. 250 Jsm^{-1}
9. 6 WA^{-1}
11. $t = 10$ s
12. $3y = 17x - 21$
13. $(2, 6), (-2, -6), y + 3x = \pm 12$
14. $y = x + 3$; $(0, 3)$
15. $y = 8x - 21$

Exercise 7.9.1

1. $-2x^3(2 - x^4)^{-1/2}$
2. $2(4 - x)^{-3}$
3. $3(2x + 1)(x^2 + x + 1)^2$
4. $0.3x^2(x^3 - 1)^{-0.9}$
5. $\frac{1}{2}(3t^2 - 6t + 1)(t^3 - 3t^2 + t - 1)^{-1/2}$

6. $-\dfrac{3}{2}(1 + 3t)^{-1/2}$

7. $1 - (1 + t)^{-2}$

8. $2t - \dfrac{1}{2}(1 - t)^{-1/2}$

9. $-2.4t(1 - t^2)^{0.2}$

10. $-v^{-2} + \dfrac{2}{3}v(v^2 - 2)^{-2/3}$

Exercise 7.9.2

1. $94.2\ \text{cm}^2\text{s}^{-1}$

2. $0.00796\ \text{cms}^{-1}$

3. $-\dfrac{2RI}{(t + 1)^2}$

4. $0.0111\ \text{ms}^{-1}$

5. $\dfrac{500}{\pi w^2}$

6. $\dfrac{-2}{(2s + 1)^3}$; $-0.0741\ \text{ms}^{-2}$, $-0.000216\ \text{ms}^{-2}$; $-2.46 \times 10^{-7}\ \text{ms}^{-2}$

Exercise 7.10.1

1. -4; decreasing
2. -1; decreasing
3. 39; increasing
4. 0; stationary
5. $-1/\sqrt{2}$; decreasing
6. $\left(\dfrac{1}{6}, \dfrac{1}{12}\right)$
7. $(-\frac{1}{2}, -1)$
8. $(0, 1)$
9. $(0, 1)$
10. $(0, 0)$
11. None
12. $(1, -3)$ and $(-1, 5)$

Exercise 7.10.2

1. $(-1, 2)$ minimum
2. $(0, 1)$ inflection
3. None
4. $(1, 2)$, $(-1, 2)$ minima

Exercise 7.10.3

1. Max $\left(\dfrac{1}{2}, \dfrac{17}{4}\right)$
2. Min $(-2, -16)$, max $(2, 16)$
3. Max $(-3, -27\frac{1}{3})$, min $(3, 27\frac{1}{3})$
4. Min $(0, 1)$
5. Max $(-1, 3)$, min $(0, 0)$, max $(1, 3)$

Exercise 7.11.1

1. $2x(2x + 1)^2(5x + 1)$
2. $\frac{1}{2}(x + 1)^{-1/2}(3x + 2) + 2x$
3. $-x(1 - x^2)^{1/2}(1 + 5x^2)$
4. $\frac{2x^2(x + 6)}{(4 + x)^2}$
5. $\frac{44}{(10 - 7x)^2}$
6. $\frac{2x(1 + x^2)}{(1 + 2x)^{3/2}} + 2$
7. $\frac{1}{2}(1 - t)^{-1/2}(1 - 6t)$
8. $\frac{1}{3}t(6 - t)^{-2/3}(36 - 7t)$
9. $1 + \frac{1 - t^2}{(1 + t^2)^2}$
10. $3t^2 + \frac{t^{1/2}}{2}\left[\frac{9 + 2t}{(3 + 2t)^2}\right]$
11. $(-1, -\frac{1}{2})$ minimum, $(1, \frac{1}{2})$ maximum
12. $(1, \frac{1}{4})$ maximum

Exercise 7.12.1

1. 1.633 m from A, deflection of 2.2 mm
2. $R_1 = R_2 = 50$ kΩ; $P = 0.1$ W
3. 62.8 mpg at 51.8 mph
4. 37.5 m × 25 m
5. $h = 85.4$ cm, $r = 39.8$ cm
6. One 1 m diameter; other zero diameter

Exercise 7.13.1

1. $\cos x$
2. $-\sin x$ (i.e. the sine graph inverted)

Exercise 7.13.3

1. $-\text{cosec}^2 x$
2. $\sec x \tan x$
3. $-\text{cosec}\, x \cot x$
4. $-5 \sin 5x$
5. $\frac{1}{2}\sec^2(\frac{1}{2}x)$
6. $-\sin x + 2 \cos x$
7. $2 \sin x \cos x$
8. $6 \tan x \sec 2x$
9. $2x \cos(x^2)$
10. $-4 \cos 2x \sin 2x$
11. $-\sin \alpha \sin x$
12. $\cos 2x$
13. $2\omega \cos \omega t$, $-2\omega^2 \sin \omega t$
14. $-2\omega\, \text{cosec}\, 2\omega t \cot 2\omega t$, $4\omega\, \text{cosec}\, 2\omega t\, (\cot^2 2\omega t + \text{cosec}^2 2\omega t)$
15. $\cos \omega t - \omega t \sin \omega t$, $-\omega\,(2 \sin \omega t + \omega t \cos \omega t)$
16. $-2\omega \sin(2\omega t + \alpha)$, $-4\omega^2 \cos(2\omega t + \alpha)$
17. $2\omega \sin(\omega t - \alpha) \cos(\omega t - \alpha) \equiv \omega \sin[2(\omega t - \alpha)]$, $2\omega^2 \cos[2(\omega t - \alpha)]$
18. $\omega \cos 2\omega t$, $-2\omega^2 \sin 2\omega t$

19. $\theta = \frac{\alpha}{2} + \frac{\pi}{2}$

20. $\theta = \arctan \sqrt{2}$ ($\simeq 54.7°$), $V = 2\pi l^3/9\sqrt{3}$

Exercise 7.14.1

3. 1.00
4. 1.10

Exercise 7.14.2

1. $-6e^{-6x}$
2. $\frac{1}{2}(e^x - e^{-x})$
3. $-2xe^{-x^2}$
4. $e^{2x}(1 + 2x)$
5. $-e^{1-x}$
6. $xe^{-x}(2 - x)$
7. $e^{-2t}(\cos t - 2 \sin t)$
8. $e^{at}(a \cos \omega t - \omega \sin \omega t)$
9. $\dfrac{4}{(e^t + e^{-t})^2}$
10. $e^{-t}[\omega\cos(\omega t + \alpha) - \sin(\omega t + \alpha)]$
11. $t^2e^{t^2}(2t^2 + 3)$
12. $\dfrac{e^{at}(at - 1)}{t^2}$
13. $y = ex$, $ey + x = e^2 + 1$, $A(0, 0)$, $B(e^2 + 1, 0)$, $\frac{1}{2}e(e^2 + 1)$
14. $(-1, -e)$, minimum
15. $5e^{-3t}$ $(2 \cos 2t - 3 \sin 2t)$, $5e^{-3t}(5 \sin 2t - 12 \cos 2t)$; 0.294 s

Exercise 7.14.3

1. $a^x \ln a$
2. $2^x \ln 2$, $3^{2x} \ln 9$

Exercise 7.15.1

1. $\dfrac{1}{x}$
2. $\dfrac{2}{x}$
3. $\dfrac{2}{x}$
4. $\dfrac{2}{x}$
5. $\dfrac{3}{x}$
6. $\dfrac{1}{2x}$
7. $1 + \dfrac{1}{t}$
8. $\dfrac{1}{1 + t}$
9. $-\dfrac{1}{t^2} - \dfrac{1}{t}$
10. $\dfrac{2at}{1 + at^2}$
11. $\dfrac{a^2}{t}$
12. $2t \ln(2t) + t$
13. $\dfrac{2 - \ln t^2}{t^2}$
14. $\dfrac{2t \ln t - t}{(\ln t)^2}$
15. 1
16. $\dfrac{6 \ln t}{t}$
17. $y = \frac{1}{2}x + \ln\left(\dfrac{2}{e}\right)$
18. (0, 0), minimum

Exercise 7.16.1

(1)

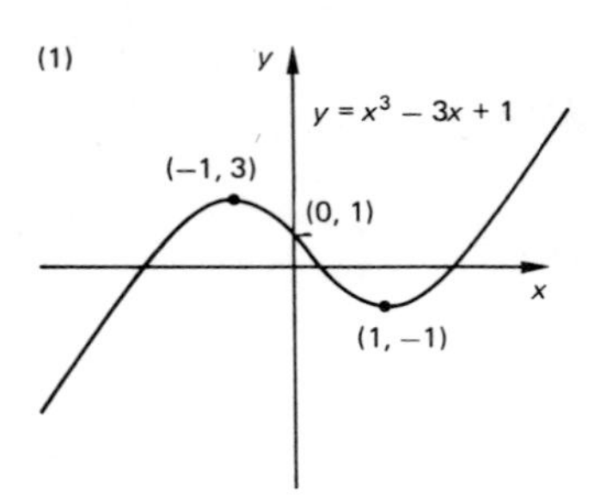

(2)

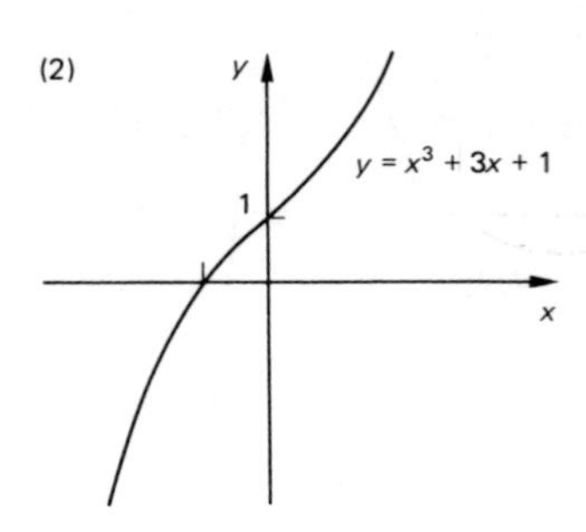

(3)

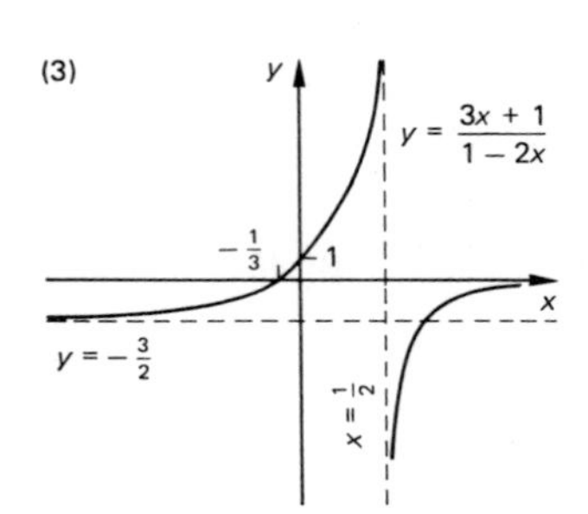

(4)

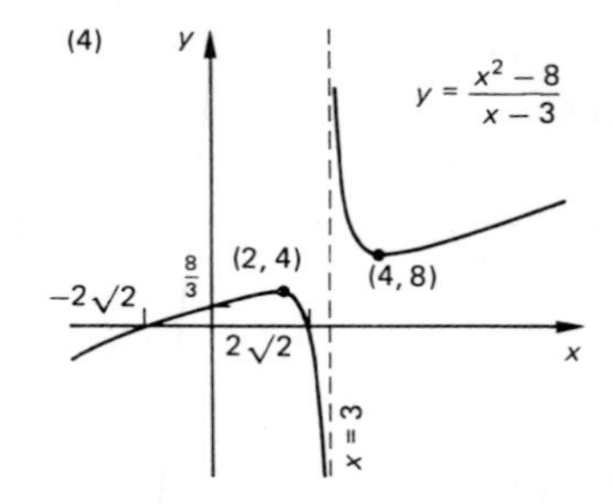

(5)

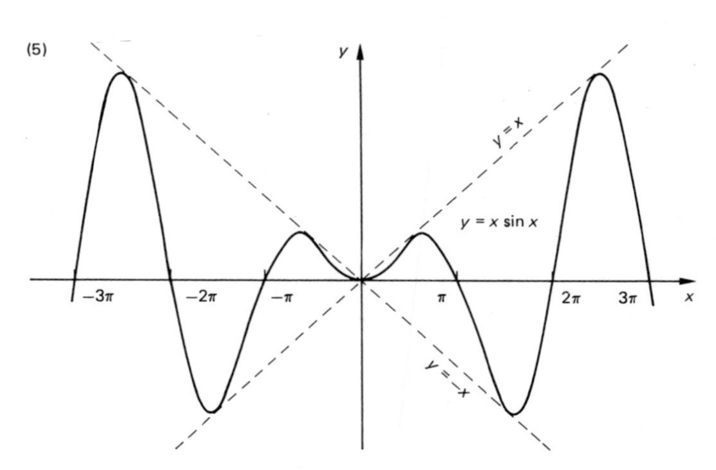

(6)

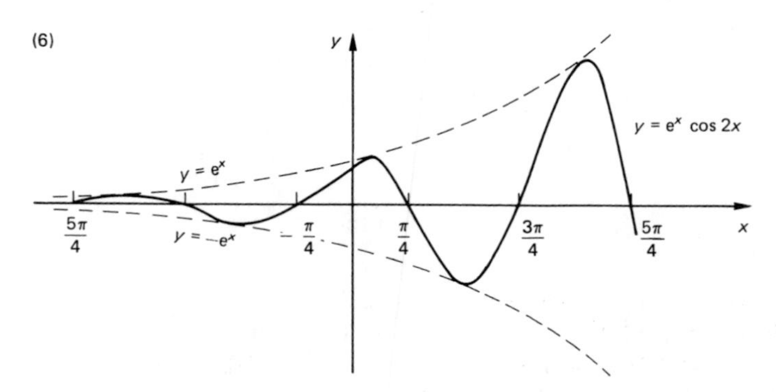

(7)

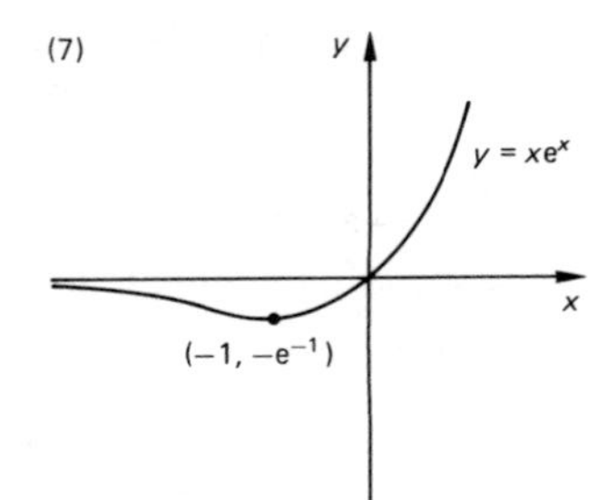

(8)

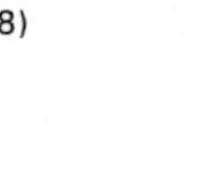
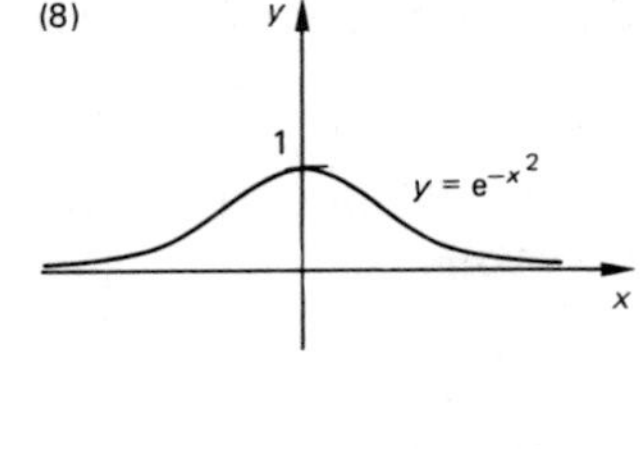

(9)

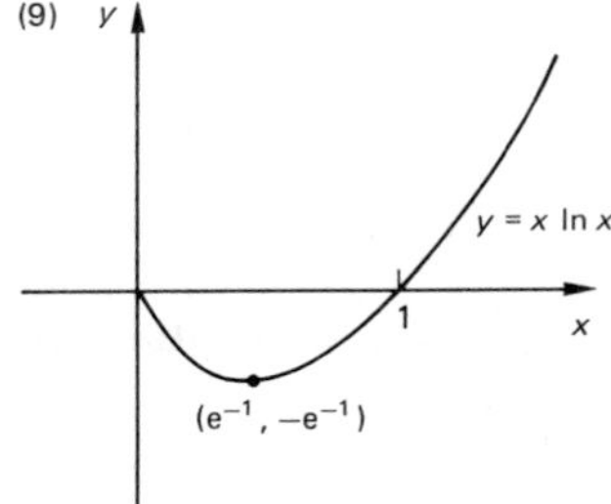

(10)

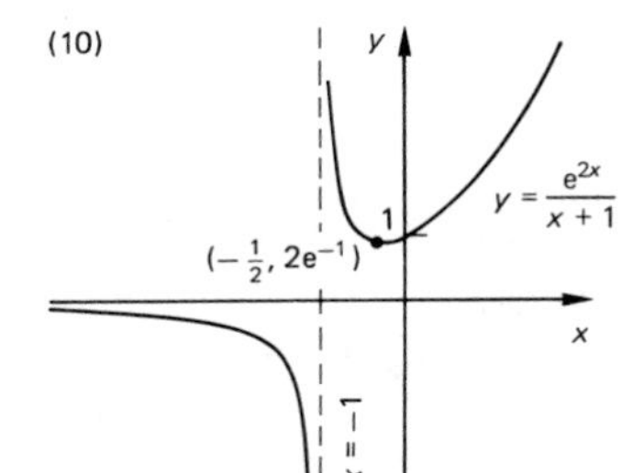

Exercise 7.17.1

1. $6 \cos(2x)$
2. $-\frac{1}{2} \sec^2 \left(\frac{\pi}{4} - \frac{x}{2}\right)$
3. 0
4. $(2x \, e^{-2x} - 2x^2 e^{-2x}) \sin(2x) + 2x^2 \, e^{-2x} \cos(2x)$
5. $\dfrac{-2x^2 \sin(2x) - 2x \cos(2x)}{x^4}$
6. $2 \cos(2x) \cos(3x) - 3 \sin(2x) \sin(3x)$
7. $3 \, e^{3x} \ln(2x + 1) + \dfrac{2e^{3x}}{(2x + 1)}$
8. $\sec^2(x) \sin(2x) + 2 \tan(x) \cos(2x)$
9. $12 \sin(1 - 4x)$
10. $20 \, e^{4t}$
11. $-14 \, e^{-2t}$
12. $e^{2t} + 2te^{2t}$
13. $2t \, e^{t^2}$
14. $2e^{2t} \sin(4t) + 4e^{2t} \cos(4t)$
15. $\dfrac{2t \, e^{4t} - 4t^2 \, e^{4t}}{e^{8t}}$
16. $\dfrac{2}{t}$
17. $-\dfrac{4}{(3 - 4t)}$
18. $v = \dfrac{ds}{dt} = 2 + 10t$

    (a) $t = 0, v = 2$ (b) $t = 2, v = 22$ (c) $t = 4, v = 42$
19. $dT/dt = -500 \, e^{-5t}$

    (a) $t = 0, dT/dt = -500$ (b) $t = 1, dT/dt = -500 \, e^{-5}$
20. $dM/dt = 4.3(-2.1 \, e^{-2.1t}) = -9.03 \, e^{-2.1t}$

21. $v = \mathrm{d}s/\mathrm{d}t = 0.21\cos(0.7t)$ $\quad a = \mathrm{d}v/\mathrm{d}t = -0.147\sin(0.7t)$
    (a) $t = 0$, $v = 0.21$, $a = 0$
    (b) $t = 1$, $v = 0.21\cos(0.7) = 0.161$,
    $a = -0.147\sin(0.7) = -0.095$
22. (a) $r(t) = 9$ cm
    (b) $r = 3$ when $t = 1$ second
    (c) $\mathrm{d}r/\mathrm{d}t = 9/t^2 = 9$ when $t = 1$ (or $r = 3$)
    (d) Largest value of r is 12 cm $[\lim_{t\to\infty} r(t)]$
23. Speed $v = \mathrm{d}h/\mathrm{d}t = 21 - 10t$. The table shows the results.

    (a)

t (s)	1	2	3	4
v ($\mathrm{ms^{-1}}$)	11	1	−9	−19

    (b) The stone comes to rest when $v = 0$, i.e. when $t = 2.1$ seconds.
    (i) Initially v is positive so that the stone is climbing for $t < 2.1$.
    (ii) For times $t > 2.1$, the value of v is negative, so that the stone is falling.

24. $\dfrac{1 - t^2}{(1 + t^2)^2}$; $t = 1$; $\dfrac{9}{2}$; it approaches (0, 4) along the axis

25. $t = \dfrac{5}{2}$; $v = -\dfrac{1}{8}\mathrm{e}^{-5/4} \simeq -0.0358$

26. $-8.08\ \mathrm{ms^{-2}}$
27. (3, 1), $x + 3y = 6$
28. (a) 41 (b) 36 (c) $y = 36x - 31$ (d) $x + 36y = 1478$
29. (a) P is a maximum when $X = R$
    (b)

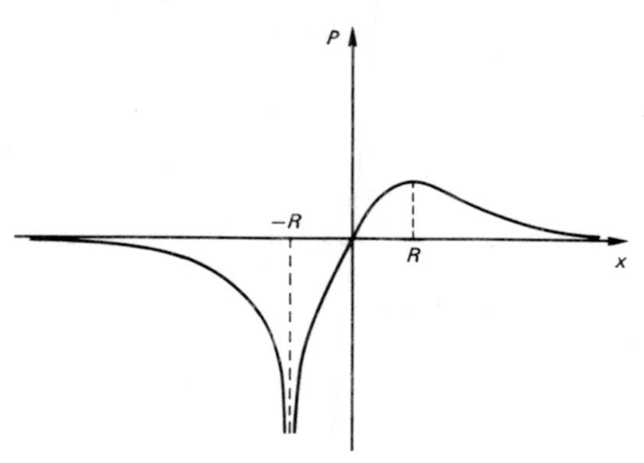

30. (b)

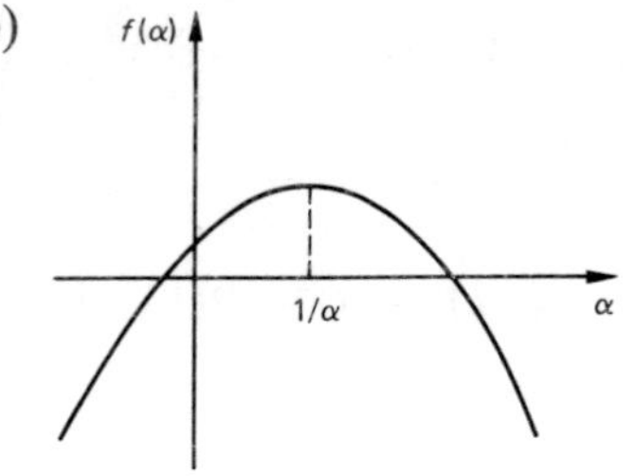

31.

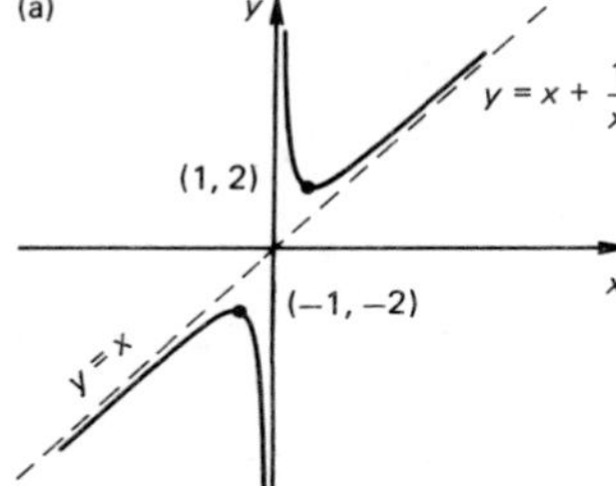

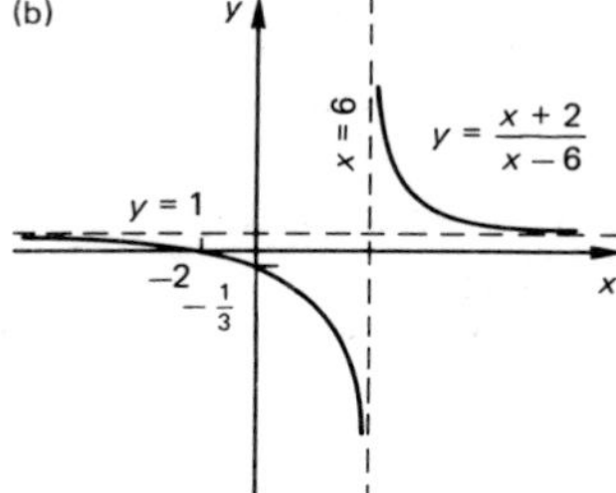

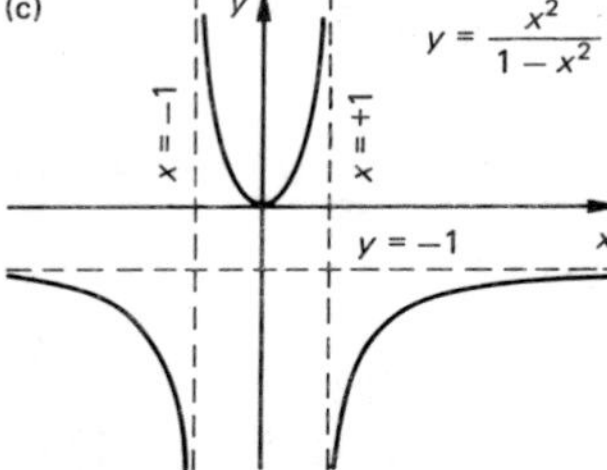

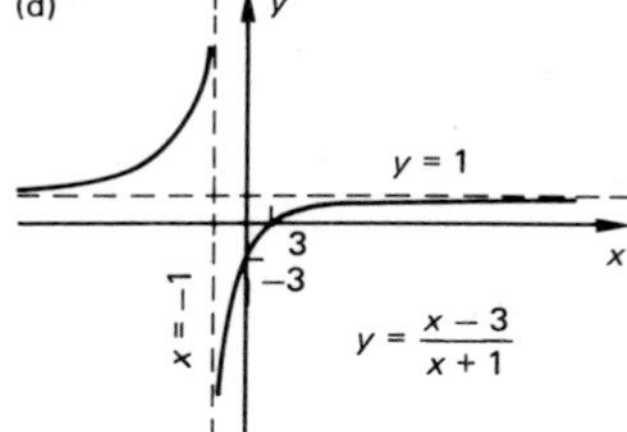

32. (a) $V_0 = \sqrt{\left(\frac{T}{3a}\right)}$

(b)

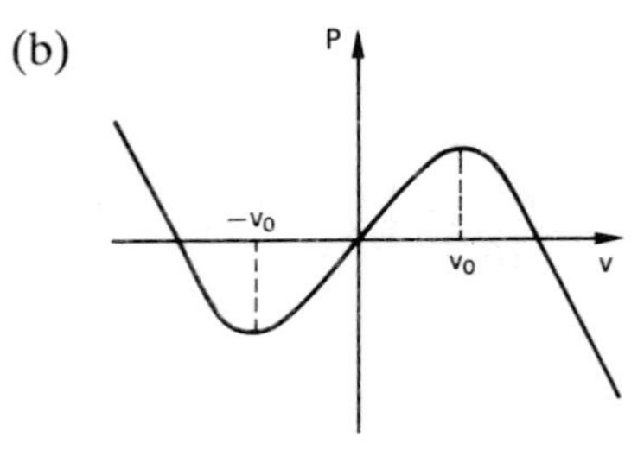

Objective Test 7

1. 15.08, 15.08 (three decimal places)
2. -2.297
3. $\frac{1}{x} - x - 2x \ln x$
4. $(1 - p)^{-2}, 2(1 - p)^{-3}$
5. $y = 1$
6. 36191 mm^3s^{-1}
7. -0.081
8. $\left(e^{-1/2}, -\frac{1}{2} e^{-1}\right)$
9. $\frac{e^{x^2}(1 + 2x + 2x^2)}{(1 + x)^2}$
10. $\frac{1}{x} + \cot x$

CHAPTER 8

Exercise 8.1.1

s should settle down to 35.33 m, but there may be trouble with rounding errors

Exercise 8.1.2

1. 3 m
2. $5\frac{1}{3}$ m, $-2\frac{1}{3}$ m
3. $\sqrt{12} \approx 3.46$ s

Exercise 8.3.3

1. $(q^6 - p^6)/6$
2. 20
3. 7.5
4. $2(2\sqrt{2} - 1)/3$
5. $\frac{1}{2}$
6. 12.4
7. 1.5125, $12\frac{2}{3}$, $1\frac{3}{4}$
8. $-4, 4, 0, 8$

Exercise 8.4.1

In each case c is an arbitrary constant.

1. $x + c$
2. $\frac{1}{3}x^3 + c$
3. $e^x + c$
4. $\frac{3}{16}x^4 + c$
5. $-\cos x + c$
6. $\frac{1}{3}x^3 + \frac{3}{2}x^2 + c$

Exercise 8.5.1

1. $-\cos t + e^t + c$
2. $\frac{u^3}{3} + \frac{3}{2}u^2 + 8u + c$

3. $-1/v + c$
4. $2p^{1/2} + c$ or $2\sqrt{p} + c$
5. $2\sqrt{x} + c$
6. $-e^{-x} + c$
7. $\tan x + c$
8. $-\cot x + c$

Exercise 8.6.1

1. 12.4
2. $\frac{2}{3}$
3. $3\frac{2}{3}$
4. $1\frac{1}{8}$
5. $1\frac{1}{3}$
6. $\frac{5}{6}$
7. $26\frac{2}{3}$
8. $25\frac{2}{3}$
9. $1\frac{11}{15}$
10. $3\frac{19}{24}$
11. $21\frac{1}{3}$
12. 53.99 (two decimal places)
13. $3\frac{2}{3}$
14. 12
15. $12\frac{2}{3}$
16. Integral $= -\frac{1}{6}$, area $= \frac{1}{6}$
17. Integral $= -1\frac{1}{3}$, area $= 1\frac{1}{3}$
19. $4\frac{1}{2}$
20. $1\frac{1}{3}$
21. $10\frac{2}{3}$
22. $1\frac{1}{3}$
25. $2 + e^{-1} - e^{-3} \simeq 2.318$
26.2
27.e − 1

Exercise 8.7.1

1. $-\frac{1}{3}\cos 3x + c$
2. $\frac{2}{3}(x-1)^{3/2} + c$
3. $\frac{1}{3}(2t-1)^{3/2} + c$
4. $-\cos\left(x + \frac{\pi}{3}\right) + c$
5. $\frac{1}{2}\sin\left(2x - \frac{\pi}{3}\right) + c$
6. $\frac{1}{2}e^{x^2} + c$
7. $\frac{1}{12}(x^2 + 2x - 1)^6 + c$
8. $-\frac{1}{1+t} + c$
9. $\frac{1}{4}(3-2v)^{-2} + c$
10. $\frac{1}{7}(2t+1)^{7/2} + c$
11. $-\frac{3}{4}(1-e^y)^{4/3} + c$
12. $\frac{5}{6}(1+v)^{1.2} + c$
13. $\frac{5}{12}(1+2v)^{1.2} + c$

Exercise 8.7.2

A is a positive arbitrary constant.

1. $\ln(Ax^{1/4})$ 2. $\ln(Ax^2)$ 3. $\ln(Ax^9)$

4. $-\frac{7}{2}\ln[A(1-2x)]$ 5. $\ln[A(3t+2)^{1/3}]$

6. $\ln[A(3t-2)^{1/3}]$ 7. $\frac{1}{a}\ln(Ax)$ 8. $a\ln(AX)$

9. $\frac{1}{a}\ln[A(ax+b)]$ 10. $\frac{a}{b}\ln[A(bx+c)]$

11. $\frac{1}{3}\ln[A(1+t^3)]$ 12. $\frac{2}{3}\ln[A(4+t^{3/2})]$

Note that in each case the constant of integration could be added on; i.e. in 1 the answer could have been written

$\ln(x^{1/4}) + C$ or $\frac{1}{4}\ln(x) + C$.

Exercise 8.7.3

1. $\frac{1}{3}$ 2. $\frac{1}{5}\ln\left(\frac{9}{4}\right)$ 3. $\frac{1}{5}\ln\left(\frac{9}{4}\right)$

4. $\frac{1}{3}(3\sqrt{3}-1) \simeq 1.399$ 5. $\frac{1}{3}\ln\left(\frac{9}{2}\right)$

6. 9 7. $\frac{1}{4}$ 8. -0.575 (three decimal places)

9. $\frac{1}{4} - \frac{1}{2(1+e^2)}$ 10. $\frac{\sqrt{3}}{2\pi}$

Exercise 8.8.1

1. $\ln(\sin\theta) + c$ 2. $\frac{2}{3}\ln(2+x^{3/2}) + c$ 3. $\frac{1}{2}\ln(\sec 2\theta) + c$

4. $-\ln(1+\cos\theta) + c$ 5. $\ln(\sqrt{6})$

6. $\frac{1}{3}\ln\left(\frac{7}{4}\right)$ 7. $\ln(1+e^x) + c$

8. $\frac{1}{2}\ln(1+2x) + c$

Exercise 8.8.2

1. $\frac{1}{3}\arctan(3x) + c$ 2. $\frac{1}{3}\arctan\left(\frac{x}{3}\right) + c$

3. $\frac{1}{6}\arctan\left(\frac{3x}{2}\right) + c$

4. $\frac{1}{ab}\arctan\left(\frac{bx}{a}\right) + c$

5. $\arcsin x + c$

6. $\arcsin\left(\frac{x}{5}\right) + c$

7. $\frac{1}{b}\arcsin\left(\frac{bx}{a}\right) + c$

8. $\frac{\pi}{6}$

9. $\frac{\pi}{24}$

10. $\frac{\pi}{12}$

Exercise 8.8.3

1. $\ln(\text{cosec}\,\theta - \cot\theta) + c$
2. $-2(1 + \tan\theta/2)^{-1} + c$

Exercise 8.8.4

1. $\frac{1}{2}x - \frac{1}{4}\sin 2x + c$

2. $\frac{\pi}{8} - \frac{1}{12}$

3. $\frac{1}{2}(x + \sin x) + c$

4. $\frac{1}{16} + \frac{1}{8\pi}$

Exercise 8.9.1

1. $x\sin x + \cos x + c$

2. $(x - 1)e^x + c$

3. $\frac{1}{2}x\sin 2x + \frac{1}{4}\cos 2x + c$

4. $\frac{1}{4}e^{2x}(2x - 1) + c$

5. $\frac{2}{15}(x + 1)^{3/2}(3x - 2) + c$

6. $\frac{1}{27}(2 - 9x^2)\cos 3x + \frac{2x}{9}\sin 3x + c$

7. $-(x^2 + 2x + 2)e^{-x} + c$

8. $\frac{x^2}{2}\ln x - \frac{x^2}{4} + c$

Exercise 8.9.2

1. $\frac{4}{15}(11\sqrt{2} - 4)$

2. $6(2 - e^{1/2})$

3. $\frac{1}{15}(297 - 56\sqrt{7})$

4. $-\frac{2}{9}$

5. $8(\pi - 2)$

6. $8(5e^2 - 1)$

7. $2\ln 2 - 1$

Exercise 8.9.3

1. $\frac{1}{2}e^{x}(\sin x - \cos x) + c$
2. $\frac{1}{2}e^{-x}(\sin x - \cos x) + c$
3. $\frac{1}{10}e^{2x}(\cos 4x + 2\sin 4x) + c$
4. $\frac{2}{5}(e^{\pi} - 2)$
5. $\frac{2}{5} - \frac{1}{10}(\sqrt{3} + 2)e^{-\pi/6}$
6. $\frac{1}{10}(3\sqrt{2}e^{\pi} - 4e^{\pi/2})$

Exercise 8.10.1

$\ln A$ and c are arbitrary constants.

1. $\ln\left[\frac{A(x+1)^3}{(x-1)^2}\right]$ or $3\ln(x+1) - 2\ln(x-1) + c$
2. $\ln[A(x+3)^4(2x-1)^{3/2}]$ or $4\ln(x+3) + \frac{3}{2}\ln(2x-1) + c$
3. $\frac{1}{4}\ln[A(2x+3)(2x-1)]$ or $\frac{1}{4}\ln[(2x+3)(2x-1)] + c$
4. $\ln\left[\frac{Ax^5(x-1)}{(x+3)^2}\right]$ or $\ln[x^5(x-1)] - 2\ln(x+3) + c$
5. $\frac{2}{3}\ln(3x+1) + 3\arctan x + c$
6. $\ln[(x+3)^2\sqrt{x^2+1}] - \arctan x + c$

   or $2\ln(x+3) + \frac{1}{2}\ln(x^2+1) - \arctan x + c$
7. $-\frac{1}{x+1} + \frac{2}{3}\ln(3x-2) + c$
8. $\ln(2x+3) + \frac{3}{2}\arctan\left(\frac{x}{2}\right) + c$
9. $\ln(2.7)$
10. $\ln\left(\frac{21}{16}\right)$
11. $\ln 160 - \frac{3}{2}$

Exercise 8.11.1

1. $\frac{\pi}{7}$
2. $\frac{\pi^2}{2}$
3. $\frac{\pi}{2}(e^8 - e^4)$
4. $\frac{1016\pi}{15}$

5. $\frac{\pi}{2}$ 6. $\frac{\pi}{2}$ 7. $\frac{\pi}{5}$ 8. $\frac{\pi}{2}(e^3 - e^2)$

9. π 10. $\frac{128\pi}{7}$

Exercise 8.12.1

1. $\left(\frac{3}{4}, \frac{3}{10}\right)$ 2. $\left(\frac{2}{5}, \frac{2}{7}\right)$ 3. $\left(\frac{\pi}{2}, \frac{\pi}{8}\right)$

4. $\left(\frac{\pi}{4} - \frac{1}{2}, \frac{\pi}{8}\right)$ 5. (0.459, 0.402) to three decimal places

Exercise 8.13.1

1. 4
2. 16.5
3. 3.1312
4. $\frac{1}{2}\ln(x^2 + 1) + c$
5. 1
6. $\frac{1}{3}te^{3t} - \frac{1}{9}e^{3t} + c$
7. $\frac{1}{8}(e^4 - 1)$
8. $\frac{1}{2}\ln\left(\frac{2 + x}{2 - x}\right) + c$
9. $-\frac{1}{9}\sqrt{1 - 9x^2} - \frac{1}{3}\arcsin(3x) + c$
10. $-\frac{1}{4}\cos 2x + c$
11. $0.5 + \ln(3/4)$
12. $\frac{\pi a^2}{4}$
13. $\ln(x^3 - 3) + c$
14. $\ln\left(\frac{x - 2}{x + 3}\right) + c$
15. $\ln\left(\frac{5}{4}\right)$
16. $\frac{1}{\pi}\sin \pi x + c$
17. $\frac{1}{4}(e^7 - e^{-1})$
18. 2
19. $(2 - x^2)\cos x + 2x \sin x + c$
20. $\frac{x^2}{4} + \frac{\sin^2 x}{4} - \frac{x \sin x \cos x}{2} + c$
21. $\frac{1}{2}\arctan\left(\frac{x}{2}\right) + c$
22. $\frac{1}{2}\ln(t^2 + 4) + c$
23. $\frac{1}{4}\ln\left[\left(\frac{t + 2}{t - 2}\right)\right] + c$
24. $-\frac{1}{2}\ln(t^2 - 4) + c$
25. $u + \ln u + c$

26. $u - \ln(u + 1) + c$

27. $2 \ln t + \dfrac{t^2}{2} + c$

28. $t - \dfrac{2}{t} + c$

29. $t - \sqrt{2} \arctan\left(\dfrac{t}{\sqrt{2}}\right) + c$

30. $\frac{1}{2} \ln(t^2 + 2) + c$

31. $x - \frac{1}{3} x^3 + c$

32. $\frac{1}{2} \arcsin x + \frac{1}{2} x\sqrt{1 - x^2} + c$

33. $-\dfrac{2}{3}(1 - x)^{3/2} + c$

34. $-\dfrac{1}{6}\ln(5 - 6t) + c$

35. $\dfrac{1}{6(5 - 6t)} + c$

36. $\dfrac{1}{3(5 - 6t)^{1/2}}$

37. $\dfrac{(u^2 + 1)^2}{4} + c$

38. $\dfrac{1}{15} u^3 (3u^2 + 5) + c$

39. $\dfrac{1}{8} [u(2u^2 + 1) \sqrt{u^2 + 1} - u - \ln \sqrt{u^2 + 1}] + c$

Objective Test 8

1. $t^3 + c$
2. $4\left(1 - \dfrac{1}{\sqrt{2}}\right)$
3. 1
4. $\dfrac{e - 1}{2}$
5. $\frac{1}{2} \ln(64 + t^2) + c$
6. $\dfrac{1}{8} \arctan\left(\dfrac{t}{8}\right) + c$
7. $\dfrac{1}{16} \ln\left(\dfrac{8 + t}{8 - t}\right) + c$
8. $\dfrac{1}{2} x + \dfrac{1}{4\pi} \sin(2\pi x) + c$

In each case c is an arbitrary constant.

CHAPTER 9

Exercise 9.2.1

1. $y = 2x^2e^x$
2. $y = \ln(2x + 3)$
3. $y = 5 - 2e^{-x}$
4. $y = \arctan t + \pi/2$
5. $x = 4(1 - t)$

Exercise 9.3.1

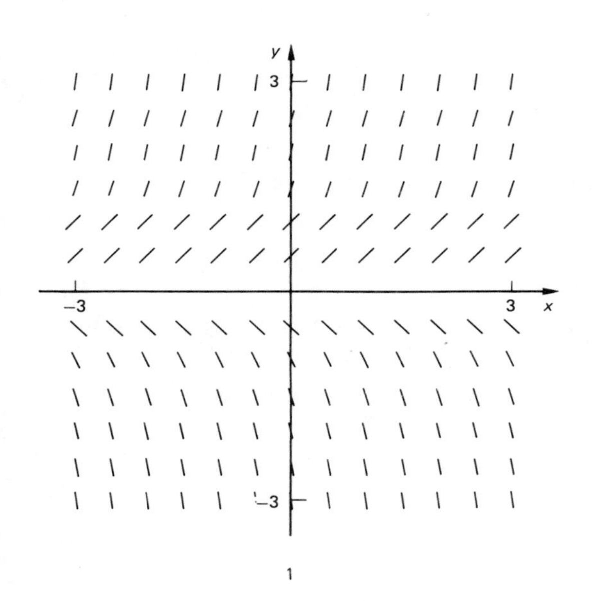

1

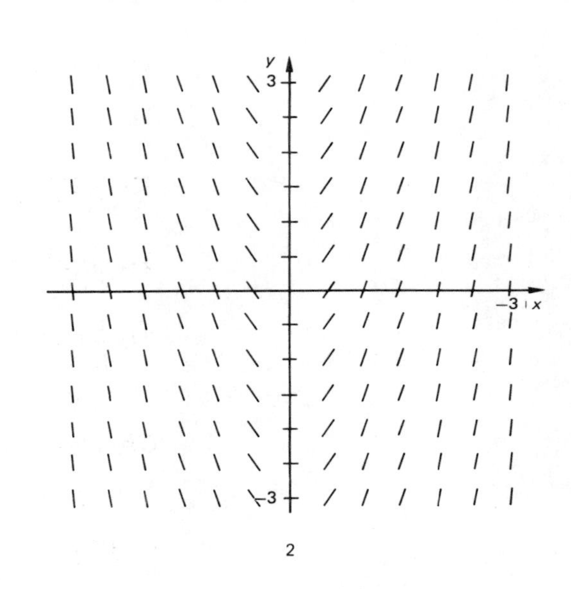

2

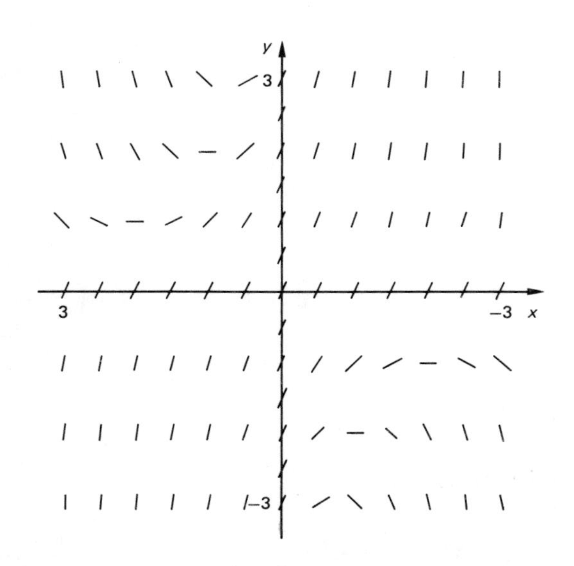

3

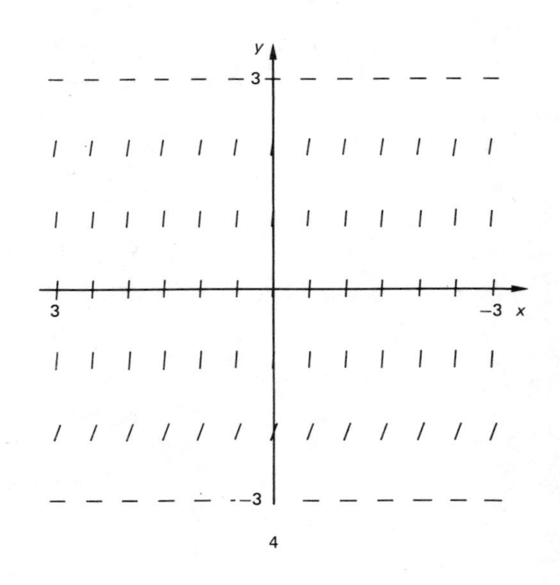

4

Exercise 9.4.1

In the following answers, A and C are arbitrary constants.

1. $y = Ax^3$
2. $y^2 - x^2 = A$
3. $y = \cos x + C$
4. $y = A\,e^{kt}$
5. $y^2 = A\,e^t$
6. $\ln y = t^2 + 4t + C$
7. $C - \cos x = \frac{1}{2}t^2$
8. $e^{-x} = C - \frac{1}{3}t^3$
9. $-\dfrac{1}{x} = e^t + C$
10. $\cot y = C - \frac{1}{2}\sin 2x$
11. $y = 4x^2 + 1$
12. $y = 3e^{-2x}$
13. $y = 1 - \dfrac{1}{2\omega}\sin 2\omega t$
14. $3t^2 + 2e^{-3y} = 2$
15. $\arctan y = t^2 + \dfrac{\pi}{4}$
16. $x^2 = \arcsin(2t)$
17. $e^{2x} = 2[2 - e^{-t}(1 + t)]$
18. $\sin 2\omega y + 2\cos \omega y = 2$
19. $1 + x^2 = \dfrac{4(1 + t)^2}{(1 - t)^4}$
20. $y \sin y + \cos y = e^x$

Exercise 9.5.1

1. 0.711 m, 147 s
2. $m = e^{-0.0127t}$

Exercise 9.6.1

1. 3.003
2. 1.084
3. 1.096, 1.095

Objective Test 9

1. $y = \frac{1}{2}\sin 2x + c$
2. $v = Ae^{1.5t^2}$
3. $i = \sqrt{(e^{t^2} + c)}$
4. $y = e^{1/(x+1)}$
5. 1.136

CHAPTER 10

Exercise 10.2.1

1. (a) 10 gram
   (b) 0.1, 10%; 0.01, 1%; 0.00001, 0.001%
   (c) $127 < m < 147$
2. $3.135 < m < 3.145$

Exercise 10.2.2

(a) $|\text{error}| = 0.0015; \quad 5.4975 < n < 5.5005$
(b) $|\text{error}| = 0.0056; \quad 23.8914 < n < 23.9026$
(c) $|\text{error}| = 0.0090; \quad 2.4099 < n < 2.4278$
(d) $|\text{error}| = 0.0058; \quad 6.8007 < n < 6.8123$
(e) $|\text{error}| = 0.9 \quad ; \quad 331.50 < n < 333.30$
(f) $|\text{error}| = 0.0045; \quad 2.3923 < n < 2.4013$

Exercise 10.2.3

(a) $\epsilon = 0.019$; $e^{1.32} = 3.743 \pm 0.019$; 4.0 to appropriate accuracy
(b) $\epsilon = 0.050$; $\cos(1.5) = 0.0707 \pm 0.0499$; 0.0
(c) $\epsilon = 0.00019$; $\ln(2.634) = 0.9685 \pm 0.00019$; 0.97
(d) $\epsilon = 17.28$; $e^{(2.1)2} = 82 \pm 17$; 100
(e) $\epsilon = 0.08$; $3.42e^{1.32} = 12.80 \pm 0.08$; 13

Exercise 10.3.1

1. For root near $x = 3.4$, method (d) only. For root near $x = 0.6$, methods (a) and (c) could be used. (c) would converge more quickly.
2. Use method (a); $x = 0.567$ to three decimal places.
3. (a) $x_{n+1} = \sin^{-1}\left(\dfrac{1 - x_n}{2}\right)$; $x = 0.338$

   (b) $x_{n+1} = \sqrt{1 + 2e^{-x_n}}$; $x = 1.253$

   (c) $x_{n+1} = \dfrac{1 + 2x_n^2 - x_n^3}{3}$; $x = 0.430$

Exercise 10.3.2

(a) $x = 0.5858$ (b) $x = 0.5671$ (c) $x = 0.3376$ (d) $x = 1.2534$
(e) $x = 0.4302$

Exercise 10.4.1

1. (a) 2.927 (b) 0.2438 (c) 0.6932 (d) 3.142 (e) 0.1054
2. 0.62 km
3. 35.8 ms^{-1}
4. $\bar{\theta} = 49.03$ (simple average = 49.45)

Objective Test 10

1. $\epsilon = 0.389$; appropriate answer 63
2. For the root near 0.3, methods (a) and (c) cannot be used. Quickest method is (d). For the root near 5, methods (b) and (d) cannot be used. Quickest method is (c). Roots are 4.9571 and 0.3429.
4. (a) −0.39 (b) 0.83

CHAPTER 11

Exercise 11.1.1

force, velocity, displacement.

Exercise 11.1.2

1.

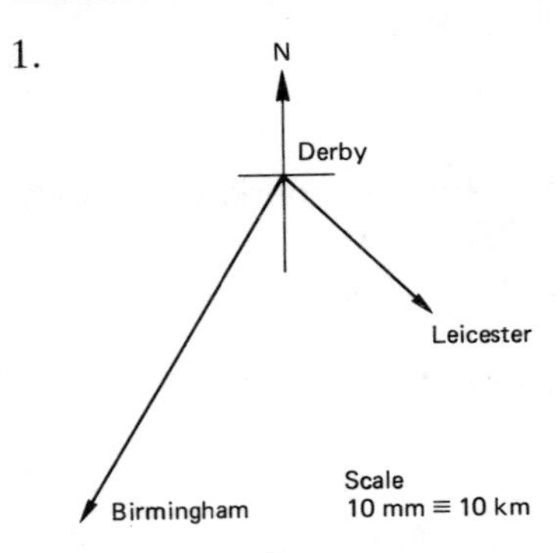

2.

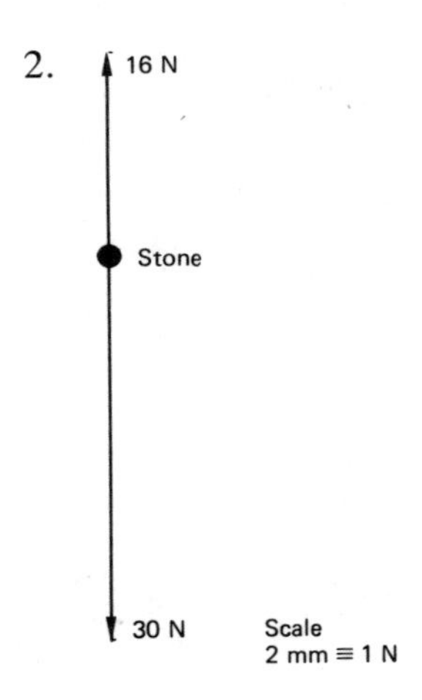

3.

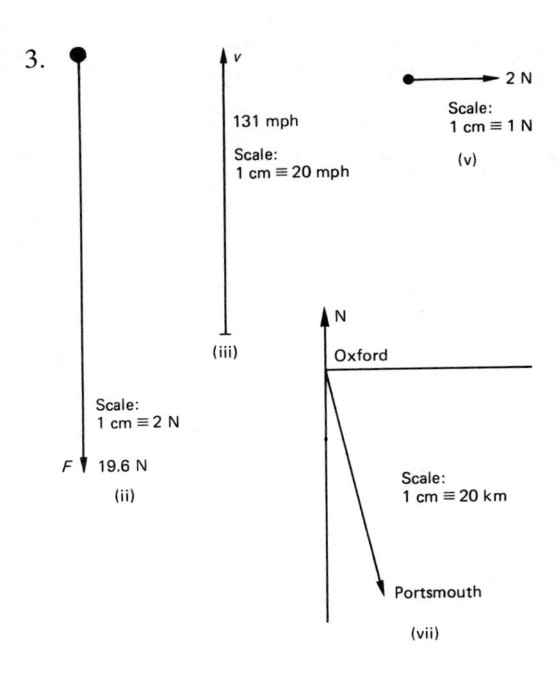

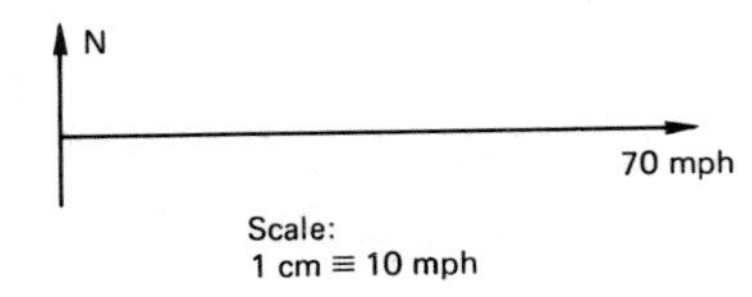

Exercise 11.2.1

1. $\boldsymbol{c} = \boldsymbol{k}$, $\boldsymbol{d} = \boldsymbol{f}$, $\boldsymbol{e} = \boldsymbol{n}$
2. (a) $\boldsymbol{e} = -2\boldsymbol{a}$ (b) $\boldsymbol{h} = -\boldsymbol{a}$ (c) $\boldsymbol{n} = -2\boldsymbol{a}$
   (d) $\boldsymbol{m} = 2\boldsymbol{d}$ (e) $\boldsymbol{g} = -\boldsymbol{d}$ (f) $\boldsymbol{i} = -2\boldsymbol{d}$
   (g) $\boldsymbol{l} = 1\frac{1}{2}\boldsymbol{d}$ (h) $\boldsymbol{j} = -\boldsymbol{b}$
3. (i) = (vi), (ii) = (v)

4.

Vector	*Magnitude*	*Direction (bearing)*
a	1N	270°
b	1.4N	135°
c	1.1N	117°
d	1.4N	045°
e	2N	090°
f	1.4N	045°
g	1.4N	225°
h	1N	090°
i	2.8N	215°
j	1.4N	315°
k	1.1N	117°
l	2.1N	045°
m	2.8N	045°
n	2N	090°

Giving the direction of a vector as a bearing provides a unique answer.

Exercise 11.2.2

1.

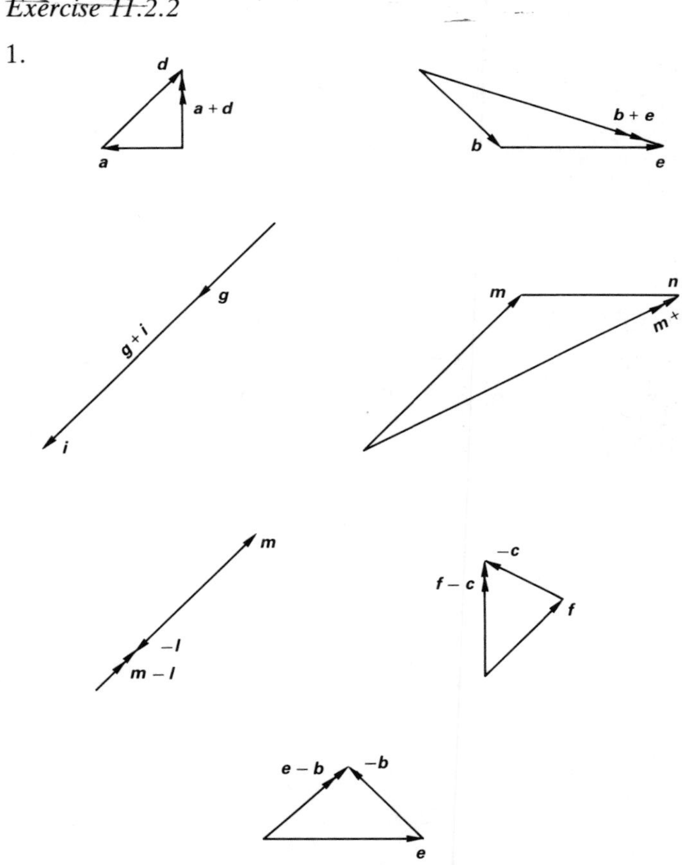

2.

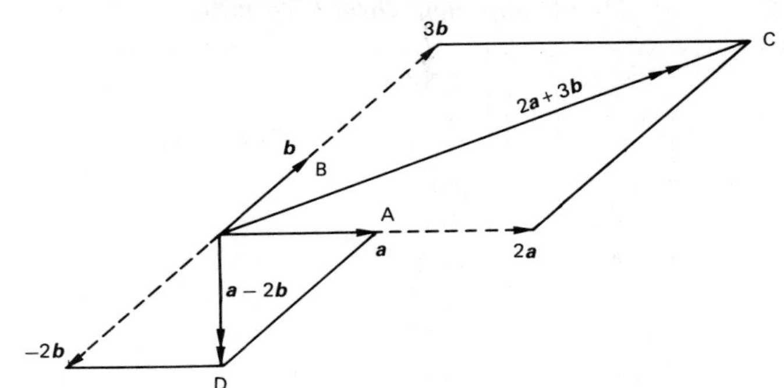

3. 13 newtons; angle = 22.6° to larger force.
4. 15.5 newtons; angle = 29.6° to smaller force.
5. 100.9 miles on bearing 112°

Exercise 11.3.1

1. $\begin{bmatrix} 20 \cos 30° \\ 20 \cos 60° \end{bmatrix} = \begin{bmatrix} 10\sqrt{3} \\ 10 \end{bmatrix}$

2. $\begin{bmatrix} 250 \cos 70° \\ 250 \cos 20° \end{bmatrix} = \begin{bmatrix} 85.5 \\ 234.9 \end{bmatrix}$

3. $\boldsymbol{F} = \begin{bmatrix} 0 \\ 100 \end{bmatrix}$; $\boldsymbol{G} = \begin{bmatrix} 75 \\ 0 \end{bmatrix}$

Exercise 11.3.2

1. $\begin{bmatrix} -30 \cos 60° \\ -30 \cos 30° \end{bmatrix} = \begin{bmatrix} -15 \\ -15\sqrt{3} \end{bmatrix}$

2. $\begin{bmatrix} -30 \cos 70° \\ -30 \cos 20° \end{bmatrix} = \begin{bmatrix} -10.26 \\ -28.18 \end{bmatrix}$

3.

Vector	*Component form*
$\boldsymbol{a}$	$\begin{bmatrix} -1 \\ 0 \end{bmatrix}$
$\boldsymbol{b}$	$\begin{bmatrix} 1 \\ -1 \end{bmatrix}$
$\boldsymbol{c}$	$\begin{bmatrix} 1 \\ -0.5 \end{bmatrix}$
$\boldsymbol{d}$	$\begin{bmatrix} 1 \\ 1 \end{bmatrix}$
$\boldsymbol{e}$	$\begin{bmatrix} 2 \\ 0 \end{bmatrix}$
$\boldsymbol{f}$	$\begin{bmatrix} 1 \\ 1 \end{bmatrix}$
$\boldsymbol{g}$	$\begin{bmatrix} -1 \\ -1 \end{bmatrix}$
$\boldsymbol{h}$	$\begin{bmatrix} 1 \\ 0 \end{bmatrix}$
$\boldsymbol{i}$	$\begin{bmatrix} -2 \\ -2 \end{bmatrix}$
$\boldsymbol{j}$	$\begin{bmatrix} -1 \\ 1 \end{bmatrix}$
$\boldsymbol{k}$	$\begin{bmatrix} 1 \\ -0.5 \end{bmatrix}$
$\boldsymbol{l}$	$\begin{bmatrix} 1.5 \\ 1.5 \end{bmatrix}$
$\boldsymbol{m}$	$\begin{bmatrix} 2 \\ 2 \end{bmatrix}$
$\boldsymbol{n}$	$\begin{bmatrix} 2 \\ 0 \end{bmatrix}$

Exercise 11.3.3

Vector	Magnitude	Direction bearing	α (see p.418)
$\boldsymbol{a}$	5	036.9°	53.1°
$\boldsymbol{b}$	13	022.6°	67.4°
$\boldsymbol{c}$	5	323.1°	126.9°
$\boldsymbol{d}$	$\sqrt{5}$	206.6°	−116.6°
$\boldsymbol{e}$	$\sqrt{2}$	225°	−135°
$\boldsymbol{f}$	$\sqrt{10}$	161.6°	−71.6°

Exercise 11.3.4

2.

Vector	Magnitude	Direction bearing	or α (see p.418)
$\boldsymbol{a}+\boldsymbol{b}=\begin{bmatrix}8\\16\end{bmatrix}$	$8\sqrt{5}$	026.6°	63.4°
$\boldsymbol{b}+\boldsymbol{c}=\begin{bmatrix}2\\16\end{bmatrix}$	$2\sqrt{65}$	007.1°	82.9°
$\boldsymbol{a}+\boldsymbol{c}=\begin{bmatrix}0\\8\end{bmatrix}$	8	0°	90°
$\boldsymbol{c}+\boldsymbol{f}=\begin{bmatrix}-2\\1\end{bmatrix}$	$\sqrt{5}$	296.6°	153.4°
$\boldsymbol{d}-\boldsymbol{e}=\begin{bmatrix}0\\-1\end{bmatrix}$	1	180°	−90°
$\boldsymbol{c}-\boldsymbol{a}=\begin{bmatrix}-6\\0\end{bmatrix}$	6	270°	180°
$\boldsymbol{b}+\boldsymbol{e}=\begin{bmatrix}4\\11\end{bmatrix}$	$\sqrt{137}$	020°	70°
$\boldsymbol{f}-\boldsymbol{e}=\begin{bmatrix}2\\-2\end{bmatrix}$	$2\sqrt{2}$	135°	−45°

3. (a) $\begin{pmatrix}5\cos 45^\circ\\5\cos 45^\circ - 4\end{pmatrix}$; 3.6N, −7.5°

(b) $\begin{pmatrix}2 - 3\cos 30^\circ\\3\cos 60^\circ - 2\end{pmatrix}$; 0.78N, −140.1°

(c) $\begin{pmatrix} 5.3 \cos 30° - 4.7 \cos 60° \\ 5.3 \cos 60° + 4.7 \cos 30° - 6 \end{pmatrix}$; 2.35N, 17.8°

(d) $\begin{pmatrix} 4 \cos 45° + 5 \cos 30° - 3 \\ 4 \cos 45° - 5 \cos 60° \end{pmatrix}$; 4.17N, 4.5°

4.

	Vector	Magnitude	Direction bearing	or α (see p.418)
$\boldsymbol{a} + \boldsymbol{b} =$	$\begin{bmatrix} 3 \\ -2 \end{bmatrix}$	$\sqrt{13}$	123.7°	−33.7°
$\boldsymbol{b} + \boldsymbol{c} =$	$\begin{bmatrix} 0 \\ -1 \end{bmatrix}$	1	180.°	−90°
$\boldsymbol{a} - \boldsymbol{b} =$	$\begin{bmatrix} -1 \\ 8 \end{bmatrix}$	$\sqrt{65}$	352.9°	97.1°
$3\boldsymbol{a} + 2\boldsymbol{b} =$	$\begin{bmatrix} 7 \\ -1 \end{bmatrix}$	$\sqrt{50}$	098.1°	−8.1°
$-2\boldsymbol{a} - 3\boldsymbol{b} + 4\boldsymbol{c} =$	$\begin{bmatrix} -16 \\ 25 \end{bmatrix}$	29.7	327.4	123.6°

Exercise 11.3.5

1. $2\boldsymbol{i}, 5\boldsymbol{i}, 4\boldsymbol{i}, \boldsymbol{i}, 7.3\boldsymbol{i}$
2. $\boldsymbol{a} = 3\boldsymbol{i} + \boldsymbol{j}$
   $\boldsymbol{b} = -\boldsymbol{i} + 3\boldsymbol{j}$
   $\boldsymbol{c} = -3\boldsymbol{i} - 2\boldsymbol{j}$
   $\boldsymbol{d} = 3\boldsymbol{i} + \boldsymbol{j}$
3. $|\boldsymbol{a}| = 5, |\boldsymbol{b}| = \sqrt{13}, |\boldsymbol{c}| = 5$

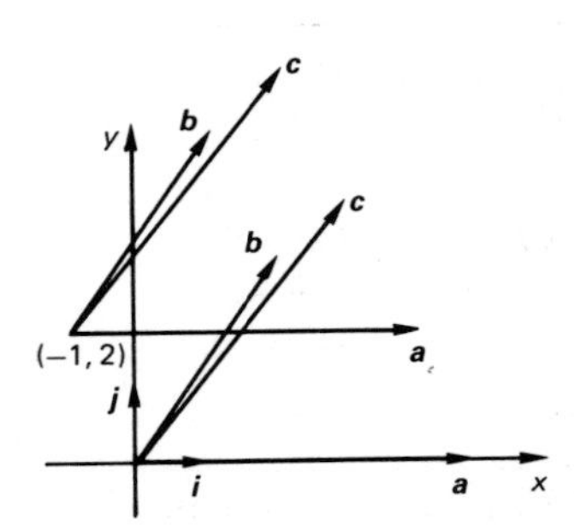

4. Same as Exercise 11.3.4, Problem 4.
5. (a) $(3 - 2.5 \cos 30°)\boldsymbol{i} + (2 - 2.5 \cos 60°)\boldsymbol{j} = 0.83\boldsymbol{i} + 0.75\boldsymbol{j}$
   (b) $(5.3 \cos 30° - 4.7 \cos 60°)\boldsymbol{i} + (5.3 \cos 60° + 4.7 \cos 30° - 6)\boldsymbol{j} = 2.24\boldsymbol{i} + 0.72\boldsymbol{j}$

(c) $(7.1 \cos 15° - 5.4 \cos 80° - 8\cos 60° + 4.1\cos 40°)\boldsymbol{i}$
$+ (7.1 \cos 75° + 5.4 \cos 10° - 8 \cos 30° - 4.1 \cos 50°)\boldsymbol{j}$
$= 5.06\boldsymbol{i} - 2.41\boldsymbol{j}$

Exercise 11.4.1

1. $\boldsymbol{a} = \boldsymbol{j} \quad \boldsymbol{b} = \boldsymbol{i} \quad \boldsymbol{c} = -2\boldsymbol{i} + 3\boldsymbol{j}$
   $\boldsymbol{d} = -\boldsymbol{i} - 2\boldsymbol{j}$

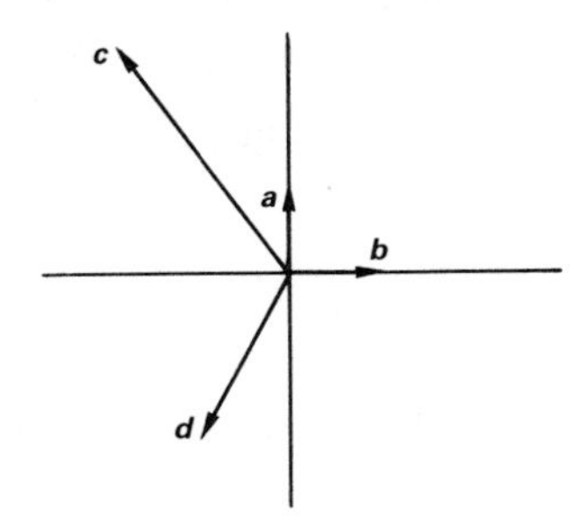

2. $\boldsymbol{c} = 3\boldsymbol{i} + 4\boldsymbol{j}$; $\boldsymbol{f} = 3\boldsymbol{i} + 4\boldsymbol{j}$; $\boldsymbol{d} = 3\boldsymbol{i} + 4\boldsymbol{j}$; $\boldsymbol{c}$ is the position vector of (3, 4)
3. $\boldsymbol{a} = \boldsymbol{i} + 2\boldsymbol{j}$; $\boldsymbol{b} = 5\boldsymbol{i} + \boldsymbol{j}$; $\boldsymbol{c} = 7\boldsymbol{i} + 8\boldsymbol{j}$
   $\overrightarrow{AB} = \boldsymbol{b} - \boldsymbol{a} = 4\boldsymbol{i} - \boldsymbol{j}$
   $\overrightarrow{BC} = \boldsymbol{c} - \boldsymbol{b} = 2\boldsymbol{i} + 7\boldsymbol{j}$
   $\overrightarrow{CA} = \boldsymbol{a} - \boldsymbol{c} = -6\boldsymbol{i} - 6\boldsymbol{j}$
4. $\boldsymbol{a} = 4\boldsymbol{i} + 2\boldsymbol{j}$; $\boldsymbol{b} = 5\boldsymbol{i} + 4\boldsymbol{j}$; $\overrightarrow{AB} = \boldsymbol{i} + 2\boldsymbol{j}$
   $|\overrightarrow{AB}| = \sqrt{5}$, direction = 63.4°

Exercise 11.4.2

1.

Case	Velocity $t = 0$	Velocity $t = 1$	Acceleration $t = 0$	Acceleration $t = 1$
(a)	$3\boldsymbol{i}$	$6\boldsymbol{i} + 8\boldsymbol{j}$	$8\boldsymbol{i}$	$6\boldsymbol{i} + 8\boldsymbol{j}$
(b)	$10\boldsymbol{i} + 10\boldsymbol{j}$	$10\boldsymbol{i} + 0.2\boldsymbol{j}$	$-9.8\boldsymbol{j}$	$-9.8\boldsymbol{j}$
(c)	$w\boldsymbol{j}$	$(w - g)\boldsymbol{j}$	$-g\boldsymbol{j}$	$-g\boldsymbol{j}$

speed at time t = 0 is (a) 3 ms^{-1} (b) $10\sqrt{2}$ ms^{-1} (c) w ms^{-1}

2. $\boldsymbol{v} = t^2\boldsymbol{i} - (3t - 2)\boldsymbol{j}$; $\boldsymbol{r} = \dfrac{t^3}{3}\boldsymbol{i} - \left(\dfrac{3t^3}{2} - 2t\right)\boldsymbol{j}$

3. $\boldsymbol{v} = -dw \sin wt\boldsymbol{i} + dw \cos wt\,\boldsymbol{j}$
   $\boldsymbol{a} = -dw^2 \cos wt\,\boldsymbol{i} - dw^2 \sin wt\,\boldsymbol{j}$
   $\boldsymbol{a} = -w^2\boldsymbol{r}$ hence $\boldsymbol{a}$ is parallel to $\boldsymbol{r}$

Objective Test 11

1. *Scalars*: area, temperature, volume, energy, time; *vectors*: velocity, force, displacement, acceleration
2.

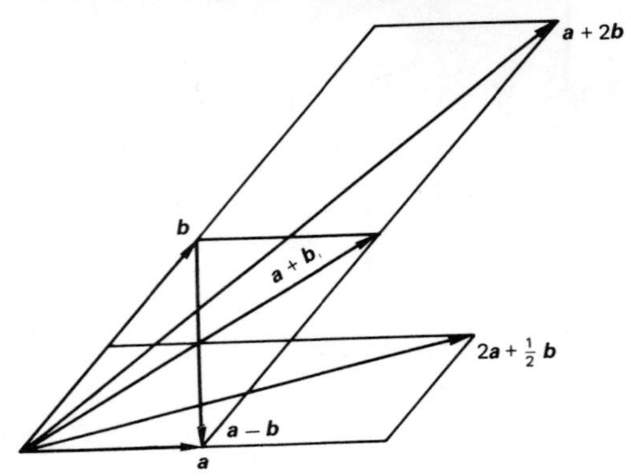

3. $\boldsymbol{a} = \begin{bmatrix} 3 \\ 0 \end{bmatrix}$ or $\boldsymbol{a} = 3\boldsymbol{i}$

$$\boldsymbol{b} = \begin{bmatrix} 2.5 \\ 5\sqrt{3}/2 \end{bmatrix} \text{ or } \boldsymbol{b} = 2.5\boldsymbol{i} + \frac{5\sqrt{3}}{2}\boldsymbol{j}$$

Vector	Component form	Magnitude	Direction bearing	Direction or α
$\boldsymbol{a} + \boldsymbol{b}$	$\begin{bmatrix} 5.5 \\ 5\sqrt{3}/2 \end{bmatrix}$ or $5.5\boldsymbol{i} + \frac{5\sqrt{3}}{2}\boldsymbol{j}$	7	052°	38.2°
$\boldsymbol{a} - \boldsymbol{b}$	$\begin{bmatrix} 0.5 \\ -5\sqrt{3}/2 \end{bmatrix}$ or $0.5\boldsymbol{i} - \frac{5\sqrt{3}}{2}\boldsymbol{j}$	4.36	173.4°	−83.4°
$2\boldsymbol{a} - 3\boldsymbol{b}$	$\begin{bmatrix} -1.5 \\ -15\sqrt{3}/2 \end{bmatrix}$ or $-1.5\boldsymbol{i} - \frac{15\sqrt{3}}{2}\boldsymbol{j}$	13.1	186.6°	−96.6°

4. $(3\cos 45° - 4\cos 60° - 2)\boldsymbol{i} + (3\cos 45° + 4\cos 30° - 8)\boldsymbol{j} = -1.879\boldsymbol{i} - 2.415\boldsymbol{j}$; magnitude = 3.06 N; direction 217.9° or $\alpha = -127.9°$
5. 18.5 knots bearing 240.3°
6. 12.4 km bearing 108°
7. $\boldsymbol{r}(t) = (t^3 + t)\boldsymbol{i} - (2t^2 + 2t)\boldsymbol{j}$, $\boldsymbol{a}(t) = 6t\boldsymbol{i} - 4\boldsymbol{j}$

CHAPTER 12

Exercise 12.2.1

1. $\begin{bmatrix} 3 & -1 \\ 8 & 0 \end{bmatrix}$
2. $\begin{bmatrix} -1 & 1 \\ 6 & 1 \end{bmatrix}$
3. $\begin{bmatrix} 3 & 0 \\ 11 & -1 \end{bmatrix}$
4. $\begin{bmatrix} 5 & -1 \\ 2 & -4 \end{bmatrix}$
5. $\begin{bmatrix} -5 & 1 \\ -2 & 4 \end{bmatrix}$
6. $\begin{bmatrix} -4 & 2 \\ -2 & 1 \end{bmatrix}$

Exercise 12.3.1

$\begin{bmatrix} 3 \\ 6 \\ -3 \end{bmatrix}$

2. $\begin{bmatrix} 0.4 & -0.1 \\ -0.3 & -0.7 \end{bmatrix}$
3. $\begin{bmatrix} cp & cq \\ cr & cs \end{bmatrix}$
4. $3\begin{bmatrix} -2 & 1 \\ 3 & -4 \end{bmatrix}$
5. $\frac{1}{8}[4 \quad 1 \quad -2]$
6. $x\begin{bmatrix} 3 & x \\ a & -x^2 \\ 10 & 2 \end{bmatrix}$
7. $\begin{bmatrix} -19 \\ -12 \end{bmatrix}$
8. $\begin{bmatrix} 17 \\ 1 \\ 7 \end{bmatrix}$
9. $\begin{bmatrix} 3x + 8y \\ 2y - x \end{bmatrix}$
10. $[a + 2b \quad 2a + 3b \quad 3a + 4b]$
11. $\begin{bmatrix} 0 & -0.5 \\ 2.3 & -2.4 \end{bmatrix}$
12. $\begin{bmatrix} 0 & -5 \\ 23 & -24 \end{bmatrix}$
13. $\begin{bmatrix} x \\ y \\ z \end{bmatrix}$
14. $\begin{bmatrix} -8 & 1 & 0 \\ 3 & 3 & 0 \\ 2 & 5 & 0 \end{bmatrix}$

16. $a = 6,\ b = -\frac{5}{3}$
17. $x = 5,\ m = 2$

Exercise 12.4.1

1. $\begin{bmatrix} 3 & 1 \\ -1 & 2 \end{bmatrix}\begin{bmatrix} u \\ v \end{bmatrix} = \begin{bmatrix} 6 \\ 5 \end{bmatrix}$

2. $\begin{bmatrix} 2 & 1 & 1 \\ 0 & 1 & -1 \end{bmatrix} \begin{bmatrix} p \\ q \\ r \end{bmatrix} = \begin{bmatrix} 0 \\ -2 \end{bmatrix}$

3. $\begin{bmatrix} 2 & 1 \\ -1 & 1 \end{bmatrix} \begin{bmatrix} x \\ y \end{bmatrix} = \begin{bmatrix} 1 \\ 4 \end{bmatrix}$

Exercise 12.5.1

1. 1
2. $2y - 3x$
3. 1
4. 0
5. 1
6. 3; (3, 0)
7. −10; (2, 3)
8. 4; −1.25, 3.75
9. 22; (2.7, 1.8)
10. 0, no unique solution

Exercise 12.5.2

1. 27
2. 1
3. −0.018
4. 4
5. −18
6. $xy(y - x)$
7. 3

Exercise 12.5.3

1. 0
2. 0
3. 0
7. −1860
8. $-\dfrac{2}{15}$
9. 0
10. $-y(z - x)^2$

Exercise 12.5.4

1. $x = 0$, $y = 0.7$, $z = 0.3$
2. det $M = 0$, so no unique solution
3. $a = -2$, $b = 2$, $c = 1$

Exercise 12.6.1

1. $-12, \dfrac{1}{12}\begin{bmatrix} -6 & 2 \\ 3 & 1 \end{bmatrix}$
2. $4, \dfrac{1}{4}\begin{bmatrix} \sqrt{3} & -1 \\ 1 & \sqrt{3} \end{bmatrix}$
3. 0, no inverse
4. $-1, \begin{bmatrix} 0 & 1 & -1 \\ 2 & -2 & 3 \\ -1 & 1 & -1 \end{bmatrix}$
5. $abc, \begin{bmatrix} 1/a & 0 & 0 \\ 0 & 1/b & 0 \\ 0 & 0 & 1/c \end{bmatrix}$
6. $1, \begin{bmatrix} 1 & -k & kn-m \\ 0 & 1 & -n \\ 0 & 0 & 1 \end{bmatrix}$

Exercise 12.6.2

1. $x = 2, y = 3$ 2. $s = 2, t = 3$ 3. $x = 5, y = -4$
4. $x = -1, y = 5$ 5. $x = 0, y = 2, z = -1$
6. $p = 0.3, q = 0.5, r = 0.2$

Exercise 12.7.1

1. $x = 2, y = -1, z = 4$ 2. $x = -0.7, y = 0.2, z = 0.5$
3. $u = 1, v = 0.5, w = 2.5$ 4. $p = 5, q = 1, r = 0$

Exercise 12.8.1

1. inconsistent, no solutions
2. Infinitely many solutions: $u = \dfrac{3 + \lambda}{6}, v = \lambda$
3. Inconsistent, no solutions
4. Infinitely many solutions: $x = \dfrac{4 + \lambda}{6}, y = -\left(\dfrac{1 + 4\lambda}{3}\right), z = \lambda$
5. $u = 8 - 4\lambda, v = 3(\lambda - 1), w = \lambda$

Exercise 12.9.1

1. Currents (A): 0.581, −0.452, 1.032;
   Voltages (V): 2.903, −0.903, 3.097 across AB, BC and BD.
2. Currents (mA): −0.243, 0.054, −0.297
   Voltages (V): −1.216, 0.216, −1.784 across AB, BC and BD.
3. Currents (mA): 0.415, −0.418, −0.727, 0.564, 0.309;
   Voltages (V): 0.873, −0.418, −1.455, 1.127, 1.545 across AB, BC, CD, BF and CE.
4. Currents (A): 0.343, 0.214, 0.129, 1.000;
   Voltages (V): 1.714, 1.286, 1.288, 4.000 across AB, BC, BD, EF
5. Currents (A): 0.75, 0.25, 0.5,1.5, −0.75;
   Voltages (V): 1.5, 1.5, 1.5, 3.0, −3.0 across AB, BC, BD, DF, EF.

Objective Test 12

1. (a) $\begin{bmatrix} -0.6 & 1.2 \\ 0.8 & -0.6 \end{bmatrix}$ (b) $\begin{bmatrix} 1.5 & -1.2 \\ -0.7 & -0.3 \end{bmatrix}$
   (c) $\begin{bmatrix} 0.13 & 0.04 \\ -0.46 & 0.29 \end{bmatrix}$ (d) $\begin{bmatrix} 0.17 & -0.68 \\ 0.20 & 0.25 \end{bmatrix}$
2. $\begin{bmatrix} 1 & 0 & 0 \\ 0 & 1 & 0 \\ 0 & 0 & 1 \end{bmatrix}$

3. $\begin{bmatrix} 0 & -1 & 7 \\ -4 & 6 & 8 \end{bmatrix}$
4. $\begin{bmatrix} 8 & -2 \\ -3 & 1 \end{bmatrix} \begin{bmatrix} p \\ q \end{bmatrix} = \begin{bmatrix} 4 \\ 9 \end{bmatrix}$
5. $p = 11, q = 42$
6. -15
7. $\det M = 0.2; M^{-1} = 5 \begin{bmatrix} 0.1 & 0.3 \\ -0.6 & 0.2 \end{bmatrix}$
8. $x = 1.5, y = 11$
9. $x = -3, y = 0, z = 4$
10. $u = \dfrac{3 - 5\lambda}{2}, v = \lambda$

Index